AF887298

# Yang-Baxter Equations, Conformal Invariance and Integrability in Statistical Mechanics and Field Theory

# Yang-Baxter Equations, Conformal Invariance and Integrability in Statistical Mechanics and Field Theory

Proceedings of a Conference

Centre for Mathematical Analysis
Australian National University, Canberra, Australia
July 10-14, 1989

Editors
**Michael N. Barber**
*Department of Mathematics & Centre for Mathematical Analysis*
*Australian National University*

**Paul A. Pearce**
*Department of Mathematics, University of Melbourne*

**World Scientific**
*Singapore • New Jersey • London • Hong Kong*

*Published by*

World Scientific Publishing Co. Pte. Ltd.
P O Box 128, Farrer Road, Singapore 9128
*USA office:* 687 Hartwell Street, Teaneck, NJ 07666
*UK office:* 73 Lynton Mead, Totteridge, London N20 8DH

**YANG-BAXTER EQUATIONS, CONFORMAL INVARIANCE AND INTEGRABILITY IN STATISTICAL MECHANICS AND FIELD THEORY**

Copyright © 1990 by World Scientific Publishing Co. Pte. Ltd.

*All rights reserved. This book, or parts thereof, may not be reproduced in any form or by any means, electronic or mechanical, including photocopying, recording or any information storage and retrieval system now known or to be invented, without written permission from the Publisher.*

ISBN 981-02-0067-6

Printed in Singapore by JBW Printers and Binders Pte. Ltd.

# PREFACE

During 1989 the Centre for Mathematical Analysis (a Commonwealth of Australia funded Special Research Centre) sponsored a Special Year in Mathematical Physics under the direction of Professor R. J. Baxter FAA, FRS and Professor M. N. Barber. The research activities of this program centred on:

i. solvable models in statistical mechanics,

ii. conformal and modular invariance in field theory and statistical mechanics,

iii. the Yang-Baxter equations and integrability,

iv. related mathematics including applications to knot theory, Virasoro and Kac-Moody algebras, Riemann surfaces and strings, and integrable systems.

These themes formed the rationale for the conference,whose proceedings form this volume. The conference was, in fact, the major event of the Special Year and featured 25 lecturers, 15 of whom came from overseas. Many of the lecturers were world leaders in their fields and there was considerable interaction between all the participants. This breadth of expertise and the fruitfulness of the resulting interactions are reflected in these Proceedings. The Editors hope that the volume captures some of the excitement of the Conference and serves as an introduction to some of the latest developments in these very active areas of research.

We would like to thank the Director of the Centre for Mathematical Analysis, Professor Neil Trudinger for the support of the Centre and indeed for the initial vision that the Centre should host a program in Mathematical Physics. The success of any conference in the Centre relies heavily on the dedication of the staff of the Centre; this conference was no exception. In particular, the Editors thank Mrs. Marilyn Gray, the Secretary of the the Centre, for her apparently effortless organisation. We are also appreciative of the assistance that Dr Murray Batchelor gave in the production of this volume. However, in the end the success of a conference and of its Proceedings depends on the contributors. We thank all our contributors for their efforts.

Michael N. Barber and Paul A. Pearce, editors

# CONTENTS

Some Participants in the Workshop on Yang-Baxter Equations, Conformal Invariance and Integrability in Statistical Mechanics and Field Theory.

**First (Front) Row:** V. Pasquier, M. Jimbo, P.A. Pearce, R. Crewther, R.J. Baxter, G.R.W. Quispel, T. Miwa, H.J. de Vega.

**Second Row:** V. Jones, C.J. Hamer, A. Klümper, M.N. Barber, T.T. Truong, D.E. Evans, H. He, M.T. Batchelor, I.R. Aitchison.

**Third Row:** A.J. Bracken, P.J. Forrester, B.M. McCoy, J. Oitmaa, B. Nienhuis, B. Davies, M. Murray, A.L. Carey, V. Rittenberg, V. Dotsenko.

**Fourth (Back) Row:** R.B. Zhang, V. Korepin, C.A. Tracy, J.H.H. Perk, N. Joshi, K.A. Seaton, P. Jarvis, C.A. Hurst.

International Journal of Modern Physics B, Vol. 4, No. 5 (1990) 701–713
© World Scientific Publishing Company

# BAXTERIZATION

V.F.R. Jones

*Mathematics Department, University of California, Berkeley, Berkeley CA* 94720 *USA*

In statistical mechanics one is interested in the globally observable properties of a system with an enormous number of degrees of freedom. One approach to the analysis of such systems is to define simplistic mathematical models which can be calculated exactly. The first example is the Ising model, solved by Onsager in 1944. A state of the system is defined by a configuration of + and – signs (spins) at the vertices of a square lattice in the plane. Each edge of the lattice is to be thought of as an interaction and contributes an energy $E(\sigma,\sigma')$ to the total energy where $\sigma, \sigma'$ are the spins at the ends of the edge. Thus in the Ising model the total energy of a state $\sigma$ is

$$\sum_{i,j} k_1\sigma_{i,j}\sigma_{i+1,j} + k_2\sigma_{i,j}\sigma_{i,j+1} .$$

The partition function is then defined (for any system) to be

$$Z = \sum_{\sigma} \exp(-E(\sigma)/kT) .$$

Note that Z depends on some parameters $(k_1, k_2, kT)$. We will adopt the convention of summing up this dependence of Z (in any model) by a single "parameter" called the "spectral parameter" $\lambda$. Thus $Z = Z(\lambda)$.

The partition function does not make sense for an infinite system so we consider rectangular approximations to the system with N vertices and let $Z_N$ be the corresponding partition function (the boundary conditions need to be treated in some way, e.g. periodic). Then the free energy per site is defined as $F(\lambda) = \lim_{N\to\infty} \frac{1}{N} \log(Z_n)$, where the union of the approximating rectangles is the whole lattice. One would like to calculate $F(\lambda)$ explicitly as a function of $\lambda$. This is called "solving" a model.

## A BAXTER STRATEGY

The most successful technique in solving models is the method of *transfer matrices.* One concocts a maxtrix $T(\lambda)$ (depending on the horizontal dimension m of the approximating rectangle) such that trace $(T(\lambda)^n)$ is the partition function for the $m \times n$ rectangle. The largest eigenvalue of $T(\lambda)$ then gives the limit of $F(\lambda)$ as $n \to \infty$. Unfortunately the size of $T(\lambda)$ grows exponentially with m so diagonalization is not straightforward.

Baxter's idea was to consider systems for which the transfer matrices commute among themselves for different values of $\lambda$, thus $T(\lambda)T(\lambda') = T(\lambda')T(\lambda)$. Then the $T(\lambda)$'s have a common eigenvector v and $T(\lambda)v = f(\lambda)v$. By a detailed analysis of the symmetries and analytic dependence of the matrix entries of $T(\lambda)$ on $\lambda$, the Baxter program is then able to nail down a finite list of possible functions $f(\lambda)$, in particular the largest. By an act of faith this $f_{max}(\lambda)$ does actually occur and the model is solved.

But how is one to find systems for which the transfer matrices commute? It is natural to factorize the matrix $T(\lambda)$ as $R_m(\lambda)R_{m-1}(\lambda)\ldots R_1(\lambda)$ where each $R_i$ corresponds to adding an energy contribution at the ith position. Then one can ask the question: is there a *local* condition, involving the $R_i$'s, which guarantees that the $T(\lambda)$'s commute? Suppose that for each $\lambda$ and $\lambda'$ there is a third $\lambda''$ such that (letting $R_i(\lambda') = R_i'$ etc)

(YBE) $$R_i R'_{i+1} R''_i = R''_{i+1} R'_i R_{i+1}$$

and

$$R_i R'_j = R'_j R_i \quad \text{for } |i-j| \geq 2.$$

Then an easy computation (see [Ba]) shows that

$$\{R_m R''_m (R'_m)^{-1}\}\ TT' = T'T\ \{R_1^{-1}(R''_1)^{-1}R_1^1\}.$$

Thus up to boundary terms the transfer matrices commute. It is to be hoped, and indeed true in a host of examples, that the boundary conditions can be defined so that these terms disappear and we have commuting transfer matrices.

Using this method Baxter defined and solved a model (the "8 vertex model") that generalized the "six vertex method" solved using a Bethe ansatz technique by Lieb. The theory of "quantum groups" was designed by Jimbo and Drinfeld to produce examples of $R_i$'s satisfying YBE. They certainly find such examples for any simple Lie algebra $\mathfrak{g}$ and any finite dimensional representation of $\mathfrak{g}$ (see [Ji] and [Dr]). Note that there is a free parameter $q$ in these $R_i(\lambda)$ solutions.

## THE BRAID GROUPS

A geometric braid on $n$ strings is a way of tying two horizontal sticks together with the strings so that the tangent vector to each string always has a non-zero vertical component. Thus

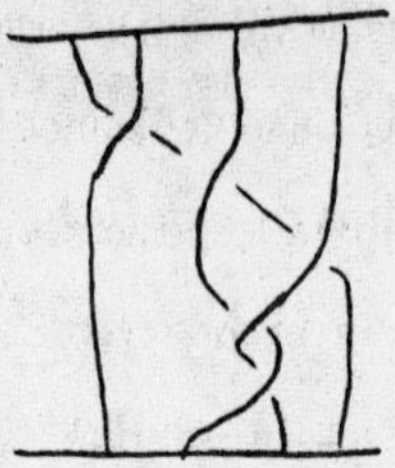

is an example of a 4-string braid. Braids form a *group* with concatenation as operation.

Thus if $\alpha =$  and $\beta =$ 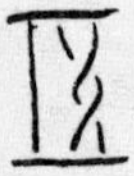, then $\alpha\beta =$ . The identity element is

obviously the braid , and the inverse of a braid is obtained by reflecting it in a mirror and turning it upside down.

Thus for each n we have a group $B_n$ which has a natural surjection onto the symmetric group $S_n$ given by permutation of the end points.

If we define $\sigma_i$, for $1 \le i < n$ as the braid

$\sigma_i$ :

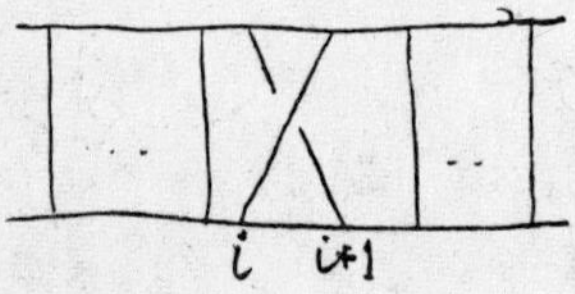


then the following relations are easy to see

(Br)
$$\sigma_i\sigma_{i+1}\sigma_i = \sigma_{i+1}\sigma_i\sigma_{i+1}$$
$$\sigma_i\sigma_j = \sigma_j\sigma_i \quad \text{if } |i-j| \ge 2 .$$

E. Artin proved that the braid group is actually equal to the group presented on $\sigma_1,\ldots,\sigma_{n-1}$ with relations (Br). Thus to define a representation of the braid group it suffices to find matrices $\sigma_i$ satisfying (Br).

The similarity between (Br) and (YBE) is striking and one could look for representations of $B_n$ using the $R_i(\lambda)$'s of solvable models. The actual way to get rid of the spectral parameter to get a nontrivial representation may be subtle. Typically one will have to let the $\lambda$'s go to infinity in some controlled way. But in the cases of models obtained from quantum groups this always works so we get, for every simple Lie algebra $\mathfrak{g}$, and every finite dimensional representation of it, a representation of $B_n$, for every $n$, depending on a parameter $q$.

## KNOTS AND LINKS

A knot is a smooth non-self intersecting curve in $S^3$. Two knots are equivalent if there is a diffeomorphism of $S^3$ taking one onto the other. Knots always admit projections onto $\mathbb{R}^2$ whose singularities are at worst double points so one may always draw a picture which adequately, but highly non-uniquely, represents the knot.

e.g.

the unknot

trefoil

figure 8

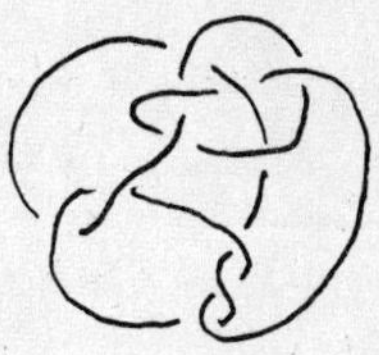

Conway knot

Links are "knots with several components",

e.g.

unlink on 2 components

Hopf Link

Whitehead Link

Borromean rings

It is in general difficult to tell, from looking at pictures of knots, whether they are the same or not. What one needs are invariants, i.e., entities associated with the knot in 3-space rather than just the picture. Most invariants are obtained by looking at the topological space obtained when the knot is removed from 3-space and doing some algebraic topology.

Consider a knot $K$ in the three sphere and remove a tubular neighbourhood $T$ of it. This tubular neighbourhood is abstractly just a torus $D^2 \times S^1$ and, for each integer $p > 0$ there is a $p$ fold covering space of $D^2$, branched over the centre, given by the map $z \to z^p$ of $D^2(=\{z \in \mathbb{C} \mid |z| \leq 1\})$. This defines a covering of $D^2 \times S^1$, branched over the centre of the solid torus, inducing a regular $p$-fold covering on the boundary. The fundamental group of the complement of $T$ contains a copy of $\mathbb{Z}$ coming from a path that circles the knot with linking number one. Thus the complement of $T$ admits a $p$-fold covering space which matches, on the boundary, the covering of $D^2 \times S^1$. Glueing the branched cover of $D^2 \times S^1$ back in, one obtains a compact 3-manifold which is a $p$-fold branched cover of $S^3$, branched precisely along the knot. This manifold is obviously an invariant of the knot. For $p = 2$, the homology of this manifold was used very early on as an invariant in knot theory.

Another old invariant is the Alexander polynomial $\Delta(t)$, discovered in 1928 by J. Alexander. It can be defined, up to normalisation, simply as the characteristic polynomial of the action of a generator of $\mathbb{Z}$ on the homology of the infinite cyclic cover of the knot complement for the group $\mathbb{Z}$ described above. The normalisation ambiguity can be eliminated by requiring $\Delta(t) = \Delta(t^{-1})$, $\Delta(1) = 1$. For the trefoil $\Delta(t) = t^{-1} - 1 + t$ and for the Conway knot $\Delta(t) \equiv 1$.

Given a braid $\alpha \in B_n$, it may be turned into an oriented link by "closing" it as below .

$\alpha =$  closure = $\hat{\alpha}$ 

It is a result of Alexander that any oriented link may be obtained in this way. So it was natural to look for invariants of links by studying representations of the braid groups. In particular one may look at representations coming from solvable models as described above. Since the partition function is given by a trace, it is natural to take the trace of the matrix given by the braid. The remarkable result is that, if the trace is slightly modified (corresponding to special boundary conditions), all models coming from quantum groups give polynomial invariants (in the variable q) of links in this way. This was proved by Rosso and Reshetikhin. Thus to every simple Lie algebra $\mathfrak{g}$ and every finite dimensional representation of it there is a knot polynomial.

The simplest example is for $s\ell_2$ in its 2-dimensional representation. The corresponding polynomial is called $V_L(q)$, and the corresponding model is a 6-vertex model. As examples, $V_{trefoil} = q+q^3-q^4$, $V_{fig.8} = q^{-2}-q^{-1}+1-q+q^2$, $V_{conway}$ = complicated. The n-dimensional representations of $s\ell_n$ combine to form a 2-variable polynomial $P_L$ known as the HOMFLY polynomial and the symplecic and orthogonal algebras (in their defining representations) combine to form the Kauffman polynomial $F_L$.

It is indeed a bizarre way to obtain knot invariants: we took a link, turned it into a braid, used some infinite limit in the spectral parameter of a solvable model to obtain a braid group representation and then took a modified trace of the braid to get our invariant! It must be said that this process is not very well understood at this stage. It is possible to avoid the use of braids by using the statistical mechanical model directly on an arbitrary link diagram and calculating its partition function. In the infinite limit of the spectral parameter one obtains an invariant. But the role of solvability and the reason for the infinite limit remain obscure. (see also [W]).

## BAXTERIZATION

In a small attempt to understand the correspondence between solvable models and knot theory I propose the following procedure: take a knot invariant, turn it (if possible) into a coherent sequence of braid group representations, and then Baxterize (if possible) by inserting a spectral parameter so that YBE is satisfied and so that the original braid group representation is the infinite limit in the spectral parameter of the Baxterized version.

The simplest way to Baxterize the $\sigma_i$'s is what I call *self-Baxterization* where $R_i(\lambda)$ actually belongs to the algebra generated by $\sigma_i$. I know of many cases where this is possible, the first two I shall give are *universal Baxterizations*

1) Hecke case

If the $\sigma_i$'s satisfy a quadratic equation it is possible to renormalize them so that

$$\sigma_i + \sigma_i^{-1} = x1$$

Then we simply put

$$R_i(\lambda) = e^{\lambda}\sigma_i + e^{-\lambda}\sigma_i^{-1}$$

Equation (YBE) follows immediately.

**2) BIRMAN-MURAKAMI-WENZL CASE** (see [BW])

Suppose that $\sigma_i$'s have the following property for some numbers x and a:

$$\text{(define } E_i = \tfrac{1}{x}(\sigma_i+\sigma_i^{-1})-1)$$

$$E_i^2 = (a+a^{-1}-x)x^{-1}E_i$$

(BMW) $$E_i\sigma_{i-1}^{\pm1}E_i = a^{\mp1}E_i\,, \qquad E_i\sigma_{i+1}^{\pm1}E_i = a^{\mp1}E_i$$

$$E_i\sigma_{i\mp1}\sigma_i = E_iE_{i\pm1}$$

(the reader should understand these relations by representing $E_i$ by a "braid like" object  , an idea due to Kauffman).

Then one may show with some pain that

$$R_i(\lambda) = (e^{\lambda}-1)k\sigma_i + x(k+k^{-1})1 + (e^{-\lambda}-1)k^{-1}\sigma_i^{-1}$$

satisfies (YBE). The result can also be deduced with a different kind of pain, from Jimbo's paper [Ji].

Thus whenever a braid group representation factors through the Hecke or BMW algebras it can be Baxterized.

**ABELIAN BAXTERIZATION** occurs when the $R_i(\lambda)$'s commute among themselves for different values of $\lambda$. Self-Baxterization is necessarily abelian. Non-abelian Baxterization occurs in quantum group theory. For the adjoint representation of $s\ell_3$, the $R_i(\lambda)$'s do not commute. I will return to this later.

To see an example of abelian Baxterization which is not self-Baxterization I turn to the main example of this talk (also seen in [KMM]), where we see that the simplest of all knot invariants, the homology of the double cover mod N, can be Baxterized.

Let $D_n$ be the algebra on generators $u_1,u_2,\dots,u_{n-1}$ with relations $u_\ell^N = 1$, $u_\ell u_{\ell+1} = \omega u_{\ell+1}u_\ell$ ($\omega = -e^{2\pi i/N}$), $u_ku_\ell = u_\ell u_k$ if $|\ell-k| \geq 2$. Then it is easy to show that if we define $\sigma_k = \sum_{j=0}^{N-1} \omega^{j^2}u_k^j$, that the braid group relations are satisfied. It is equally easy to see that the trace on $D_n$, defined by $tr(\omega) = 0$ if $\omega$ is a non-scalar product of

$u_\ell$'s, has the right property (Markov) to define a link invariant. In [GJ] it is shown that the absolute value of this invariant is, at least for N odd, the order of the homology of the 2-fold branched cover mod N. One can attempt to Baxterize these representations but if one looks in [FZ] one finds that the models already exist. The Fateev Zamolodchikov model has transfer matrices

$$R_k(\lambda) = \sum_{j=0}^{N-1} \left[ \prod_{a=0}^{j-1} \frac{\sin(\pi a/N + \lambda/2N)}{\sin(\pi(a+1)/N - \lambda/2N)} \right] u_k^j$$

Taking the limit $\lambda \to i\infty$ we obtain $\sigma_k$ as above.

If N is a prime (or 6) it is easy to see that $R_k(\lambda)$ is in the algebra generated by $\sigma_k$, so we have self-Baxterization. For other values of N this is not true but the Baxterization is clearly abelian.

## FURTHER EXAMPLES

There are many known knot invariants that can be turned into traces of braid group representations. The number of representations of the fundamental group of the complement into a finite group is one due to Kuperberg. But the one that is most tempting to try at this stage is the homology of higher branched covers mod N. These may be obtained from the Burau representation as in [GJ]. To develop the theory in an analogous way to the mod 2 case one may proceed as follows (suppose N is an odd prime).

Let $\mathbb{F}$ be a quadratic extension of $\mathbb{Z}/N\mathbb{Z}$ with involution $*$ and let $S \in \mathbb{F}$ be any element with $S \neq S^* = S^{-1}$, write $\mathrm{Im}(a) = \frac{a-a^*}{s-s^*}$ for $a \in \mathbb{F}$. Then to replace the algebra $D_n$ previously defined we let $D_{n,s}$ be the (complex) algebra on generators $u_\ell^a$, $a \in \mathbb{F}$, $\ell = 1,\ldots,n-1$ with relations $(\omega = e^{2\pi i/N})$

$$u_\ell^a u_\ell^b = \omega^{(s+s^{-1})\mathrm{Im}(ab^*)} u_\ell^{a+b}$$

$$u_\ell^a u_{\ell+1}^b = \omega^{2\mathrm{Im}(\bar{a}b)} u_{\ell+1}^b u_\ell^a$$

$$u_k^a u_\ell^b = u_\ell^b u_k^a \quad \text{for } |k-\ell| \geq 2$$

Then it may be deduced from [GJ] or shown directly that if $\sigma_i = \sum_{a \in \mathbb{F}} \omega^{a^* a} u_i^a$, then the braid relations are satisfied.

We conjecture that these representations can be Baxterized.

## CONCLUDING REMARKS

Besides the probability of new solvable models, what are the lessons to be learned from Baxterization? I believe the biggest is that YBE is in some sense more natural than the braid relations. Let me give two reasons for thinking this, based on our examples.

1) In the Fateev Zamolodchikov model one may calculate the structure of two algebras: one the algebra generated by $R_1(\lambda)$, $R_2(\lambda), \ldots, R_n(\lambda)$, the other the algebra generated by $\sigma_1, \sigma_2, \ldots, \sigma_n$. When N is a prime (or 6) these algebras are the same. In the other cases the *first* algebra is natural and simple to describe: it is the algebra of fixed points for the involution on $D_n$ given by $u_i \to u_i^{-1}$. The *second* algebra seems quite unpleasant.

We give the structure of the first algebra as a semisimple algebra over $\mathbb{C}$. For this it suffices to give the "Bratteli diagram" which displays the various matrix algebra direct summands as natural numbers giving their size, and lines which show, according to the only reasonable convention, how the algebra generated by $R_1(\lambda), \ldots, R_n(\lambda)$ is included in that generated by $R_1, \ldots, R_{n+1}(\lambda)$. Let N be odd, then we have the following diagrams

N odd:

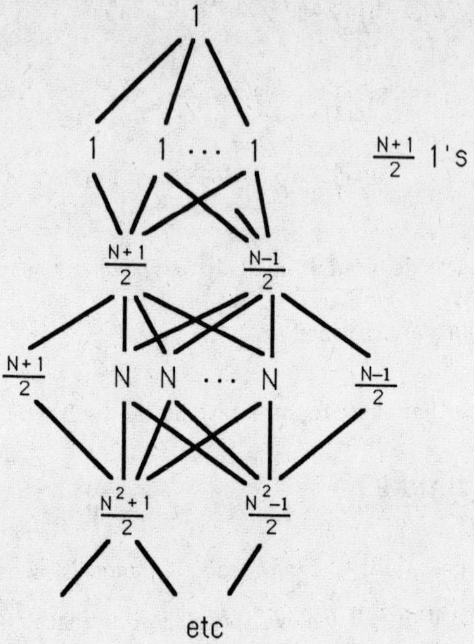


N even:

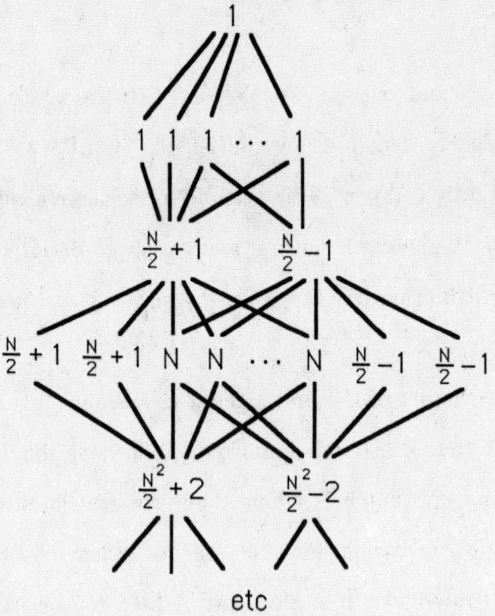


Here the number of repetitions in any row is always determined by the simple rule that a number is always the sum of the numbers connected to it on the previous row.

2) In the theory of quantum groups one may consider the commutant of the representations of $U_h(\mathfrak{g})$ in the tensor powers of an irreducible representation. It is known that, unlike in the classical limit where it is only true for $s\ell_n$, the braid group generates the commutant for all representations of $s\ell_2$ and the smallest representations of the A,B,C,D series. On the other hand this is not true for the adjoint representation of $s\ell_3$. But it is precisely in this case that the $R_i(\lambda)$ matrices generate more than the braid group. One is naturally led to the question: is the commutant of $U_h(\mathfrak{g})$ always equal to the algebra generated by the $R_i(\lambda)$ with $\lambda$ varying with i .

## REFERENCES

[A] J. Alexander. "*Topological Invariants of Knots and Links*", Trans. AMS, Vol.30 (1928) 275.

[Ba] R. Baxter "*Exactly solved models in statistical mechanics*", New York Academic Press 1982.

[BW] J. Birman, H. Wenzl "*Braids, link polynomials and a new algebra*" Trans. AMS (to appear).

[Dr] V. Drinfeld "*Quantum groups*" Proc. ICM 1986.

[FZ] V. Fateev and A. Zamolodchikov "*Self-dual solutions of the star-triangle relations in* $\mathbb{Z}_N$ *models*" Phys. Lett. 92A (1982) 37-39.

[GJ] D. Goldschmidt and V. Jones "*Metaplectic link invariants*" Geometric Dedicata (to appear).

[Ji] M. Jimbo "*Quantum R-matrix for the generalized Toda system*" Comm. Math. Phys. 102 (1986) 537-547.

[KMM] T. Kobayashi, H. Murakami, J. Murakami "*Cyclotomic invariants for links*" Proc. Japan Acad. (1988) 235-238.

[W] E. Witten "*Gauge theories an integrable lattice models*" IAS Preprint 1989.

International Journal of Modern Physics B, Vol. 4, No. 5 (1990) 715–734
© World Scientific Publishing Company

# TEMPERLEY-LIEB OPERATORS AND CRITICAL *A–D–E* MODELS

Paul A. PEARCE

*Centre for Mathematical Analysis, Australian National University, Canberra ACT 2601, Australia*

*and*

*Mathematics Department, University of Melbourne, Parkville Vic. 3052, Australia*

A review is given of recent work connecting exactly solvable lattice models with Temperley-Lieb and Hecke algebras, Dynkin diagrams of simply-laced Lie algebras, conformal and modular invariance and Virasoro characters.

## 1. Introduction

These notes are a summary of lectures presented during 1989 as part of the special year on mathematical physics at the Centre for Mathematical Analysis, Australian National University. The intent of these lectures was to introduce to an audience well versed in statistical mechanics some of the notions, ideas and results flowing from the recent advances connecting two-dimensional critical lattice models with conformal field theory. The exposition is for the most part elementary and designed to whet appetites for further research. There are some very good reviews elsewhere [1,2,3,4,5,6] which cover much the same area. Indeed, some have appeared since these lectures were given. I refer the interested reader to these for more details of some of the topics discussed here. Of course, ultimately, there is no substitute for the original papers.

## 2. Solvable *A–D–E* Lattice Models

One of the overriding goals of modern statistical mechanics is to understand how symmetry determines the various observed universality classes of critical systems and to classify them accordingly. In two dimensions, the application of conformal field theory has ushered in the first major step in this direction. Many of the theoretical advances have been either motivated by, or corroborated by, direct calculations on exactly solvable two-dimensional lattice models [7]. So recent developments on solvable models and conformal theory have been closely linked. It is known that Yang-Baxter equations give a sufficient condition for solvability of lattice models. These same equations arise, albeit in different guises, in other branches of physics as well. An important open problem is to classify all solutions to the Yang-Baxter equations. This is a difficult problem in general but large classes of solutions are known. We will start by considering a particularly interesting class of solutions leading to the critical $A$–$D$–$E$ models. The rest of the review will then focus on the application of conformal theory as it pertains to this class of critical lattice models.

### 2.1 Temperley-Lieb interaction models

The Temperley-Lieb interaction (TLI) models [8,9] are a large class of exactly solvable interaction-round-a-face or IRF models [7]. These models were studied by Owczarek and Baxter [9] in 1987 and will be used in this review to construct solvable lattice models corresponding to arbitrary graphs. In particular, we show that these models include Pasquier's critical $A$–$D$–$E$ models [10,11], which were also introduced in 1987.

The face weights of the TLI models we consider are given by

$$W\begin{pmatrix} d & c \\ a & b \end{pmatrix} = \rho[\delta(a,c) + x h_{ab} h_{cd} \delta(b,d)]. \tag{2.1}$$

Here $a, b, c, d$ are discrete $L$-state spins around an elementary square of the lattice, $\delta$ is the Kronecker delta function, $\rho$ is a normalization factor and $x$ is a free parameter. The face edges $(a,b), (b,c), (c,d)$ and $(d,a)$ are restricted to certain allowed pairs. The interactions $h_{ab}$ are in general not symmetric $h_{ab} \neq h_{ba}$ and should be thought of as three-spin interactions, that is, $h_{ab} = h_{abd}\delta(b,d)$. The TLI face weights can be used to construct two diagonal transfer matrices $V_1$ and $V_2$ corresponding to the black and white faces of the square lattices. Explicitly, the elements of $V_1$ and $V_2$ are given by

$$\begin{aligned} V_1(\sigma,\sigma') &= \prod_{i=1}^{N} W\begin{pmatrix} \sigma_{2i-2} & \sigma'_{2i-1} \\ \sigma_{2i-1} & \sigma_{2i} \end{pmatrix} \delta(\sigma_{2i}, \sigma'_{2i}) &(2.2)\\ V_2(\sigma,\sigma') &= \prod_{i=1}^{N} W\begin{pmatrix} \sigma_{2i-1} & \sigma'_{2i} \\ \sigma_{2i} & \sigma_{2i+1} \end{pmatrix} \delta(\sigma_{2i-1}, \sigma'_{2i-1}) &(2.3)\end{aligned}$$

where $\sigma = \{\sigma_0, \sigma_1, \ldots, \sigma_{2N+1}\}$ and $\sigma' = \{\sigma'_0, \sigma'_1, \ldots, \sigma'_{2N+1}\}$ are consecutive diagonal rows of spins. The dimension of the transfer matrices $V_1$ and $V_2$ is given by the number of configurations or allowed paths of spins along a diagonal row with the end spins $\sigma_0$ and $\sigma_{2N+1}$ fixed. The partition function of a lattice with $M$ diagonal rows closed on a cylinder can then be written as

$$Z = \mathrm{Tr}(V_1 V_2)^M. \tag{2.4}$$

It is convenient to factorize the transfer matrices $V_1$ and $V_2$ by introducing face transfer matrices

$$X_j = \rho(I + xU_j) \tag{2.5}$$

where $I$ is the identity matrix and

$$U_j(\sigma,\sigma') = h_{\sigma_j,\sigma_{j+1}} h_{\sigma'_j,\sigma_{j-1}} \delta(\sigma_{j-1},\sigma_{j+1}) \prod_{i \neq j} \delta(\sigma_i, \sigma'_i). \tag{2.6}$$

The transfer matrices $V_1$ and $V_2$ are then given by the ordered matrix products

$$\begin{aligned} V_1 &= \rho^N(1+xU_1)(1+xU_3)\dots(1+xU_{2N-1}) && (2.7)\\ V_2 &= \rho^N(1+xU_2)(1+xU_4)\dots(1+xU_{2N}). && (2.8)\end{aligned}$$

Let us consider a graph whose vertices are the $L$ states of the spins and whose edges are the allowed nearest-neighbour pairs. The adjacency matrix of this graph is

$$A_{\sigma,\sigma'} = \begin{cases} 1, & (\sigma,\sigma') \text{ allowed} \\ 0, & \text{otherwise.} \end{cases} \tag{2.9}$$

Owczarek and Baxter [9] have shown that if

$$V_{ab} = h_{ba}^2 \tag{2.10}$$

satisfies the conditions

$$\begin{aligned} V_{ab}V_{ba} &= 1, \quad (a,b) \text{ allowed} && (2.11)\\ \sum_b A_{ab}V_{ab} &= \sqrt{q} && (2.12)\end{aligned}$$

then the matrices $\{U_j, j=1,\dots,2N\}$ satisfy the Temperley-Lieb algebra [8]

$$\begin{aligned} U_j^2 &= \sqrt{q}U_j && (2.13)\\ U_jU_{j\pm1}U_j &= U_j \\ U_iU_j &= U_jU_i, \quad |i-j|\geq 2. \end{aligned}$$

A Temperley-Lieb algebra $\{U_j\}$ gives rise to a lattice model satisfying the Yang-Baxter equations

$$\sum_g W\begin{pmatrix} a & g \\ b & c\end{pmatrix} W'\begin{pmatrix} f & e \\ a & g\end{pmatrix} W''\begin{pmatrix} e & d \\ g & c\end{pmatrix} = \sum_g W''\begin{pmatrix} f & g \\ a & b\end{pmatrix} W'\begin{pmatrix} g & d \\ b & c\end{pmatrix} W\begin{pmatrix} f & e \\ g & d\end{pmatrix}. \tag{2.14}$$

Following Baxter [12] we define face transfer matrices

$$\begin{aligned} X_j &= \rho(I+xU_j) && (2.15)\\ X_j' &= \rho'(I+x'U_j) \\ X_j'' &= \rho''(I+x''U_j). \end{aligned}$$

The Yang-Baxter equation then becomes

$$X_jX_{j+1}'X_j'' = X_{j+1}''X_j'X_{j+1} \tag{2.16}$$

and is satisfied provided only that the parameters $x, x', x''$ are related by

$$x'' = \frac{x'-x}{1+\sqrt{q}x+xx'}. \tag{2.17}$$

To see this just expand the Yang-Baxter equation in the $U_j$ and equate the coefficients of $I, U_j, U_{j+1}, U_j U_{j+1}$ and $U_{j+1}U_j$. In fact, the $U_j$ need only satisfy the conditions

$$U_j^2 - \sqrt{q}U_j = U_{j+1}^2 - \sqrt{q}U_{j+1} \tag{2.18}$$

$$U_j U_{j+1} U_j - U_j = U_{j+1} U_j U_{j+1} - U_{j+1} \tag{2.19}$$

so it is sufficient that the $U_j$ satisfy a Hecke algebra.

The most convenient way to ensure that $x, x', x''$ satisfy the required identity is to set

$$\sqrt{q} = \Delta = \begin{cases} 2\cos\lambda, & q < 4 \\ 2, & q = 4 \\ 2\cosh\lambda, & q > 4 \end{cases} \tag{2.20}$$

where $\lambda$ is the crossing parameter and set

$$x = \frac{s}{s_-}, \quad x' = \frac{s'}{s'_-}, \quad x'' = \frac{s''}{s''_-} \tag{2.21}$$

where

$$s = \begin{cases} \dfrac{\sin u}{\sin\lambda}, & q < 4 \\ u, & q = 4 \\ \dfrac{\sinh u}{\sinh\lambda}, & q > 4 \end{cases} \qquad s_- = \begin{cases} \dfrac{\sin(\lambda - u)}{\sin\lambda}, & q < 4 \\ 1 - u, & q = 4 \\ \dfrac{\sinh(\lambda - u)}{\sinh\lambda}, & q > 4 \end{cases} \tag{2.22}$$

and similarly for the primed and double primed variables with $u$ replaced by $u'$ and $u''$. If the spectral parameters satisfy the difference property

$$u'' = u' - u \tag{2.23}$$

then the required relation follows from the generalized trigonometric identity

$$ss'_-s''_- + s_-s'_-s'' + ss's'' + \Delta ss'_-s'' = s_-s's''_-. \tag{2.24}$$

Physically, the spectral parameter $u$ is related to spatial anisotropy. This anisotropy acts to distort a square face into a rhombus with angle $\theta$ given by

$$\theta = \begin{cases} \dfrac{\pi u}{\lambda}, & q \neq 4 \\ \pi u, & q = 4. \end{cases} \tag{2.25}$$

The difference property of the Yang-Baxter equation then becomes

$$\theta'' = \theta' - \theta \tag{2.26}$$

which has the meaningful geometric interpretation shown in Fig. 1.

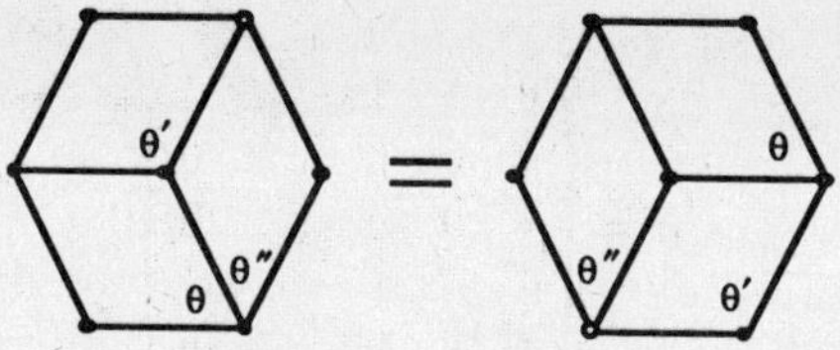


Figure 1. Diagrammatic representation of the Yang-Baxter equation showing the anisotropy angles satisfying the geometric constraint $\theta'' = \theta' - \theta$. The central spin is summed out and equality holds for all values of the six perimeter spins.

### *2.2 Critical A–D–E models*

The TLI models contain the critical $A$–$D$–$E$ models [10,11] as a special case. To see this notice that the second Owczarek-Baxter condition (2.12) looks like an eigenvalue equation. In particular, if we set

$$V_{ab} = S_b/S_a \tag{2.27}$$

then the first condition (2.11) is automatically satisfied and the second condition takes the form

$$\sum_b V_{ab} S_b = \sqrt{q} S_a. \tag{2.28}$$

Clearly this is satisfied if $S_a$ is an eigenvector of the adjacency matrix $A$ with eigenvalue $\sqrt{q}$. It follows that if $S_a$ is any fixed eigenvector of $A$, then the matrices

$$U_j(\sigma, \sigma') = \sqrt{\frac{S\sigma_j S\sigma'_j}{S\sigma_{j-1} S\sigma_{j-1}}}\, \delta(\sigma_{j-1}, \sigma_{j+1}) \prod_{i \neq j} \delta(\sigma_i, \sigma'_i). \tag{2.29}$$

satisfy the Temperley-Lieb algebra.

Putting everything together, it has been shown that if $S_a$ are the components of an eigenvector of the adjacency matrix $A$ with corresponding eigenvalue $\sqrt{q}$, then the face weights

$$W\begin{pmatrix} d & c \\ a & b \end{pmatrix} = \rho \left[ s_- \delta(a, c) + s \sqrt{\frac{S_a S_c}{S_b S_d}}\, \delta(b, d) \right] \tag{2.30}$$

satisfy the Yang-Baxter equations and so the corresponding lattice model is exactly solvable. It is usual, but not necessary, to take $S_a$ to be the positive Perron-Frobenius vector of $A$. The face weights are then real and positive for $0 \le u \le \lambda$ (or $0 \le u \le 1$ in the case $q = 4$).

To make the discussion more concrete at this point let us consider the $q$-state Potts model. The $q$-state Potts model has an adjacency graph consisting of a $q$-pointed star. The spins assume the states $\sigma = 0, 1, 2, \ldots, q$ and the $(q+1) \times (q+1)$

adjacency matrix

$$A = \begin{pmatrix} 0 & 1 & 1 & \dots & 1 \\ 1 & 0 & 0 & \dots & 0 \\ 1 & 0 & 0 & \dots & 0 \\ \vdots & \vdots & \vdots & \ddots & \vdots \\ 1 & 0 & 0 & \dots & 0 \end{pmatrix} \tag{2.31}$$

has largest eigenvalue $\sqrt{q}$ with Perron-Frobenius vector

$$S = (\sqrt{q}, 1, 1, \dots, 1). \tag{2.32}$$

The TLI model constructed starting with the star graph is precisely the self-dual $q$-state Potts model. This model is critical for $q \leq 4$ and exhibits first-order phase coexistence for $q > 4$. The Temperley-Lieb equivalence [8,7,12] implies that the free energy of a Temperley-Lieb interaction model is given by the free energy of the Potts model with the corresponding value of $q$. This equivalence can also be extended to certain correlation functions. It therefore follows that a Temperley-Lieb interaction model is critical if the corresponding Potts model is critical, that is, if the largest eigenvalue $\Lambda_{\text{max}}$ of the adjacency matrix $A$ satisfies

$$\Lambda_{\text{max}} \leq 2. \tag{2.33}$$

### *2.3 Dynkin diagrams*

In statistical mechanics the object of central study is critical behaviour. It is therefore natural to ask which graphs lead via the Temperley-Lieb construction to critical lattice models. More precisely, the question is which undirected connected graphs without loops have adjacency matrices with $\Lambda_{\text{max}} \leq 2$. The answer is an old result in the spectra of graphs [13] and surprisingly has links to Lie algebras. The only connected graphs for which $\Lambda_{\text{max}} = 2$ are the Dynkin diagrams of the simply-laced affine Lie algebras. The allowed connected graphs for which $\Lambda_{\text{max}} < 2$ are the Dynkin diagrams of the simply-laced classical Lie algebras. These graphs are shown in Fig. 2. The Cartan matrix $C$ of the Lie algebra is related to the adjacency matrix by $C = 2I - A$. Both the classical and affine Dynkin diagrams consist of an infinite $A$ and $D$ series plus a few exceptional cases. These results can be proved from the following lemmas [13] which we state without proof:

**Lemma 1.** *The maximal eigenvalue of an irreducible nonnegative matrix $A$ is a strictly increasing function of its matrix elements.*

**Lemma 2.** *If $G$ is a connected graph then either (i) $G$ is an affine Dynkin diagram (ii) $G$ is a subgraph of an affine Dynkin diagram, that is, $G$ is a classical Dynkin diagram or (iii) $G$ has a subgraph which is an affine Dynkin diagram.*

Another surprise is that the eigenvalues of the Dynkin diagrams can all be written in the form

$$\Lambda_j = 2\cos(m_j \pi / h) \tag{2.34}$$

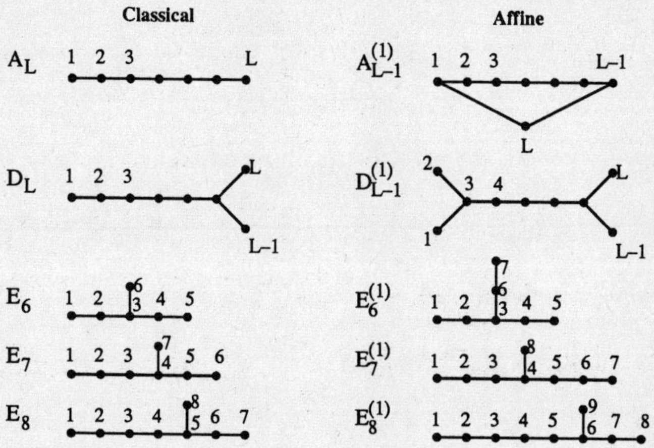


Figure 2. Dynkin diagrams of the simply-laced classical and affine Lie algebras.

**Coxeter Exponents**

| Lie Algebra | h | $m_j$ |
|---|---|---|
| $A_L$ | L+1 | 1, 2, 3,..., L |
| $D_L$ | 2L–2 | L–1, 1, 3, 5,..., 2L–3 |
| $E_6$ | 12 | 1, 4, 5, 7, 8, 11 |
| $E_7$ | 18 | 1, 5, 7, 9, 11, 13, 17 |
| $E_8$ | 30 | 1, 7, 11, 13, 17, 19, 23, 29 |
| $A^{(1)}_{L-1}$ | L | 0, 2, 4,..., 2L–2 |
| $D^{(1)}_{L-1}$ | 2(L–3) | 0, 2, 4,..., 2(L–3), L–3, L–3 |
| $E^{(1)}_6$ | 6 | 0, 2, 2, 3, 4, 4, 6 |
| $E^{(1)}_7$ | 12 | 0, 3, 4, 6, 6, 8, 9, 12 |
| $E^{(1)}_8$ | 30 | 0, 6, 10, 12, 15, 18, 20, 24, 30 |

Table 1. The Coxeter number $h$ and the Coxeter exponents $m_j$ for simply-laced classical and affine Lie algebras.

where $h$ and $m_j$ are integers called the Coxeter number and Coxeter exponents respectively. These numbers are listed in Table 1. The corresponding Perron-Frobenius eigenvectors are listed in Table 2.

| Algebra | Perron-Frobenius Eigenvector |
|---|---|
| $A_L$ | $\left(\sin\frac{\pi}{L+1}, \sin\frac{2\pi}{L+1}, \ldots, \sin\frac{L\pi}{L+1}\right)$ |
| $D_L$ | $\left(2\cos\frac{(L-2)\pi}{2L-2}, \ldots, 2\cos\frac{2\pi}{2L-2}, 2\cos\frac{\pi}{2L-2}, 1, 1\right)$ |
| $E_6$ | $\left(\sin\frac{\pi}{12}, \sin\frac{\pi}{6}, \sin\frac{\pi}{4}, \sin\frac{\pi}{3} - \frac{\sin \pi/4}{2\cos \pi/12}, \sin\frac{5\pi}{12} - \sin\frac{\pi}{4}, \frac{\sin \pi/4}{2\cos \pi/12}\right)$ |
| $E_7$ | $\left(\sin\frac{\pi}{18}, \sin\frac{\pi}{9}, \sin\frac{\pi}{6}, \sin\frac{2\pi}{9}, \sin\frac{5\pi}{18} - \frac{\sin 2\pi/9}{2\cos \pi/18}, \sin\frac{\pi}{3} - \sin\frac{2\pi}{9}, \frac{\sin 2\pi/9}{2\cos \pi/18}\right)$ |
| $E_8$ | $\left(\sin\frac{\pi}{30}, \sin\frac{\pi}{15}, \sin\frac{\pi}{10}, \sin\frac{2\pi}{15}, \sin\frac{\pi}{6}, \sin\frac{\pi}{5} - \frac{\sin \pi/6}{2\cos \pi/30}, \sin\frac{7\pi}{30} - \sin\frac{\pi}{6}, \frac{\sin \pi/6}{2\cos \pi/30}\right)$ |
| $A_{L-1}^{(1)}$ | (1, 1,..., 1) |
| $D_{L-1}^{(1)}$ | (1, 1, 2, 2,...,2, 2, 1, 1) |
| $E_6^{(1)}$ | (1, 2, 3, 2, 1, 2, 1) |
| $E_7^{(1)}$ | (1, 2, 3, 4, 3, 2, 1, 2) |
| $E_8^{(1)}$ | (1, 2, 3, 4, 5, 6, 4, 2, 3) |

Table 2. The Perron-Frobenius eigenvectors for simply-laced classical and affine Lie algebras. The order of the components is fixed by the labelling of sites in the Dynkin diagrams of Fig. 2

In the $A$–$D$–$E$ classification, the critical $q = 2, 3, 4$ state Potts models correspond to Lie algebras $A_3, D_4$ and $D_4^{(1)}$ respectively. More generally, the classical $A$ series corresponds to the ABF or RSOS models [14] and the affine $A$ series corresponds to the cyclic solid-on-solid or CSOS models [15,16,17]. Classical $D$ lattice models have been introduced by Pasquier [18] and affine $D$ models by Akutsu, Kuniba and Wadati [19] and Choi, Kwon and Kim [20]. The $A$ and $D$ models admit off-critical elliptic extensions to the solution of the Yang-Baxter equations but no elliptic solutions are known for the exceptional cases.

## 3. $c < 1$ Theories

### *3.1 Conformal invariance*

It has been known for some time that statistical mechanics models are scale invariant at criticality. The principle of scale invariance has led to the formulation of the renormalization group and implies, among other things, scaling relations among the critical exponents that characterize critical behaviour. Recently, the notion of scale invariance of critical systems has been refined by considering invariance under more general conformal [21,22] and modular [23,24] transformations. In two dimensions, the theory of conformal invariance predicts that the critical exponents fall into various universality classes classified by the central charge $c$ of the corresponding Virasoro algebra. In particular, for unitary theories with $c < 1$, the critical exponents are quantized and a complete classification of critical behaviours can be given in terms of the unitary series of conformal field theories with central charge [25]

$$c = 1 - \frac{6}{h(h-1)} \tag{3.1}$$

where $h = 4, 5, 6, \ldots$. For these theories the scaling dimensions $x$ and spins $s$ of the various primary scaling operators are given by

$$x = \Delta_{r,s} + \bar{\Delta}_{r,s} \tag{3.2}$$

$$s = \Delta_{r,s} - \bar{\Delta}_{r,s} \tag{3.3}$$

in terms of the conformal weights $(\Delta_{r,s}, \bar{\Delta}_{r,s})$ of two copies of the Virasoro algebra, each given independently by the Kac formula [26]

$$\Delta_{r,s} = \frac{[hr-(h-1)s]^2 - 1}{4h(h-1)} \qquad 1 \le r \le h-2, 1 \le s \le h-1, r \ge s. \tag{3.4}$$

with $\bar{\Delta}_{r,s} = \Delta_{\bar{r},\bar{s}}$. The allowed values of the conformal weights for $h = 4, 5$ and 6 are listed in Table 3.

h = 4

| s \ r | 1 | 2 |
|---|---|---|
| 2 | | 1/16 |
| 1 | 0 | 1/2 |

h=5

| s \ r | 1 | 2 | 3 |
|---|---|---|---|
| 3 | | | 1/10 |
| 2 | | 3/80 | 3/5 |
| 1 | 0 | 7/16 | 3/2 |

h=6

| s \ r | 1 | 2 | 3 | 4 |
|---|---|---|---|---|
| 4 | | | | 1/8 |
| 3 | | | 1/15 | 2/3 |
| 2 | | 1/40 | 21/40 | 13/8 |
| 1 | 0 | 2/5 | 7/5 | 3 |

Table 3. Grids of conformal weights $\Delta_{r,s}$ for $h = 4, 5$ and 6.

The simplest possibility is $h = 4, c = 1/2$ with three spinless operators given by $(\Delta, \bar{\Delta}) = (0,0)$, $(1/2, 1/2)$ and $(1/16, 1/16)$. In two dimensions, the thermal and magnetic scaling dimensions are related to the usual critical exponents by

$$2-\alpha=\frac{2}{2-x_\epsilon},\qquad \frac{2\beta}{2-\alpha}=x_\sigma. \tag{3.5}$$

Hence these three scaling fields are easily recognized as the identity (1), energy ($\epsilon$) and magnetization ($\sigma$) operators of the Ising model which has critical exponents $\alpha=0$ and $\beta=1/8$. More generally, it was pointed by Huse [27] that the generic critical points of the ABF or $A_L$ models, possessing $\mathbb{Z}_2$ symmetry, realize the entire unitary series with $h=L+1$. More specifically, the scaling dimensions calculated from the known critical exponents of the order parameters [14] agree with allowed zero-spin values given by the Kac formula. Thus, for $h=4,5,6,\ldots$ the unitary series describes the critical point of the Ising model, the tricritical Ising model, the tetracritical Ising model and so on.

The application of conformal invariance gives a list of possible scaling dimensions that can appear in a conformal theory or critical lattice model with central charge $c<1$. It does not give the precise combinations of scaling operators allowed nor does it give any information on the degeneracy of operators. The answers to these questions are found by applying the stronger requirement of modular invariance on a torus.

### *3.2 Partition function on a torus*

Let us consider a conformal field theory or critical lattice model on a finite $\ell\times\ell'$ periodic lattice or torus. The partition function can be written as

$$Z_{\ell,\ell'}=\exp(-\ell\ell' f)Z(q) \tag{3.6}$$

where $f$ is the bulk free energy and $Z(q)$ is a universal term describing the leading finite-size corrections in the limit of $\ell,\ell'$ large with the aspect ratio $\delta=\ell'/\ell$ fixed. The argument $q$ is the modular parameter. For a spatially isotropic model, it is simply related to the aspect ratio $\delta$ by $q=\exp(-2\pi\delta)$. The partition function on a torus can be calculated from the eigenvalues of the row transfer matrix $T$ of a periodic row of $\ell$ faces by use of the formula

$$Z_{\ell,\ell'}=\mathrm{Tr}\,T^{\ell'}=\sum_n\Lambda_n^{\ell'}=\sum_n\exp(-\ell' E_n) \tag{3.7}$$

where

$$\Lambda_n=\exp(-E_n) \tag{3.8}$$

are the eigenvalues of $T$ and $E_n$ are corresponding energy levels. But now conformal invariance dictates [28,29] that the leading finite-size corrections to the energy levels take the form

$$E_0 = \ell f-\frac{\pi c}{6\ell}\sin\theta \tag{3.9}$$

$$E_n-E_0 = \frac{2\pi}{\ell}(x_n\sin\theta+is_n\cos\theta) \tag{3.10}$$

where $E_0$ is the groundstate energy, $n = 1, 2, \ldots$, and

$$x_n = \Delta + \bar{\Delta} + k + \bar{k}, \quad s_n = \Delta - \bar{\Delta} + k - \bar{k}, \quad k, \bar{k} \in \mathbb{N}. \tag{3.11}$$

The angle $\theta$ is determined by the spatial anisotropy [30] and is given by (2.25).

The energy spectrum consists of a groundstate and an excited state for each primary operator. Above each of these is a tower of equally spaced levels or descendents. It follows that

$$Z(q) = (q\bar{q})^{-c/24} \sum_{\Delta,\bar{\Delta}} \sum_{k,\bar{k}} d_\Delta(k) d_{\bar{\Delta}}(\bar{k}) q^{\Delta+k} \bar{q}^{\bar{\Delta}+\bar{k}}. \tag{3.12}$$

The first sum is over all conformal weights allowed by the Kac formula including multiplicities. The second sum is over all nonnegative integers $k, \bar{k}$ and the factors $d_\Delta(k), d_{\bar{\Delta}}(\bar{k})$ are integers giving the degeneracy of the levels. The modular parameter is

$$q = \exp(2\pi i \tau), \qquad \tau = \frac{\ell'}{\ell} \exp[i(\pi - \theta)] \tag{3.13}$$

and $\bar{q}$ is the complex conjugate. Hence the universal finite-size partition function on a torus can be written as a sesquilinear form in Virasoro characters

$$Z(q) = \sum_{\Delta,\bar{\Delta}} \chi_\Delta(q) \mathcal{N}(\Delta, \bar{\Delta}) \chi_{\bar{\Delta}}(\bar{q}) \tag{3.14}$$

where the sum is now over distinct conformal weights and the integer $\mathcal{N}(\Delta, \bar{\Delta})$ gives the multiplicity of the primary operator $(\Delta, \bar{\Delta})$. Here

$$\chi_\Delta(q) = q^{-c/24} \sum_{k=0}^{\infty} d_\Delta(k) q^{\Delta+k} \tag{3.15}$$

is the character of the representation of the Virasoro algebra. It is given explicitly by

$$\chi_\Delta(q) = \chi_{\Delta_{r,s}}(q) = \chi_{r,s}(q) \tag{3.16}$$

$$= \frac{q^{-c/24}}{\prod_{n=1}^{\infty}(1-q^n)} \sum_{n=-\infty}^{\infty} \left\{ q^{\frac{[2h(h-1)n + hr - (h-1)s]^2 - 1}{4h(h-1)}} - q^{\frac{[2h(h-1)n + hr + (h-1)s]^2 - 1}{4h(h-1)}} \right\}. \tag{3.17}$$

The set of numbers $\mathcal{N}(\Delta, \bar{\Delta}) \in \mathbb{N}$ is called the operator content of the theory. For the Ising model we have seen that there are three scaling operators. In this case the operator content is given by

$$\mathcal{N}(0,0) = \mathcal{N}(1/2, 1/2) = \mathcal{N}(1/16, 1/16) = 1, \qquad \mathcal{N}(\Delta, \bar{\Delta}) = 0, \Delta \neq \bar{\Delta} \tag{3.18}$$

and so

$$Z(q) = |\chi_0(q)|^2 + |\chi_{1/2}(q)|^2 + |\chi_{1/16}(q)|^2 \tag{3.19}$$

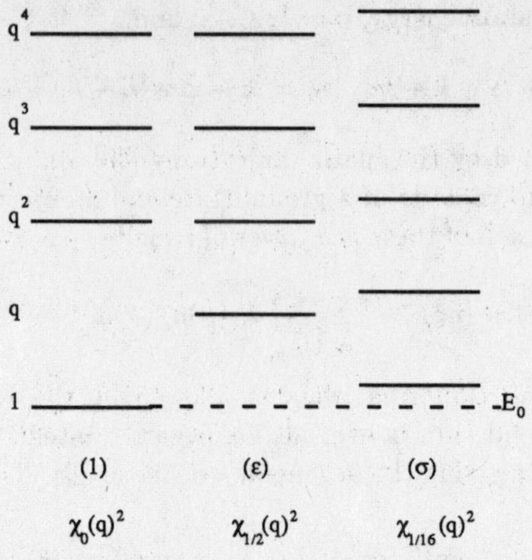


Figure 3. Equally spaced towers of levels for the isotropic Ising model above the identity, energy and magnetization operators. The degeneracies of the levels are given by the coefficients in the expansion of the squares of the Virasoro characters.

where the $c = 1/2$ Virasoro characters can be written as

$$\chi_0(q) = \tfrac{1}{2}q^{-\frac{1}{48}}\left\{\prod_{n=1}^{\infty}(1+q^{n-\frac{1}{2}}) + \prod_{n=1}^{\infty}(1-q^{n-\frac{1}{2}})\right\} \tag{3.20}$$
$$= q^{-\frac{1}{48}}(1+q^2+q^3+2q^4+\ldots)$$
$$\chi_{1/2}(q) = \tfrac{1}{2}q^{-\frac{1}{48}}\left\{\prod_{n=1}^{\infty}(1+q^{n-\frac{1}{2}}) - \prod_{n=1}^{\infty}(1-q^{n-\frac{1}{2}})\right\} \tag{3.21}$$
$$= q^{-\frac{1}{48}+\frac{1}{2}}(1+q+q^2+q^3+\ldots)$$
$$\chi_{1/16}(q) = q^{\frac{1}{24}}\prod_{n=1}^{\infty}(1+q^n) \tag{3.22}$$
$$= q^{-\frac{1}{48}+\frac{1}{16}}(1+q+q^2+2q^3+\ldots).$$

This situation is shown schematically, for the isotropic Ising model, in Fig. 3. Amazingly, the same Virasoro characters and analogous tower structures occur in the critical spectrum of the Ising model corner transfer matrices [31].

### *3.3 Modular invariance*

The universal partition function $Z(q)$ must be invariant under transformations of the modular group generated by

$$T : \quad \tau \mapsto 1+\tau \tag{3.23}$$
$$S : \quad \tau \mapsto -\frac{1}{\tau}. \tag{3.24}$$

Here the torus is formed by identifying opposite sides of a parallelogram in the complex plane $\mathbb{C}$ with

$$q = e^{2\pi i\tau}, \qquad \operatorname{im}\tau > 0. \tag{3.25}$$

The vertices of the torus are $0, 1, \tau$ and $1+\tau$. The generators $T$ and $S$ map this torus into itself. Under the generators of the modular group the Virasoro characters transform as

$$T : \quad \chi_\Delta(q) \mapsto e^{2\pi i(\frac{c}{24}-\Delta)} \tag{3.26}$$

$$S : \quad \chi_\Delta(q) \mapsto \sum_{\Delta'} \mathcal{A}(\Delta, \Delta')\chi_{\Delta'}(q) \tag{3.27}$$

where the elements of the $\frac{1}{2}(h-1)(h-2) \times \frac{1}{2}(h-1)(h-2)$ matrix $\mathcal{A}$ are given by

$$\mathcal{A}(\Delta_{r,s}, \Delta'_{r',s'}) = \sqrt{\frac{8}{h(h-1)}}(-1)^{(r+s)(r'+s')} \sin\frac{\pi r r'}{h-1} \sin\frac{\pi s s'}{h}. \tag{3.28}$$

The matrix $\mathcal{A}$ is real symmetric, orthogonal and an involution

$$\mathcal{A}^{\mathrm{T}} = \mathcal{A} = \mathcal{A}^{-1}. \tag{3.29}$$

Hence the requirement that the sesquilinear form in Virasoro characters giving the partition function is modular invariant becomes

$$\mathcal{A}\mathcal{N} = \mathcal{N}\mathcal{A}, \qquad \Delta - \bar{\Delta} \in \mathbb{N}. \tag{3.30}$$

Since $\mathcal{N}$ is a nonnegative integer matrix this is a system of diophantine equations. One family of solutions to these equations for $h = 4, 5, \ldots$ is given by

$$\mathcal{A}(\Delta, \bar{\Delta}) = \delta(\Delta, \bar{\Delta}). \tag{3.31}$$

Clearly, this gives the operator content of the ABF [14] or classical $A_L$ models with $h = L+1$. For $h = 4$ and 5 these are the only solutions corresponding to the Ising and tricritical Ising models. For $h = 6$, however, there are two solutions corresponding to the tetracritical Ising model and the three-state Potts models. The modular invariant partition function of the three-state Potts or $D_4$ model is

$$Z(q) = |\chi_0(q) + \chi_3(q)|^2 + |\chi_{2/5}(q) + \chi_{7/5}(q)|^2 + 2|\chi_{1/15}(q)|^2 + 2|\chi_{2/3}(q)|^2. \tag{3.32}$$

This solution, which involves only a subset of the allowed conformal weights, exhibits both degenerate primary operators and operators with spin.

The diophantine equations have in fact been solved in complete generality by Capelli, Itzykson and Zuber [32,33] and Kato [34]. Remarkably, they found that the modular invariant sesquilinear forms in Virasoro characters, giving the universal partition functions of unitary conformal theories with $c < 1$, are in a one-to-one correspondence with the classical $A$–$D$–$E$ Lie algebras. More specifically they found two infinite hierarchies corresponding to the $A$ and $D$ series and three exceptional solutions. Their results are tabulated in Table 4. The occurrence of the Coxeter numbers and Coxeter exponents in these expressions is striking. This classification has been derived from purely theoretical coniderations based on conformal and modular invariance. However, Pasquier [11] obtained precisely the same operator content and modular invariant partition functions by direct computation on the critical $A$–$D$–$E$ lattice models.

| Algebra | Modular Invariant Partition Function |
|---|---|
| $A_L$ | $Z=\frac{1}{2}\sum_{r=1}^{h-2}\sum_{s=1}^{h-1}\lvert\chi_{r,s}\rvert^2$ |
| $D_L$ ($L$ even) | $Z=\frac{1}{2}\sum_{r=1}^{h-2}\Big\{\sum_{s=1,\ s\ \mathrm{odd}}^{\frac{h}{2}-2}\lvert\chi_{r,s}+\chi_{r,h-s}\rvert^2+2\lvert\chi_{r,h/2}\rvert^2\Big\}$ |
| $D_L$ ($L$ odd) | $Z=\frac{1}{2}\sum_{r=1}^{h-2}\Big\{\sum_{s=1,\ s\ \mathrm{odd}}^{h-1}\lvert\chi_{r,s}\rvert^2+\lvert\chi_{r,h/2}\rvert^2+\sum_{s=2,\ s\ \mathrm{even}}^{\frac{h}{2}-2}(\chi_{r,s}\bar\chi_{r,h-s}+\bar\chi_{r,s}\chi_{r,h-s})\Big\}$ |
| $E_6$ | $Z=\frac{1}{2}\sum_{r=1}^{h-2}\Big\{\lvert\chi_{r,1}+\chi_{r,7}\rvert^2+\lvert\chi_{r,4}+\chi_{r,8}\rvert^2+\lvert\chi_{r,5}+\chi_{r,11}\rvert^2\Big\}$ |
| $E_7$ | $Z=\frac{1}{2}\sum_{r=1}^{h-2}\Big\{\lvert\chi_{r,1}+\chi_{r,17}\rvert^2+\lvert\chi_{r,5}+\chi_{r,13}\rvert^2+\lvert\chi_{r,7}+\chi_{r,11}\rvert^2+\lvert\chi_{r,9}\rvert^2$ |
| | $+\,[(\chi_{r,3}+\chi_{r,15})\bar\chi_{r,9}+(\bar\chi_{r,3}+\bar\chi_{r,15})\chi_{r,9}]\Big\}$ |
| $E_8$ | $Z=\frac{1}{2}\sum_{r=1}^{h-2}\Big\{\lvert\chi_{r,1}+\chi_{r,11}+\chi_{r,19}+\chi_{r,29}\rvert^2+\lvert\chi_{r,7}+\chi_{r,13}+\chi_{r,17}+\chi_{r,23}\rvert^2\Big\}$ |

Table 4. Modular invariant partition functions of classical $A$–$D$–$E$ lattice models expressed as sesquilinear forms in Virasoro characters. Here $c=1-\frac{6}{h(h-1)}$ where $h$ is the Coxeter number. The index $s$ is ranges over the Coxeter exponents.

## 4. $c=1$ Theories

It is natural to ask whether it is possible to classify the $c = 1$ two-dimensional conformal field theories [35] and critical lattice models [11] along the same lines as for the $c < 1$ theories. In particular, we would like to know if the critical affine $A$–$D$–$E$ models give an exhaustive classification of $c = 1$ theories. The $c < 1$ theories possess a finite number of scaling operators with fixed rational scaling dimensions. By contrast, the $c = 1$ theories admit continuously varying scaling dimensions and an infinite number of scaling operators. The $c = 1$ theories are therefore intrinsically more difficult. Some progress has be made, however, for the $c = 1$ rational conformal field theories which we consider in this section.

### 4.1 Gaussian partition functions

The canonical example of a $c = 1$ theory is the Gaussian model (Coulomb gas, six-vertex model, scalar free boson compactified on a circle) [36,37] with the modular invariant partition function on a torus given by

$$Z(q,r) = \frac{1}{\eta(q)\eta(\bar{q})} \sum_{m,n\in\mathbb{Z}} q^{\Delta_{m,n}} \bar{q}^{\bar{\Delta}_{m,n}} \tag{4.1}$$

where

$$\eta(q) = q^{1/24} \prod_{n=1}^{\infty} (1-q^n) \tag{4.2}$$

is the Dedekind eta function and $r$ is a free parameter called the radius of compactifiation. The conformal weights are

$$\Delta_{m,n} = \tfrac{1}{2}(\frac{m}{2r}+nr)^2, \qquad \bar{\Delta}_{m,n} = \tfrac{1}{2}(\frac{m}{2r}-nr)^2. \tag{4.3}$$

so the scaling dimensions and spins are

$$x_{m,n} = \Delta_{m,n} + \bar{\Delta}_{m,n} = \frac{m^2}{4r^2} + n^2r^2 \tag{4.4}$$

$$s_{m,n} = \Delta_{m,n} - \bar{\Delta}_{m,n} = mn. \tag{4.5}$$

The theory is called rational if $r^2$ is rational. In this case all the conformal weights and scaling dimensions are also rational. It is straightforward to verify that $Z(q,r)$ is indeed modular invariant and satisfies the duality relation

$$Z(q,r) = Z(q, \frac{1}{2r}) \tag{4.6}$$

so that the Gaussian partition function is self-dual at $r = 1/\sqrt{2}$.

The $c = 1$ Virasoro characters are given by

$$\chi_\Delta(q) = \begin{cases} q^\Delta/\eta(q), & \Delta \neq n^2/4, \quad \Delta \geq 0 \\ \left[q^{n^2/4} - q^{(n+2)^2/4}\right]/\eta(q), & \Delta \neq n^2/4, \quad n = 1,2,3,\ldots \end{cases} \tag{4.7}$$

It is therefore clear that $Z(q,r)$ can only be decomposed as a sesquilinear form in Virasoro characters with an infinite number of terms. Fortunately, the $c = 1$ theories have a chiral or Kac-Moody symmetry in addition to the Virasoro symmetry. For the rational theories, it is then possible [38] to decompose the modular invariant partition function as a *finite* sesquilinear form in characters of the larger chiral or Kac-Moody algebra. Suppose that

$$r^2 = \frac{p}{2p'}, \quad p,p' \text{ coprime.} \tag{4.8}$$

The chiral or Kac-Moody characters at level $N = pp'$ are then defined by

$$\chi^{(1)}_{N,k}(q) = \frac{1}{\eta(q)} \sum_{n=-\infty}^{\infty} q^{N(n+\frac{k}{2N})^2}, \quad N,k \in \mathbb{Z}. \tag{4.9}$$

| Algebra | Modular Invariant Partition Function |
|---|---|
| $A_{L-1}^{(1)}$ | $Z_L$ (L odd), $Z_{L/2}$ (L even) |
| $D_{L-1}^{(1)}$ | $\frac{1}{2}(Z_{L-3}+2Z_2-Z_1)$ |
| $E_6^{(1)}$ | $\frac{1}{2}(2Z_3+Z_2-Z_1)$ |
| $E_7^{(1)}$ | $\frac{1}{2}(Z_4+Z_3+Z_2-Z_1)$ |
| $E_8^{(1)}$ | $\frac{1}{2}(Z_5+Z_3+Z_2-Z_1)$ |
| $A_L$ | $\frac{1}{2}(Z_{L+1}-Z_1)$ |
| $D_L$ | $\frac{1}{2}(Z_{2(L-1)}-Z_{L-1}+Z_2-Z_1)$ |
| $E_6$ | $\frac{1}{2}(Z_{12}-Z_6-Z_4+Z_3+Z_2-Z_1)$ |
| $E_7$ | $\frac{1}{2}(Z_{18}-Z_9-Z_6+Z_3+Z_2-Z_1)$ |
| $E_8$ | $\frac{1}{2}(Z_{30}-Z_{15}-Z_{10}-Z_6+Z_5+Z_3+Z_2-Z_1)$ |

Table 5. Modular invariant partition functions of critical classical and affine $A$–$D$–$E$ lattice models expressed as linear combinations of Gaussian partition functions. The modular parameter $q$ is given by $q=\exp(2\pi i\tau), \tau=(\ell'/\ell)\exp[i(\pi-\theta)]$ with $\theta=\pi u$ for the affine models and $\theta=\pi u/\lambda$ for the classical models.

and the result of Kiritsis [38] states that the modular invariant partition functions of $c=1$ rational conformal field theories can all be written as

$$Z(q)=\sum_{\substack{(N,k)\\(\bar{N},\bar{k})}} \mathcal{N}_{N,k}^{\bar{N},\bar{k}}\chi_{N,k}^{(1)}(q)\chi_{\bar{N},\bar{k}}^{(1)}(\bar{q}). \tag{4.10}$$

where $\mathcal{N}_{n,k}^{\bar{N},\bar{k}}$ are nonegative integers giving the operator content of the theory. In addition, Kiritsis also showed that the only modular invariants at level $N$ are the Gaussian partition functions of the form

$$Z(q,\sqrt{\frac{p}{2p'}})=\sum_{a=0}^{p-1}\sum_{b=0}^{p'-1}\left\{\chi_{N,ap'+bp}^{(1)}(q)\chi_{N,ap'-bp}^{(1)}(\bar{q})+\chi_{N,N+ap'+bp}^{(1)}(q)\chi_{N,N+ap'-bp}^{(1)}(\bar{q})\right\}. \tag{4.11}$$

It therefore follows that the modular invariant partition function of a $c=1$ rational conformal field theory can also be written as a finite linear combination of Gaussian partition functions. Indeed, if we set

$$Z_n=Z(q,n/\sqrt{2}) \tag{4.12}$$

then the modular invariant partition functions of the critical affine $A$–$D$–$E$ models are given by direct calculation [37,11] as the linear combinations of Gaussian partition functions shown in Table 5 where the modular parameter $q$ is given by (3.13) with

$$\theta=\pi u. \tag{4.13}$$

Dijkgraaf, Verlinde and Verlinde [39] have shown that the $A$–$D$–$E$ models exhaust all the conformal field theories that can be built from Gaussian partition functions. Under the further assumptions, that the operator algebra is consistent and that the identity operator has multiplicty 1, Kiritsis [38] has shown the subset of $c = 1$ rational conformal field theories given by the critical affine $A$–$D$–$E$ models is complete. A detailed discussion of the operator algebra of the rational conformal field theories can be found in Dijkgraaf, Vafa, Verlinde and Verlinde [40].

The $c < 1$ characters can be written simply as differences of chiral or Kac-Moody characters at level $N = h(h-1)$

$$\chi_{r,s}(q) = \chi^{(1)}_{h(h-1),hr-(h-1)s}(q) - \chi^{(1)}_{h(h-1),hr+(h-1)s}(q). \tag{4.14}$$

From this identity it follows that the modular invariant partition functions of the classical $A$–$D$–$E$ models can also be written [37] as a linear combinations of Gaussian partition functions as shown in Table 5.

## 5. $c > 1, SU(n)$ and Beyond

It is apparent from the previous sections that the critical behaviour of lattice models with central charge $c \leq 1$ can be understood in terms of the line of critical points of the six-vertex model or coulomb gas [41]. This points to a common structure underlying these models. In fact, all the $A$–$D$–$E$ models considered here are connected in some way with the affine Lie algebra $\hat{SU}(2)$. The quantum Hamiltonians for example, obtained from these models in the extreme anisotropic limit $u \to 0$, describe $SU(2)$ quantum spin chains. The generalization [42,43,44] of the construction of solvable lattice models to the higher rank algebras $SU(n)$ with $n \geq 3$ leads to conformal theories with central charge $c > 1$. Specifically, for $SU(n)$ at level $l$ the central charge of these theories is

$$c = (n-1)\left[1 - \frac{n(n+1)}{(n+l)(n+l-1)}\right]. \tag{5.15}$$

Another process which builds models with $c > 1$ is fusion [45,46,47,48,49]. In fusion, $p \times p$ blocks of faces of a solvable model are fused together to form a new solvable model with an increased value of the central charge. For example, fusion of the $SU(2)$ models considered here leads to models with central charge [50]

$$c = \frac{3p}{p+2}\left[1 - \frac{2(p+2)}{h(h-p)}\right]. \tag{5.16}$$

Having given some of the basic references, we will not proceed to review the study of these more complicated theories. In any case, the classification and content of these theories is largely *terra incognita*. Instead we recommend it to the intrepid reader as a fascinating area to explore.

## Acknowledgements

It is pleasure to thank the Centre for Mathematical Analysis and my friends and colleagues at the Australian National University, particularly Michael Barber and Rodney Baxter, for their interest and encouragement.

## References

1. J. L. Cardy, Conformal Invariance and Statistical Mechanics, in Les Houches, Session XLIV, Fields, Strings and Critical Phenomena, eds. E. Brézin and J. Zinn-Justin, 1989
2. P. Ginsparg, Applied Conformal Field Theory, in Les Houches, Session XLIV, Fields, Strings and Critical Phenomena, eds. E. Brézin and J. Zinn-Justin, 1989
3. P. Ginsparg, Some Statistical Mechanical Models and Conformal Field Theories, Trieste Spring School Lectures, HUTP-89/A027
4. H. Saleur, Lattice Models and Conformal Field Theories, in Cargèse School on Common Trends in Condensed Matter and Particle Physics, 1988
5. H. Saleur and P. di Francesco, Two Dimensional Critical Models on a Torus, in Brasov Summer School on Conformal Invariance and String Theory, 1987
6. A. Capelli, Recent Results in Two-Dimensional Conformal Field Theory, in Proceedings of the XXIV International Conference on High Energy Physics, eds. R. Kotthaus and J. Kühn, Springer, Berlin, 1988
7. R. J. Baxter , "Exactly Solved Models in Statistical Mechanics", Academic Press, London, 1982.
8. H. N. V. Temperley and E. H. Lieb, *Proc. Roy. Soc. (London) A* **322** (1971), 251
9. A. L. Owczarek and R. J. Baxter , *J. Stat. Phys.* **49** (1987), 1093
10. V. Pasquier, *Nucl. Phys. B* **285** **[FS19]** (1987), 162
11. V. Pasquier, *J. Phys. A* **20** (1987), L1229, 5707
12. R. J. Baxter , *J. Stat. Phys.* **28** (1982), 1
13. D. M. Cvetkovic, M. Doob and H. Sachs, "Spectra of Graphs", Academic Press, London 1980
14. G. E. Andrews, R. J. Baxter and P. J. Forrester, *J. Stat. Phys.* **35** (1984), 193
15. P. A. Pearce and K. A. Seaton, *Phys. Rev. Lett.* **60** (1988), 1347
16. P. A. Pearce and K. A. Seaton, *Ann. Phys. (N.Y.)* **193** (1989), 326

17. A. Kuniba and T. Yajima, *J. Stat. Phys.* **52** (1988), 829

18. V. Pasquier, *J. Phys. A* **20** (1987), L217, L221

19. Y. Akutsu, A. Kuniba and M. Wadati, *J. Phys. Soc. Japan* **55** (1986), 1466

20. J.-Y. Choi, K. Kwon and D. Kim, *Europhysics Lett.* **xx** (1990), to appear

21. A. A. Belavin, A. M. Polyakov and A. B. Zamolodchikov, *Nucl. Phys. B* **241** (1984), 333

22. J. L. Cardy, "Phase Transitions and Critical Phenomena, Vol.11", eds. C. Domb and J.L. Lebowitz, Academic Press, London 1988

23. J. L. Cardy, *Nucl. Phys. B* **270** **[FS16]** (1986), 186

24. J. L. Cardy, *Nucl. Phys. B* **275** **[FS17]** (1986), 200

25. D. Friedan, Z. Qiu and S. Shenker, *Phys. Rev. Lett.* **52** (1984), 1575 ; in "Vertex Operators in Mathematics and Physics", eds. J. Lepowsky, S. Mandelstam and I.M. Singer, Springer, 1984

26. V. G. Kac, *Lecture Notes in Physics* **94** (1979), 441-445

27. D. A. Huse, *Phys. Rev. B* **30** (1984), 3908

28. H. W. J. Blöte, J. L. Cardy and M. P. Nightingale, *Phys. Rev. Lett.* **56** (1986), 742

29. I. Affleck, *Phys. Rev. Lett.* **56** (1986), 746

30. D. Kim and P. A. Pearce , *J. Phys. A* **20** (1987), L451-6

31. B. Davies and P. A. Pearce , *Int. J. Mod. Phys. B* (1990), this issue

32. A. Capelli, C. Itzykson and J.-B. Zuber, *Nucl. Phys. B* **280** **[FS18]** (1987), 445-65

33. A. Capelli, C. Itzykson and J.-B. Zuber, *Comm. Math. Phys.* **113** (1987), 1-26

34. A. Kato, *Mod. Phys. Lett. A* **2** (1987), 585

35. P. Ginsparg, *Nucl. Phys. B* **295** **[FS21]** (1988), 153-70

36. L. P. Kadanoff and A. C. Brown, *Ann. Phys. (N.Y.)* **121** (1979), 318-42

37. P. di Francesco, H. Saleur and J.-B. Zuber, *J. Stat. Phys.* **49** (1987), 57-79

38. E. B. Kiritsis, *Phys. Lett. B* **217** (1989), 427

39. R. Dijkgraaf, E. Verlinde and H. Verlinde, in Proceedings of the 1987 Copenhagen Conference, World Scientific, 1988

40. R. Dijkgraaf, C. Vafa, E. Verlinde and H. Verlinde, *Comm. Math. Phys.* **123** (1989), 485

41. B. Nienhuis, in Phase Transitions and Critical Phenomena, Vol.11, eds. C. Domb and J.L. Lebowitz, Academic Press, 1987

42. V. Pasquier, *Nucl. Phys. B* **B295** **[FS21]** (1988), 491-510

43. M. Jimbo, T. Miwa and M. Okado, *Comm. Math. Phys.* **116** (1988), 507-25

44. P. di Francesco and J.-B. Zuber, $SU(N)$ Lattice Models Associated with Graphs, Saclay preprint SPhT/89-92, 1989

45. P. P. Kulish, N. Yu. Reshetikhin and E. K. Sklyanin, *Lett. Math. Phys.* **5** (1981), 393

46. E. Date, M. Jimbo, T. Miwa and M. Okado, *Lett. Math. Phys.* **12** (1986), 209

47. E. Date, M. Jimbo, A. Kuniba, T. Miwa, and M. Okado, *Nucl. Phys. B* **B290** (1987), 231

48. E. Date, M. Jimbo, A. Kuniba, T. Miwa, and M. Okado, *Advanced Studies in Pure Mathematics* **16** (1988), 17

49. Y. Akutsu, A. Kuniba and M. Wadati, *J. Phys. Soc. Japan* **55** (1986), 2907

50. V.V. Bazhanov and N. Yu. Reshetikhin , *Int. J. Mod. Phys. A* **4** (1989), 115-42

International Journal of Modern Physics B, Vol. 4, No. 5 (1990) 735 – 801
© World Scientific Publishing Company

# YANG-BAXTER ALGEBRAS, INTEGRABLE THEORIES AND BETHE ANSATZ

H.J. DE VEGA

Laboratoire de Physique Théorique et Hautes Energies* - Paris **

**Abstract :**

The Yang-Baxter algebras (YBA) are introduced in a general framework stressing their power to exactly solve the lattice models associated to them. The algebraic Bethe Ansatz is developed as an eigenvector construction based on the YBA. The six-vertex model solution is given explicitly. The generalization of YB algebras to face language is considered. The algebraic BA for the SOS model of Andrews, Baxter and Forrester is described using these face YB algebras.

It is explained how these lattice models yield both solvable massive QFT and conformal models in appropiated scaling (continuous) limits within the lattice light-cone approach. This approach permit to define and solve rigorously massive QFT as an appropiate continuum limit of gapless vertex models.

The deep links between the YBA and Lie algebras are analyzed including the quantum groups that underly the trigonometric/hyperbolic YBA. Braid and quantum groups are derived from trigonometric/hyperbolic YBA in the limit of infinite spectral parameter.

To conclude, some recent developments in the domain of integrable theories are summarized.

---

* Laboratoire associé au CNRS UA 280.

** Postal address :
Université Pierre et Marie Curie
Tour 16 - 1er étage
4, place Jussieu
75252 PARIS CEDEX 05 - FRANCE -

# I. YANG-BAXTER ALGEBRAS

A Yang-Baxter (YB) algebra consists of a set of operators $T(\theta)$. They depend on the complex variable $\theta$ (the spectral parameter). Each operator $T(\theta)$ acts on two vector spaces, say $\mathcal{A}$ and $\mathcal{V}$. This means that they have two couples of indices of different kind, in general. The best way to work with Yang-Baxter algebras is to use graphical notation. It is defined as follows :

a) a line of different type is associated to each vector space

———— = $\mathcal{A}$, ~~~~~~ = $\mathcal{V}$, etc.

*Figure 1*

b) The intersection of two lines is associated to an operator $T(\theta)$ where $\theta$ is the angle between the two lines.

$$[\, T_{ab}{}^{(\mathcal{A},\mathcal{V})}(\theta)\,]_{\alpha\beta} \;=$$

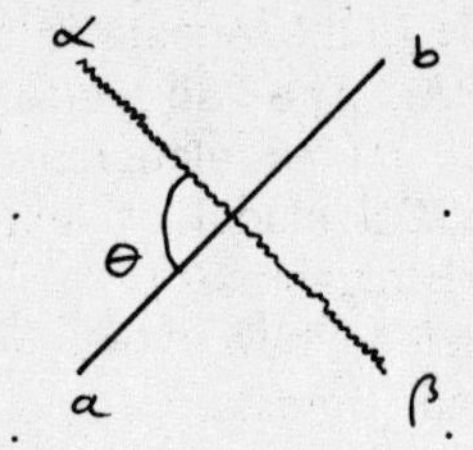


*Figure 2*

c) There is summation over all states in the vector spaces associated to the lines between two vertices ["internal lines"].

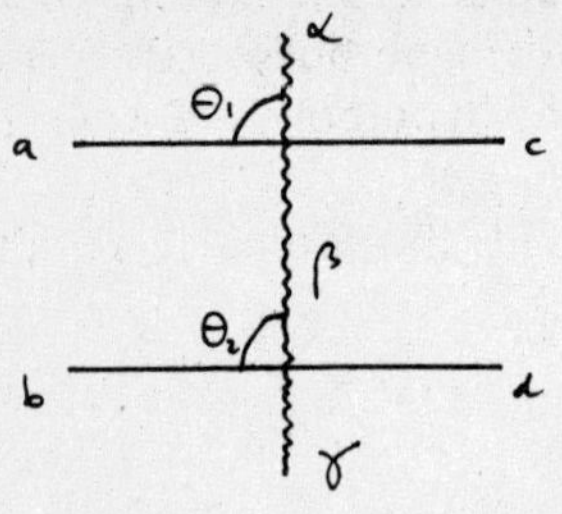


$$= \sum_{\beta} [\, T_{ac}(\theta_1)\,]_{\alpha\beta}\; [\, T_{bd}(\theta_2)\,]_{\beta\gamma}$$

*Figure 3*

d) Left to right order in the formulas correspond to up to down in the pictures ( see fig. 3 ).

This definition originates in two dimensional vertex models where the links can be in different states spanning a vector space, say $\mathcal{V}$ Here we consider the general case when the lattice is formed by different types of bonds (they are associated to vector spaces $\mathcal{A}$, $\mathcal{V}$, ... ).

Let us call $\mathcal{S}$ the set of all vector spaces where the YB algebra (YBA) generators act

$$\mathcal{J} = \{ V^I \}$$

$\mathcal{J}$ is also the set of different types of lines. The basic equation that characterizes the YBA is

$$T^{(K,I)}(\theta - \theta')\, T^{(K,J)}(\theta)\, T^{(I,J)}(\theta') = T^{(I,J)}(\theta')\, T^{(K,J)}(\theta)\, T^{(K,I)}(\theta - \theta') \tag{1.1}$$

for all spaces $V^I$, $V^J$, $V^K \in \mathcal{J}$. Eq. (1.1) is called Yang-Baxter equation (YBE) or triangular relation or factorization equation. It can be represented graphically as

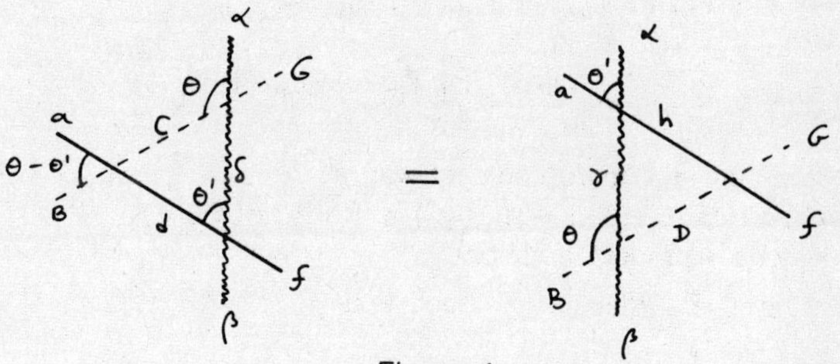


*Figure 4*

Here ——— $= V^I$, ~~~~~~ $= V^J$, - - - - - $= V^K$. We used repeatedly the rules of Fig. 2 and 3 to write down Fig. 4. Eq. (1.1) writes putting all indices explicitly

$$\sum_{C,d,\delta} [\, T_{BC}{}^{(K,I)}(\theta - \theta')\,]_{ad}\, [\, T_{CG}{}^{(K,J)}(\theta)\,]_{\alpha\delta}\, [\, T_{df}{}^{(I,J)}(\theta')\,]_{\delta\beta} =$$

$$= \sum_{D,h,\gamma} [\, T_{ah}{}^{(I,J)}(\theta')\,]_{\alpha\gamma}\, [\, T_{BD}{}^{(K,J)}(\theta)\,]_{\gamma\beta}\, [\, T_{DG}{}^{(K,I)}(\theta - \theta')\,]_{hf} \, . \tag{1.2}$$

As one sees in fig. 4 the YBE says that one can displace any line through the intersection of other two provided the angles are kept fixed. This is called sometimes Z-invariance since it leaves the partition function unchanged [see below][1]. That is, we have invariance under parallel displacements in the lattice. Eqs.(1.1)-(1.2) [or fig. 4] shows the general YBE.

Eqs.(1.1) or (1.2) hold for all values of the external indices ( a, B, $\beta$, f, G and $\alpha$ ) and all values of the spectral parameters $\theta$ and $\theta'$.

The YB generator associated to the intersection of two lines of the same type is called a R-matrix : $R^I(\theta) \equiv T^{(I,I)}(\theta)$ ,

$$R^{ab}{}_{cd}(\theta) = [\,T_{bc}{}^{(I,I)}(\theta)\,]_{ad} =$$

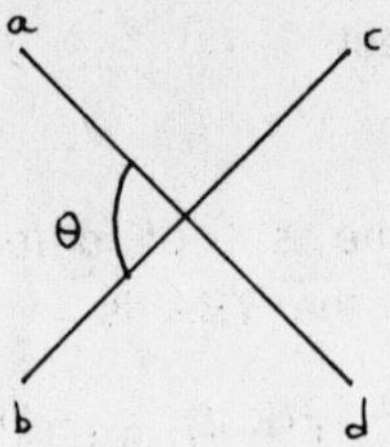


*Figure 5*

In the particular case when two of the vector spaces are identical, say $V^I = V^K = \mathcal{A}$ and $V^J = \mathcal{V}$, eqs. (1.1)-(1.2) become

$$\sum_{c,d,\delta} R^{ab}{}_{cd}(\theta-\theta')\ [\,T_{cg}(\theta)\,]_{\alpha\delta}\ [\,T_{df}(\theta')\,]_{\delta\beta} =$$
$$= \sum_{c,d,\gamma} [\,T_{ac}(\theta')\,]_{\alpha\gamma}\ [\,T_{bd}(\theta)\,]_{\gamma\beta}\ R^{cd}{}_{gf}(\theta-\theta').$$

This can be rewritten in a more compact way as[3]

$$R(\theta-\theta')\,[\,T(\theta)\otimes T(\theta')\,] = [\,T(\theta')\otimes T(\theta)\,]\,R(\theta-\theta') \qquad (1.3)$$

where we use tensor product notation

$$(A\otimes B)_{ab,cd} \equiv A_{ac}\ B_{bd}\ ,$$

and

$$R(\theta) = R^I(\theta) \qquad , \qquad T(\theta) = T^{(I,J)}(\theta)\,.$$

In eq. (1.3) an operator product in the space $\mathcal{V}$ is understood. The $\otimes$ means tensor product of the space $\mathcal{A}$ multiplied by itself. R acts in $\mathcal{A}\otimes\mathcal{A}$ as a matrix. The R-matrix associated to the space $V_o$ of lowest dimensionality in $\mathcal{I}$ as called the fundamental R-matrix. The fundamental R-matrix characterizes the YB algebra.

A YB algebra is then, generally speaking , a set of operators $T^{(K,I)}(\theta)$ fulfilling eq.(1.1) identically. Usually, this is an infinite set. Non-trivial examples of such algebras exist and simple cases will be considered below.

Let us see why YB algebras are connected deeply with integrable theories. Eq. (1.1) can be written as

$$T^{(K,J)}(\theta)\,T^{(I,J)}(\theta') = \{T^{(K,I)}(\theta-\theta')\}^{-1}\,T^{(I,J)}(\theta')\ T^{(K,J)}(\theta)\,T^{(K,I)}(\theta-\theta') \qquad (1.4)$$

Taking the trace of eq. (1.4) in the space $V^K\otimes V^I$ yields

$$\tau_K(\theta)\ \tau_I(\theta') = \tau_I(\theta')\ \tau_K(\theta) \qquad (1.5)$$

where we use the cyclic property of the trace and

$$\mathrm{Tr}_{V^I\otimes V^K}(\,T^{(I,J)}\otimes T^{(K,J)}) = \mathrm{Tr}_{V^I}(\,T^{(I,J)}\,)\ \mathrm{Tr}_{V^K}(\,T^{(K,J)}) \qquad (1.6)$$

Here an operatorial product in the space $V^J$ is understood. We denote by $\tau_I(\theta)$ and $\tau_K(\theta)$ the transfer matrices

$$\tau_I(\theta) = Tr_{V^I}( T^{(I,J)} ) = \sum_a T_{aa}(\theta)^{(I,J)} \tag{1.7}$$

$$\tau_K(\theta) = Tr_{V^K} ( T^{(K,J)}) = \sum_A T_{AA}(\theta)^{(K,J)} \tag{1.8}$$

$\tau_I(\theta)$ and $\tau_K(\theta)$ are operators acting on $V^J$. They form a **set** of families of commuting transfer matrices

$$[ \tau_I(\theta) , \tau_K(\theta') ] = 0 , \quad \forall \theta, \theta' \in \mathbb{C} , \forall I, K \in \mathcal{I} \tag{1.9}$$

Moreover, series expanding in $\theta$ yields an infinite number of commuting operators acting on $V^J$.

$$[ c_n{}^I , c_m{}^K ] = 0 \quad , \forall n , m \geq 0 , \quad \forall I , K \in \mathcal{I} \tag{1.10}$$

Here $c_n{}^I$ are the expansion coefficients of $\tau_I(\theta)$ or $\log \tau_I(\theta)$ in powers of $\theta$. The existence of an infinite number of commuting operators is the necessary condition to have a quantum integrable system with an infinite number of degrees of freedom. Actually, only in the thermodynamic limit this number of degrees of freedom is attained. In addition, the hamiltonian of the system (H) must commute with $\tau_I(\theta)$. Actually, H often expresses itself in terms of $\tau_I(\theta)$ either as the logarithmic derivative at $\theta = 0$ (spin chains, see eq.(2.13)) or in terms of $\log \tau_I(\theta)$ at some special value of $\theta$ (field theories in the light cone approach, see sec.IV).

Since the operators $\tau_I(\theta)$ are mutually commuting for all $\theta$ and $V^I$, one can expect to be able to diagonalize all of them simultaneously. This is actually possible through the algebraic Bethe Ansatz (BA). Moreover, in the BA, the eigenvectors and eigenvalues can be constructed using the YB algebra itself. This is probably the main application of YBA. They permit to built eigenvectors and eigenvalues of all $\tau_I(\theta)$ and operators $c_n{}^I$ derived from them in a purely algebraic framework.

A specially important YB equation follows when the three vector spaces in eq. (1.1) are equal : $V^I = V^J = V^K = \mathcal{A}$. In this case eqs.(1.1)-(1.2) can be written as

$$\begin{aligned} &[ 1 \otimes R( \theta - \theta' ) ] [ R( \theta ) \otimes 1 ] [ 1 \otimes R( \theta' ) ] = \\ &= [ R( \theta' ) \otimes 1 ] [ 1 \otimes R( \theta ) ] [ R( \theta - \theta' ) \otimes 1 ] \end{aligned} \tag{1.11}$$

In explicit notation this reads

$$\sum_{1\le c,d,e\le q} R^{cd}{}_{a_2a_1}(\theta-\theta')\; R^{b_1e}{}_{a_3c}(\theta)\; R^{b_2b_3}{}_{ed}(\theta') =$$

$$= \sum_{1\le m,n,p\le q} R^{mn}{}_{a_3a_2}(\theta')\; R^{pb_3}{}_{na_1}(\theta)\; R^{b_1b_2}{}_{mp}(\theta-\theta') \quad , \qquad (1.12)$$

where $q \equiv \dim \mathcal{A}$ . This equation can be depicted as

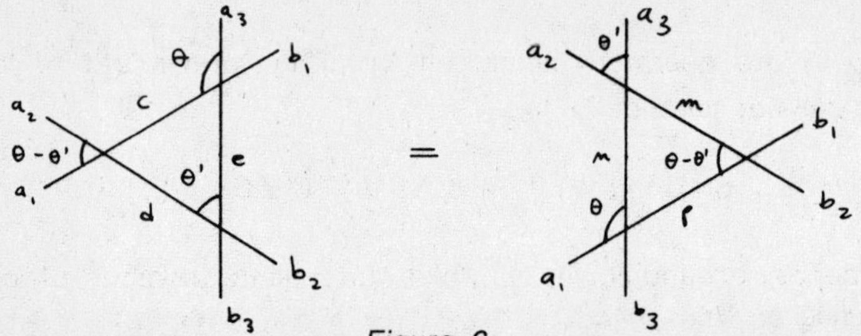


*Figure 6*

We see that eq. (1.11) or (1.12) is a system of $q^6$ equations ( $q = \dim \mathcal{A}$ ) with $q^4$ unknowns (the functions $R^{ab}{}_{cd}(\theta)$, $1 \le a, b, c, d \le q$). That is, one finds a heavily over-determined set of equations. The existence of a solution is clearly a necessary condition to have a YBA. Actually it is also a sufficient condition since one can define a YB generator acting on $\mathcal{A} \otimes \mathcal{A}$ as ( see fig. 5 )

$$[\, t_{ab}{}^{(\mathcal{A},\mathcal{A})}(\theta)\, ]_{cd} = R^{ca}{}_{bd}(\theta) \qquad (1.13)$$

It obeys

$$R(\theta-\theta')\, [\, t^{(\mathcal{A},\mathcal{A})}(\theta) \otimes t^{(\mathcal{A},\mathcal{A})}(\theta')\, ] = [\, t^{(\mathcal{A},\mathcal{A})}(\theta') \otimes t^{(\mathcal{A},\mathcal{A})}(\theta)\, ]\, R(\theta-\theta')$$

which just follows by rewriting eq. (1.12) with the help of eq. (1.13).

The most remarkable fact in integrable theories is that eqs. (1.11) or (1.12) do admit a rich set of non-trivial solutions. Actually each solution exhibits some invariance which probably explains its very existence. That is, thanks to the presence of an invariance the number of actual independent equations is largely reduce from $q^6$.

A YB algebra is invariant [see eq. (1.3)] under a transformation $g \in \mathcal{G}$ in $\mathcal{A}$

$$T_{ab}(\theta) \;\rightarrow\; \sum_c g_{ac}\, T_{cb}(\theta) \qquad (1.15)$$

provided [4,5]

$$[\, g \otimes g\, , R(\theta)\, ] = 0 \;,\; \forall\, \theta\, \varepsilon\, \mathcal{C} \;,\; \forall\, g\, \varepsilon\, \mathcal{G} \qquad (1.16)$$

This can be proven as follows from eq.(1.3). Let us multiply eq.(1.3) by $g \otimes g$ from the left. We find

$$( g \otimes g ) R( \theta - \theta' ) ( g^{-1} \otimes g^{-1} ) \ [ gT(\theta) \otimes gT(\theta') ] = \\ = [ gT( \theta' ) \otimes gT( \theta ) ] R( \theta - \theta')$$

Therefore $gT(\theta)$ obey a YB algebra with the same $R(\theta)$ as $T(\theta)$ provided

$$( g \otimes g ) R( \theta - \theta' ) ( g^{-1} \otimes g^{-1} ) = R( \theta - \theta' ) ,$$

which is just eq.(1.16). This is clearly a sufficient condition of invariance. More generally, we find

$$[ g_I \otimes g_J , T^{(I,J)}( \theta ) ] = 0 , \ \forall \theta \in \mathbb{C} , \ \forall g \in \mathfrak{G} \tag{1.17}$$

where $g_I$ and $g_J$ are the representation of $g \in \mathfrak{G}$ acting on the vector spaces $V^I$ and $V^J$ respectively. For an infinitesimal transformation

$$g_I = 1 + i \varepsilon S_I \quad , \quad g_J = 1 + i \varepsilon S_J$$

where $\varepsilon \ll 1$ and $S_I$ and $S_J$ are the generators representation in $V^I$ and $V^J$ respectively. Hence eq. (1.17) yields

$$[ S_I , T^{(I,J)}( \theta ) ] + [ S_J , T^{(I,J)}( \theta ) ] = 0 \tag{1.18}$$

The invariance of a YB algebra under a group $\mathfrak{G}$ can be formalized as follows :

Let us define the transformed YB operator $T_g{}^{(I,J)}( \theta )$ as follows :

$$T_g{}^{(I,J)}( \theta ) \equiv g_I \ T^{(I,J)}( \theta ) \ g_J{}^{-1} \tag{1.19}$$

This definition includes an additional transformation $g^{-1}$ in the vertical space compared with eq.(1.15). In this way $T_g{}^{(I,J)}( \theta )$ also obeys the YBE (1.1). This is easy to check , taking into account that

$$g_I \ T^{(I,J)}( \theta ) \ g_J{}^{-1} = g_J{}^{-1} \ T^{(I,J)}( \theta ) \ g_I \tag{1.20}$$

There exists a direct connection between the kind of symmetry group $\mathfrak{G}$ of the YBA and the functional dependence on $\theta$. This connection is displayed in table I. This is a first hint about the deep connection between YB algebras and Lie algebras and their deformations (quantum groups).

Another important invariance of YB algebras is the shift invariance. That is, if $T(\theta)$ is a YB generator, so it is

$$T( \theta - \alpha )$$

with fixed $\alpha$. A look to eq. (1.3) shows that this is true since R depends on the difference $\theta-\theta'$, $\alpha$ must be the same so it drops.

Let us now discuss the most important property of YBA : the reproduction property. It can be stated as follows : if $t(\theta)$ obeys the YBA

$$R(\theta-\theta')\,[\,t(\theta)\otimes t(\theta')\,] \;=\; [\,t(\theta')\otimes t(\theta)\,]\,R(\theta-\theta')$$

with horizontal space $\mathcal{A}$ and vertical $\mathcal{V}$, so does

$$T_{ab}^{[N]}(\theta) \;=\; \sum_{a_1,\ldots a_{N-1}=1}^{N} t_{aa_1}(\theta)\otimes t_{a_1a_2}(\theta)\otimes \ldots\ldots \otimes t_{a_{N-1}b}(\theta)\;, \qquad (1.21)$$

with the same R-matrix. The auxiliary space for $T_{ab}^{[N]}(\theta)$ is also $\mathcal{A}$, the vertical one being

$$\mathcal{V}^{(N)} \;=\; \bigotimes_{1\leq j\leq N} \mathcal{V}_j \qquad (1.22)$$

In order to show that $T_{ab}^{[N]}(\theta)$ [eq.(1.21)] fulfils the YBE (1.3) , the best is to use graphical methods.

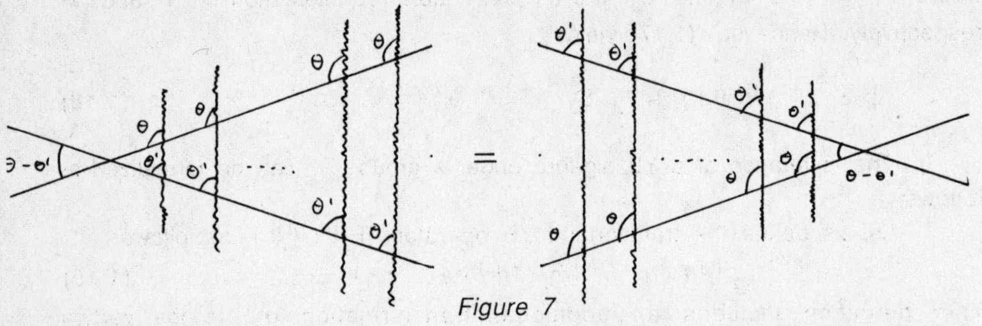


*Figure 7*

We must show that the equal sign holds in fig. 7. Remembering fig. 4, we can push one by one to the left the wavy lines in the lhs of fig. 7 through the R-matrix vertices leaving the expression invariant. After displacing all vertical wavy lines, we precisely get the rhs. This ends the proof. Of course, it can be also done analytically inserting eq.(1.20) in the lhs of eq.(1.3) and using eq.(1.19) repeatedly.

For N = 2 eq. (1.19) can be considered as a way of multiplying YB generators yielding new YB generators. This can be called a coproduct and shows that we have a Hopf algebra structure. More generally, if the YB generators are invariant under a group $\mathcal{G}$ [eq. (1.15)-(1.17)] we have as generator in $\mathcal{A}\otimes\mathcal{V}^N$,

$$T_{ab}^{[N]}(\theta,\bar{\mu},\mathbf{g}) \;=\; \sum_{a_1,\ldots a_{N-1}=1}^{N} [\,g_1\, t(\theta-\mu_1)]_{aa_1} \otimes [\,g_2\, t(\theta-\mu_2)]_{a_1a_2} \otimes \ldots\ldots \otimes [\,g_N\, t(\theta-\mu_N)]_{a_{N-1}b} \qquad (1.23)$$

It obeys the YB eq. (1.3) for any **fixed** transformations $\mathbf{g} = (g_1, \ldots, g_N)$ and $\bar{\mu} = (\mu_1,\ldots, \mu_N)$, with $g_i \in \mathcal{G}$ , $\mu_i \in \mathbb{C}$ , $1 \leq i \leq N$. The introduction of the parameters $\mu_i$ leads to integrable inhomogeneous vertex models[6].

There exists in addition, another coproduct multiplying the generators from right to left

$$\tilde{T}_{ab}^{[N]}(\theta,\bar{\mu},\mathbf{h}) = \sum_{a_1,\ldots a_{N-1}=1}^{N} [\, h_1\, t(\theta-\mu_1)]_{a_1 b} \otimes [\, h_2\, t(\theta-\mu_2)]_{a_2 a_1} \otimes \ldots\ldots \otimes [\, h_N\, t(\theta-\mu_N)]_{a a_{N-1}} \tag{1.24}$$

That is, $\tilde{T}_{ab}^{[N]}(\theta,\bar{\mu},\mathbf{h})$ obeys the same YBA [eq. (1.3)] as $t_{ab}(\theta)$ does.

As we see, YBA are not Lie algebras since the sum of two generators $T(\theta)$ is **not** a YB generator. However, one finds for the YBA the analogous for most of the features of Lie algebras. The YBE (1.12) plays the role of the Jacobi identity in Lie algebras. The fundamental R-matrix being the analogue of the structure constants. There exists for YBA an "adjoint representation" [eq. (1.13)] provided by the R-matrix. We also have a "Cartan algebra" formed by the commuting transfer matrices $\tau_I(\theta)$ [eq. (1.7)]. A representation theory for YBA has been developped. That is, the construction of $T(\theta)$ for different spaces $(\mathscr{A},\mathscr{V})$ given a fundamental R-matrix[3].

Actually there exist more general commuting transfer matrices than (1.7). It follows from eqs. (1.1) and (1.17) that the following operators on $V^J$ :

$$\begin{aligned} \tau_{g_I}(\theta) &= \mathrm{Tr}_{V^I}(\, g_I T^{(I,J)}\,) = \sum_{a,b} (g_I)_{ab}\, T_{ba}(\theta)^{(I,J)} \\ \tau_{g_K}(\theta) &= \mathrm{Tr}_{V^K}(\, g_K\, T^{(K,J)}) = \sum_{A,B} (g_K)_{AB}\, T_{BA}(\theta)^{(K,J)} \end{aligned} \tag{1.25}$$

commute

$$[\ \tau_{g_I}(\theta)\ ,\ \tau_{g_K}(\theta')\ ] = 0\ ,\quad \forall\, \theta,\theta' \in \mathbb{C}\ , \forall\, g \in \mathfrak{G} \tag{1.26}$$

Notice in eq. (1.24) that the transformation $g \in \mathfrak{G}$ is the **same** in both transfer matrices.

It is legitime to call $T(\theta, \mathbf{g}, \bar{\mu})$ [eq. (1.23)] a gauge transformation of $T(\theta)$ [eq. (1.22)]. We apply in eq.(1.23) a group symmetry transformation ( $g_i$ , $\mu_i$ , $i = 1,\ldots,N$ ) that depends upon the site. This is a one-dimensional **local** gauge transformation on the lattice. Since the gauge transformed operator (1.22) or (1.23) obeys a YB algebra, these gauge transformations respect integrability[7].

Actually, a YB gauge transformed generator under $\mathfrak{G}$ can be related with the untransformed one as follows [4]. Let us call $g_{\mathscr{A}}$ and $g_{\mathscr{V}}$ the group element representations acting on the horizontal and vertical spaces respectively and we set $\alpha_i = 0$ , $i = 1, \ldots ,N$. Then eq.(1.20) implies

$$g_{\mathcal{A}} \; t^{(\mathcal{A},\mathcal{V})}(\theta) \; = \; g_{\mathcal{V}}^{-1} \; t^{(\mathcal{A},\mathcal{V})}(\theta) \; [g_{\mathcal{A}}^{-1} \otimes g_{\mathcal{A}}^{-1}] \tag{1.27}$$

Now, we recognize in the l.h.s. of eq.(1.27) each of the operator factors in the r.h.s. of eq.(1.23). Inserting eq.(1.27) in eq.(1.23) yields after a little calculation:

$$T_{ab}^{[N]}(\theta, g) = \sum_{c} \bigotimes_{1\leq i\leq N} G_i^{-1} \; T_{ac}^{[N]}(\theta) \bigotimes_{1\leq i\leq N} G_i \; J_{cb} \tag{1.28}$$

where $G_i^{-1}$ and $G_i$ here act on the i-th vertical space with

$$G_i = \prod_{j=1}^{i} g_j \quad \text{and} \quad J = \prod_{j=1}^{N} g_j$$

That is, the gauge transformed $T_{ab}^{[N]}(\theta, \mathbf{g})$ can be obtained from the untransformed $T_{ab}^{[N]}(\theta)$ by a similarity transformation $\otimes_{1\leq i\leq N} G_i$ on the vertical space plus a right transformation $J$ on the horizontal space. Hence, the gauge transformed transfer matrix,

$$\tau(\theta, \mathbf{g}) = \sum_{a} T_{aa}(\theta, \mathbf{g}) \tag{1.29}$$

can be written as

$$\tau(\theta,\mathbf{g}) = K \sum_{a,c} T_{ac}(\theta) \, J_{ca} \, K^{-1} = K \, \tau_J(\theta) \, K^{-1} \tag{1.30}$$

where $K \equiv \otimes_{1\leq i\leq N} G_i$ .

$\tau_J(\theta)$ is precisely the generalized transfer matrix introduced by eq.(1.25). As we shall see in sec. II, it corresponds to twisted boundary conditions. The matrix $J$ defines the twist in the vertical space [see eq.(2.2)]. Eq.(1.29) has a deep implication : local gauge transformations only affect the physical operators like the transfer matrix and the spin hamiltonian (generated by it) through a twist $J$ on the boundary conditions and a similarity transformation $K$ .

When $\theta = \theta'$ eq. (1.3) naturally suggests that R(0) is a multiple of the unit matrix in $\mathcal{A} \otimes \mathcal{A}$. When this happens the corresponding R-matrix is called regular. That is

$$R(0) = c \; 1 \tag{1.35}$$

where c is a numerical constant. This property can be represented graphically as follows (cf. fig. 5)

$$R^{ab}{}_{cd}(0) \;=\; c\,\delta_{ac}\,\delta_{bd} \;= \qquad (1.36)$$

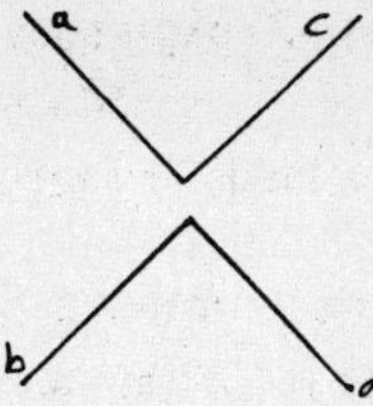


This property plays a key role in the theory of integrable models. First, it implies that the transfer matrices $\tau(\theta)$ built from R-matrices are generating functionals of local lattice operators. That is, those $\tau(\theta)$ following from eqs. (1.7) and (1.21) when $t(\theta)$ is given by eq. (1.13) (see eqs. (2.12)-(2.13)).

## II. PHYSICAL REALIZATIONS OF YANG-BAXTER ALGEBRAS

In this section we shall describe YB algebras in two-dimensional statistical models, field theories and S-matrix theory.

We associate in sec. I a vector space $V^I$ to each type of lines and a YB generator $T^{(I,J)}(\theta)$ to a pair of lines (I,J) intersecting with an angle $\theta$. This can be immediately applied to a two-dimensional lattice of lines [fig. 8] intersecting at the sites. The vector spaces describe the possible local states of the bonds and the $t(\theta)$ describe the statistical weights of the different link configurations.

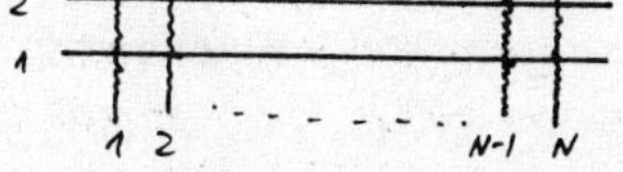


*Figure 8*

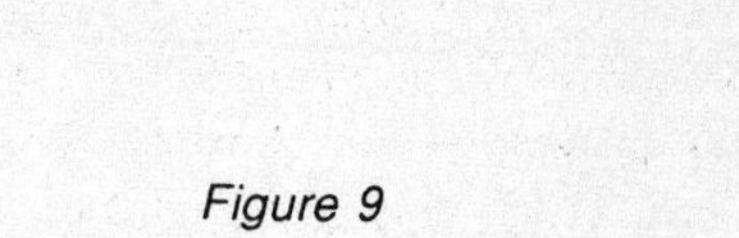


*Figure 9*

That is the matrix element $[t_{ab}(\theta)]_{\alpha\beta}$ defines the probability for the local configuration depicted in fig. 9. The product of the local weights over all sites in the lattice yields the probability for such configuration of the whole system. Finally summing over all possible configurations gives the partition function Z. When periodic boundary conditions are used in both horizontal and vertical directions, Z expresses as

$$Z = \mathrm{Tr}_{\mathcal{V}} [\, \tau(\theta)^M \,] \tag{2.1}$$

Actually eq. (2.1) holds irrespective of the YB equations.

The transfer matrices $\tau_g(\theta)$ [5] [eq. (1.22)] correspond to twisted boundary conditions. That is, when the operators at sites N+1 and 1 are related by the transformation g :

$$S_{N+1} = g\, S_1\, g^{-1} \;, \quad \forall\, S \,. \tag{2.2}$$

Here g acts in the appropriate representation of $\mathfrak{G}$. Then, $\tau_{g_{\mathcal{A}}}(\theta)$ is the transfer matrix. If we also impose twisted b.c in the vertical direction with a twist $h_{\mathcal{V}}$, Z writes

$$Z_{g,h} = \mathrm{Tr}_{\mathcal{V}} [\, \tau_{g_{\mathcal{A}}}(\theta)^M\, h_V \,] \tag{2.3}$$

Eq. (2.1) and (2.3) show how important is the knowledge of the eigenvalues of $\tau(\theta)$. Actually, just the largest eigenvalue $\Lambda^{[N]}{}_{MAX}(\theta)$ gives the free energy in the thermodynamic limit

$$f = -\lim_{N,M\to\infty} (\, 1 \,/\, NM \,) \log Z = -\lim_{N\to\infty} \{(1/N\,) \log \Lambda^{[N]}{}_{MAX}(\theta)\} \tag{2.4}$$

(The dependence on the b.c. drops in the N = M = ∞ limit).

The lattice model here described is called a vertex model. It is homogeneous but not isotropic since horizontal and vertical lines are of different nature. Horizontal bonds live in states of the space $\mathcal{A}$ and vertical bonds in states of the space $\mathcal{V}$ . One can even generalize these integrable vertex models taking lines at arbitrary intersection angles[9]. Also taking inhomogeneous weights $g^{\mathcal{A}}(x)\ \ h^{\mathcal{V}}(y)\ \ t(\,\theta - \alpha(x) - \beta(y)\,)$ that depend upon the horizontal (x) and the vertical (y) coordinates. Moreover, one could take the lattice lines from all possible vector spaces $V^I \in \mathcal{J}$ at will. All these models are integrable and solvable although inhomogeneous and anisotropic.

Let us now study the transfer matrices $\tau(\theta)$ as generating functionals of commuting local operators on the lattice. This is the case for R-matrix models (where $\mathcal{A} = \mathcal{V}$) when R is a regular R-matrix [eq. (1.30)]. We find from eqs. (1.13) and (1.30)-(1.31)

$$[\, t_{ab}(\,0\,)\,]_{cd} = c\ \delta_{bc}\, \delta_{ad} \tag{2.5}$$

Then for a N-site transfer matrix as defined by (1.7) and (1.19)

$$\tau(0)^{[N]}{}_{\mathbf{a}|\mathbf{b}} = c^N \prod_{i=1}^{N} \delta_{a_i,b_{i+1}} \tag{2.6}$$

$$\mathbf{a} \equiv (a_1, a_2, \dots, a_N) \quad ; \quad \mathbf{b} \equiv (b_1, b_2, \dots, b_N)$$

where $b_{N+1} \equiv b_1$ . Here we assume periodic boundary conditions (PBC). The operator in the rhs of (2.6) is just the lattice unit shift operator to the right. Therefore, we can define the momentum operator as

$$\mathcal{P} = i \operatorname{Log}[\ c^{-N}\, \tau(0)^{[N]}\ ] \tag{2.7}$$

Let us now show that the logarithmic derivative of $\tau^{[N]}(\theta)$ at $\theta = 0$ gives an operator coupling nearest neighbors.

Using eq. (1.31), $T_{ab}{}^{[N]}(0)_{\mathbf{c}|\mathbf{d}}$ and $\tau^{[N]}(0)_{\mathbf{c}|\mathbf{d}}$ can be drawn as follows

$T_{ab}{}^{[N]}(0)_{\mathbf{c}|\mathbf{d}} =$

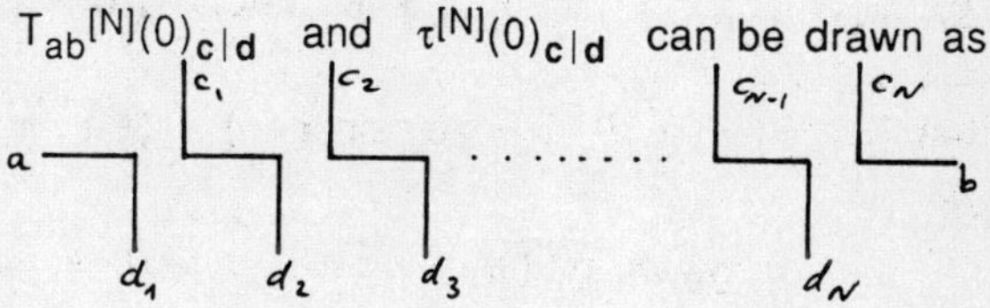

(2.8)

$\tau^{[N]}(0)_{\mathbf{c}|\mathbf{d}} =$

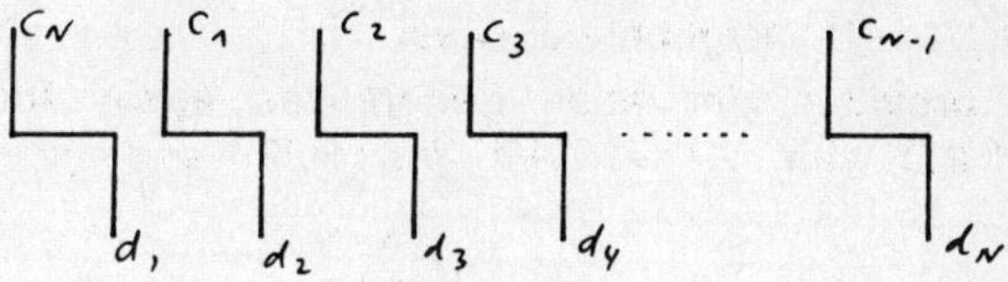

(2.9)

Similarly,

$\{\tau^{[N]}(0)\}^{-1}{}_{\mathbf{c}|\mathbf{d}} =$

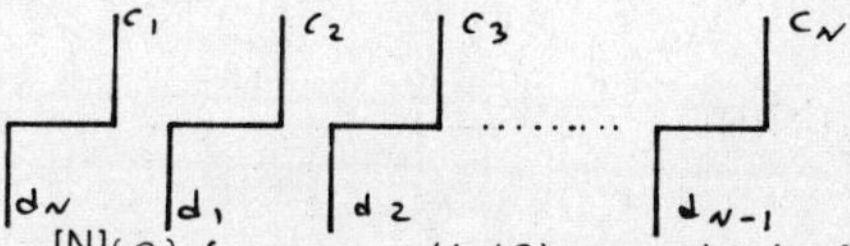

(2.10)

Now, if we compute $d/d\theta\ \tau^{[N]}(\theta)$ from eq. (1.19) we obtain N terms, each one containing $d/d\theta\ t^{(h)}(\theta)$, $1 \leqslant h \leqslant N$ and the others $t^{(\ell)}(\theta)$ $(\ell \neq h)$ not derived. Hence, setting $\theta = 0$ yields

$\dot{\tau}^{[N]}(0) =$

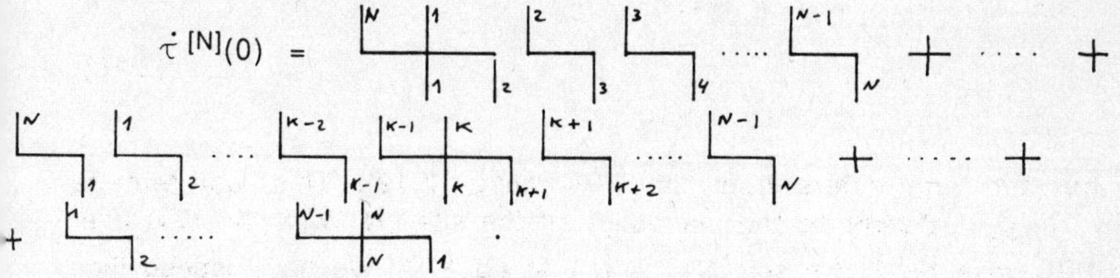

(2.11)

Here 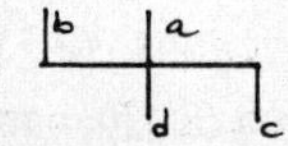 stands for $\dot{R}^{ab}{}_{cd}(0)$

It is now very simply to perform the product $\tau^{[N]}(0)^{-1}\ \dot{\tau}^{[N]}(0)$ just combining eqs. (2.10) and (2.11) with the result

$$\{\tau^{[N]}(0)\}^{-1}\,\dot{\tau}^{[N]}(0) \;=\; \sum_{K=1}^{N} \qquad (2.12)$$

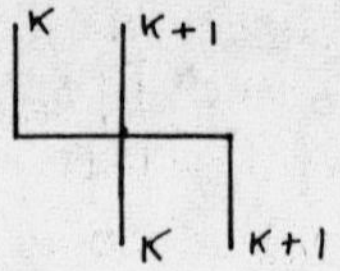


Therefore $\tau^{[N]}(0)^{-1}\,\dot{\tau}^{[N]}(0)$ is a sum of terms each one acting as an operator on two neighboring sites. Now, putting all factors

$$H \;=\; \partial/\partial\theta \,\log \tau^{[N]}(\theta)\,\Big|_{\theta=0} \;=\; \sum_{n=1}^{N} h_{n,n+1} \qquad (2.13)$$

where the matrix elements of $h_{n,n+1}$ read

$$< c_n\, c_{n+1} \mid h_{n,n+1} \mid d_n\, d_{n+1} > \;=\; (1/c)\;\dot{R}(\,0\,)^{c_n\, c_{n+1}}{}_{d_n\, d_{n+1}} \qquad (2.14)$$

Notice that $[\,\tau(\theta)\,,\;\dot{\tau}(\theta)\,] = 0$ since $[\,\tau(\theta)\,,\,\tau(\theta')\,] = 0\;,\;\forall\,\theta,\,\theta'$. Therefore,

$$\partial/\partial\theta\,\log \tau^{[N]}(\theta) \;=\; \dot{\tau}(\theta)\;\tau^{-1}(\theta) \;=\; \tau^{-1}(\theta)\,\dot{\tau}(\theta)\,.$$

More generally the $n^{th}$ derivative of $\log \tau(\theta)$ at $\theta = 0$ is an operator that couples $(n+1)$ neighboring sites[10].

The previous derivation generalizes easily for the twisted transfer matrices (1.25) with $\mathcal{V}^{I} = \mathcal{V}^{(N)}$ [eq.(1.22)] and $\mathcal{V}^{J} = \mathcal{A}$ . We find in this case

$$\tau_g(0)^{[N]}{}_{a\,|\,b} \;=\; c^N \prod_{i=1}^{N-1} \delta_{a_i,b_{i+1}}\; g_{a_N b_1} \qquad (2.15)$$

and

$$\{\tau_g{}^{[N]}(0)\}^{-1}\,\dot{\tau}_g{}^{[N]}(0) \;=$$

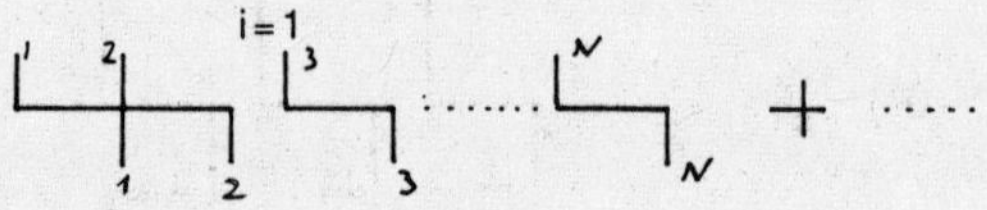


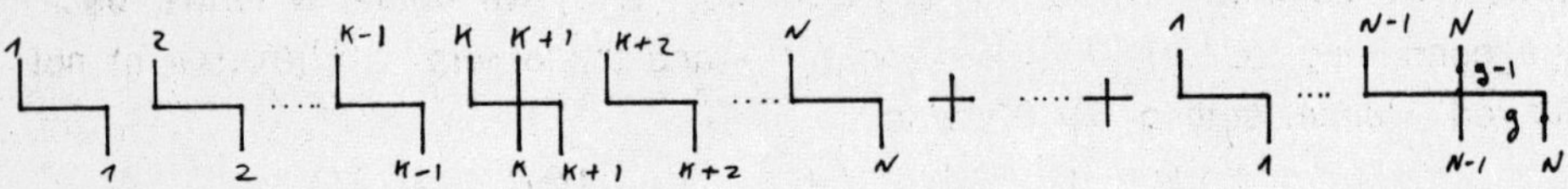


(2.16)

Only the last term differs from the PBC case (2.12). This last term is actually $h_{N,N+1}$ describing the interaction of the site N with the next one. In the PBC case (2.12) we set $N+1 \equiv N$ . In eq.(2.15) we have indeed the $h_{N,N+1}$ as given by eq.(2.13) provided we make the identification (2.2) for the operators acting on $\mathcal{V}_1$ and $\mathcal{V}_{N+1}$ . The factor $g_{a_N b_1}$ in eq.(2.15) has the same explanation. It is equivalent through eq.(2.2) to $\delta_{a_N b_{N+1}}$ .

The operator H can be interpreted as a one-dimensional quantum hamiltonian. It is an operator coupling neighboring q-component "spins". The word spins only applies, rigorously speaking, when the fundamental R-matrix corresponds to the six or eight vertex model. That is, the underlying Lie algebra in $A_1$ and we have true SU(2) spins. Otherwise one finds SU(q) spins, O(q) spins, etc. Eq. (2.12) suggest that $\theta$ may be the imaginary time variale. This possibility has not been fully explored yet. Anyway, it must be noticed that $\tau(\theta) \neq e^{\theta H}$.

Let us expand the gauge transformed transfer matrix $\tau(\theta,\mathbf{g})$ [eq.(1.29)] around $\theta = 0$ . We find :

$$\tau(0,\mathbf{g})_{\mathbf{a}|\mathbf{b}} = c^N \prod_{i=1}^{N} (g_i)_{a_{i-1},b_i} =$$

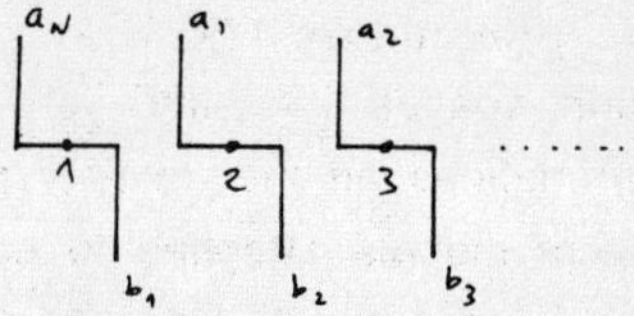


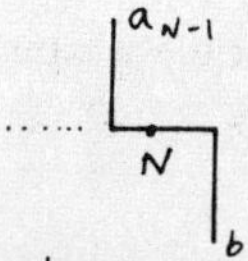


(2.17)

where 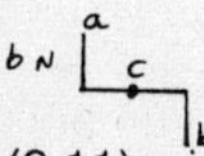

$= g_c : \mathcal{V}_b \to \mathcal{V}_a$. A calculation analogous to that from eq.(2.10)-(2.11) yields for $\tau(\theta,\mathbf{g})$ [7]

$$H[\mathbf{g}] = \partial/\partial\theta \log \tau^{[N]}(\theta,\mathbf{g}) \Big|_{\theta=0} = \sum_{n=1}^{N} (g^{-1})_{n+1}\, h_{n,n+1}\, g_{n+1} \tag{2.18}$$

We see from eq.(2.18) that the way $\mathbf{g} = (g_1,\ldots\ldots, g_N)$ acts on the spin hamiltonian can be considered as a gauge transformation on the lattice. From the analysis in sec. I for $\tau(\theta,\mathbf{g})$ we conclude that $H[\mathbf{g}]$ is, up to a similarity transformation K, identical to $H[\mathbf{1}]$ with boundary conditions twisted by $J = \prod_{1\leqslant i\leqslant N} g_i$ [4].

One can decide to expand $\tau(\theta)$ around a point $\theta = \theta_0 \neq 0$ to generate commuting operators. The trouble is that these operators are in general non-local. That is, they couple all sites in the line. Only when $\tau(\theta_0)$ is the shift operator (2.6) the logarithmic derivative of $\tau(\theta)$ at $\theta_0$ is local and the higher order derivatives multilocal. It is nevertheless interesting to expand or around $\theta = \infty$ Besides the quantum group generators [see sec.V], this expansion yields non-local integrable hamiltonians.

To conclude this section, let us sketch the classification of the known YB algebras. That is, solutions of eq.(1.1). All known solutions posses symmetries in the sense of eq.(1.15) [see table I]. Probably, the presence of such symmetries is the basic reason why a so heavily overdetermined set of equations [ see eq.(1.12) ] has non-trivial solutions.

Elliptic solutions depend on three continuous parameters $\theta$ (spectral parameter), $\gamma$ (anisotropy parameter), and $k$ (elliptic modulus). The trigonometric and hyperbolic solutions depend upon two parameters : $\theta$ and $\gamma$ . They can be obtained from each other by "Wick rotation" : $\gamma \to i\gamma$,

$\theta \to i\,\theta$ . The degenerate limit $k \to 0$ ( $k \to 1$ ) of the elliptic solutions yield trigonometric (hyperbolic) solutions. Trigonometric/hyperbolic solutions associated to all simple Lie algebras are known in vertex language[4,46,63,64].

The trigonometric/hyperbolic solutions are not invariant under the full Lie group $\mathfrak{G}$ [in the sense of eq.(1.17)] but only under its Cartan subalgebra. In addition they are invariant under the corresponding quantum group as we discuss in sec.V. The rational YB algebras follow from the trigonometric/hyperbolic ones in the isotropic limit $\gamma \to 0$. Besides these continuous parameters $(k, \gamma, \theta)$ , the YB operators are labeled by the representation spaces $\mathcal{A}$ and $\mathcal{V}$ where they act. As we have seen, each $T^{(\mathcal{A},\mathcal{V})}(\theta)$ defines an inequivalent physical vertex model. Actually there is a set of physical models attached to each $T^{(\mathcal{A},\mathcal{V})}(\theta)$: a vertex model, a spin hamiltonian, a conformal theory and a massive quantum field theory. The last two cases are exposed in ref.[4] and sec. IV.

Algebraic Bethe Ansatz solutions for a significant number of vertex models are known[17-22] and we believe that such constructions should exist for **any** solution of the YB equations. In sec. 3 we describe the simplest case : the six vertex model solution. Further algebraic Bethe Ansatz constructions (for richer models) can be found in refs.[4,17-22]. Let us recall some important features of the BA solutions.

The vertex (and spin hamiltonians) turn to have a non-zero gap in the elliptic and hyperbolic regimes. This gap is identically zero in the trigonometric and rational regimes. Therefore, interesting scaling limits exist in these gapless regimes. Two inequivalent continuous limits exist : a) a massless limit yielding conformal invariant models b) a massive limit leading to integrable massive quantum field theories. These scaling behaviors are reviewed in refs.[4]. Let us just recall that all (known) conformal field theories derive from some integrable model in the continuous limit. The anisotropy parameter appears in the conformal dimensions (which vary continuously in general) but not in the central charge. Both depend on the YB algebra considered as well as on $\mathcal{A}$ and $\mathcal{V}$. The spectral parameter turns to be irrelevant (in the sense of crical phenomena) for this massless continuous limit. For massive scaling limits (yielding QFTs) the spectral parameter undergoes a dimensional transmutation and generates the mass scale together with the lattice spacing. The anisotropy parameter provides the (single) continuos coupling constant. The elliptic modulus vanishes and leaves no trace in the scaling limits[22].

We consider here solutions of the YB algebra where the operators $T^{(\mathcal{A},\mathcal{V})}(\theta)$ depend on a single argument $\theta$ . More generally, they can be functions of a pair of variables $\theta$ and $\theta'$ associated to each of the lines intersecting at the vertex, respectively. In refs.[23] there has been found solutions of such type associated to higher genus curves. These solutions are actually related to the YB algebra associated to the six vertex model[24]. For YB solutions of genus zero or one depending on $(\theta,\theta')$ there always exist parametrizations where the dependence becomes on $\theta-\theta'$ and hence reduces to the one variable case treated here[1].

Besides the symmetry transformations (1.19) leaving invariant the YB algebras, one can also look for transformations not involving linear combinations of YB operators like

$$[T_{ab}(\theta)]_{\alpha\beta} \quad \rightarrow \quad C_{ab\alpha\beta}(\theta)\,[T_{ab}(\theta)]_{\alpha\beta} \qquad (2.19)$$

Let us analyze these abelian transformations for the R-matrix,

$$R^{ab}{}_{cd}(\theta) \rightarrow \mathring{R}^{ab}{}_{cd}(\theta) = \exp[F_{abcd}(\theta)]\; R^{ab}{}_{cd}(\theta) \qquad (2.20)$$

in the trigonometric/hyperbolic cases where

$$a+b=c+d \qquad (2.21)$$

due to the U(1) symmetry(ies). One can seek for $F_{abcd}(\theta)$ an expansion of the type

$$F_{a_1a_2a_3a_4}(\theta) = \sum_{1\le i\le 4} a_i\, f_i(\theta) + \sum_{1\le i,j\le 4} a_i\, a_j\, f_{ij}(\theta) + \ldots\ldots\ldots \qquad (2.22)$$

Inserting eqs.(2.34)-(2.36) in eq.(1.12) and requiring $\mathring{R}^{ab}{}_{cd}(\theta)$ to be also a solution of the YBE (1.12) leads to a set of constraints on $f_i(\theta)$, $f_{ij}(\theta)$ ,.....We find after a long but straightforward algebra

$$F_{abcd}(\theta) = \mu\theta\,(a+c-b-d) + \nu\,(a+d-b-c) + \omega\,(ab-cd) + \ldots. \qquad (2.23)$$

where $\mu$, $\nu$ and $\omega$ are arbitrary parameters and we used also eq.(2.35). These transformations has been given in refs.(25) where they are called 'symmetry breaking transformations'. The reason of the name is that these transformations map R-matrices with P and T invariances [eqs.(2.30)] to R-matrices ($\mathring{R}$) where P and T do not hold.

## III. THE SIX VERTEX MODEL AND ITS DESCENDANTS

The six vertex model corresponds to the trigonometric and hyperbolic solutions of the YBE (1.11) for $q = 2$. That is,

$$R(\theta) = \begin{pmatrix} a(\theta,\gamma) & 0 & 0 & 0 \\ 0 & c(\gamma) & b(\theta) & 0 \\ 0 & b(\theta) & c(\gamma) & 0 \\ 0 & 0 & 0 & a(\theta,\gamma) \end{pmatrix} \tag{3.1}$$

We have here three different regimes,

I) $a(\theta,\gamma) = \mathrm{sh}(\gamma-\theta)$, $b(\theta) = \mathrm{sh}\theta$, $c(\gamma) = \mathrm{sh}\gamma$, $\gamma > \theta > 0$, $\gamma > 0$ in the antiferroelectric regime.

II) $a(\theta,\gamma) = \sin(\gamma-\theta)$, $b(\theta) = \sin\theta$, $c(\gamma) = \sin\gamma$, $0 < \gamma < \pi$, $\gamma > \theta > 0$ in the trigonometric regime. This regime is critical (gapless) and antiferroelectric.

III) $a(\theta,\gamma) = \mathrm{sh}(\theta+\gamma)$, $b(\theta) = \mathrm{sh}\theta$, $c(\gamma) = \mathrm{sh}\gamma$, $\theta > 0$, $\gamma > 0$ in the ferroelectric regime.

The parameter $\gamma$ describes the anisotropy of the model. The character of regimes I, II and III will be clear from the ground state and excitations obtained below.

This model enjoys the following symmetry group $\mathfrak{G}$ (in the sense of eq. (1.15)]

$$\mathfrak{G} = \{ \exp(i\,\alpha\,\sigma_z) \ , \ 0 \le \alpha < 2\pi \ ; \ \sigma_x \ \} \tag{3.2}$$

That is $\mathfrak{G} = U(1) \otimes Z_2$. When $\gamma = 0$ this group enlarges to $SU(2)$. This point corresponds to a Kosterlitz-Thouless type transition as we will see below from the explicit solution.

It is called six vertex model, since the non-zero elements of the R-matrix, eq.(3.1) define six allowed configurations. The integrable eight-vertex model will not be considered here[1]. The state of a bond in the six-vertex (and eight-vertex) models is usually characterized by the sense of an arrow. This corresponds here to the values 1 or 2 of the vertical and horizontal indices. In fig. 10 the allowed configurations and their respective statistical weights are depicted.

It must be recalled that the trigonometric regime of the six-vertex model (II) describes the critical (zero gap) limit of the eight-vertex model[1]. As it will be clear from the solution one describes a critical line when $\gamma$ varies from 0 to $\pi$. As a S-matrix eq.(3.1) for regime II describes the scattering of a particle and its antiparticle with a conserved U(1) charge[12]. The crossing symmetry (2.23) writes here

$$[P\, R(\theta)\,]^{t_1} = (1 \otimes \sigma)\, P\, R(-\theta-\gamma)\,(1 \otimes \sigma) \tag{3.3}$$

where
$$\sigma \equiv \begin{pmatrix} 0 & -1 \\ 1 & 0 \end{pmatrix} = -i\sigma_y$$

The YB generators read here (for one site)

$$t_{11}(\theta) = \begin{pmatrix} a(\theta,\gamma) & 0 \\ 0 & b(\theta) \end{pmatrix}, \quad t_{22}(\theta) = \begin{pmatrix} b(\theta) & 0 \\ 0 & a(\theta,\gamma) \end{pmatrix} \tag{3.4}$$

$$t_{12}(\theta) = c(\gamma)\,\sigma_- \quad , \quad t_{21}(\theta) = c(\gamma)\,\sigma_+$$

The YB generator $T_{ab}{}^{[N]}(\theta)$ for a N-sites line follows from eq. (1.19) where one inserts the $t_{ab}(\theta)$ given by eq. (3.4). One can then set

$$T_{ab}{}^{[N]}(\theta) = \begin{pmatrix} A(\theta) & , & B(\theta) \\ C(\theta) & , & D(\theta) \end{pmatrix} \tag{3.5}$$

The one-dimensional quantum hamiltonian associated to the six-vertex model is the XXZ Heisenberg hamiltonian. One finds from eqs. (2.12)-(2.13) and (3.1)

$$h = [\, \cos\gamma + \sigma_x \otimes \sigma_x + \sigma_y \otimes \sigma_y + \Delta\ \sigma_z \otimes \sigma_z\,]/(\,2\sin\gamma\,)$$

where $\Delta = -\cos\gamma$ .Then

$$H_{XXZ} = -\tfrac{1}{2}\sum_{a=1}^{N} (\,\sigma_x{}^a \otimes \sigma_x{}^{a+1} + \sigma_y{}^a \otimes \sigma_y{}^{a+1} + \Delta\ \sigma_z{}^a \otimes \sigma_z{}^{a+1}\,) \tag{3.6}$$

where we droped the terms proportional to the unit operator and redefined H by a factor $-\sin\gamma$ .

The six vertex model enjoys gauge invariance (see sec.I) under the symmetry group (3.2). That is we can perform U(1) transformations $g_a = \exp(\,i\,\mu_a\,\sigma_z{}^a\,)$ at each site a of the line and $T(\theta,\mathbf{g})$ will continue to obey the YB algebra. In particular, the gauge transformed XXZ Heisenberg hamiltonian reads from eq.(2.18) and (3.6)[7]

$$H_{XXZ}[\,\bar{\mu}\,] = -\tfrac{1}{2}\sum_{1\le a\le N} \exp(\,-i\,\mu_{a+1}\,\sigma_z{}^{a+1}\,)\,(\,\sigma_x{}^a \otimes \sigma_x{}^{a+1} + \sigma_y{}^a \otimes \sigma_y{}^{a+1} +$$

$$+ \Delta\, \sigma_z{}^a \otimes \sigma_z{}^{a+1} )\exp( i\, \mu_{a+1}\, \sigma_z{}^{a+1} ) \tag{3.7}$$

where $\bar{\mu} = ( \mu_1, \ldots, \mu_r )$.

As discussed in sec.II, the expansion of $\tau(\theta)$ around any point $\theta = \theta_0 \neq 0$ yields non-local operators. Nevertheless, expanding around $\theta_0 = \infty$ leads to interesting objects : generators of the $SU(2)_q$ quantum group as we shall see in sec.VI ( $q = \exp(i\gamma)$ or $\exp(-\gamma)$ ). Let us start to expand $A(\theta,\bar{\mu})$ and $D(\theta,\bar{\mu})$ around $\theta = \infty$. We find in the regime I[5]

$$\begin{aligned} A(\theta,\bar{\mu}) \underset{\theta\to\infty}{=} & \; y^N \exp( \gamma S_z - \alpha ) [ 1 + Q_+(\bar{\mu}) / y^2 + O(1/y^4) ] \\ D(\theta,\bar{\mu}) \underset{\theta\to\infty}{=} & \; y^N \exp( -\gamma S_z - \alpha ) [ 1 + Q_-(\bar{\mu}) / y^2 + O(1/y^4) ] \end{aligned} \tag{3.8}$$

where $\quad y \equiv \frac{1}{2}\exp( \theta + \gamma/2 )\ , \quad \alpha \equiv \sum_{a=1}^{N} \mu_a\ , \quad S_z = \frac{1}{2} \sum_{a=1}^{N} \sigma_z{}^a .$

and

$$\begin{aligned} Q_\pm(\bar{\mu}) = & \; \mathrm{sh}2\gamma \sum_{1\le a<b\le N} \sigma_-{}^a \exp(\mu_a + \mu_b) \prod_{c=a+1}^{b-1} \exp(\pm\gamma\, \sigma_z{}^c)\, \sigma_\pm{}^b \\ & - \frac{1}{4} \sum_{a=1}^{N} \exp(2\mu_a) [ \cosh\gamma - \mathrm{sh}\gamma\, \sigma_z{}^a ] \end{aligned} \tag{3.9}$$

For simplicity we have set **g = 1** in eq.(1.23).

The operator

$$Q_\mu( \bar{\mu} ) \equiv \exp(i\mu)\, Q_+( \bar{\mu} ) + \exp(-i\mu)\, Q_-( \bar{\mu} ) \tag{3.10}$$

belongs to the infinite sequence of operators generated by the transfer matrix

$$\tau_\mu(\theta) = \exp( -\gamma\, S_z + i\mu )\, A(\theta) + \exp( \gamma S_z - i\mu )\, D(\theta) \tag{3.11}$$

since

$$\tau_\mu(\theta) = y^N \exp( - \alpha ) [ 1 + Q_\mu( \bar{\mu} ) / y^2 + O(1/y^4) ] \quad \text{for } y \to \infty,$$

and

$$[ \tau_\mu(\theta) , \tau_\mu(\theta') ] = 0 \quad , \quad [ \tau_\mu(\theta) , Q_\mu( \bar{\mu} ) ] = 0 \ .$$

These operators simplify in the isotropic limit $\gamma \to 0$, $\mu_a \to \kappa_a\, \gamma$, $\theta \to \gamma\, \lambda$ . We find from eqs.(3.9)-(3.10)

$$\underset{\gamma\to 0}{Q_\mu(\bar{\mu})} = -\tfrac{1}{2}N\cos\mu - \gamma(\nu\cos\mu + 2i\sin\mu\, S_z) - \tfrac{1}{4}\gamma^2 [\, 3N + \sum_{a=1}^{N} \kappa_a^2 - 4(S_x^2 + S_y^2)\,]\cos\mu + \gamma^2 \sin\mu\; h(\bar{\kappa}) + O(\gamma^3)$$

where $\nu = \alpha / \gamma$ for $\gamma \to 0$ and $h(\bar{\kappa})$ is the hermitean operator

$$h(\bar{\kappa}) \equiv \tfrac{1}{4}\sum_{1\le a\ne b\le N} \mathrm{sign}(a\text{-}b)\,(\varepsilon_{zij}\,\sigma_i^a\,\sigma_j^b) - \sum_{a=1}^{N}\kappa_a\,\sigma_z^a \tag{3.12}$$

Here i,j = x , y , $\varepsilon_{123} = +1$ and $\bar{\kappa} = (\kappa_1, \dots, \kappa_N)$ . $h(\bar{\kappa})$ can be interpreted as a non-local Dzialozhinski-Moriya interaction hamiltonian plus a coupling with an external magnetic field $\nu_a$ pointing in the z direction. That is a magnetic field varying from site to site on the line. The properties of $h(\bar{\kappa})$ are analyzed in ref.[5]. This exactly solvable hamiltonian can describe disordered systems since the $\kappa_a$ are completely arbitrary. For example they can follow some probability distribution.

The YB algebra defined by the R-matrix (3.1) yields some number of bilinear algebraic relations between the $T_{ab}^{[N]}(\theta)$. Let us just write down the more useful ones for the subsequent derivations

$$A(\theta)\, B(\theta') = g(\theta' - \theta)\, B(\theta')\, A(\theta) - h(\theta' - \theta)\, B(\theta)\, A(\theta') \tag{3.13}$$

$$D(\theta)\, B(\theta') = g(\theta - \theta')\, B(\theta')\, D(\theta) - h(\theta - \theta')\, B(\theta)\, D(\theta')$$

$$B(\theta)\, B(\theta') = B(\theta)\, B(\theta') \tag{3.14}$$

$$[\, C(\theta)\, , B(\theta')\, ] = h(\theta - \theta')\,\{ A(\theta')\, D(\theta) - A(\theta)\, D(\theta') \}$$

where $g(\theta) = a(\theta,\gamma) / b(\theta)$ and $h(\theta) = c(\gamma) / b(\theta)$.

Let us now proceed to construct the exact eigenvectors and eigenvalues of

$$\tau^{[N]}(\theta) = \mathrm{Tr}_{\mathcal{A}}\, T^{[N]}(\theta) = A(\theta) + D(\theta)\ , \tag{3.15}$$

using the algebraic Bethe Ansatz[17]. We shall assume N to be even. One notices that the ferromagnetic state

$$|\Omega\rangle = \begin{pmatrix}1\\0\end{pmatrix} \otimes \begin{pmatrix}1\\0\end{pmatrix} \otimes \cdots\cdots \otimes \begin{pmatrix}1\\0\end{pmatrix} \tag{3.16}$$

is an eigenvector of $A(\theta)$ and $D(\theta)$

$$A(\theta) \mid \Omega> = a(\theta,\gamma)^N \mid \Omega> \quad , \quad D(\theta) \mid \Omega> = b(\theta)^N \mid \Omega> \tag{3.17}$$

In addition

$$C(\theta) \mid \Omega> = 0 \tag{3.18}$$

whereas $B(\theta)|\Omega>$ is non-zero and not proportional to $|\Omega>$. The algebraic Bethe ansatz consist in looking for eigenvectors of $\tau(\theta)$ with the form

$$\Psi(\theta_1,...,\theta_r) = B(\theta_1)\, B(\theta_2)........ B(\theta_r) \mid \Omega> \tag{3.19}$$

Here, the complex number $\theta_1$, ..., $\theta_r$ will be determined by requiring that $\Psi(\theta_1,...,\theta_r)$ is an eigenvector of $\tau(\theta)$.

In order to do that one applies $A(\theta) + D(\theta)$ to the r.h.s. of eq. (3.20) and pushes $A(\theta) + D(\theta)$ through the $B(\theta_j)$ with the help of eqs. (3.13). After using eqs. (3.13) $r$ times, $A(\theta)$ and $D(\theta)$ reach $\mid \Omega>$ where their action is known from eqs. (3.17). These operations produced a lot of terms. Let us first write down explicitely those generated by the first term in eqs. (3.13) :

$$A(\theta)\, \Psi(\theta_1,...,\theta_r) = a(\theta,\gamma)^N \prod_{j=1}^{r} g(\theta_j - \theta)\, B(\theta_1)\, B(\theta_2)........ B(\theta_r) \mid \Omega> + \tag{3.20}$$
$$+ \text{ unwanted terms } = \Lambda_+(\theta)\, \Psi(\theta_1,...,\theta_r) + \text{ unwanted terms}$$

and an analogous formula for $D(\theta)\, \Psi(\theta_1,...,\theta_r)$. The remaining terms are called "unwanted" since they are not proportional to $\Psi(\theta_1,...,\theta_r)$ and hence they must finally cancel in order that $\Psi(\theta_1,...,\theta_r)$ be an eigenvector of $\tau(\theta)$.

Now, let us concentrate in terms containing the vector

$$B(\theta)\, B(\theta_2)........ B(\theta_r) \mid \Omega>$$

They originate when the second term in eq. (3.13) is used to express $A(\theta)\, B(\theta_1)$ and the first term for the rest when $A(\theta_1)$ is pushed through $B(\theta_j)$ $(2 \leq j \leq r)$. Hence, one finds

$$- h(\theta_1 - \theta)\, B(\theta)\, A(\theta_1)\, B(\theta_2)........ B(\theta_r) \mid \Omega> =$$
$$= - h(\theta_1 - \theta)\, a(\theta_1,\gamma)^N \prod_{j=2}^{r} g(\theta_j - \theta_1)\, B(\theta)\, B(\theta_2)........ B(\theta_r) \mid \Omega> + \text{ other types of terms.} \tag{3.21}$$

It is now very easy to determine the remaining coefficients since $\Psi(\theta_1,...,\theta_r)$ is a symmetric function of $\theta_1,...,\theta_r$ due to eq. (3.14). Therefore one can permute $\theta_1$ by $\theta_j$ in eq. (3.21) with the result

$$A(\theta)\,\Psi(\theta_1,\ldots,\theta_r) = \Lambda_+(\theta,\theta_1,\ldots,\theta_r)\,\Psi(\theta_1,\ldots,\theta_r) + \sum_{k=1}^{r} \Lambda_k{}^+(\theta,\theta_1,\ldots,\theta_r)\,\Psi_k(\theta,\theta_1,\ldots,\theta_r) \tag{3.22}$$

where

$$\Psi_k(\theta,\theta_1,\ldots,\theta_r) \equiv B(\theta) \prod_{1\le j\le r,\, j\ne k} B(\theta_j)\,|\,\Omega> \tag{3.23}$$

and

$$\Lambda_+(\theta,\theta_1,\ldots,\theta_r) = a(\theta,\gamma)^N \prod_{j=1}^{r} g(\theta_j-\theta)$$

$$\Lambda_k{}^+(\theta,\theta_1,\ldots,\theta_r) = h(\theta_k-\theta)\;\; a(\theta_k,\gamma)^N \prod_{1\le \ell\le r,\, \ell\ne k} g(\theta_\ell-\theta_k) \tag{3.24}$$

One analogously finds

$$D(\theta)\,\Psi(\theta_1,\ldots,\theta_r) = \Lambda_-(\theta,\theta_1,\ldots,\theta_r)\,\Psi(\theta_1,\ldots,\theta_r) + \sum_{k=1}^{r} \Lambda_k{}^-(\theta,\theta_1,\ldots,\theta_r)\,\Psi_k(\theta,\theta_1,\ldots,\theta_r) \tag{3.25}$$

where

$$\Lambda_-(\theta,\theta_1,\ldots,\theta_r) = b(\theta)^N \prod_{j=1}^{r} g(\theta-\theta_j)$$

$$\Lambda_k{}^-(\theta,\theta_1,\ldots,\theta_r) = h(\theta-\theta_k)\;\; b(\theta_k)^N \prod_{1\le \ell\le r,\, \ell\ne k} g(\theta_k-\theta_\ell) \tag{3.26}$$

Now, in order to get an eigenvector of $\tau(\theta)$ we must require

$$\Lambda_k{}^+(\theta,\theta_1,\ldots,\theta_r) + \Lambda_k{}^-(\theta,\theta_1,\ldots,\theta_r) = 0 \tag{3.27}$$

This yields a set of r algebraic equation in $\theta_k$ $(1 \le k \le r)$ usually called Bethe Ansatz equations (BAE). Notice that the dependence on $\theta$ drops in eq.(3.27) thanks to the fact that $h(\theta)$ is an odd function. This could be expected since the commutativity of $\tau(\theta)$ for different $\theta$ suggest that its eigenvectors can be chosen $\theta$-independent. More explicitely these BAE read,

$$[\,\mathrm{sh}(\lambda_j + i\gamma/2)\,/\,\mathrm{sh}(\lambda_j - i\gamma/2)\,]^N =$$

$$= -\prod_{k=1}^{r} [\mathrm{sh}(\lambda_j-\lambda_k+i\gamma)\,/\,\mathrm{sh}(\lambda_j-\lambda_k-i\gamma)] \qquad \text{regime II} \tag{3.28}$$

$$[\,\sin(\lambda_j + i\gamma/2)\,/\,\sin(\lambda_j - i\gamma/2)\,]^N =$$

$$= - \prod_{k=1}^{r} [\sin( \lambda_j - \lambda_k + i\gamma) / \sin( \lambda_j - \lambda_k - i\gamma )] \qquad (3.29)$$

regime I and III

Here we have introduced $\lambda_j \equiv i(\theta_j + \gamma/2)$, for regime III and $\lambda_j \equiv -i(\theta_j - \gamma/2)$ for regimes I and II, $1 \leq j \leq r$. Once the $\lambda_j$ are found by solving eqs. (3.28) or (3.29) the eigenvalues $\Lambda(\theta)$ of $\tau(\theta)$ follow from eqs.(3.24) and (3.26) as

$$\Lambda(\theta,\theta_1,\ldots,\theta_r) = \Lambda_+(\theta,\theta_1,\ldots,\theta_r) + \Lambda_-(\theta,\theta_1,\ldots,\theta_r) \qquad (3.30)$$

We can assume $|\mathrm{Re}\ \lambda_j| \leqslant \pi/2$ in regimes I and III, whereas $-\infty < \mathrm{Re}\ \lambda_j < +\infty$ for regime II. The r.h.s. of eq. (3.30) would seem to have poles at $\theta = +i\lambda_j + \gamma/2$. However the corresponding residues identically vanish due to eqs. (3.28)-(3.29). Actually one can use this property as a short-cut to derive the BAE when the construction of the explicit eigenvectors is more involved.

Eq. (1.18) gives for the six-vertex model symmetry (3.2) (rotations around z)

$$[ A(\theta) , S_z ] = [ D(\theta) , S_z ] = 0$$
$$[ S_z , B(\theta) ] = - B(\theta) \quad , \quad [ S_z , C(\theta) ] = C(\theta) \qquad (3.31)$$

where $S_z = \frac{1}{2} \sum_{a=1}^{N} \sigma_z^a$ acts in the vertical space. Therefore $B(\theta)$ [$C(\theta)$] lowers [raises] the z-component of the spin in one unit.

In particular we find that the state (3.19) is an eigenvectors of $S_z$

$$S_z \Psi(\theta_1,\ldots,\theta_r) = ( N/2 - r ) \Psi(\theta_1,\ldots,\theta_r) \qquad (3.32)$$

This algebraic BA construction of the eigenvectors of $\tau(\theta)$ has some analogies with the angular momentum states obtained from the maximal weight state applying lowering operators $S_-$ . Here $|\Omega\rangle$ is also a maximal weight vector for the total spin and $B(\theta)$ lowers the spin by one unit [eq.(3.31)]. Nevertheless, to obtain exact eigenvectors of $\tau(\theta)$ the arguments $\theta_1,\ldots,\theta_r$ must fulfil the BAE (3.28)-(3.29). This notion is absent in Lie algebra constructions of eigenvectors.

The resolution of the BAE (3.28)-(3.29) is possible analytically for small r and N . For large r and N this is a formidable task. Only in the $N = \infty$, $r = \infty$ limit , things become easier since the roots become closer and closer for large N. The separation is $O(1/N)$ between neighboring $\lambda_j$ in the real axis. The density of roots is defined as

$$\rho(\lambda_j) = \lim_{N\to\infty} 1/[N(\lambda_{j+1} - \lambda_j)] \tag{3.33}$$

This function obeys a linear integral equation that follows from eqs.(3.28)-(3.29). Since it has a difference kernel, it can be solved in closed form by Fourier integrals or Fourier series. Once $\rho(\lambda)$ is known, the calculation of physical quantities like $\Lambda(\theta)$ (eigenvalue of $\tau(\theta)$ ) reduce to quadratures. That is

$$\lim_{N\to\infty} (1/N)\Lambda(\theta) = \int d\lambda\ \phi(\lambda + i\theta, \gamma/2)\ \rho(\lambda) \tag{3.34}$$

where

$$\phi(\lambda,\alpha) = i \operatorname{Log}\{\sin(\lambda + i\alpha) / \sin(\lambda - i\alpha)\} \quad \text{, regime I}$$
$$\phi(\lambda,\alpha) = i \operatorname{Log}\{\sinh(\lambda + i\alpha) / \sinh(\lambda - i\alpha)\} \quad \text{, regime II} \tag{3.35}$$

Besides real roots the BAE posses complex roots (see, for example refs.[26]). There exist an important literature on numerical solutions of the BAE[27]. Moreover, it is possible to solve the BAE asymptotically for N large but finite using the method proposed in ref.[28] , extended and generalized in refs.[29-33]. The finite size corrections turn out to be exponiantially small in N for regime I and power like in the gapless regime II .

In the QFT associated to vertex models, the vacuum (ground state) corresponds precisely to the antiferroelectric ground state. Let us concentrate on this state and excitations around it from now on. The operators $B(\theta_j)$ play here the rôle of creation operators of excitations over the bare vacuum $|\Omega\rangle$. That is pseudo-particles or "bare" particles. The antiferroelectric ground state is the analog of the filled Dirac sea for free fermions. However, the pseudoparticles are here not free, they interact through two-body interactions. The functions $\phi(\lambda_i - \lambda_j, \gamma)$ describe the two-body phase-shift associated to such interactions. These functions are equal to $\pi$ when $\gamma = \pi/2$ in regime II : the free fermion (or Ising) case.

The BAE (3.28)-(3.29) can be rewritten as

$$\exp i[N\,\phi(\lambda_j, \gamma/2) - \sum_{k\neq j} \phi(\lambda_j - \lambda_k, \gamma)] = 1 \tag{3.36}$$

The first term in the exponent, $N\phi(\lambda_j, \gamma/2)$, is just the momentum of the $j^{th}$ pseudoparticle times the number of sites. That is the phase for a free particle moving around this ring of length N. The second term can be interpreted as the phase shifts induced in the wave function of the $j^{th}$ pseudoparticle by the (pair) interaction with the rest of them. In other words eq. (3.29) ensures the periodicity of the "wave function" when turning aroung the ring. This interpretation of the BAE extends for more general models[4].

This concludes our exposition of the Bethe Ansatz solution of the six-vertex model. A more general Bethe Ansatz construction provides the eigenvectors of the eight-vertex model[17,34]. It has been shown recently that these eight-vertex eigenvectors become the six-vertex eigenvectors in the limit where the eight-vertex weights become those of the six-vertex model[35].

The Bethe Ansatz has been also generalized for multi-state vertex models. That is when dim$\mathcal{A}$ and/or dim $\mathcal{V}$ is larger than two. The resulting construction is a set of nested Bethe Ansatz. It is reviewed in ref.[4] .

## IV - FACE MODELS (SOS and RSOS) and FACE ALGEBRAS

We have exposed up to here statistical models in the vertex language. Let us now start with face models. That is, models where the degrees of freedom are attached to faces or points of the lattice instead of being attached to the links joining them. This is a language dual to the vertex language. One can always associate a face model to a vertex model through the Wu-Kadanoff-Wegner duality (see fig.16). The converse is also true but more involved . A face model with $q^2$ states per face associates to a vertex model with q states per bond[36].

The local state of the face is defined by a number $\alpha \in \mathcal{Z}$ and depicted as a dot. Neighboring sites are in local states differing by plus or minus one. More generally, the local states are described by weights of a given simple Lie algebra. For these generalized SOS and RSOS models see ref.[62]. The case consider here ( $\alpha - \alpha_{\text{neighbor}} = \pm 1$ ) corresponds to the Lie algebra $A_1$ and they are the face partners of the six and eight vertex models.
We consider both RSOS and SOS type models where $1 \leq \alpha \leq Q$ and when $\alpha$ is unrestricted, respectively. The local (microscopic) statistical weights depend on the states of four neighboring faces as depicted in fig.17. The variable $\theta$ ( $\theta \in \mathbb{C}$ ) is called spectral parameter and is associated with an angle as in fig 2.

As integrability constraint we assume that the $w(\alpha, \beta, \gamma, \delta \mid \theta)$ obeys the star-triangle relation:

$$\sum_{\nu} w(\nu, \gamma, \delta, \varepsilon \mid \theta')\, w(\alpha, \beta, \gamma, \nu \mid \theta)\, w(\mu, \alpha, \nu, \varepsilon \mid \theta-\theta') = \sum_{\nu} w(\nu, \beta, \gamma, \delta \mid \theta-\theta')\, w(\mu, \nu, \delta, \varepsilon \mid \theta)\, w(\alpha, \beta, \nu, \mu \mid \theta') \tag{4.1}$$

This relation can be conveniently represented in graphical form as in fig.18. Notice that the picture is consistent (mod $\pi$ ) with the interpretation of $\theta$, $\theta'$ and $\theta-\theta'$ as angles.

As it is known the face language is very convenient to define many integrable models [37-40]. It is therefore a fundamental problem for the theory of integrable models to generalize the YBA and the BA to the face language. This is what we do in the present paper.

The statistical weights $w(\alpha, \beta, \gamma, \delta \mid \theta)$ are said to be regular if

$$w(\alpha, \beta, \gamma, \delta \mid 0) = c\, \delta_{\alpha\gamma} \tag{4.2}$$

where c is a (non-zero) constant.

Setting $\theta = 0$ in eq (1.1) and using eq. (4.2) yields

$$\delta_{\mu\delta}\, P_{\beta\gamma\mu\alpha}(\theta') = \delta_{\alpha\gamma}\, P_{\alpha\delta\varepsilon\mu}(-\theta') \tag{4.3}$$

where

$$P_{\beta\gamma\mu\alpha}(\theta) = \sum_{\nu} w(\nu, \beta, \gamma, \mu \mid -\theta)\, w(\alpha, \beta, \nu, \mu \mid \theta) \tag{4.4}$$

Therefore, $P_{\beta\gamma\mu\alpha}(\theta)$ must have the index structure

$$P_{\beta\gamma\mu\alpha}(\theta) = \delta_{\alpha\gamma}\, Q_{\beta\mu}(\theta) \tag{4.5}$$

Moreover, inserting eq.(4.5) in eq.(4.3) shows that $Q_{\beta\mu}(\theta)$ is actually independent of the indices $\beta\,\mu$ and that it is an even function of $\theta$. We finally obtain

$$\sum_{\gamma} w(\alpha, \beta, \gamma, \delta \mid \theta)\, w(\gamma, \beta, \sigma, \delta \mid -\theta) = \rho(\theta)\, \delta_{\alpha\sigma} \tag{4.6}$$

where $\rho(\theta) = \rho(-\theta)$ . Eq.(4.6) is depicted in fig. 17.

Eq. (6) can be interpreted as a unitarity property of the weights. Its derivation closely parallels the vertex language derivation (sec.II).

We consider up to now the **local** face algebra (1.1) associated to one site. Let us consider N sites. We define the YB operator for a line of N sites as[41]

$$t_{\alpha\alpha',\beta\beta'}(\underline{\sigma},\underline{\sigma}' \mid \theta) = \delta_{\beta\sigma_N}\, \delta_{\beta'\sigma'_N} \prod_{\ell=0}^{N} w(\sigma_\ell, \sigma_{\ell+1}, \sigma'_{\ell+1}, \sigma'_\ell \mid \theta) \tag{4.7}$$

where $\sigma_0 \equiv \alpha$, $\sigma'_0 \equiv \alpha'$ , $\underline{\sigma} \equiv (\sigma_1, \sigma_2, \ldots, \sigma_N)$ and $\underline{\sigma}' \equiv (\sigma'_1, \sigma'_2, \ldots, \sigma'_N)$. We depict $t_{\alpha\alpha',\beta\beta'}(\underline{\sigma},\underline{\sigma}' \mid \theta)$ in fig 19. Eq.(4.7) defines the analogue of the operator $T_{ab}(\theta)_{\underline{\sigma}|\underline{\sigma}'}$ used in vertex language. As in the vertex case the variables $\underline{\sigma}, \underline{\sigma}'$ are here vertical (or quantum) indices whereas the pair of indices $\alpha\,\alpha'$ and $\beta$

$\beta'$ play the rôle of horizontal (or auxiliary) indices. This is actually consistent with the pictures (fig.19). A look to them shows that in the face YB operators (fig.19) the horizontal indices are forcely a couple $\alpha\ \alpha'$ and $\beta\ \beta'$ on left and right respectively. By consistency, we must set at the corners $\beta = \sigma_N$ and $\beta' = \sigma'_N$.

The transfer matrix for faces is defined as a trace of the YB operator (4.7)

$$\tau(\underline{\sigma},\underline{\sigma}'\,|\,\theta) \equiv \sum_{\alpha\alpha'} t_{\alpha\alpha',\alpha\alpha'}(\underline{\sigma},\underline{\sigma}'\,|\,\theta) \tag{4.8}$$

analogously to the the vertex language case.

We introduce **shifted** transfer matrices as[41]

$$\tau^{(K)}(\underline{\sigma},\underline{\sigma}'\,|\,\theta) \equiv \sum_{\alpha\alpha'} t_{\alpha,\alpha';\alpha+K,\alpha'+K}(\underline{\sigma},\underline{\sigma}'\,|\,\theta) \tag{4.9}$$

These expressions (4.8) and (4.9) are analogous to the vertex language transfer matrices and twisted transfer matrices (see secs. I and II).

A twist $\exp(iKS_z)$ in vertex. language corresponds to a shift K in the face variables.

Notice that the sums in eqs.(4.8) and (4.9) only contain one term due to the Kronecker deltas on the right corner [cfr. eq. (4.7)] . That is

$$\begin{aligned} \tau(\underline{\sigma},\underline{\sigma}'\,|\,\theta) &= t_{\sigma_N,\sigma'_N;\sigma_N,\sigma'_N}(\underline{\sigma},\underline{\sigma}'\,|\,\theta) \\ \tau^{(K)}(\underline{\sigma},\underline{\sigma}'\,|\,\theta) &= t_{\sigma_N-K,\sigma'_N-K;\sigma_N,\sigma'_N}(\underline{\sigma},\underline{\sigma}'\,|\,\theta) \end{aligned} \tag{4.10}$$

An important difference between vertex YB operators and the face YB operators eq.(4.7) is that **no summation** appear in our definitions in face language. $T_{ab}(\theta)$ [eq.(1.21)] is in vertex language the sum of $2^{N-1}$ terms (when there are two possible states per bond).

Now, the partition function $\mathcal{Z}$ for a N x M lattice with periodic boundary conditions (PBC) writes in terms of the transfer matrix (4.8) as usual

$$\mathcal{Z} = \mathrm{Tr}[\ \tau(\theta)^M\ ] \tag{4.11}$$

where Tr means trace over the indices $\underline{\sigma}$. If we impose as boundary condition on the face variables a shift K in the horizontal direction. That is,

$$\alpha_N = \alpha_1 + K \tag{4.12}$$

Then, the partition function writes

$$\mathcal{Z}_K = \mathrm{Tr}[\ \tau^{(K)}(\theta)^M\ ] \tag{4.13}$$

Let us now derive a YB algebra for the operators $t_{\alpha\alpha',\beta\beta'}(\underline{\sigma},\underline{\sigma}'\,|\,\theta)$ [eq.

(4.7)] . Using the star-triangle eq.(1) repeatedly, we find

$$\sum_{\delta} w(\beta, \alpha, \delta, \gamma \mid \theta-\theta')\, t_{\alpha\delta,\lambda\nu}(\theta)\, t_{\delta\gamma,\nu\mu}(\theta') = \tag{4.14}$$
$$= \sum_{\delta} t_{\alpha\beta,\lambda\delta}(\theta')\, t_{\beta\gamma,\delta\mu}(\theta)\, w(\delta, \lambda, \nu, \mu \mid \theta-\theta')$$

This equation is depicted in fig. 20. We use here the convention that left to right in the formulas correspond to up to down in the figures. The derivation of eq. (4.14) is actually easier graphically than analytically. Applying fig. 18 repeatedly to the lhs of fig. 20 yields the rhs.

It should be noticed that the regularity condition (1.2) is naturally suggested by eq. (4.14) at $\theta = \theta'$.

As in vertex language a tensor product notation can be introduced for the horizontal indices,

$$[t(\theta) \otimes t(\theta')]_{\alpha\delta\delta'\gamma,\lambda\nu\nu'\mu} = t_{\alpha\delta,\lambda\nu}(\theta)\; t_{\delta'\gamma,\nu'\mu}(\theta') \tag{4.15}$$

Then eq. (4.14) can be written as[41]

$$W(\theta - \theta')\, [t(\theta) \otimes t(\theta')]\, Q = Q\, [t(\theta') \otimes t(\theta)]\, W(\theta - \theta') \tag{4.16}$$
or
$$W(\theta - \theta')\, Q[t(\theta) \otimes t(\theta')]\, Q = Q\, [t(\theta') \otimes t(\theta)]\, Q\, W(\theta - \theta')$$

where

$$Q_{\alpha\beta\mu\nu,\alpha'\beta'\mu'\nu'} \equiv \delta_{\beta\mu}\, \delta_{\alpha\alpha'}\, \delta_{\beta\beta'}\, \delta_{\mu\mu'}\, \delta_{\nu\nu'}$$
$$W_{\alpha\beta\beta'\gamma,\alpha'\delta\delta'\gamma'}(\theta) = \delta_{\alpha\alpha'}\, \delta_{\beta\beta'}\, \delta_{\delta\delta'}\, \delta_{\gamma\gamma'}\; w(\beta, \alpha, \delta, \gamma \mid \theta) \tag{4.17}$$
$$QW = WQ = W$$

Eq. (4.16) has the same structure as the vertex-language YBA (sec.I)

$$R(\theta-\theta')\, [T(\theta) \otimes T(\theta')] = [T(\theta') \otimes T(\theta)]\, R(\theta-\theta') \tag{4.18}$$

except for the constant matrix Q which is essentially a projector, since

$$Q^2 = Q \tag{4.19}$$

The matrix Q makes the job of getting equal between them the middle indices of the t's which are not summed up. That is $\nu = \nu'$ and $\beta = \beta'$ in the l.h.s. and the r.h.s of eq.(4.14) [fig.20] respectively.

Let us now prove the commutativity of the transfer matrices (1.10) from the face YBA (4.14). Multiplying both sides of eq. (4.14) by $w(\alpha, \beta, \gamma, \delta \mid \theta)$ and summing over $\beta$ yields.

$$t_{\alpha\sigma,\lambda\nu}(\theta)\ t_{\sigma\gamma,\nu\mu}(\theta')\ \rho(\theta-\theta') = \tag{4.20}$$

$$= \sum_{\beta\delta} w(\ \sigma, \alpha, \beta, \gamma \mid \theta'-\theta\ )\ t_{\alpha\beta,\lambda\delta}(\theta')\ t_{\beta\gamma,\delta\mu}(\theta)\ w(\ \delta, \lambda, \nu, \mu \mid \theta-\theta'\ )$$

Now setting $\mu = \gamma + K$, $\lambda = \alpha + K$ and $\nu = \sigma + K$ and summing over $\alpha$, $\sigma$ and $\gamma$ gives the desired relation

$$[\ \tau^{(K)}(\theta)\ ,\ \tau^{(K)}(\theta')\ ] = 0 \tag{4.21}$$

In particular, for $K = 0$ we find the commutativity of the PBC transfer matrices

$$[\ \tau(\theta)\ ,\ \tau(\theta')\ ] = 0 \tag{4.22}$$

It must be noticed that the transfer matrix $\tau(\theta)$ at $\theta = 0$ becomes

$$\tau(\underline{\sigma},\underline{\sigma}'\,|\,0) = \prod_{i=1}^{N} \delta_{\sigma_{i-1}\sigma_i'} \tag{4.23}$$

where we used eq. (4.2). That is, $\tau(0)$ is the unit shift operator to the left.

The regularity equation (1.2) gave us the inversion (or unitarity) property (4.6) for the microscopic weights $w(\ \alpha, \beta, \gamma, \delta \mid \theta\ )$. Using eq. (4.6) repeatedly yields for the YB face operator (4.7)

$$\sum_{\alpha''\,\beta''\underline{\sigma}''} t_{\alpha\alpha'',\beta\beta''}(\underline{\sigma},\underline{\sigma}''\,|\,\theta)\ t_{\underline{\sigma}'',\underline{\sigma}'}\,(\beta''\beta,\alpha''\alpha'\,|\,-\theta) = \rho(\theta)^N\ \delta_{\alpha\alpha'}\,\delta_{\underline{\sigma},\underline{\sigma}'} \tag{4.24}$$

where the operator $t_{\underline{\sigma},\underline{\sigma}'}\,(\beta\beta',\alpha\alpha'\,|\,\theta)$ is depicted in fig 19. Eq(4.24) tells us that the YB face operator possess an inverse given by

$$t^{-1}{}_{\alpha\alpha',\beta\beta'}(\underline{\sigma},\underline{\sigma}'\,|\,\theta) = \rho(\theta)^{-N}\ t_{\underline{\sigma},\underline{\sigma}'}\,(\alpha'\alpha,\beta'\beta\,|\,-\theta) \tag{4.25}$$

The weights $w(\ \alpha, \beta, \gamma, \delta \mid \theta\ )$ possess in addition the crossing property in many cases :

$$w(\ \alpha, \beta, \gamma, \delta \mid \theta\ ) = w(\ \delta, \alpha, \beta, \gamma \mid -\theta-\eta\ )$$

then,

$$t_{\underline{\sigma},\underline{\sigma}'}\,(\beta\beta',\alpha\alpha'\,|\,-\theta) = t_{\alpha\alpha',\beta\beta'}(\underline{\sigma},\underline{\sigma}'\,|\,\theta-\eta) \tag{4.26}$$

Let us consider explicit solutions of the star-triangle relations (1.1). There exist a solution of eq.(1.1) for each two dimensional graph. In the graph the nodes describe the allowed local states of the faces. Two nodes linked by a bond in the graph mean that this a couple of allowed states for two neighboring faces. An unrestricted SOS model has associated an infinite one-dimensional graph. The periodic SOS model where states $\alpha$ and $\alpha + Q$ are identified corresponds to a ring with $Q$ nodes. The ABF RSOS model is described by a Q node linear graph identical to the $A_Q$ Dynkin diagram. The solution of eq.(1.1) associated to a graph $\Gamma$ can be written as [40]

$$w(\alpha, \beta, \gamma, \delta \mid \theta) = \delta_{\alpha\gamma} \sin(\gamma+\theta) + \delta_{\beta\delta} \sin(\theta) \sqrt{[\Psi(\alpha)\Psi(\gamma)/\Psi(\beta)\Psi(\delta)]} \tag{4.27}$$

where $\Psi(\alpha)$ is an eigenvector of the incidence matrix of $\Gamma$ with eigenvalue $2\cos\gamma$. That is,

$$\sum_{\beta} K_{\alpha\beta}\,\Psi(\beta) = 2(\cos\gamma - 1)\ \Psi(\alpha) \tag{4.28}$$

Here $K_{\alpha\alpha} = -2$ and $K_{\alpha\beta} = +1$ if $\alpha$ and $\beta$ are neighbors in $\Gamma$. Since the Perron-Frobenius theorem holds for the matrix $(-K_{\alpha\beta})$, there exist an eigenvector with all its components $\Psi(\alpha) \geq 0$ and with positive eigenvalue $1 - \cos\gamma > 0$. This eigenvector lead to real weights through eq.(4.27).

One recognizes that K is the Cartan matrix if $\Gamma$ is a simply-laced Dynkin diagram. In that case one recovers the models of refs.[39] in their critical regime. To summarize: for each graph $\Gamma$, eq.(4.27) provides a solution of eq.(4.1) provided $\Psi(\alpha)$ fulfils eq.(4.28). As it is clear a finite $\Gamma$ gives a RSOS type model and an infinite $\Gamma$ a SOS type model. In the first case the parameter $\gamma$ will take some discrete values determined by eq.(4.28).

For an infinite one-dimensional $\Gamma$, one finds

$$\Psi(\alpha) = \sin(\alpha\gamma + w_0) \tag{4.29}$$

where $w_0$ is an arbitrary constant and $0 \leq \gamma \leq \pi$. Eq.(4.29) inserted in (4.27) yields the critical ABF SOS model[37]. One finds as non-zero weights :

$$\begin{aligned} w(\alpha, \alpha \pm 1, \alpha, \alpha-1 \mid \theta) &= a(\theta) \\ w(\alpha \pm 1, \alpha, \alpha-1, \alpha \mid \theta) &= \beta_\alpha(\theta) \\ w(\beta \pm 1, \beta, \beta\pm1, \beta \mid \theta) &= \alpha_\beta^{\pm}(\theta) \end{aligned} \tag{4.30}$$

where

$$\begin{aligned} a(\theta) &= \sin(\theta + \gamma) \\ \beta_\alpha(\theta) &= b(\theta) \sqrt{\phantom{x}} s_{\alpha+1}\, s_{\alpha-1} / s_\alpha \\ \alpha_\beta^{\pm}(\theta) &= \sin\gamma\ \sin(\beta\gamma + w_0 - \theta) / s_\beta \end{aligned} \tag{4.31}$$

where $s_\alpha \equiv \sin(\alpha\gamma + w_0)$.

When

$$\gamma/(2\pi) = Q'/Q \text{ with } Q, Q' \varepsilon \mathscr{Z},$$

we have

$$\Psi(\alpha + Q) = \Psi(\alpha)$$

and we obtain the periodic SOS model. Setting in addition $w_0 = 0$ yields the RSOS (Q–1)-state ABF model since then,

$$\Psi(Q) = \Psi(0) = 0$$

It is always possible to map a solution of the vertex language YBE (1.12) into a solution of the star-triangle eq.(4.1) through the Wu-Kadanoff-Wegner duality

$$w(\alpha, \beta, \gamma, \delta \mid \theta) = R^{\alpha-\delta,\beta-\alpha}_{\gamma-\delta,\beta-\gamma}(\theta) \qquad (4.32)$$

Inserting eq.(4.30) in eq.(4.1) yields an identity after using (4.12). Notice that eq.(4.30) provides a particular class of solutions of traslational invariant solutions.

$$w(\alpha+\lambda, \beta+\lambda, \gamma+\lambda, \delta+\lambda \mid \theta) = w(\alpha, \beta, \gamma, \delta \mid \theta) \qquad (4.33)$$

This suggests that the face language is richer than the vertex description.

The dual of the six vertex model through eq.(4.32) gives a SOS model with weights

$$\begin{aligned} a(\theta) &= \sin(\theta+\gamma) \\ \beta_\alpha(\theta) &= b(\theta) \\ \alpha_\beta{}^{\pm}(\theta) &= \sin\gamma \end{aligned} \qquad (4.34)$$

In this SOS model [model I] as well as in the model defined by eqs.(4.31) [model II], we have an infinite number of states per site, whereas there are only two per bond in the six-vertex model.

We shall construct the eigenvectors of $\tau(\theta)$ by a Bethe-Ansatz following ref.[41] . We start to find eigenvectors of $\tau^{(N)}(\theta)$ where N, the number of sites, in assumed to the even. For even N, the shift K in the shifted transfer matrix (4.9) must also be even, and we cannot exceed N/2 in absolute value : $|K| < N/2$ . It will be useful to write this matrix $\tau^{(2K)}(\theta)$ as a sum of two terms

$$\tau^{(2K)}(\theta) = \tau_+{}^{(K)}(\theta) + \tau_-{}^{(K)}(\theta) \qquad (4.35)$$

Here ± stands for projections in the subspaces where $\sigma'_N = \sigma_N \pm 1$, respectively. If we call

$$\ell \equiv \sigma'_N - 2K - 1 \qquad (4.36)$$

then

$$\begin{aligned} \tau_+{}^{(K)}(\theta) &= t_{\ell+1,\ \ell+2,\ 2K+\ell+1,\ 2K+\ell+2}(\theta) \\ \tau_+{}^{(K)}(\theta) &= t_{\ell+1,\ \ell,\ 2K+\ell+1,\ 2K+\ell}(\theta) \end{aligned} \qquad (4.37)$$

Define the vectors

$$\Omega_\ell \equiv \delta_{\sigma_1,\ell}\, \delta_{\sigma_2,\ell+1} \cdots\cdots \delta_{\sigma_N,\ell+N-1} \qquad (4.38)$$

Let us compute

$$\tau^{(N)}(\theta)\, \Omega_\ell = \tau_+{}^{(N/2)}(\theta)\, \Omega_\ell + \tau_-{}^{(N/2)}(\theta)\, \Omega_\ell \qquad (4.39)$$

Each term in the rhs has only one non-zero component. We find that the resulting vectors are proportional to $\Omega_{\ell+1}$ and $\Omega_{\ell-1}$ respectively. That is,

$$\tau_-{}^{(N/2)}(\theta)\, \Omega_\ell = a(\theta)^N\, \Omega_{\ell+1}$$

$$\tau_+^{(N/2)}(\theta)\,\Omega_\ell = [\prod_{n=1}^{N} \beta_{\ell+n-2}(\theta)]\,\Omega_{\ell-1} \tag{4.40}$$

For model I, we trivially find from eq. (4.34)

$$\prod_{n=1}^{N} \beta_{\ell+n-2}(\theta) = b(\theta)^N \tag{4.41}$$

For model II, we have from eq. (4.30)

$$\prod_{n=1}^{N} \beta_{\ell+n-2}(\theta) = b(\theta)^N \sqrt{[s_{\ell-2}\, s_{\ell+N-1}/\, s_{\ell-1}\, s_{\ell+N-2}]} \tag{4.42}$$

It is easy to see that the rhs of (4.42) is $\ell$ independent for $0 \le \gamma < \pi$ if

$$N\gamma/\pi = \text{integer} \tag{4.43}$$

We shall assume that this is the case for model II in what follows [later on, we shall need to assume this integer to be even and $\gamma/(2\pi) = Q'/Q$, where Q and Q' are positive integers]. Therefore, eq.(4.41) holds for models I **and** II and we find

$$\tau^{(N)}(\theta)\,\Omega_\ell = a(\theta)^N\,\Omega_{\ell+1} + b(\theta)^N\,\Omega_{\ell-1} \tag{4.44}$$

Then, an eigenvector of $\tau^{(N)}(\theta)$ is given by

$$\Omega_\alpha = \sum_{\ell\,\varepsilon\,\mathcal{M}} \exp(-i\alpha\ell)\,\Omega_\ell \tag{4.45}$$

where $\mathcal{M} = \mathcal{Z}$ for model I and $\mathcal{M} = \{1, 2, \ldots, Q-1\}$ for model II. It fulfills

$$\tau^{(N)}(\theta)\,\Omega_\alpha = [\,\exp(i\alpha)\, a(\theta)^N + \exp(-i\alpha)\, b(\theta)^N\,]\,\Omega_\alpha \tag{4.46}$$

In addition $\alpha$ must obey $\exp(iN\alpha) = 1$ for model II.

Following the analogy with the algebraic Bethe Ansatz in vertex models, we need creation operator (B-type operators, see sec.III). Let us define

$$B^{(m,\ell)}(\theta) = t_{\ell+2,\,\ell+1,\,2m+\ell,\,2m+\ell+1}(\theta)$$
$$C^{(m,\ell)}(\theta) = t_{\ell-2,\,\ell-1,\,2m+\ell+2,\,2m+\ell+1}(\theta) \tag{4.47}$$

They enjoy the properties

$$C^{(N/2,\ell)}\Omega_k = 0\ ,\ \forall\, k, \ell\ ; \qquad B^{(N/2,\ell-2)}\Omega_\ell \neq 0 \tag{4.48}$$

In ref.[41] we show that the eigenvectors of $\tau(\theta)$ are given by

$$\Psi_\alpha(\underline{v}) = \sum_{\ell\varepsilon\mathcal{M}} \exp(-i\alpha\ell)\,\Psi_\ell(\underline{v}) \tag{4.49}$$

where $\underline{v} \equiv (v_1, v_2, \ldots\ldots, v_{N/2})$ and

$$\Psi_\ell(\underline{v}) \equiv B^{(1,\ell)}(v_1)\, B^{(2,\ell-1)}(v_2) \ldots\ldots B^{(N/2,\ell-N/2+1)}(v_1)\, \Omega_{\ell-N/2+3} \tag{4.50}$$

The numbers $v_1, v_2, \ldots\ldots, v_{N/2}$ must fulfil the BAE

$$[\, sh(\lambda_j + i\gamma/2) / sh(\lambda_j - i\gamma/2)\,]^N =$$

$$= - \exp(-2i\alpha) \prod_{k=1}^{N/2} [sh(\lambda_j - \lambda_k + i\gamma) / sh(\lambda_j - \lambda_k - i\gamma)] \tag{4.51}$$

where $\lambda_j \equiv i(v_j + \gamma/2)$ and $1 \le j \le N/2$. In conclusion, for any set of roots $\underline{\lambda} \equiv (\lambda_1, \lambda_2, \ldots\ldots, \lambda_{N/2})$ of eqs.(4.51), eqs.(4.49)-(4.50) provide an eigenvector of $\tau(\theta)$ with eigenvalue

$$\Lambda(\theta) = \exp(i\alpha)\, \Lambda_+(\theta, \underline{\lambda}) + \exp(-i\alpha)\, \Lambda_-(\theta, \underline{\lambda}) \tag{4.52}$$

where

$$\Lambda_+(\theta, \underline{\lambda}) = \sin^N(\theta + \gamma) \prod_{j=1}^{N/2} [sh(\lambda_j - i\theta + i\gamma/2) / sh(\lambda_j - i\theta - i\gamma/2)] \tag{4.53}$$

$$\Lambda_-(\theta, \underline{\lambda}) = \sin^N\theta \prod_{j=1}^{N/2} [sh(\lambda_j - i\theta - 3i\gamma/2) / sh(\lambda_j - i\theta - i\gamma/2)]$$

$\alpha$ is an arbitrary real number for model I and $\alpha = 2\pi s/Q$, $s \in \mathcal{M}$ for model II.

Eqs. (4.51) look very similar to those of the six-vertex model BA. In particular they look identical to the solution of the six-vertex given in ref.[42]. There is, however, a crucial difference. As shown in ref.[35], non-zero and independent eigenvectors of the six vertex model are associated only to solutions where $N/2 - r = |S_3|$ roots are at infinity ( with $\alpha = |S_3|\gamma$ ). This is not the case for the SOS model as shown in an example (see Appendices of ref.[41]). In the thermodynamic limit ( $N = \infty$ ), the solutions of the SOS BAE (4.51) for the free energy and excitations are identical to those of the six-vertex model. For finite N , the vertex model eigenvalues and eigenvectors are a subset of those of the SOS model whose Bethe Ansatz equations look identical. We see explicitely that face models have a richer content than their vertex associates.

## V. THE LIGHT-CONE LATTICE APPROACH

This approach starts by discretizing the two-dimensional Minkowski space-time in light-cone coordinates $x_{\pm} = x \pm t$. Space time is thus approximated by a diagonal lattice. This discretization scheme turns to be an useful regularization method for quantum field theories ( integrable or not ) since they become naturally connected with vertex models in their scaling limit[43-44].

The sites in the light-cone lattice (fig. 11) are considered as world events. Each site (event) is joined by light-like links to its four nearest neighbours along $x_+$ and $x_-$ . There diagonal links are possible world lines for the propagation upwards in time of "bare" massless particles. Particles on right-oriented (R) and left-oriented (L) links are called respectively right and left-movers.

One then associates microscopic amplitudes to each site (world event) where two oppositely oriented world lines cross. These amplitudes describe the different processes that can take place, and must verify general invariance properties like unitarity.

Let us start for the simplest case where each link describes only two different configurations. The general case of any number of states per link is considerezd later on. We assume that these two cases correspond to the presence or absence of a bare fermion without internal degrees of freedom. In general, there can be 16 different amplitudes per site corresponding to the 16 configurations (occupied/empty) of the four links joining there. Only U(1) invariant microscopic amplitudes will be considered here such that the number of particles is conserved at each site. U(1) transformations act on the link states by

$$|0\rangle \rightarrow e^{i\lambda}|0\rangle$$
$$|1\rangle \rightarrow e^{-i\lambda}|1\rangle \qquad (5.1)$$

where $|0\rangle$ = (empty) and $|1\rangle$ = (occupied). Therefore, there are only six non-zero amplitudes as depicted in fig. 12. The correspondence with the general (non-symmetric) six-vertex model is evident.

Of course, space-time translational invariance implies that the amplitudes are the same in all sites of the lattice. It is natural (and causes no loss of generality) to set the nothing-to-nothing amplitude to be 1. Unitarity then requires

$$\Omega\,\Omega^{\dagger} = 1 \;, \qquad \Omega = \begin{pmatrix} \omega_3 & \omega_5 \\ \omega_6 & \omega_4 \end{pmatrix} \;, \quad |\omega_2|^2 = 1 \qquad (5.2)$$

While $\omega_3$ and $\omega_4$ are naturally interpreted as amplitudes for free propagation (being therefore related to kinetic energies in the continuum limit), $\omega_5$ and $\omega_6$ play the role of mass terms since they couple right and left movers.

Symmetry under parity transformation holds if

$$\omega_3 = \omega_4 = b \quad , \quad \omega_5 = \omega_6 = c \tag{5.3}$$

This corresponds now to an integrable six vertex model. Unitarity now reads

$$|b|^2 + |c|^2 = 1 \quad , \quad b\bar{c} + \bar{b}c = 0 \tag{5.4}$$

One can organize these microscopic amplitudes at a site into a 4x4 unitarity "bare" S-matrix

$R^{ab}{}_{cd}(\theta) =$

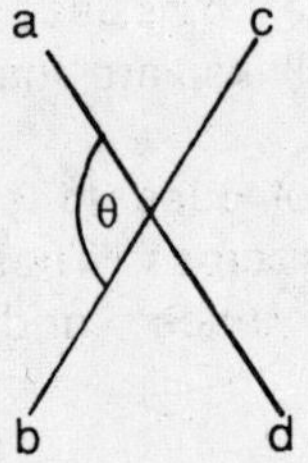


$$= \begin{pmatrix} 1 & 0 & 0 & 0 \\ 0 & c & b & 0 \\ 0 & b & c & 0 \\ 0 & 0 & 0 & \omega \end{pmatrix} \tag{5.5}$$

where $\omega \equiv \omega_2$ and $\alpha, \beta, \alpha', \beta'$ take the values 0 or 1 for empty or occupied links like in eq. (1.1).

The amplitude for a global process, from a given state at $t = t_0$ to another given state at a later time, is obtained by summing over the amplitudes of all allowed vertex configurations compatible with initial and final conditions and with boundary conditions. Each of these is given by a product of microscopic amplitudes $\omega_i$. It clearly corresponds to the sum over all possible paths of an arbitrary, but constant in time number of particles. At any instant, a particle can move to the left or to the right at the speed of light. We are thus dealing with a discretization of Feynman path integral for fermions.

It is convenient to parametrize b and c following the constraints (5.4) as

$$b = b(\theta,\gamma) = \operatorname{sh}\theta / \operatorname{sh}(\theta - i\gamma) \quad , \quad c = c(\theta,\gamma) = \operatorname{sh}\gamma / \operatorname{sh}(\theta - i\gamma)$$
$$0 < \theta < \infty \quad , \quad 0 < \gamma < \pi \tag{5.6}$$

This makes (5.5) identical the six-vertex model R matrix (3.1) up to an overall factors $\operatorname{sh}(\theta - i\gamma)$ and a redefinition of $\theta \to i\theta$ when $\omega = 1$. Actually $\omega \neq 1$ corresponds to a six-vertex model in an external field.

Let us now describe the operator formalism for the light-cone approach[43]. The unit evolution operators in the light-cone direction (R or L) are given by simply juxtaposition of the microscopic S-matrices (5.5) at the same horizontal level. That is

$$U_L = \begin{matrix} {}^{1\ 2}_{\ } \\ \mathrm{X} \\ {}_{2N\ 1} \end{matrix} \begin{matrix} {}^{3\ 4} \\ \mathrm{X} \\ {}_{2\ 3} \end{matrix} \begin{matrix} {}^{5\ 6} \\ \mathrm{X} \\ {}_{4\ 5} \end{matrix} \begin{matrix} {}^{7\ 8} \\ \mathrm{X} \\ {}_{6\ 7} \end{matrix} \cdots\cdots\cdots \begin{matrix} {}^{2N-1\ 2N} \\ \mathrm{X} \\ {}_{2N-2\ 2N-1} \end{matrix} \tag{5.7}$$

$$U_R = \begin{matrix} {}^{1\ 2} \\ \mathrm{X} \\ {}_{2\ 3} \end{matrix} \begin{matrix} {}^{3\ 4} \\ \mathrm{X} \\ {}_{4\ 5} \end{matrix} \begin{matrix} {}^{5\ 6} \\ \mathrm{X} \\ {}_{6\ 7} \end{matrix} \begin{matrix} {}^{7\ 8} \\ \mathrm{X} \\ {}_{8\ 9} \end{matrix} \cdots\cdots\cdots \begin{matrix} {}^{2N-1\ 2N} \\ \mathrm{X} \\ {}_{2N\ 1} \end{matrix} \tag{5.8}$$

where the numbers 1, 2, 3,...,2N label the sites. Here time evolves upward, according to rule d) of sec. I. Here N is assumed to be even and $\alpha_{j+N} \equiv \alpha_j$. Notice that there are no summations in eq. (5.7)-(5.8). One can now define the two light-cone lattice evolution generators as

$$H \pm P = (2i/a) \operatorname{Log} U_{R_L}(\theta) \tag{5.9}$$

where H and P stand for lattice hamiltonian and momentum and a is the latttice spacing.

Eq. (5.9) is extremely suggestive since it provides a lattice version of field-theoretic H and P in terms of lattice vertex transfer matrices $U_R$ and $U_L$. The natural question is now to find the eigenvectors of them. It will be shown now that this is possible using the techniques of sec. III and IV (and their generalizations) provided $R(\theta)$ verifies the YB algebra (1.12)[43].

Let us consider the row-to-row transfer matrix $\tau^{[N]}(\theta,\alpha)$ of eq. (1.23) [with $g_1 = \ldots = g_N = 1$] with the particular choice of inhomogeneities (recall N = even)

$$\mu_j = (-1)^{j+1}\ \theta$$
$$\bar{\theta} = (+\theta, -\theta, + \ldots\ldots, +\theta, -\theta) \tag{5.10}$$

It then follow from eqs.(1.23) and (1.35) using eq. (1.13) that

$$\tau(\theta,\bar{\theta}) = U_L(\theta) \tag{5.11}$$
$$\tau(-\theta,\bar{\theta}) = U_R(\theta)^{\dagger} \tag{5.12}$$

Let us check (5.12). Setting (5.10) in (1.13) and (1.23) yields

$$
\begin{aligned}
\tau(\theta,\bar{\theta})_{\bar{\alpha}|\bar{\sigma}} = \sum_{\gamma_1,\ldots,\gamma_N} \delta^{\alpha_1}{}_{\gamma_1}\, \delta^{\gamma_1}{}_{\sigma_1}\, R(\theta)^{\alpha_2\gamma_3}{}_{\gamma_2\sigma_2} \ldots\ldots \delta^{\alpha_{N-1}}{}_{\gamma_{N-1}}\, \delta^{\gamma_N}{}_{\sigma_{N-1}} \\
R(\theta)^{\alpha_N\gamma_1}{}_{\gamma_N\sigma_N} = \prod_{j=1}^{N/2} R(\theta)^{\alpha_{2j},\alpha_{2j+1}}{}_{\sigma_{2j-1},\sigma_{2j}} = U_L(\theta)_{\bar{\alpha}|\bar{\sigma}}
\end{aligned}
\tag{5.13}
$$

after using eq.(1.35) repeatedly.

The key relations (5.9), (5.11)-(5.12) connect the lattice H and P with the row-to-row transfer matrices whose eigenvectors and eigenvalues can be constructed by the algebraic Bethe Ansatz developed in sec. III and IV. The light-cone or diagonal-to-diagonal transfer matrices result to be particular cases of the inhomogeneous row-to-row transfer matrices. The commutativity property (1.9) gives in addition :

$$
\begin{aligned}
[\, \tau(\lambda,\bar{\theta}) \,,\, U_L(\theta) \,] &= 0 \\
[\, \tau(\lambda,\bar{\theta}) \,,\, U_R(\theta)^{\dagger} \,] = 0 \,,\; [\, U_L(\theta) \,,\, U_R(\theta)^{\dagger} \,] &= 0
\end{aligned}
\tag{5.14}
$$

One can consider the infinite sequence of commuting operators ($0 \leqslant K < \infty$)

$$
c_K = \partial^K/\partial\lambda^K \, \mathrm{Log}\, \tau(\lambda,\bar{\theta})\big|_{\lambda=\theta} \tag{5.15}
$$

They all commute with $U_L(\theta)$, $U_R^{+}(\theta)$ and with each other.

Let us now consider the continuum limit ($a \to 0$) of the lattice models through eq. (5.9). The antiferroelectric ground state (regimes I and II) of $\tau(\lambda, \theta)$ corresponds just to the physical vacuum (filled Dirac sea) of the QFT defined by H and P. The bare vaccuum is the ferromagnetic state $|\Omega\rangle$. The physical vaccuum follows here by BA by filling the bare one [eq.(3.19)] with **interacting** pseudoparticles (their phase shift is given by $\phi(\lambda_i-\lambda_j,\gamma)$ ) and not free particles as for the free Dirac field. Only when $\gamma = \pi/2$ the phase shift is a constant ($\pi$).

The particle states follow from the lowest excitations. Since a factor $a^{-1}$ appears in H $\pm$ P [see eq. (5.9)] only gapless models yield finite energy states in the scaling limit. Moreover, in order to compute the energy and momentum in the scaling limit it is enough to know the eigenvalues of $\tau(\pm\theta,\bar{\theta})$ close to the bottom of the spectrum. The low-lying excitations are associated to holes and complex solutions with large (real) rapidity. Moreover, their eigenvalues normalized to the vacuum ones [as in eq. (3.55)] are independent of the inhomogeneities

Let us start by the fermion model (with $\omega = 1$) associated to the six-vertex models (eqs. (5.1)-(5.6)). We need the eigenvalues $\Lambda(\theta)$ of the transfer matrix $\tau(\theta)$. The solution of the six vertex BAE derived in sec.III yields for the excited states in the thermodynamic limit

$$\lim_{N\to\infty} \Lambda(\theta,\phi) / \Lambda_0(\theta) = \exp[-ig(\theta,\phi)] \tag{5.16}$$

where $g(\theta,\phi)$ reads for hole excitations

$$g(\theta,\phi) = 2 \operatorname{arctg}[\exp\pi(\phi + i\theta)/\gamma] \tag{5.17}$$

and $\phi$ is the hole position. Combining eqs.(5.16)-(5.17) with eqs.(5.9)-(5.12) yields

$$\epsilon \pm p = -(1/a)\ g(-iv) \qquad , v \in \mathcal{R} \tag{5.18}$$

Let us start by a hole excitation. We find for large $v$ from eq. (3.55) and (3.61) of ref.[4]

$$g(\pm iv) \underset{v\to+\infty}{=} \pm 2 \exp[\pm \pi\theta_h/\gamma - \pi v/\gamma] + O(\exp[-2\pi v/\gamma]) \tag{5.19}$$

after discarding an irrelevant $\pi$ (It does not contribute to the eigenvalue of $\tau(\theta)$ since the holes appear always by pairs). We find a relativistic spectrum provided $v \to +\infty$ when $a \to 0$ keeping fixed the renormalized mass

$$m = (4/a) \exp[-\pi v/\gamma] \tag{5.20}$$

The dispersion law results

$$\epsilon = m \cosh(\pi\theta_h/\gamma)$$
$$p = m \sinh(\pi\theta_h/\gamma) \tag{5.21}$$

So, $\pi\theta_h/\gamma$ is the physical rapidity of the particle [see eq. (2.14)]. The particles in eq.(5.21) correspond to the fermion or antifermion in the massive Thirring model or the sine-Gordon solitons. Besides these holes one finds complex solutions of the BAE disposed as strings (that is, with imaginary parts equally spaced by $\gamma$ or $\pi-\gamma$). They provide relativistic particles in the **same** scaling limit (4.20) with masses

$$m_n = 2m \sin\{ \tfrac{1}{2} n\pi( \pi/\gamma - 1 ) \} \quad , \quad 1 \le n < [\![ \pi/(\pi-\gamma) ]\!] \tag{5.22}$$

where $[\![x]\!]$ stands for integer part of x. Eq.(5.22) follows from eqs. (5.18) and (3.61) of ref.[4]. This set of particles are fermion-antifermion bound states. They relate semiclassically to the breathers of sine-Gordon as the fermions or holes relate to the sine-Gordon solitons. The S-matrix describing the two-body interaction of all these particles follows by direct calculation from the BAE using standard methods[49].

The preceding exposition of the light-cone lattice method applies to all gapless vertex models. In ref.[43] the models with rational R-matrices associated to simple Lie algebras are analysed. The q(2q−1) vertex model associated to the deformed $A_{q-1}$ algebra is also considered in its gapless regime[4,43].

Within this light-cone approach it is possible to construct explicitly the canonical bare fields on the lattice and to show that in the scaling limit (5.20) the massive Thirring model emerges[43].

Eq.(5.20) defines the renormalized scaling limit yielding rigorously all physical quantities. There is in addition the **bare** scaling limit giving the bare quantum fields in the continuum[43].

We use the word "rigorous" since we solve in this approach a lattice model exactly, then we take the infinite volume limit and finally the $a \to 0$ (scaling) limit. In other words, here one solves (exactly) a model with both UV and volume cutoffs and then lets the cutoffs to infinity in a precise way. This is clearly much better than the coordinate Bethe-Ansatz (CBA) where the UV cutoff is introduced after the obtention of the solution. For the MTM and the chiral Gross-Neveu model the results of the CBA coincide with the light-cone approach for on-shell magnitudes. Hence the CBA works well in these cases. This is not the case for the multiflavor Chiral fermion model treated in ref.[45] by CBA. As it is shown in ref.[43] the results of ref.[45] are not correct.

Starting from richer vertex models than the six-vertex a large set of QFTs arises[43]. Let us first summarize the integrable vertex models classification in terms of simple Lie Algebras. There exist an inequivalent vertex model for each representation space of each simple Lie algebra. In the rational case these models are invariant unde the corresponding Lie group in the sense of eq.(1.17). In the trigonometric/hyperbolic case the invariance under the Cartan algebra survives and for the rest we have the invariance under the quantum group in the sense of eq.(6.28).

In these models, the links can be in any of $q$ different local states where $q$ is the dimension of the respective representation space.As an illustration let us take the model associated to the $A_{q-1}$ Lie algebra in the fundamental representation. The R-matrix reads for this case[4,46]:

$$R^{ab}_{ij}(\theta) = \delta_{ia}\ \delta_{jb}\ \mathrm{sh}\gamma \exp[\ \theta\ \mathrm{sign}(a-b)] + \mathrm{sh}\theta\ \delta_{ib}\ \delta_{ja} \qquad \text{for} \quad i \neq j,$$
$$R^{aa}_{ii}(\theta) = \delta_{ia}\ \mathrm{sh}(\gamma \pm \theta) \quad , \quad \pm \text{ for regime I / III.} \tag{5.23}$$

Here $1 \leq i, j, a, b, \leq q$ where $q$ can take any value $\geq 2$ . In the trigonometric regime (regime II), some weights are complex. The R-matrix reads:

$$R^{ab}_{ij}(\theta) = \delta_{ia}\ \delta_{jb}\ \sin\gamma \exp[\ i\theta\ \mathrm{sign}(a-b)] + \sin\theta\ \delta_{ib}\ \delta_{ja} \qquad \text{for} \quad i \neq j,$$
$$R^{aa}_{ii}(\theta) = \delta_{ia}\ \sin(\gamma + \theta) \quad , \tag{5.24}$$

This is a regular R-matrix, since

$$R^{ab}_{ij}(0) = \delta_{ia}\ \delta_{jb}\ \mathrm{sh}\gamma \quad (\text{or} \quad \delta_{ia}\ \delta_{jb}\ \sin\gamma) \tag{5.25}$$

The YB algebra defined by the R-matrix (5.34) is invariant under the group

$$\mathcal{G} = [U(1)]^{q-1} \otimes Z_q \qquad (5.26)$$

The cyclic group $Z_q$ is generated by powers of the matrix $h$ defined as

$$h_{ab} = \delta_{a+1,b} \quad \text{with} \quad \delta_{a+q,b} = \delta_{a,b+q} = \delta_{a,b} \qquad (5.27)$$

We have $h^q = \mathbf{1}$.

The R-matrix (5.23) enjoys PT invariance but not P or T separately :

$$R^{ab}{}_{ij}(\theta) = R^{ij}{}_{ab}(\theta) \quad , \quad R^{ab}{}_{ij}(\theta) \neq R^{ba}{}_{ji}(\theta) \quad , \quad R^{ab}{}_{ij}(\theta) \neq R^{ji}{}_{ba}(\theta) \qquad (5.28)$$

There are q(2q-1) non-vanishing weights in the vertex model defined by this R-matrix. Eq. (5.28) tells that the model is invariant under a 180° rotation of the whole lattice. For $q = 2$ we recover the six-vertex model.

In this model the links can be in q different states. In regime III, for $\gamma > 1$, $2\theta + \gamma$ fixed, the model has a long-range generalised ferroelectric order and the dominant configurations are formed mostly by vertices of type $R^{a1}{}_{a1}$ (for $\theta > 0$) and some $R^{a,q-1}{}_{a,q-1}$ . There are q different predominant patterns following one from each other by shifting by one the state of all links. This generalises the six vertex ($q = 2$) situation [1]. Regimes I and III map into each other through $\gamma \to -\gamma + i\pi$.

The exact solution of this q(2q-1) vertex model can be found in ref.[4,18]. That is the eigenvectors and eigenvalues of the transfer matrix $\tau^{[N]}(\theta, \alpha)$ associated to the R-matrix (5.23)-(5.24).

The structure of the nested Bethe Ansatz (NBA) that solves each of these vertex models looks like the one of their respective Dynkin diagram. It must be noticed that a proof that these BAE lead to the eigenvectors and eigenvalues of the transfer matrix has been explicited only for a subset of models : those associated to U(N)[4,18], Sp(2N)[19], and SO(2N) [20] and some others. However, these statements are extremely likely to hold for all semisimple Lie algebras.

Let us describe the BAE for the trigonometric models. The derivation of these equations (for a subset of Lie algebras) is in refs.[19,20]. The eigenvalues of the transfer matrix can be written as a sum of terms. The dominant one in the infinite volume limit ($N \to \infty$) is

$$\lambda_\omega(\theta, \underline{\lambda}^{(j)}) =$$

$$= \prod_{a=1}^{N} \prod_{k=1}^{r} \prod_{j_k=1}^{p_k} \operatorname{sh}[\, i(\theta - \theta_a) + \lambda^{(k)}_{j_k} - i\,\gamma\,(\omega_a, \alpha_k)\,] / \operatorname{sh}[\, i(\theta - \theta_a) + \lambda^{(k)}_{j_k} + i\,\gamma\,(\omega_a, \alpha_k)\,] \qquad (5.29)$$

for $\theta$ in the vicinity of $\theta = 0$, $|\theta| < \theta_0$ . Here $\theta_1, \theta_2, \ldots\ldots, \theta_N$ are given numbers describing inhomogeneities of the lattice as discussed in sec.II [1,4,5] and $\underline{\lambda}^{(j)} = \{\lambda_{j_k}{}^{(k)}, 1 \leqslant j_k \leqslant p_j, 1 \leqslant k \leqslant r\}$. The $\omega_a$ are fundamental weights and $\alpha_k$ are the simple roots of G whose rank is r.

$(\alpha, \beta)$ stands for the usual inner product in root space. The $\lambda_{j_k}^{(k)}$ $(1 \leqslant j_k \leqslant p_j,\ 1 \leqslant k \leqslant r)$ are solutions of the nested BAE (NBAE) :

$$\prod_{a=1}^{N} \mathrm{sh}[\lambda^{(k)}_{j_k} - i\theta_a - i\gamma(\omega_a,\alpha_k)] / \mathrm{sh}[\lambda^{(k)}_{j_k} - i\theta_a + i\gamma(\omega_a,\alpha_k)] =$$

$$= -\prod_{i=1}^{r} \prod_{j_i=1}^{p_i} \mathrm{sh}[\lambda^{(k)}_{j_k} - \lambda^{(i)}_{j_i} - i\gamma(\alpha_k,\alpha_i)] / \mathrm{sh}[\lambda^{(k)}_{j_k} - \lambda^{(i)}_{j_i} + i\gamma(\alpha_k,\alpha_i)] \tag{5.30}$$

$$1 \leq j_k \leq p_k\ ,\quad 1 \leq k \leq r\ .$$

Here the upper indices $^{(i)}$ label the steps in the NBA. Each step is associated to a simple root $\alpha_i$. The structure of eq.(5.30) coincides with the respective Dynkin diagram: when two roots, say $\alpha_\ell$ and $\alpha_i$, are orthogonal, their associated parameters $\lambda_{j_\ell}^{(\ell)}$ and $\lambda_{j_i}^{(i)}$ $(1 \leqslant j_\ell \leqslant p_\ell,\ 1 \leqslant j_i \leqslant p_i)$ are not directly coupled through (5.30) since $(\alpha_i, \alpha_\ell) = 0$. It must be noticed that due to the orthogonality of fundamental weights and simple roots [47]

$$(\omega_a,\alpha_k) = \tfrac{1}{2}\,\delta_{ak}\ (\alpha_k,\alpha_k) \tag{5.31}$$

The normalization of the simple roots can be absorbed as a multiplicative factor on the $\lambda_{j_i}^{(i)}$

We refer to the original references and to the reviews [4] for the resolution of the NBAE. Let us just say that there is a ferroelectric regime (regime III) where the ground state ( eigenvalue of $\tau(\theta)$ with maximal modulus) is a maximal weight vector. In regimes I and II the ground state is antiferroelectric with non-zero and zero gap respectively. There exists $r$ branches of excitations formed by holes in the NBAE. In addition complex roots that can be interpreted as hole bound states appear in regime II for $0 < \gamma < \pi/2$ yielding further excitations. When non-fundamental representations or non-simply laced algebras (even in their fundamental representation) are considered the ground state turns to be formed by complex roots (strings for $N = \infty$ ). For example, this happens in the $A_1$ case for spin $S \geq 1$[48]. In order to derive a QFT in the continuum limit from these lattice models in the light-cone approach it is enough to know the excitation eigenvalues $\Lambda_k(\theta, \phi)$ of $\tau(\theta)$ . We find

$$\lim_{N\to\infty} \Lambda_k(\theta, \phi) / \Lambda_0(\theta) = \exp[-i\, g_k(\theta, \phi)] \quad ,\ 1 \leq k \leq r \tag{5.32}$$

where $\Lambda_0(\theta)$ is the ground state eigenvalue of $\tau(\theta)$ , $k$ labels the branch of excitation, $\phi$ the hole position in rapidity and

$$g_k(\theta, \phi) \underset{i\theta \to -\infty}{=} (m_k / \pi\gamma)\ \exp[\pm \kappa(\phi + i\theta)/\gamma]\ \{ 1 + o(\exp(-|\theta\delta|)) \} \tag{5.33}$$

where $\delta > 0$. The parameters $\kappa$ and $m_\ell$ are given in Table II. $\kappa$ is just $2\pi$ times the length squared of the shortest simple root in the normalization where [47]

$$B(E_\alpha , E_{-\alpha}) = 1 ,$$

and $B(x, y)$ is the Killing form.

Light-cone evolution operators can be defined through eqs.(5.7)-(5.9) for any R-matrix. Let us see that a relativistic dispersion law arises from any excitation spectrum as given by eq.(5.32). Let us call $E_\ell(\varphi)$ and $p_\ell(\varphi)$ the eigenvalues of H and P, respectively. Eqs. (5.9) and (5.32) yield

$$\begin{aligned} E_\ell(\theta_h) &\underset{i\theta \to \infty}{=} [\exp(-i\kappa\theta/\gamma)/(\pi a\gamma)]\, m_\ell \cosh(\kappa\theta_h/\gamma) + O(\exp(-2i\kappa\theta/\gamma)) \\ p_\ell(\theta_h) &\underset{i\theta \to \infty}{=} [\exp(-i\kappa\theta/\gamma)/(\pi a\gamma)]\, m_\ell \sinh(\kappa\theta_h/\gamma) + O(\exp(-2i\kappa\theta/\gamma)) \end{aligned} \tag{5.34}$$

It is then natural to define the scaling limit according to

$$a \to 0 , \quad i\theta \to \infty , \mu = \exp(-i\kappa\theta/\gamma)/(\pi a\gamma) = \text{fixed} \tag{5.35}$$

is the renormalised or physical mass scale and the particle mass spectrum of these integrable QFTs is given by

$$M_\ell = \mu \, m_\ell \tag{5.36}$$

We recognize in eq.(5.34) $\kappa\theta_h/\gamma$ as the physical particle rapidity.

This is a very general way of constructing integrable QFTs. The operators H and P given by eq. (5.9) are well defined on the lattice as well as all the higher conserved charges. In the continuum limit $a \to 0$, they provide the energy and momentum of a relativistic invariant QFT, as long as the spectrum of the initial vertex model is gapless. This is usually the case for rational or trigonometric weights. In addition to the particle spectrum , the S-matrix is exactly calculable from the BAE by standard methods[49,50].

As it was the case for the MTM, the evolution operators $U_R$ and $U_L$ are much simpler than H and P on the lattice. This was exploited before in ref.[43] to obtain the lattice field equations for the fermionic fields of the MTM regularized by the lattice. An analogous local construction would be very interesting to obtain in the general case of a Lie algebra G. We present in ref.[43] a lattice construction for the current operators for all rational models. The H and P are always given by eqs. (5.7) and (5.9). The renormalized scaling limit (5.35) yields the mass spectrum (5.36) [see Table II] .

In refs.[43] it is shown that we obtain in this scaling limit of rational vertex models associated to $\mathcal{G}$ the non-abelian Thirring model associated to the group $\mathcal{G}$. This theory, also called Chiral Gross-Neveu model, has as Lagrangian,

$$\mathcal{L} \doteq i\bar{\Psi}\partial\!\!\!/\Psi - g(\bar{\Psi}\gamma_\mu t^\alpha \Psi)(\bar{\Psi}\gamma^\mu t^\beta \Psi) K_{\alpha\beta} \tag{5.37}$$

Here $\Psi$ transforms under the irreducible representation $\rho$ of $\mathcal{G}$, $t^\alpha$ are the $\mathcal{G}$-generators in that representation and $K_{\alpha\beta}$ is proportional to the inverse of the Killing form. Actually the H and P constructed from eqs. (4.7)-(4.9) describe the zero-chirality sector of the model (5.37).

The field theoretic models discussed up to here correspond to finite dimensional $\mathcal{V}$ and $\mathcal{A}$. Namely , a finite dimensional vector space at each link in the light-cone lattice. This is clearly appropiate for fermion or parafermion fields. Since there exists infinite dimensional representations of YB algebras, also bosonic QFTs may be described in this framework.The SU(2) principal chiral model (PCM) is defined by the lagrangian

$$\mathcal{L} = \mathrm{Tr}[(g^{-1}\partial_\mu g)^2] \tag{5.38}$$

where $g \in SU(2)$. The lagrangian (5.38) exhibits a $SU(2) \otimes SU(2)$ invariance.

The infinite spin representation of the SU(2) invariant R-matrix (rational limit of the six-vertex model) relates to the PCM as it is investigated in ref. [51]. For arbitrary spin S, this R-matrix writes[52]

$$R_{12}(\theta) = \Gamma(2S+1+i\theta)\,\Gamma(\mathbf{J}+1-i\theta) / [\Gamma(2S+1-i\theta)\,\Gamma(\mathbf{J}+1+i\theta)] \tag{5.39}$$

where the operator $\mathbf{J}$ is defined by

$$\mathbf{J}(\mathbf{J}+1) = 2S(S+1) + 2\,\vec{S}_1 \otimes \vec{S}_2$$

$\vec{S}_1$ and $\vec{S}_2$ are spin S operators acting on the spaces $\mathcal{A}$ and $\mathcal{V}$ respectively $[(\vec{S}_1)^2 = (\vec{S}_2)^2 = S(S+1)]$. The light-cone hamiltonian (5.7)-(5.9) provides particle states that yield all particle masses and S-matrix amplitudes for the PCM letting $S = \infty$.

The conserved currents in the PCM read

$$J_\mu{}^L = g\,\partial_\mu(g^{-1}) \quad \text{and} \quad J_\mu{}^R = g^{-1}\partial_\mu g$$

They transform under the left and right SU(2) groups, respectively.

One obtains left or right transforming states depending upon we identify $J_\pm{}^L$ or $J_\pm{}^R$ with the spin operators in the lattice:

$$(1/ga\theta) \sum_{1\le i\le 3} \sigma^i S^i{}_{2n} \rightarrow J_+{}^L(x) \text{ or } J_+{}^R(x) \quad , \tag{5.40}$$

$$(1/ga\theta) \sum_{1 \le i \le 3} \sigma^i S^i_{2n-1} \rightarrow J\_^R(x) \quad \text{or} \quad J\_^L(x)$$

with $x = na$ for $a \rightarrow 0$ .That is, the H obtained through eq.(4.9) is not the full hamiltonian of the PCM as it is proven in refs.(51) and (53). There is a very simple explanation for this, the physical particle states for this model transform under the group $SU(2)_L \otimes SU(2)_R$ and from the present construction only left or right operators can be obtained. Therefore all states obtained in this way are left (or right) singlets. The detailed counting of states in ref.(51) is confirmed by the simple proof of ref. (53).

This whole construction generalizes to the SU(N) PCM. It also applies for Chiral fermion models and PCM with one anisotropy axis (trigonometric YB algebras)[54].

## VI. BRAID GROUPS AND QUANTUM GROUPS FROM YANG-BAXTER ALGEBRAS

Let us see first how braid groups follow from trigonometric/hyperbolic YBA. In the limit $\theta \rightarrow \pm \infty$ $[\pm i\infty]$ the hyperbolic/trigonometric generators behave as $e^{\pm K\theta}(e^{+K\theta})$ times a well defined operator (K being a constant). Since such exponential factor can be absorbed in $T(\theta)$ respecting the YBE, we can in general assume that the limit

$$\lim_{\theta \rightarrow \pm\infty} T(\theta) = T_\pm \qquad (6.1)$$

is finite and non-trivial for hyperbolic (trigonometric) YB generators. Actually a symmetry breaking transformation (2.23) may be necessary in order to find a non-trivial limit. These limiting operators can be graphically represented as follows

$T_+$ $T_-$

fig. 15

Letting $\theta \rightarrow \pm\infty$, $\theta' \rightarrow \pm\infty$ with $\theta - \theta' \rightarrow \pm\infty$ in the hyperbolic regime of eq. (1.1) yields

$$T_{\pm}^{(K,I)}\, T_{\pm}^{(K,J)}\, T_{\pm}^{(I,J)} \;=\; T_{\pm}^{(I,J)}\, T_{\pm}^{(K,J)}\, T_{\pm}^{(K,I)} \qquad (6.2)$$

In addition eq.(1.42) tells us that $T_+$ and $T_-$ are inverses of each other

$$T_+\, T_- \;=\; T_-\, T_+ \;=\; \mathbf{1} \quad , T_{\pm} \;=\; T_{\pm}^{(I,J)} \qquad (6.3)$$

A factor $\sqrt{\rho(\theta)}$ has been absorbed in the definition of $T(\theta)$ as in eq. (5.1).

If one consider R matrices instead of general YB operators $T^{(I,J)}(\theta)$, eqs. (5.2)-(5.3) read

$$R_{\pm}^{23}\, R_{\pm}^{12}\, R_{\pm}^{23} \;=\; R_{\pm}^{12}\, R_{\pm}^{23}\, R_{\pm}^{12} \qquad (6.4)$$

$$R_+ R_- \;=\; R_-\, R_+ \;=\; \mathbf{1} \qquad (6.5)$$

where now $V^I = V^J = V^K = \mathcal{A}$, $R_{12} = R \otimes 1$, $R_{23} = 1 \otimes R$ and $R_{\pm} = T_{\pm}^{(\mathcal{A},\mathcal{A})}$.

Lines intersect in just two ways in the ultrarelativistic limit as depicted in fig.15. ( For finite $\theta$ we have a continuous family of possibilities parametrized by the angle $\theta$ ). We can interpret geometrically this two possibilities as one line being over the other or viceversa. This fact naturally connects YBA in the $\theta = \pm\infty$ limit with braids, knots and links. The YB property allowing to push lines through intersections (Z-invariance, see sec. I) means now simply continuous deformation of lines without tearing them. More technically this connects with the Reidemeister moves. An important problem is to compute link and knot invariants which are intrinsic and therefore the same for topologically invariant objects. This can be achieved by computing the partition function Z for $\theta = \pm\infty$ of an integrable model taking as lattice a given knot, braid or link. Since Z is precisely Z-invariant it will be a topological invariant.

Let us now study the connection of the ultrarelativistic limit of hyperbolic/trigonometric YB algebras with braids, knots, links and quantum groups. The matrices $R_+$ and $R_-$ will give a representation of a braid group in the following way. Let us consider the operators $X_i(\theta)$ acting in the tensor produict of n auxiliary spaces $\mathcal{A}$[55]

$$X_i(\theta) = \underset{1}{1} \otimes \ldots\ldots\ldots \otimes \underset{(i,i+1)}{R(\theta)} \otimes \ldots\ldots\ldots\ldots \otimes \underset{n}{1} \qquad (6.6)$$

That is,

$$(a_1\, a_2 \ldots a_n \mid X_i(\theta) \mid b_1\, b_2 \ldots b_n) \;=\; R^{b_i b_{i+1}}{}_{a_i a_{i+1}}(\theta) \prod_{k\neq i,i+1}^{n} \delta_{a_k, b_k} \qquad (6.7)$$

They fulfil the relations

$$[X_i(\theta), X_j(\theta')] = 0 \qquad \text{if} \quad |i-j| \geq 2, \quad \forall\, \theta, \theta'$$

$$X_i(\theta)\, X_{i+1}(\theta+\theta')\, X_i(\theta') = X_{i+1}(\theta')\, X_i(\theta+\theta')\, X_{i+1}(\theta) \tag{6.8}$$

$$X_i(\theta)\; X_i(-\theta) = 1$$

The last two equations being a consequence of the YB equation (1.11) and the unitarity eq.(1.42), respectively.

These operators are clearly of "light-cone" type. They are closely related to the light-cone evolution operators discussed in sec. IV. We find

$$U_+ = X_1\, X_3 \,\ldots\, X_{N-1}$$
$$U_- = X_2\, X_4 \,\ldots\, X_N$$

where[43] (see eqs.(5.7)-(5.8))

$$U_R = V\, U_-$$
$$U_L = U_+\, V^\dagger$$

Here $V$ ( $V^\dagger$ ) is the shift operator affecting one-half translation to the right (to the left). There are two different limits for hyperbolic (trigonometric) YB algebras : $\theta \to \pm\infty$ ( $\pm\, i\,\infty$ ) . Upon adequately normalizing $R(\theta)$ and eventually applying a symmetry breaking transformation (2.23) in order to have a finite and non-trivial limit, we find

$$b_i = \lim_{\theta\to\infty} X_i(\theta) \qquad , \; b_i^{-1} = \lim_{\theta\to-\infty} X_i(\theta) \tag{6.9}$$

and analogous relations for trigonometric YBA. The operators $b_i$ ( $1 \leq i \leq n$ ) fulfil:

$$b_i\, b_j = b_j\, b_i \qquad \text{when} \quad |i-j| \geq 2 \;,$$
$$b_i\, b_{i+1}\, b_i = b_{i+1}\, b_i\, b_{i+1} \tag{6.10}$$

Let us briefly recall the notion of a braid group[56]. Braids are formed when n points in a straight line are connected by n lines with othe n points on a paralell line as shown in fig.13. When the lines connecting the points have no intersections, the braid is called trivial . A general n-braid is obtained from the trivial one applying succesively the operations $b_i$ and / or the inverses $b_i^{-1}$ ( $1 \leq i \leq n-1$ ). The operations $b_i$ and $b_{i-1}$ are depicted in fig.16. Then each topologically equivalent class of braids is identified with an element in $B_n$. Eq. (6.10) shows that the $\theta = \pm\infty$ limit of hyperbolic R-matrices provide a representation of $B_n$. This connection between YB algebras and braid groups revealed recently very fruitful to obtain knot invariants and link polynomials [57,58].

We want to remark that sometimes eq.(6.4) is called YB equation in the literature. This is somehow misleading since eq.(6.4) is only a particular case of the the YBE with spectral parameter eq.(1.1) or (1.11). It would be more appropiate to call eq.(6.4) 'braid equation'.

A braid invariant is directly obtained by taking the trace of the product of $b_i$'s and $b_i^{-1}$'s that represent the given braid. Moreover, a knot or a link can be associated to many different braids (Alexander's theorem). The so-called Markov moves relate equivalent braids associated to the same link. Remarkably enough, it is possible to define a trace which is invariant under Markov moves and therefore yields a link invariant. This trace is analogous to Z with twisted boundary conditions.

The exchange of points in the n-point conformal blocks forming the conformal invariant correlation functions yields a representation of a braid group (6.10) [59]. We want to remark that the R-matrix associated to such braid groups defines a lattice statistical model whose critical behavior is described precisely by the conformal theory yielding this braid group.

Let us now discuss the quantum groups. They are related to trigonometric/hyperbolic YBA in the $\theta = \pm\infty$ limit (as the braid groups). Let us start by the six-vertex case, where [eq.(3.4)]. In the $\theta = \pm\infty$ limit, the YB operators $T_{ab}(\theta)$ $(1 \le a,b \le 2)$ yield for regime I [5]

$$T_{11}(\theta) \underset{\theta\to\pm\infty}{=} (y_\pm)^N \exp(\pm\gamma S_z)\,[1 + O(y_\pm^{-2})]\ , \tag{6.11}$$

$$T_{22}(\theta) \underset{\theta\to\pm\infty}{=} (y_\pm)^N \exp(\mp\gamma S_z)\,[1 + O(y_\pm^{-2})]\ , \tag{6.12}$$

$$T_{12}(\theta) \underset{\theta\to\pm\infty}{=} (y_\pm)^{N-1}\,\mathrm{sh}\,\gamma\, J_-(\mp\gamma)\,[1 + O(y_\pm^{-2})]\ , \tag{6.13}$$

$$T_{21}(\theta) \underset{\theta\to\pm\infty}{=} (y_\pm)^{N-1}\,\mathrm{sh}\,\gamma\, J_+(\pm\gamma)\,[1 + O(y_\pm^{-2})]\ , \tag{6.14}$$

where

$$y_\pm = \pm\tfrac{1}{2}\exp[\pm(\theta+\gamma/2)]\quad ,\qquad S_Z = \tfrac{1}{2}\sum_{a=1}^{N}(\sigma_a)_Z\quad ,$$

and

$$J_\pm(\gamma) = \sum_{k=1}^{N}\prod_{j=1}^{k-1}\exp\{\tfrac{1}{2}\gamma(\sigma_j)_z\}\,(\sigma_\pm)_k\prod_{l=k+1}^{N}\exp\{-\tfrac{1}{2}\gamma(\sigma_l)_z\} \tag{6.15}$$

(Compare with eqs.(3.8)-(3.9)).

The ultrarelativistic limit of the YB algebra relations (3.13) that followed from the R-matrix (3.1) yields the algebra of the operators $J_\pm(\gamma)$ and $S_Z$. We find [1]

$$[\, J_+(\gamma)\, ,\, J_-(\gamma)\, ] \;=\; \mathrm{sh}(2\gamma S_z) / \mathrm{sh}\,\gamma \tag{6.16}$$

$$[\, S_z\, ,\, J_\pm(\gamma)\, ] \;=\; \pm J_\pm(\gamma) \tag{6.17}$$

Analogous relations hold with $\gamma \to i\gamma$ in the trigonometric regime. Eqs.(6.13)-(6.14) define the so-called $SU_\gamma(2)$ quantum group. That is, a one-parameter deformation of the SU(2) Lie algebra. Note that eqs.(6.13)-(6.14) for $\gamma \to 0$ reduces to the usual angular momentum algebra. The operators $J_+(-\gamma)$, $J_-(-\gamma)$ and $S_z$ obey the same algebra than $J_+(\gamma)$, $J_-(\gamma)$ and $S_z$. In regime II these representations are complex conjugate of each other.

We want to notice that $J_\pm(\gamma)$ and $S_z$ have a coproduct structure heritated from the YB algebra [eq.(1.19)]. Let us call $J_\pm^{(N-1)}(\gamma)$, $S^{(N-1)}{}_z$ ( $J_\pm^{(1)}(\gamma)$, $S^{(1)}{}_z$) the operators acting on (N-1) sites (the Nth site) of the line. Then, we find from eq.(6.15)

$$J_\pm^{(N)}(\gamma) \;=\; \Delta^{(N-1,1)}(\, J_\pm\,) \;\equiv$$

$$\equiv \quad J_\pm^{(N-1)}(\gamma) \otimes \exp[\,-\gamma S^{(1)}{}_z] \;+\; \exp[\,\gamma S^{(N-1)}{}_z] \otimes J_\pm^{(1)}(\gamma), \tag{6.18}$$

$$S^{(N)}{}_z \;=\; \Delta^{(N-1,1)}(\, S_z\,) \;\equiv$$

$$\equiv \quad S^{(N-1)}{}_z \otimes 1^{(1)} \;+\; 1^{(N-1)} \otimes S^{(1)}{}_z\,, \tag{6.19}$$

where $S^{(1)}{}_z \equiv \frac{1}{2}\sigma^{(1)}{}_z$ and $J_\pm^{(1)} \equiv \sigma^{(1)}{}_\pm$ . Eqs.(6.18)-(6.19) define a coproduct endowing (6.16)-(6.17) with a Hopf algebra structure. That is, if $J_\pm^{(\mathcal{A})}(\gamma)$, $S^{(\mathcal{A})}{}_z$ and $J_\pm^{(\mathcal{V})}(\gamma)$, $S^{(\mathcal{V})}{}_z$ separately obey eqs.(5.16)-(5.17) as operators on $\mathcal{A}$ and $\mathcal{V}$, respectively, then $\Delta^{(\mathcal{A},\mathcal{V})}(\, J_\pm\,)$ and $\Delta^{(\mathcal{A},\mathcal{V})}(\, S_z\,)$ as given by eqs.(5.18)-(5.19) **also** obey eqs.(6.16)-(6.17). Notice that $\Delta^{(\mathcal{A},\mathcal{V})}(\, J_\pm\,) \neq \Delta^{(\mathcal{V},\mathcal{A})}(\, J_\pm\,)$ . That is this coproduct is not cocommutative (although $\Delta^{(\mathcal{A},\mathcal{V})}(\, J_z\,) = \Delta^{(\mathcal{V},\mathcal{A})}(\, J_z\,)$ ).

The physical interpretation of the coproduct definition for YB [eq.(1.19)] is transparent : it is the way to combine the operators associated to two independent sites in order to have the same mathematical structure (YB, there) for the combined operator as for each one separately. In most physical situations the coproduct is a simple direct sum. For example, the spin $\vec{S}$ of a two site lattice reads,

$$\vec{S} \;=\; \vec{S}^{(1)} \otimes 1^{(2)} \;+\; 1^{(1)} \otimes \vec{S}^{(2)}$$

where $\vec{S}^{(1)}$ ( $\vec{S}^{(2)}$) is the spin of site 1 (2). For YBA the coproduct is depicted in fig.3 for two sites and in eq.(1.19) for N sites. This coproduct

(1.19) ensures that $T_{ab}^{(N)}$ fulfils YB when each $t_{ab}^{(k)}$ ( $1 \leq k \leq N$ ) obeys YB . The coproduct (6.19) is the consequence of the YB coproduct (1.19) for infinite $\theta$.

One parameter deformations of all simple Lie algebras are known. They appear as ultrarelativistic limits of hyperbolic YB algebras. They can also be defined directly as deformations of classical Lie algebras using the Weyl basis.

It is possible to relate the operators $J_\pm(\gamma)$ with the usual spin operators $S_\pm$, $S_z$ obeying

$$[S_+ , S_- ] = 2S_z \quad , \quad [ S_z , S_\pm ] = \pm S_\pm \tag{6.20}$$

Inserting the ansatz

$$J_+(\gamma) = S_+ f( \gamma, S_z, S) \quad , \quad J_-( \gamma ) = f( \gamma, S_z, S) S_- \tag{6.21}$$

in eqs. (6.16)-(6.17) yields the recursion relation

$$[f( \gamma, S_z-1, S)]^2( S_z -1)[ S(S+1) - S_z( S_z - 1)] - [f( \gamma, S_z, S)]^2 [ S(S+1) - S_z( S_z + 1)] = \sin( 2\gamma S_z ) / \sin\gamma \tag{6.22}$$

where $S ( S + 1 ) = (S_+ S_- + S_- S_+)/2 + (S_z)^2$, as usual. This has as solution[60]

$$f( \gamma, S_z, S) = \frac{1}{\sin\gamma} \sqrt{\frac{\sin[\gamma(S-S_z)] \sin[\gamma(S+S_z+1)]}{(S-S_z ) (S+S_z+1)}} \tag{6.23}$$

The quadratic ("Casimir") operator commuting with $J_\pm(\gamma)$ and $S_z$ writes here

$$\mathcal{C} = \tfrac{1}{2}[ J_-(\gamma) J_+(\gamma) + J_+(\gamma) J_-(\gamma) ] + \cos\gamma \sin^2( \gamma S_z ) / \sin^2\gamma \tag{6.24}$$

This deformation of $\vec{S}^2$ has the value

$$\mathcal{C} = \sin[(S+1)\gamma] \sin[S\gamma] / \sin^2\gamma \tag{6.25}$$

In summary eqs. (6.21)-(6.23) explicitely display the $SU(2)_\gamma$ quantum group generators in terms of the usual SU(2) generators $S_\pm$, $S_z$.

The six-vertex model is invariant only under z-rotations and not under the full SU(2) group as long as $\gamma \neq 0, \pi$ . Let us see how this a deformed SU(2) invariance appears using $J_+( \gamma )$ , $J_-( \gamma )$ and $S_z$ .This analysis is easier upon a symmetry breaking transformation (2.23) with $\mu = -\frac{1}{2}$ , $\nu = \omega = 0$. Only the off-diagonal elements of $R^{ab}{}_{cd}(\theta)$ are affected by this transformation :

$$\mathring{R}^{12}{}_{12}(\theta) = \exp(\theta)\, c \quad , \quad \mathring{R}^{21}{}_{21}(\theta) = \exp(-\theta)\, c$$

(It must be noticed that this form $\mathring{R}$ naturally emerges from the Toda field theory[46]). Then, the generalization of the invariance under $S_\pm$ for $\gamma \neq 0$

is[61]

$$[\, R(\theta)\,,\ \sigma_\pm \otimes \exp(-\tfrac{1}{2}\gamma\sigma_z) + \exp(\tfrac{1}{2}\gamma\sigma_z) \otimes \sigma_\pm \,] = 0 \qquad (6.26)$$

This implies for the N-site YB operator $T_{ab}{}^{(\mathcal{A},\mathcal{V})}(\theta)$ using eqs. (1.13) and (1.21) :

$$T^{(\mathcal{A},\mathcal{V})}(\theta)\ [\, J_\pm(\gamma) \otimes \exp(-\tfrac{1}{2}\gamma\sigma_z) + \exp(\gamma S_z) \otimes \sigma_\pm \,] =$$

$$= [\, \sigma_\pm \otimes \exp(-\gamma S_z) + \exp(\tfrac{1}{2}\gamma\sigma_z) \otimes J_\pm(\gamma) \,]\ T^{(\mathcal{A},\mathcal{V})}(\theta) \qquad (6.27)$$

where $\otimes$ stands for tensor product between the auxiliary space $\mathcal{A} = \mathbb{C}^2$ (spin $\frac{1}{2}$ ) and the vertical space

$$\mathcal{V} = \bigotimes_{1\le i\le N} (\, \mathbb{C}^2 \,)_i$$

Eq.(6.27) can also be derived from the asymptotic behavior of the YBE (1.3) using the eqs.(6.11)-(6.13) transformed for the $\mathcal{T}(\theta)$.

In eq.(6.27) we recognize the coproduct (6.18)-(6.19). Therefore, we can express the 'quantum group invariance' in the compact form

$$T^{(\mathcal{A},\mathcal{V})}(\theta)\ \Delta^{(\mathcal{V},\mathcal{A})}(\, J_\pm \,) = \Delta^{(\mathcal{A},\mathcal{V})}(\, J_\pm \,)\ T^{(\mathcal{A},\mathcal{V})}(\theta) \qquad (6.28)$$

where $\Delta^{(\mathcal{A},\mathcal{V})}(\, J_\pm \,)$ is given by eq.(6.18). The invariance under z-rotations (the Cartan algebra here) writes in the same fashion :

$$T^{(\mathcal{A},\mathcal{V})}(\theta)\ \Delta^{(\mathcal{V},\mathcal{A})}(\, J_z \,) = \Delta^{(\mathcal{A},\mathcal{V})}(\, J_z \,)\ T^{(\mathcal{A},\mathcal{V})}(\theta) \qquad (6.29)$$

where $\Delta^{(\mathcal{A},\mathcal{V})}(\, J_z \,)$ is given by eq.(6.19).

The z-rotations invariance yield for the transfer matrix

$$[\, S_z,\ \tau_\alpha(\theta) \,] = 0$$

where $\tau_\alpha(\theta) = \exp(i\alpha/2)\ A(\theta) + \exp(-i\alpha/2)\ D(\theta)$. Now, taking trace on $\mathcal{A}$ in eq.(6.27) yields

$$\tau_{-\gamma}(\theta)\ J_+(\gamma) - J_+(\gamma)\ \tau_\gamma(\theta) + 2\exp(-\gamma/2)\ \sinh\{\, \gamma(\, S_z - \tfrac{1}{2})\}\ C(\theta) = 0$$

$$\tau_{-\gamma}(\theta)\ J_-(\gamma) - J_-(\gamma)\ \tau_\gamma(\theta) + 2\exp(\gamma/2)\ \sinh\{\, \gamma(\, S_z + \tfrac{1}{2})\}\ B(\theta) = 0$$

Here we restricted ourselves to the " $\gamma$-deformation" of the SU(2) algebra. The $\gamma$-deformations of all simple Lie algebra are known[61]. They are also connected with the $\theta = \infty$ limit of YBA[4]. Moreover the quantum group invariance (6.28)-(6.29) generalizes to all trigonometric/hyperbolic YB algebras.

To conclude, I want to stress that YB algebras are more general and powerful tools than the quantum groups. Moreover, the elliptic YB algebras provide additional structures beyond the quantum groups.

## VII. SURVEY OF RECENT PROGRESS

Let us very briefly summarize some recent progress in the domain of integrable theories. The full account can be find in the original references.

In ref.[65] we obtained the dominant finite size corrections to the free energy $f_L - f_\infty$ for a U(1)-invariant conformal 2D theory on a cilynder with arbitrarily twisted boundary conditions. We found

$$f_L - f_\infty = \pi( c - 24 \Delta_\theta ) / ( 6 L) \tag{7.1}$$

where $c$ is the central charge and $\Delta_\theta$ is the conformal weight of the operator which realizes the twist by the angle $\theta$ . As examples of the general formula, the free chiral massless field and the massless Thirring model are worked out in detail. As an application, we identify the proper twist operator, in the scaling limit, of the XXZ quantum spin chain.

Eq.(6.1) was first derived by exact finite size calculations from the BAE (see ref.[4] for a review). Later we realize that these universal results could be derived independently of the integrability of the model only using the conformal invariance. The arguments in ref.[65] can be considered as the generalization to twisted boundary conditions of refs.[66].

In ref.[67] properties of the Thirring model solution obtained from the XXZ Heisenberg chain in the continuous limit are computed exactly using the Bethe Ansatz.

The fermion fields, vector current, axial anomaly and equal-time current commutators are constructed explicitly on the lattice and then its continuous limit is obtained. An explicit coupling constant dependence is found for the Schwinger term. This result, together with the current correlation behavior and the conformal weights of the fields here, shows that this Thirring model solution is new. That is, it is a spin 1/2 solution not contained in previous families of continuous solutions. More precisely, the continuous limit of the XXZ chain does not fit into the two-parameter solution of the Thirring model described by Klaiber[68]. Moreover, field theoretic anomalies are found using the Bethe Ansatz for the first time.

In ref.[42] exact relations on the lattice are found between SOS and vertex partition functions. We find that both partition functions are identical up to boundary conditions on the four sites at the corners. Therefore, for large size their difference will be much smaller than $N^{-2}$. This shows, for example, that both models have identical conformal properties (central charge and conformal weights).

In ref.[35] the critical limit of the eight-vertex model eigenvectors obtained by means of the generalised Bethe Ansatz is shown to give the six vertex eigenvectors as constructed in ref.[42]. Furthermore an explicit mapping is established between these eigenvectors and the usual Bethe Ansatz eigenvectors of the six-vertex model. This allowed us to show that the index $\nu$ labelling the eight-vertex eigenstates in refs.[17,34] becomes exactly the third component of the total spin in the critical limit. It turned out that this endomorphism between six-vertex eigenvectors involves the quantum group generators (5.12). Refs.[35] together with ref.[42] show the remarkable mathematical richness of the Bethe Ansatz constructions of eigenvectors already in the trigonometric case.

In ref.[22] the continuous limit of the eight-vertex model is shown to give solely the Massive Thirring Model (MTM). More precisely, using the light-cone lattice approach we find a whole class of lattice spacing (a) dependences in the eight vertex parameters ( $\theta$, $\gamma$, k) yielding a relativistic field theory in the $a = 0$ limit. This turns to be the MTM with no dependence on k. The exact excitation spectrum calculation presented in ref.[22], lead us to these conclusions. Analogous results are reached from a perturbative renormalization group study of the anisotropic current-current continuum fermion field theory.

This paper[22] gives a negative answer to the question : would the light-cone eight-vertex model give a new, more general, local quantum field theory in the continum limit ? At first, one might expect a positive answer on the basis of the relativistic S-matrix interpretation put forward in ref.[69] for the eight-vertex elliptic R-matrix. Actually the fact that this answer is negative is directly connected with the c-theorem.

The finite size resolution methods for the BAE proposed in ref.[28], generalized and extended in refs.[29-32] concerns models where the ground state is formed with real roots. When the ground state is formed by complex roots (strings in the infinite volume limit) only numerical results were available[27]. This applies to models like the spin S integrable spin chains and spin S vertex models for $S \geq 1$. In these models the roots are conveniently parametrized as

Spin 1 : $\lambda^{\pm}_j = \eta_j \pm i( 1/2 + \delta_j )$

Spin S : $\lambda^m_j = \eta^m_j + i [ S - 1/2 - m ] + \delta^m_j$ (7.2)

where $m = 0, 1, 2, ...., 2S-1$ and $\eta_j$, $\delta_j$, $\eta^m_j$ and $\delta^m_j$ are real quantities. We have singled out in eq.(6.2) the imaginary parts $\pm 1/2$ and $(S - 1/2 - m)$ which are expected by the string hypothesis. $\delta_j$ and $\delta^m_j$ vanish for $N \to \infty$ and fixed j .

In ref.[33] the Bethe Ansatz equations for spin S ($S \geq 1$) integrable vertex models (and magnetic chains) are investigated for finite size N. It is shown, that the finite size corrections to the imaginary parts of the roots (Bethe strings) for $N >> 1$ are given by

$$\delta^m_j = \alpha_m / [ 2N \sigma ( \eta^m_j ) ] \tag{7.3}$$

where m is the index of the root within the string [cfr. eq.(7.2)], $\eta^m{}_j$ is the real part of the roots and

$$\sigma(\eta) = 1 / [2\cosh(\pi\eta)] \quad (7.4)$$

is the density of the real parts. The constants $\alpha_m$ are determined by a set of algebraic equations, and are given explicitely by

$$\alpha_m = (1/\pi)\ \ln\{\cos[\tfrac{1}{2}\pi(S - m - 1)/(S + 1)] / \cos[\tfrac{1}{2}\pi(S - m)/(S + 1)]\} \quad (7.5)$$

For the best known $S = 1$ case $\alpha_0 = \ln 2 / (2\pi)$. Eqs.(6.2)-(6.3) hold in the isotropic case and generalizes to the anisotropic case through a rescaling of the BAE roots by a factor $\gamma^{-1}$ for $0 \le \gamma \le \pi/(2S)$ .

These results are found through a generalisation of the Euler-Maclaurin formula including non-analytic contributions in $N^{-1}$ which turn out to be essential in the solution of the present problem. We want to notice that the explicit result for the $\alpha_m$ was obtained through an inexpected relation of the algebraic equations fulfilled by the $\alpha_m$ with the Chebyshev polynomials and their zeroes.

The connection between Conformal Field Theories (CFT) and Integrable Field Theories (IFT) is obtained working solely in the continuum in refs.[70]. It appears that for any (rational) CFT there exist, directly on the continuum, one or more relevant perturbations leading to a massive IFT[70]. This led to the costruction by bootstrap of factorizable S-matrices describing the scattering of massive particles in the perturbed CFT . In ref.[71] the exact S-matrix of the affine $E_8$ Toda field theory is found by bootstrap and checked against tree level standard perturbation theory. At a certain purely imaginary value $g_0$ of the coupling constant g, this S-matrix coincides with that associated in ref.[70] to the critical Ising model in a magnetic field. This supports the idea that the affine $E_8$ TFT at $g = g_0$ describes the scaling limit of the Ising model in a field. The calculation in ref.[71] lead to a value for $g_0$ at which the $E_8$ TFT becomes strictly renormalizable. This means that the ultraviolet fixed point of the theory at $g = g_0$ is no longer the 8-componrnt massless Bose field ( central charge $c = 8$ ). This is consistent with the expectation $c = 1/2$ , which would follow from the identification of the $g = g_0$ $E_8$ TFT with the critical Ising model in a field. This latter has obviously the $c = 1/2$ Ising field theory as UV fixed point.

## REFERENCES

[1] R.J. Baxter, Exactly solved models in Statistical Mechanics, Academic Press (1982).

[2] L.D. Faddeev, Soviet Sci. Review **C1**, 107 (1980) and Les Houches lectures, North Holland (1982) and references contained therein.

[3] P.P. Kulish and E.K. Sklyanin, in Tvärmine lectures, Springer Lectures in Physics, Vol. 151 (1981).

[4] H.J. de Vega, Int. J. Mod. Phys., **4**, 2371-2463 (1989) and in Advanced Studies in Pure Mathematics, vol.19, Kinokuniya-Academic,1989.

[5] H.J. de Vega, Nucl. Phys. **B240**, 495 (1984).

[6] R. J. Baxter, Studies in Appl. Math. **L51**, (1971).

[7] H. J. de Vega and E. Lopes, Phys. Lett. **B 186**, 180 (1987).

[8] see for example : L. D. Faddeev and L. A. Takhtadzhyan, 'Hamiltonian approach to Soliton Theory', Nauka 1986.

[9] R.J. Baxter, Phil. Trans. Roy. Soc. **289**, 315 (1978).

[10] M. Lüscher, Nucl. Phys. **B117**, 475 (1976).

[11] J.B. Mc Guire, J.M.P. **5**, 622 (1964).

[12] A.B. Zamolodchikov and Al.B. Zamolodchikov,Ann. Phys. **120**, 253 (1979).

[13] H.B. Thacker, Revs. Mods. Phys. **53**, 253 (1981).

[14] M. Lüscher, Nucl. Phys. **B135**, 1 (1978). See also: H.J. de Vega, H. Eichenherr and J.M. Maillet, Nucl. Phys. **B240**, 377 (1984) , Phys. Lett. **B132**, 337 (1983) and Comm. Math. Phys. **92**, 507 (1984).

[15] The Analytic S-Matrix. R. J. Eden, P. Landshoff, D. Olive and J.C.Polkinghorne, Cambridge Univ. Press (1966).

[16] A.B. Zamolodchikov, Comm. Math. Phys. **69**, 165 (1979). R. Shankar, Phys. Rev. Lett. **47**, 1177 (1981).

[17] L.D. Faddeev and L.A. Takhtadzhyan, Russ. Math. Surveys **34**, 11 (1979).

[18] O. Babelon, H. J. de Vega and C. M. Viallet, Nucl. Phys. **B200**, 266 (1982).

[19] L.A. Takhtadzhyan, Journal of Soviet Math. 2470 (1983). P.P. Kulish and N. Yu. Reshetikhin, J. Phys. **A16**, L591 (1983). N.Y. Reshetikhin, Theor. Math. Phys. **63**, 555 (1986).

[20] H.J. de Vega and M. Karowski, Nucl. Phys. **B280**, 225 (1987).

[21] Zhou Yu-Kui, Yan Mu-Lin and Hou Bo-Yu, J. Phys. **A 21**, L929 (1988).

[22] C. Destri and H. J. de Vega, LPTHE Paris preprint 89-20, to appear in Mod. Phys. Lett.

[23]H. Au-Yang, B. M. McCoy, J.H.H. Perk, S. Tang and M. L. Yan, Phys. Lett.**A 123**, 219 (1987).
B. M. McCoy, J.H.H. Perk, S. Tang and C. H. Sah, Phys. Lett. **A 125**, 9 (1987).
R. J. Baxter, J. H. H. Perk and H. Au-Yang, Phys. Lett. **A 128**, 138 (1988).

[24]V. V. Bazhanov and Yu. G. Stroganov, Canberra preprint, 1989.

[25]Y. Akutsu, T. Deguchi and M. Wadati, J. Phys. Soc. Japan, **56**, 3039 (1987).

[26]F. Woynarovich, J. Phys. **A15**, 2985 (1982).
C. Destri and J. H. Lowenstein, Nucl. Phys. **B205**, 369 (1982).
L. D. Faddeev and L.A. Takhtadzhyan, Zap. Nauch. Sem. LOMI, **109**, 134 (1982).
O. Babelon, H.J. de Vega and C.M. Viallet, Nucl. Phys. **B220**, 13 (1983) and 282 (1983).

[27]L. V. Avdeev and B. D. Dörfel, Theor. Math. Phys. **71**, 272 (1987).
F. C. Alcaraz, M. N. Barber and M.T. Batchelor,
Phys. Rev. Lett. **58**, 771 (1987) and Ann. Phys. **182**, 280 (1988).
F. C. Alcaraz and M. J. Martins, J. Phys. **A 22**, 1829 (1989) and **A 21**, 4397 (1988).

[28]H.J. de Vega and F. Woynarovich, Nucl. Phys. **B251**, 439 (1985).

[29]H.O. Martin and H.J. de Vega, Phys. Rev. **B32**, 5959 (1985).

[30]M.T. Batchelor et al.J. Phys.**A 20**,5677 (1987)
C. J. Hamer and M. T. Batchelor J. Phys. **A 21**, L173(1988).
F. Woynarovich and H.P. Eckle, J. Phys. **A 20**, L443 (1987).

[31]H.J. de Vega, J. Phys. **A 20**, 6023 (1987) and J. Phys. **A 21**, L1089 (1988).

[32]a) H.P. Eckle and F. Woynarovich, J. Phys. **A 20**, L97 (1987).
b) F. Woynarovich, Phys. Rev. Lett. **59**, 259 (1987).

[33]H. J. de Vega and F. Woynarovich, LPTHE Paris preprint 89-32.

[34]R. J. Baxter, Ann. Phys. **76**, 1, 25 and 48 (1973).

[35]C. Destri, H.J. de Vega and H.J. Giacomini, J. Stat. Phys. **56**, 291 (1989).

[36]J.H.H. Perk and F. Y. Wu, J. Stat. Phys. **42**, 727 (1986).

[37]G.E. Andrews, R. J. Baxter and P. J. Forrester, J. Stat. Phys. **35**, 193 (1984).

[38]M. Jimbo, T. Miwa and M. Okado, Comm. Math. Phys. **116**, 507 (1988)
and preprint RIMS-579 (1987).
E. Date, M. Jimbo, A. Kuniba T. Miwa and M. Okado,Nucl. Phys. **B 290**, 231 (1987).
A. L. Owczarek and R. J. Baxter, J. Stat. Phys. **49**, 1093 (1987).

[39]V. Pasquier, J. Phys. **A 20**, L217 and L221 (1987) and Nucl. Phys. **B 285**, 162 (1987).

[40] Y. Akutsu, T. Deguchi and M. Wadati, J. Phys. Soc. of Japan, **57**, 1173 (1988).

[41]H. J. de Vega, LPTHE preprint 89/23, to be published in Int. J. Mod. Phys.

[42] H. J. de Vega and H. J. Giacomini, J. Phys. **A 22**, 2759 (1989).

[43]C. Destri and H.J. de Vega, Nucl. Phys. **B290**, 363 (1987);

Phys. Lett. **B201**, 261 (1988) and J. Phys. **A22**, 1329 (1989).

[44]T.T. Truong and M.D. Schotte,Nucl. Phys. **B220**, 77 (1983) and **B230**, 1 (1984).
M.F. Weiss and M.D. Schotte, Nucl. Phys. **B225**, 247 (1983).
T.T. Truong in Non-linear Eqs. in Classical and QFT Springer
Lectures in Physics, Vol. **226** (1985), Ed. N. Sánchez.

[45]L.V. Avdeev and M.V. Chizov, Phys. Lett. **B184**,363 (1987).

[46]O. Babelon, H.J. de Vega and C.M. Viallet, Nucl. Phys. **B190**, 542 (1981).

[47]H. Freudenthal and H. de Vries, Linear Lie Groups, New-York, Academic Press.

[48]H. M. Babujan, Nucl. Phys. **B215**, 317 (1983).
P. P. Kulish and N. Yu. Reshetikhin, Zap. Nauk Sem. LOMI **101**, 101 (1981).
L. A. Takhtadzhyan, Phys. Lett. **A 87** , 479 (1982).

[49]V. E. Korepin, Theor. Math. Phys. **41**, N°2 (1979).
N. Andrei and J. H. Lowenstein, Phys. Lett.**B 91**, 401 (1980).

[50]E. Ogievetski, N. Y. Reshetikhin and P. B. Wiegmann, Nucl. Phys. **B280**, 45 (1987).

[51]L. D. Faddeev and N. Y. Reshetikhin, Ann. Phys. **167**,227 (1986).

[52]P. P. Kulish, N. Y. Reshetikhin and E. K. Sklianin, L.M.P. **5**, 393 (1981).

[53]C. Destri and H. J. de Vega, Phys. Lett. **B201**, 245 (1988).

[54]A. N. Kirillov and N. Y. Reshetikhin in Proceedings of the Paris-Meudon
Colloquium. H. J. de Vega and N. Sanchez , Editors. World Scientific 1987.

[55] R. J. Baxter, J. Stat. Phys. **28**, 1 (1982).

[56]E. Artin, Ann. of Math. **48**, 101 (1947).
J. S. Birman, Braids et al., Princeton Univ. Press, 1974.

[57]T. Deguchi, M. Wadati and Y. Akutsu, Tokyo Univ. preprint, 1988.
M. Wadati and Y. Akutsu, C. M. P. **117**, 243 (1988).

[58]V. G. Turaev, LOMI preprint E-3-87, Leningrad.
N. Y. Reshetikhin, LOMI preprints E-4-87 and E-17-87, Leningrad.
V. Jones (unpublished).

[59]A. Tsuchiya and Y. Kanie, LMP **13**, 303 (1987) and
in Adv. Studies Pure Math., **16**, 297 (1988), Kinokuniya, Tokyo, 1988.
J. Fröhlich, Cargèse Lectures, 1987, Plenum New York.
K.H. Rehren, CMP **116**, 675 (1988).
K. H. Rehren and B. Schroer, Nucl.Phys. **B312**, 715 (1989).

[60]H. M. Babujian and A. M. Tsevelik, Nucl. Phys. **B265**, 24 (1986).

[61]M. Jimbo, L.M.P.**11**,247 (1986) and **10**, 63 (1985).
V. G. Drinfeld Dokl. Akad. Nauk, **283**, 1060(1985) and **296**, 13 (1987).

[62]E. Date, M. Jimbo, T. Miwa and M. Okado, RIMS preprint 590,(1987)
and references therein.

[63]M. Jimbo in Springer Notes in Physics, vol. 246, H. J. de Vega and N. Sanchez, editors and Comm. Math. Phys. **102**, 537 (1986).
N. Yu. Reshetikhin, Theor. Math. Phys. **63**, 197 and 347 (1985).

[64]N. Yu. Reshetikhin and P. B. Wiegman, Phys. Lett. **B 168**, 360 (1986).
E. I. Ogievetski and P. B. Wiegman, Phys. Lett. **B 189**, 125 (1987).

[65]C. Destri and H. J. de Vega, Phys. Lett. **B223**, 365 (1989).

[66]H. W. Blöte, J. L. Cardy and M.P. Nightingale, Phys. Rev. Lett. **56**, 742 (1986).
I. Affleck, Phys. Rev. Lett. **56**, 746 (1986).

[67]H. J. de Vega and T. M. J. Simoes, Phys. Lett. **B 217**, 142 (1989).

[68]B. Klaiber, Boulder Lectures, Vol. **XA**, 141 (1968).
J. A. Swieca, Fortschr. Phys. **25**, 303 (1977).

[69]A. B. Zamolodchikov, Comm. Math. Phys. **69**, 165 (1979).

[70] A. B. Zamolodchikov in Advanced Studies in Pure Mathematics, vol.19,Kinokuniya-Academic,1989 and Int. J. Mod. Phys. **16**, 4235 (1989).

[71] C. Destri and H. J. de Vega, LPTHE Paris preprint 89-33 (to appear in Phys. Lett. B).
H. W. Braden, E. Corrigan, P. E. Dorey and R. Sasaki, Durham preprint.

## FIGURE CAPTIONS

Fig.1 To each type of line is associated a vector space.

Fig.2 A YB generator is associated to the intersection of two lines.

Fig.3 There is a summation over the states of internal lines.

Fig.4 The Yang-Baxter equation (YBE) in its general form.

Fig.5 The R-matrix.

Fig.6 The YBE for the R-matrix.

Fig.7 The YBE for the YB generator (1.19).

Fig.8 A $N \times M$ two dimensional lattice. The local states of horizontal (vertical) bonds belong to the vector space $\mathcal{A}$ ( $\mathcal{V}$ ).

Fig.9 The local statistical weights $w( \alpha\ \beta \mid a\ b )$ depend on the states of the four bonds joining at a vertex.

Fig.10 Allowed configurations in the six-vertex model and their statistical weights ( see eq.(3.1) ).

Fig.11 Discretized Minkowski space-time. Sites are world events joined by world lines of the bare particle propagation.

Fig.12 The six non-zero microscopic transition amplitudes. They coincide with the weights of fig.10.

Fig. 13 A braid from $B_n$.

Fig.14 The elementary operations $b_i$ and $b_i^{-1}$ from the braid group $B_n$.

Fig.15 The two ways that lines may intersect in the $\theta = \pm\infty$ limit.

fig. 16 Wu-Kadanoff-Wegener duality between vertex and face weights.

fig. 17 Local statistical weights associated to the depicted four sites configuration.

fig. 18 The star triangle equation. It is the integrability constraint on the face weights $w(\alpha, \beta, \gamma, \delta \mid \theta)$.

fig. 19 The faces YB operator for a N sites row.

fig. 20 The faces YB algebra for the operators of fig. 19.

## TABLE CAPTIONS

Table I Correspondence between the symmetry group of the Yang-Baxter algebras and the functional dependence of the YB operators on the spectral parameter $\theta$.

Table II Integrable QFT associated to trigonometric Yang-Baxter algebras in the light-cone approach. We indicate the underlying Lie Algebra $\mathfrak{G}$, the respective scale parameter $\kappa$ (it coincides with the one-loop beta function) and the corresponding mass spectrum.

## TABLE I

| $\mathfrak{G}$ : symmetry group | $\theta$-dependence in $R_{ab}^{cd}(\theta)$ |
|---|---|
| discrete : $Z_q$ | elliptic |
| continuos abelian : $U(1)^q$ | trigonometric or hyperbolic |
| continuous non-abelian : U(q), O(q),.... | rational |

## TABLE II

| Lie Algebra | $\kappa$ | $m_k$ |
|---|---|---|
| $A_n$ | $2\pi/(n+1)$ | $\sin(\pi k/[n+1])$ , $1 \le k \le n$ |
| $B_n$ | $\pi/(2n-1)$ | $\sin(\pi k/[2n-1])$, $1 \le k \le n-1$, $m_n = 1/2$ |
| $C_n$ | $\pi/(n+1)$ | $\sin(\pi k/2(n+1))$, $1 \le k \le n$ |
| $D_n$ | $\pi/(n-1)$ | $\sin(\pi k/2(n-1)$, $1 \le k \le n-2$<br>$m_\pm = 1/2$ |
| $E_6$ | $\pi/6$ | $m_1 = m_5 = m_6/2 = \sqrt{3}/2$<br>$m_2 = m_4 = (3+\sqrt{3})/2$<br>$m_3 = (3+\sqrt{3})/\sqrt{2}$ |
| $E_7$ | $\pi/9$ | $m_2 = 2m_1 \cos\pi/18$, $m_3 = m_1/(2\sin\pi/18)$<br>$m_4 = 2m_2\cos\pi/9$, $m_5 = 2m_2 \sin 2\pi/9$,<br>$m_6 = 2m_1 \sin 2\pi/9$, $m_7 = 2m_1 \cos\pi/9$. |
| $E_8$ | $\pi/15$ | $m_2 = 2m_1 \cos\pi/5$, $m_3 = 2m_1 \cos\pi/30$,<br>$m_4 = 2m_2 \cos 7\pi/30$, $m_5 = 2m_2 \cos 2\pi/15$,<br>$m_6 = m_2 m_3/m_1$, $m_7 = m_2 m_4/m_1$,<br>$m_8 = m_2 m_5/m_1$. |
| $F_4$ | $\pi/9$ | $m_4 = 2m_1 \cos 2\pi/9$, $m_2 = m_1/(2\sin\pi/18)$,<br>$m_3 = 2m_1 \cos\pi/18$. |
| $G_2$ | $\pi/6$ | $m_2 = m_1 [(\sqrt{6}+\sqrt{2})/2]$. |

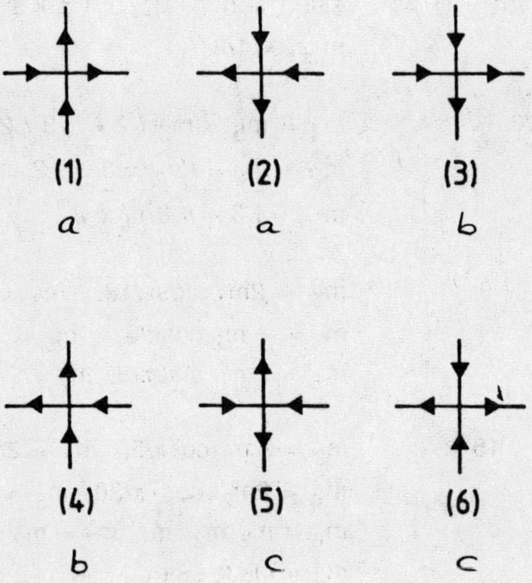


Fig.10

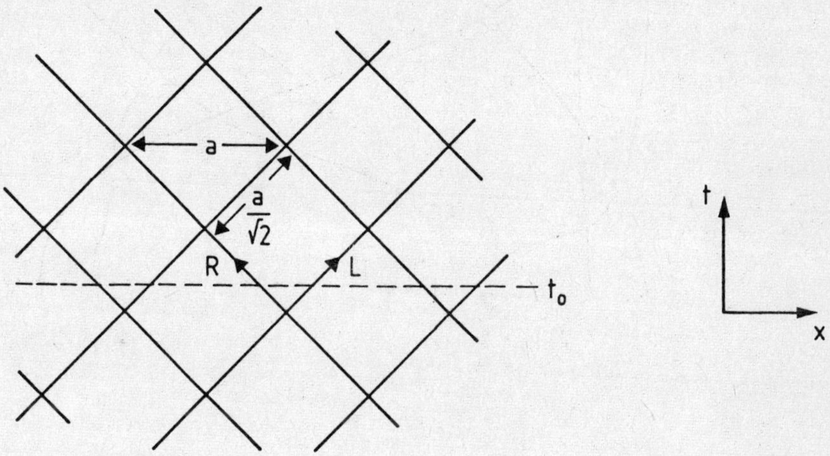


Fig.11

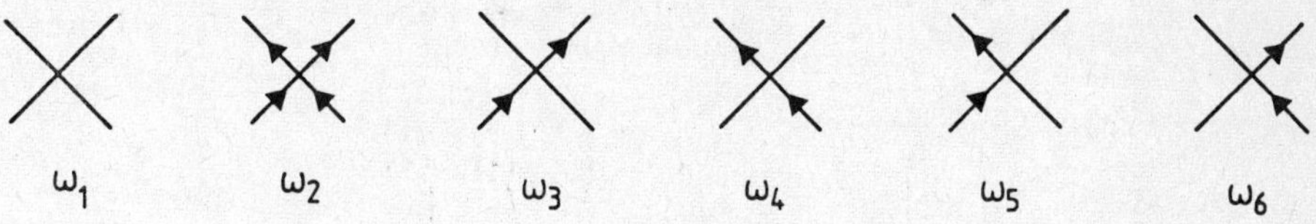


| ≡ |0 > ≡ | empty > ; ↑ ≡ |1 > ≡ | occupied >

Fig.12

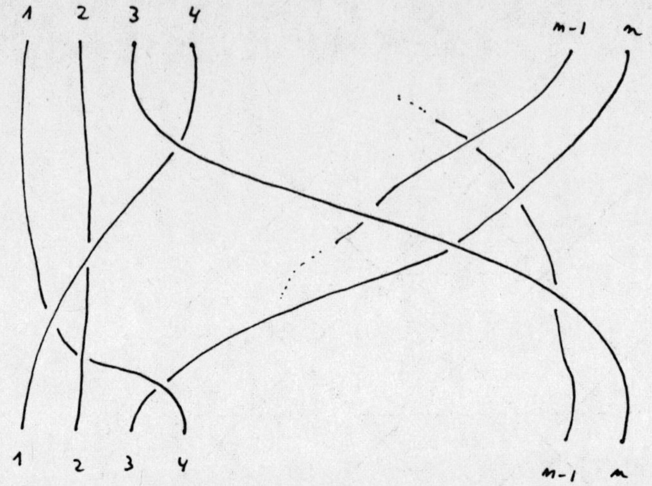


Fig.13

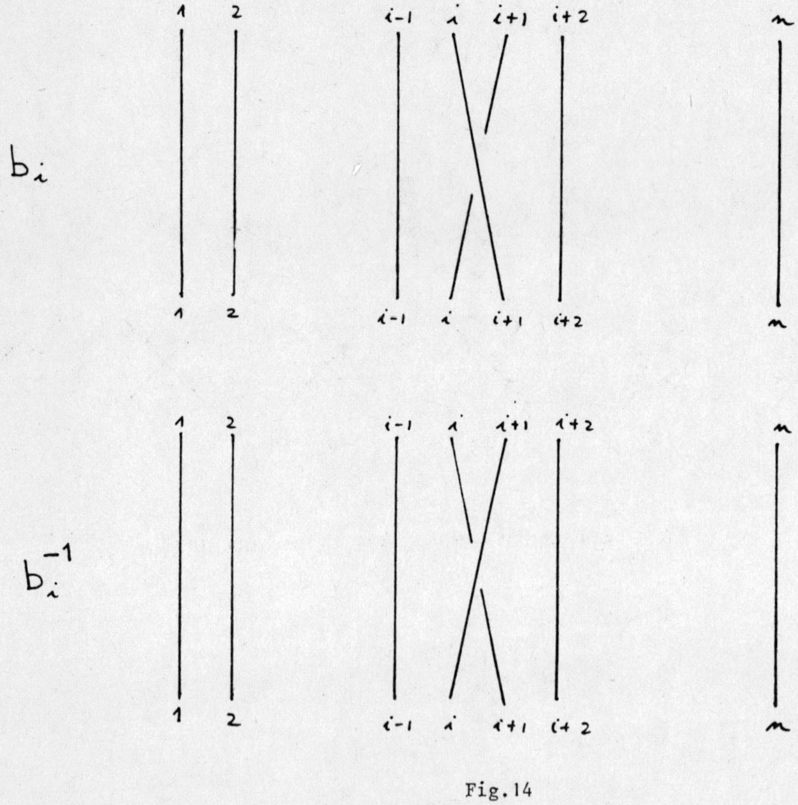


Fig.14

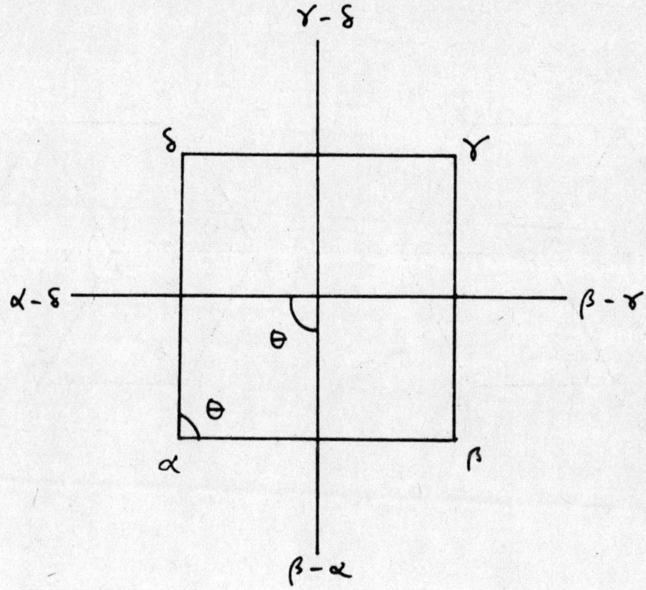


Fig.16

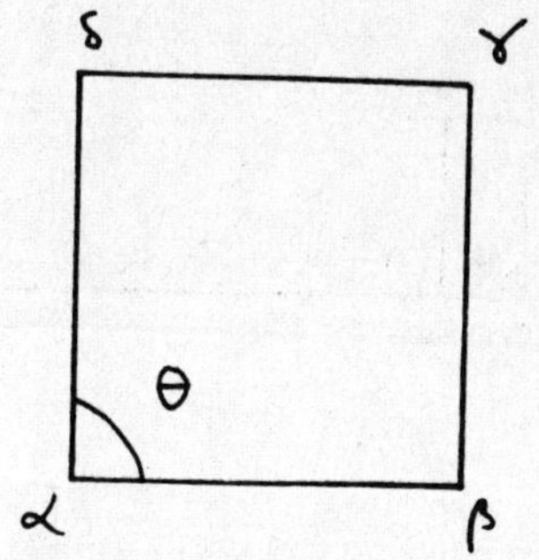


Fig.17

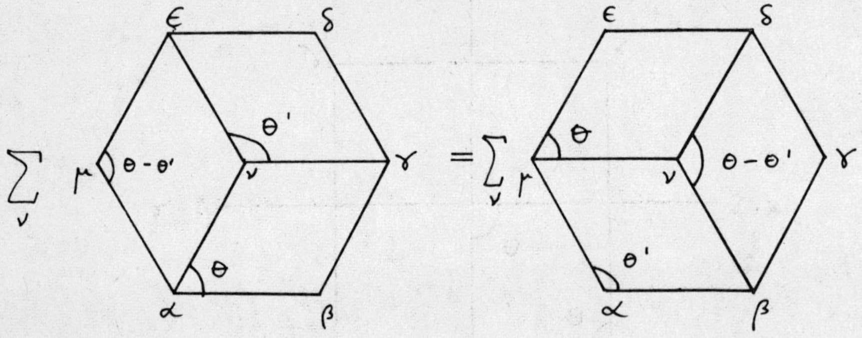


Fig.18

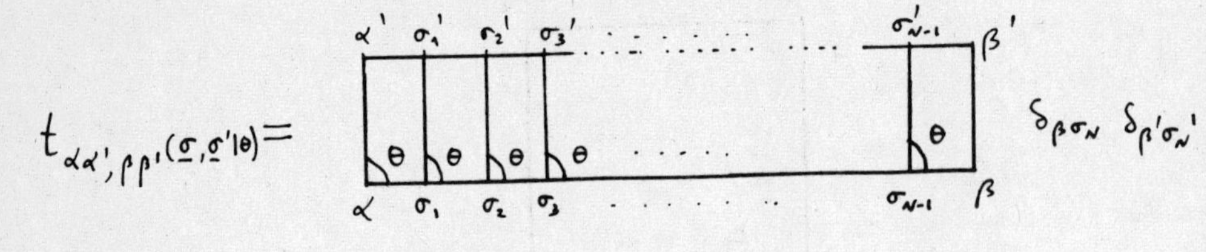


Fig.19

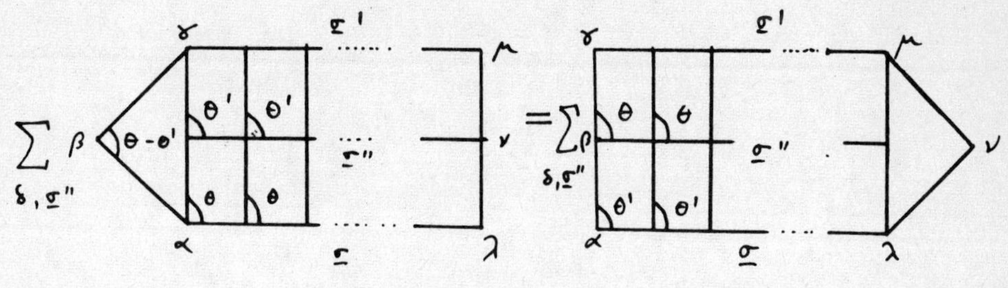

Fig.20

International Journal of Modern Physics B, Vol. 4, No. 5 (1990) 803–870
© World Scientific Publishing Company

# FUNCTIONAL RELATIONS FOR TRANSFER MATRICES OF THE CHIRAL POTTS MODEL

R.J. Baxter[1,2], V.V. Bazhanov[3] and J.H.H. Perk[4,5]

1. *Mathematics Research Section, School of Mathematical Sciences, The Australian National University, Canberra, ACT* 2601, *Australia.*

2. *Department of Theoretical Physics, Research School of Physical Sciences, The Australian National University, Canberra, ACT* 2601, *Australia.*

3. *Institute for High Energy Physics, Serpukhov, Moscow Region, USSR.*

4. *Department of Physics, Oklahoma State University, Stillwater, Ok.*74078-0444, *USA.*

5. *Supported by NSF grant DMR*-8803678.

It has recently been shown that the solvable N–state chiral Potts model is related to a vertex model with N–state spins on vertical edges, two–state spins on horizontal edges. Here we generalize this to a "j–state by N–state" model and establish three sets of functional relations between the various transfer matrices. The significance of the "super–integrable" case of the chiral Potts model is discussed, and results reported for its finite–size corrections at criticality.

## 1. INTRODUCTION

From 1985 to 1988 various studies [1-4] were made of an N-state two-dimensional lattice model (and its associated one-dimensional quantum Hamiltonian [5-7]). Guided by these, Baxter, Perk and Au-Yang [8-9] obtained in 1988 a new solution of the star-triangle (or Yang-Baxter) relation.

This solution is defined in section 2. It is a special case of the $Z_N$-symmetric, chirally asymmetric, N-state model. We shall call it simply the chiral Potts model.

Usually, the satisfying of the star-triangle relation is a first step towards exactly solving the model. In particular, for the six-vertex and eight-vertex models [10-13] one obtains n non-linear equations in n unknowns (the Bethe ansatz equations for $k_1,..,k_n$) that determine *all* the eigenvalues of the row-to-row transfer matrix T. Here n is proportional to the number of columns of the lattice. These equations can be solved explicitly for the maximum (and next maximum) eigenvalue in the large lattice limit. (For the Ising [14-15] model they can be solved completely for finite n.) This gives the bulk free energy.

We have not yet reached this stage for the chiral Potts model. One difficulty is that it lacks the "difference property" that leads naturally to a uniformizing parametrization in terms of trigonometric or elliptic functions. Indeed, it involves algebraic curves of genus greater than one, so there is no explicit elementary uniformizing parametrization.

Still, some progress has been made. Baxter [16] showed that the bulk free energy could be expressed in terms of various integrals. This result gives the critical exponent $\alpha = 1-2/N$, but is very complicated: one hopes it can be simplified.

For the N = 3 case, Albertini, McCoy and Perk [17] have postulated an exact functional relation for the row-to-row transfer matrix. For the special "superintegrable" case (discussed here in section B) they have solved this to obtain [18-20] all the eigenvalues of T for N = 3 .

Baxter [21] has studied the superintegrable case for arbitrary N, obtaining a particular set of eigenvalues and evaluating the partition function explicitly for cylindrical boundary conditions [22]. From this he deduced the eigenvalue spectrum of the column-to-column transfer matrix $T_{col}$, with fixed spin boundary conditions. This turned out to be remarkably simple: that of a direct product of N by N matrices. These matrices are themselves particularly simple when inverted, which led Baxter to consider the corresponding "inverse superintegrable" model, with transfer matrix $T_{col}$. He found that this could be expressed as a model with both N-state and 2-state spins, interacting via three-spin interactions. As a limiting case, he obtained [23] the eigenvalues of an Hamiltonian that is an N-state generalization of the one-dimensional Ising model in a transverse field (this is the Hamiltonian that commutes with the transfer matrix of the two-dimensional Ising model). All the coefficients of this Hamiltonian are arbitrary, yet its eigenvalue spectrum is always a simple direct sum.

Bazhanov and Stroganov [24] have considered a vertex model (where spins live on edges of the square lattice), where the spins on horizontal edges have 2 states, those on vertical edges have N states. The condition that row-to-row transfer matrices commute then leads to a Yang-Baxter relation with a two-state R-matrix. They take this to be the R-matrix of the six-vertex model (ref [10-12], and chapters 8, 9 of ref [15]). The 2-by-N model is then defined (to within simple gauge transformations). They then consider the condition that column-to-column transfer matrices commute and obtain a Yang-Baxter relation with an N-state R-matrix. Their remarkable observation is that this R-matrix is *precisely*, that of the N-state chiral Potts model! Further, their 2-by-N model is Baxter's inverse superintegrable model (turned through 90°)! (Strictly, it is a specialization of the row-inhomogeneous form of Baxter's model.)

Bazhanov and Stroganov write the row-to-row transfer matrix of the 2-by-N model as T, that of the chiral Potts model as Q. Let us write them here as $T_{BS}$,

$Q_{BS}$ . They establish a relation between them that is formally very similar to the "T, Q" relation of the six-vertex and eight-vertex models (eqn. 87, p.15 of ref. [13], and (10.5.32) of ref. [15]). Together with the appropriate analyticity properties it should determine all the eigenvalues of both matrices. They further go on to define a set of matrices $\tau^{(1)}$ ,..., $\tau^{(N)}$ , with $\tau^{(1)} = 1$ and $\tau^{(2)} = T_{BS}$ , satisfying an hierarchy of relations.

Here we aim to link these approaches. In section 2 we define the N-state chiral Potts model, give the star-triangle relation, and show how this leads to commutation relations for the transfer matrices T . In section 3 we consider products of two related transfer matrices. We establish a new set of relations expressing these products as a linear combination of two new matrices $\tau^{(j)}$ . These $\tau^{(j)}$ are transfer matrices of j-by-N models, and are in fact Bazhanov and Stroganov's matrices $\tau^{(j)}$ (to within scalar factors and spin translations).

In section 4 we establish a relation between $\tau^{(2)}$ and T: this is Bazhanov and Stroganov's relation between $T_{BS}$ and $Q_{BS}$ . From this and the result of section 3 we establish a closed hierarchy of relations involving only the $\tau^{(j)}$ . In one important respect these matrices are simpler than T : they are simply polynomials in an appropriate variable, so for these there is no problem in finding a uniformizing parametrization. We also obtain a functional relation between the T matrices only.

In section 5 we consider the special cases N = 2 and 3 . For the latter, the T-relation is that used by Albertini, McCoy and Perk [17].

The "superintegrable" case is discussed in section 6. It can be regarded as a general chiral Potts model, in which the vertical rapidities are alternately p and p′ , where p , p′ satisfy a simple "conjugate" relation. We discuss some of the properties of the partition function previously obtained [22] for this case, and show, for a large lattice at criticality, that the finite-size corrections do not have the usual form predicted by conformal invariance [25-26], but appear to be an interesting variant of this form.

We also remark that the column-to-column transfer matrix $T_{col}$ of the superintegrable case is the matrix $\tau^{(N)}$, and its inverse is $\tau^{(2)}$, except that we have changed from periodic to fixed-spin boundary conditions. It has the simple direct product eigenvalue spectrum mentioned above..

## 2. DEFINITIONS AND BASIC PROPERTIES

Consider the square lattice $\mathcal{L}$, drawn diagonally as in Fig. 1, with periodic (toroidal) boundary conditions. Let there be M rows (with M even); each containing L sites, numbered from 1 to L as shown.

At each site i place a "spin" $\sigma_i$, with values $0, \ldots, N-1$. Adjacent spins interact with Boltzmann weight functions $W(\sigma_i - \sigma_j)$ on SW → NE edges (i,j), $\overline{W}(\sigma_i - \sigma_k)$ on SE → NW edges (i,k), as indicated. Thus the partition function is

$$Z = \sum_{\{\sigma\}} \prod_{<i,j>} W(\sigma_i - \sigma_j) \prod_{<i,k>} \overline{W}(\sigma_i - \sigma_k) , \tag{2.1}$$

the products being over all edges of each type, the sum over all values of all the spins. We require, for all integers n, that

$$W(n+N) = W(n) , \quad \overline{W}(n+N) = \overline{W}(n) . \tag{2.2}$$

This means that the model is $Z_N$ symmetric: it is unchanged by incrementing all the spins by any fixed integer, and by incrementing any spin by N.

In general we allow

$$W(-n) \neq W(n) , \quad \overline{W}(-n) \neq \overline{W}(n) , \tag{2.3}$$

so chiral reflection symmetry is broken.

**The functions $W_{pq}$, $\overline{W}_{pq}$**

The functions $W(n)$, $\overline{W}(n)$ will also depend on certain variables $p$, $q$ (in general complex) which we call "rapidities" or "spectral parameters". We exhibit this by writing the functions as $W_{pq}(n)$, $\overline{W}_{pq}(n)$. Here we take $p$ to be a four-vector $(a_p, b_p, c_p, d_p)$ lying on the curves

$$a_p^N + k'b_p^N = kd_p^N, \quad k'a_p^N + b_p^N = kc_p^N \tag{2.4}$$
$$ka_p^N + k'c_p^N = d_p^N, \quad kb_p^N + k'd_p^N = c_p^N$$

where $k$, $k'$ are real parameters satisfying

$$k^2 + k'^2 = 1 . \tag{2.5}$$

We regard $k$, $k'$ as fixed constants; $p$ as a variable. Only 2 of the equations (2.4) are independent, and all properties depend only on the ratios $a_p : b_p : c_p : d_p$, so we have just one effective degree of freedom for $p$. We can think of $p$ as a point lying on the curve specified by (2.4).

Four related parameters that we shall also use are

$$x_p = a_p/d_p , \quad y_p = b_p/c_p , \quad \mu_p = d_p/c_p , \tag{2.6}$$
$$t_p = x_p y_p = a_p b_p / c_p d_p .$$

From (2.4), we have

$$\mu_p^N = k'/(1 - kx_p^N) = (1 - ky_p^N)/k' , \tag{2.7}$$

$$x_p^N + y_p^N = k(1 + x_p^N y_p^N) . \tag{2.8}$$

Let $\omega$ be the primitive Nth root of unity:

$$\omega = e^{2\pi i/N} . \tag{2.9}$$

We shall find it convenient to use the notation $(x, y \in \mathbb{C}, m, n \in \mathbb{Z})$

$$(x,y)_{m,n} = \prod_{j=m+1}^{n} (x - y\omega^j) , \quad n \geq m \tag{2.10a}$$

(2.10b) $$= \prod_{j=n+1}^{m} (x-y\omega^j)^{-1} , \quad n \leq m .$$

Then

(2.11) $$(x,y)_{m,n}/(x,y)_{m,r} = (x,y)_{r,n}$$

for all integers m , n , r .

With these notations, $W_{pq}(n)$ , $\overline{W}_{pq}(n)$ are defined, for all integers n , by

(2.12a) $$W_{pq}(n)/W_{pq}(o) = (d_p b_q, a_p c_q)_{o,n}/(b_p d_q, c_p a_q)_{o,n}$$

$$\overline{W}_{pq}(n)/\overline{W}_{pq}(o) = (\omega a_p d_q, d_p a_q)_{o,n}/(c_p b_q, b_p c_q)_{o,n} ,$$

or equivalently by

(2.12b) $$W_{pq}(n)/W_{pq}(o) = (c_p a_q, \omega^{-1} b_p d_q)_{o,-n}/(a_p c_q, \omega^{-1} d_p b_q)_{o,-n}$$

$$\overline{W}_{pq}(n)/\overline{W}_{pq}(o) = (b_p c_q, \omega^{-1} c_p b_q)_{o,-n}/(d_p a_q, a_p d_q)_{o,-n} .$$

Alternatively, we can express (2.12a) in terms of the $x_p$ , $y_p$ , $\mu_p$ :

(2.13) $$W_{pq}(n)/W_{pq}(o) = (\mu_p/\mu_q)^n (y_q,x_p)_{o,n}/(y_p,x_q)_{o,n}$$

$$\overline{W}_{pq}(n)/\overline{W}_{pq}(o) = (\mu_p \mu_q)^n (\omega x_p,x_q)_{o,n}/(y_q,y_p)_{o,n} .$$

The periodicity conditions (2.2) are now automatically satisfied, being a consequence of the identities

(2.14) $$J_{pq} := d_p^N b_q^N - a_p^N c_q^N = b_p^N d_q^N - c_p^N a_q^N ,$$

$$\overline{J}_{pq} := a_p^N d_q^N - d_p^N a_q^N = c_p^N b_q^N - b_p^N c_q^N ,$$

which in turn follow from (2.4).

The rapidities p ,q are associated with vertical and horizontal lines underlying the lattice $\mathcal{L}$ , shown in Fig. 1 by dotted lines. Each edge of $\mathcal{L}$ passes through the intersection of two $(p, q)$ rapidity lines, and conversely. In general we can allow each line to have its own rapidity variable, so there would be $2L$ rapidities $p_1, \ldots, p_{2L}$ associated with the $2L$ vertical lines (and $M$ rapidities $q_1, \ldots, q_M$ associated with the horizontal lines). Here we shall find it convenient to consider a "checkerboard" model, where the rapidities alternate. In particular, the vertical rapidities are $p, p', p, p', \ldots$ as indicated. (The generalization to the fully inhomogeneous case is trivial, and is briefly discussed in section 4.)

**Transfer Matrices**

Notice that there are two types of rows in sites in Fig. 1: those indicated by filled (black) circles, and those indicated by open (white) circles. Let $\sigma = \{\sigma_1, \ldots, \sigma_L\}$ be the set of spins $\sigma_1, \ldots, \sigma_L$ on a white row; and $\sigma' = \{\sigma'_1, \ldots, \sigma'_L\}$ the set on the black row immediately above it. Then the combined weight of all the interactions between the two rows is

$$T_{\sigma\sigma'} = \prod_{J=1}^{L} W_{pq}(\sigma_J - \sigma'_J)\, \overline{W}_{p'q}(\sigma_{J+1} - \sigma'_J) \tag{2.15a}$$

Similarly, if $\sigma'' = \{\sigma''_1, \ldots, \sigma''_L\}$ is the set of spins on the next white row (above $\sigma'$), the weight of the interactions between $\sigma'$ and $\sigma''$ is

$$\hat{T}_{\sigma'\sigma''} = \prod_{J=1}^{L} \overline{W}_{pq}(\sigma'_J - \sigma''_J) W_{p'q}(\sigma'_J - \sigma''_{J+1}) \tag{2.15b}$$

In the usual way (chapters 2 and 7 of ref. [15]), we can regard these combined weights as elements of $N^L$ by $N^L$ matrices $T$, $\hat{T}$, respectively. If $q$ is the same for all horizontal lines, we can write (2.1) as

$$\begin{aligned} Z &= \text{Trace}\; T\, \hat{T}\, T\, \hat{T} \ldots \hat{T} \\ &= \text{Trace}\; (T\, \hat{T})^{M/2} , \end{aligned} \tag{2.16}$$

there being one such matrix for each of the $M$ rows. Obviously these matrices depend on the vertical rapidities $p$ , $p'$ , and on the horizontal rapidity $q$ . We shall regard $p$ , $p'$ as fixed; $q$ as variable, and write $T$ , $\hat{T}$ as $T_q$ , $\hat{T}_q$ . Then (2.17) generalizes at once to the case where the horizontal rapidities are different:

$$Z = \mathrm{Trace}\; T_{q_1} \hat{T}_{q_2} T_{q_3} \ldots \hat{T}_{q_M} \quad , \tag{2.17}$$

$q_I$ being the rapidity of row $I$ .

These matrices $T_q$ , $\hat{T}_q$ are transfer matrices. The main aim of this paper is to exhibit various functional relations that exist between them, and which should enable their eigenvalues (and hence $Z$) to be calculated.

**Star-triangle Relation**

The particular form (2.12) or (2.13) of $W_{pq}(n)$ , $\overline{W}_{pq}(n)$ was obtained in ref. 8 as a solution of the star-triangle relation. [One of us (RJB) has studied the star-triangle relations for $N = 3$ in the low-temperature limit. For this solution, various additive terms in the relations are then relatively negligible, and ignoring them gives a simpler set of equations. This simpler set is easy to solve, even for an arbitrary $N = 3$ model without $Z_N$ invariance. There are actually two star-triangle relations, for both "up-pointing" and "down-pointing" triangles, that are needed to ensure commutation of transfer matrices. If one requires that both be satisfied in this simple low-temperature limit, then one finds that the only solution, to within uninteresting gauge transformations, is that given by (2.12). Thus it appears (at least for $N = 3$) that there is no solution of the star-triangle relations that continuously generalizes (2.12), containing it as a special case.]

The star-triangle relation is

$$\sum_{d=0}^{N-1} \overline{W}_{qr}(b-d)\, W_{pr}(a-d)\, \overline{W}_{pq}(d-c) = R_{pqr} W_{pq}(a-b)\, \overline{W}_{pr}(b-c)\, W_{qr}(a-c) \quad , \tag{2.18}$$

for all values of the exterior spins a , b , c , and of the rapidities p , q , r . The proof used for ref. 8, given in ref. 9, involves only ratios independent of the factor $R_{pqr}$ . It was conjectured in ref. 8, and proved very neatly by Matveev and Smirnov [27], that

$$R_{pqr} = f_{pq} f_{qr} / f_{pr} \ , \tag{2.19}$$

where

$$f_{pq} = \{\det_N[\overline{W}_{pq}(i-j)] / \prod_{n=0}^{N-1} W_{pq}(n)\}^{1/N} \ . \tag{2.20}$$

Here $\det_N[\overline{W}_{pq}(i\text{-}j)]$ is the determinant of the N-by-N matrix whose (i, j) entry is $\overline{W}_{pq}(i\text{-}j)$ . This matrix is cyclic, so

$$\det_N[\overline{W}_{pq}(i-j)] = \prod_{n=0}^{N-1} \overline{W}^{(f)}_{pq}(n) \ , \tag{2.21}$$

where

$$\overline{W}^{(f)}_{pq}(n) = \sum_{k=0}^{N-1} \omega^{nk}\, \overline{W}_{pq}(k) \tag{2.22}$$

is the Fourier transform of $\overline{W}_{pq}(k)$ . Noting from (2.12) that

$$(c_p b_q - b_p c_q \omega^k)\, \overline{W}_{pq}(k) = (\omega a_p d_q - d_p a_q \omega^k)\, \overline{W}_{pq}(k-1) \ , \tag{2.23}$$

multiplying by $\omega^{nk}$ and summing over k , it is easily seen that

$$\frac{\overline{W}^{(f)}_{pq}(n)}{\overline{W}^{(f)}_{pq}(o)} = \frac{(c_p b_q,\ a_p d_q)_{o,n}}{(b_p c_q,\ d_p a_q)_{o,n}} \ . \tag{2.24}$$

We shall find it convenient to use a Fourier transformed form of the star-triangle relation. Define

$$V_{pqr}(a, c; n) = N^{-1} \sum_{e=0}^{N-1} \omega^{ne}\, W_{pq}(a-e)\, \overline{W}_{pr}(e-c) \ . \tag{2.25}$$

Multiplying (2.18) by $\omega^{nb}$ , summing over b and interchanging sides, we obtain

$$W_{qr}(a - c)\, V_{pqr}(a, c; n) = R_{pqr}^{-1}\, V_{prq}(a, c; n)\, \overline{W}_{qr}^{(f)}(n) \ . \tag{2.26}$$

We shall make great use of this equation: it is helpful to give it a pictorial representation, as in Fig. 2.

**Commuting Transfer Matrices**

The reason the star-triangle relation is so important is that it ensures various commutation relations between transfer matrices. To see this, note from (2.15) that

$$(T_q \hat{T}_r)_{\sigma\sigma''} = \sum_{\sigma'} (T_q)_{\sigma\sigma'} (\hat{T}_r)_{\sigma'\sigma''} = \prod_{J=1}^{L} U_{qr}(\sigma_J, \sigma_{J+1}, \sigma''_{J+1}, \sigma''_J) \tag{2.27}$$

where, for all values $(0,\ldots,N-1)$ of the spins $a$, $b$, $c$, $d$:

$$U_{qr}(a, b, c, d) = \sum_{e=0}^{N-1} W_{pq}(a - e)\overline{W}_{p'q}(b - e)\, W_{p'r}(e - c)\overline{W}_{pr}(e - d) \ . \tag{2.28}$$

This $U_{qr}(a, b, c, d)$ depends implicitly on the vertical rapidities $p$, $p'$. It is the weight of the four-pointed star shown in Fig. 3. As indicated, we can write it, using (2.25), as

$$U_{qr}(a, b, c, d) = N \sum_{n=0}^{N-1} V_{pqr}(a, d; n)\, V_{p'rq}(-c, -b; n) \ . \tag{2.29}$$

Twice using (2.26), it follows that

$$\begin{aligned} W_{qr}(a - d)\, U_{qr}(a, b, c, d) &= N R_{pqr}^{-1} \sum_{n=0}^{N-1} V_{prq}(a, d; n)\, \overline{W}_{qr}^{(f)}(n)\, V_{p'rq}(-c, -b; n) \\ &= N(R_{p'qr}/R_{pqr}) \sum_{n=0}^{N-1} V_{prq}(a, d; n) V_{p'qr}(-c, -b; n) W_{qr}(b - c) \\ &= (R_{p'qr}/R_{pqr})\, U_{rq}(a, b, c, d)\, W_{qr}(b - c) \ . \end{aligned} \tag{2.30}$$

Provided $W_{qr}(n)$ is non-zero for all $n$, we can use this equation to express $U_{qr}$ as proportional to $U_{rq}$. Substituting the result into (2.27), the terms involving $W_{qr}$ cancel in pairs $W_{qr}(\sigma_J - \sigma''_J)$, $[W_{qr}(\sigma_J - \sigma''_J)]^{-1}$, leaving

$$T_q \hat{T}_r = (R_{p'qr}/R_{pqr})^L \, T_r \hat{T}_q \,, \tag{2.31}$$

or, using (2.19),

$$T_q \hat{T}_r = (f_{p'q} f_{pr}/f_{pq} f_{p'r})^L \, T_r \hat{T}_q \,. \tag{2.32a}$$

This is a commutation relation. There is a corresponding relation, obtainable at once by interchanging $T$ with $\hat{T}$, $p$ with $p'$:

$$\hat{T}_q T_r = (f_{pq} f_{p'r}/f_{p'q} f_{pr})^L \, \hat{T}_r T_q \,. \tag{2.32b}$$

The scalar f-factors are irritating but harmless. (They can be absorbed into $T$ and $\hat{T}$.) The key point is that these relations enable one to permute $q_1, \ldots, q_M$ in (2.17) in any order. To within the scalar f-factors, $Z$ is a symmetric function of $q_1, \ldots, q_M$.

In particular, for arbitrary rapidities $q$, $r$, $q'$, $r'$,

$$T_q \hat{T}_r T_{q'} \hat{T}_{r'} = T_{q'} \hat{T}_{r'} T_q \hat{T}_r \,, \tag{2.33}$$

so $T_q \hat{T}_r$, $T_{q'} \hat{T}_{r'}$ strictly commute. Provided these matrices are appropriately diagonalizable (certainly if their eigenvalues are distinct), it follows that there must exist two invertible matrices $P_B$, $P_W$ ($B$ for "black" sites, $W$ for "white") that are *independent of* $q$, such that

$$P_W^{-1} T_q P_B = \text{diagonal} \,, \tag{2.34}$$

$$P_B^{-1} \hat{T}_q P_W = \text{diagonal} \,.$$

The derivation of (2.32) breaks down if $W_{qr}(n)$ vanishes for some $n$. However, this only happens at isolated points and $T_q$, $\hat{T}_q$ are analytic functions of $q$, so by continuity (2.32) must be true even then. These points are of particular interest, and will be discussed in the next section.

**Automorphisms**

There are various automorphisms that leave the relations (2.4) unchanged [8-9], notably:

$$\begin{aligned}
R:&\quad a_{Rq}, b_{Rq}, c_{Rq}, d_{Rq} = b_q, \omega a_q, d_q, c_q \\
T:&\quad a_{Tq}, b_{Tq}, c_{Tq}, d_{Tq} = \omega a_q, b_q, \omega c_q, d_q \\
U:&\quad a_{Uq}, b_{Uq}, c_{Uq}, d_{Uq} = \omega a_q, b_q, c_q, d_q
\end{aligned} \tag{2.35}$$

We shall use the notations

$$q(k,\ell) = U^{k-\ell}R^{2\ell}q \ , \quad \bar{q}(k,\ell) = U^{k-\ell+1}R^{2\ell-1}q \ , \tag{2.36}$$

and where the meaning is obvious write $q(k,\ell)$ as $qk\ell$, $\bar{q}(k,\ell)$ as $\bar{q}k\ell$. Thus

$$\begin{aligned}
a_{qk\ell}, b_{qk\ell}, c_{qk\ell}, d_{qk\ell} &= \omega^k a_q, \omega^\ell b_q, c_q, d_q \ , \\
a_{\bar{q}k\ell}, b_{\bar{q}k\ell}, c_{\bar{q}k\ell}, d_{\bar{q}k\ell} &= \omega^k b_q, \omega^\ell a_q, d_q, c_q \ .
\end{aligned} \tag{2.37}$$

From (2.6) it follows that

$$\begin{aligned}
x_{qk\ell}, y_{qk\ell}, \mu_{qk\ell} &= \omega^k x_q, \omega^\ell y_q, \mu_q \ , \\
x_{\bar{q}k\ell}, y_{\bar{q}k\ell}, \mu_{\bar{q}k\ell} &= \omega^k y_q, \omega^\ell x_q, \mu_q^{-1} \ , \\
t_{qk\ell} = t_{\bar{q}k\ell} &= \omega^{k+\ell} t_q \ .
\end{aligned} \tag{2.38}$$

It is apparent from (2.13) and (2.15) that $T_q$ and $\hat{T}_q$ are single-valued functions of $x_q$, $y_q$ and $\mu_q$. Further, $\mu_q$ cancels out of (2.15), apart possibly from integer powers of $\mu_q^N$ that may arise through interpreting the arguments of $W$, $\overline{W}$ to modulo $N$. From (2.7), $\mu_q^N$ is a single-valued function of $x_q$ and $y_q$. Hence so are $T_q$ and $\hat{T}_q$. They are therefore invariant under any automorphism that does not change $x_q$, $y_q$; in particular $U^{-2}R^2T$. Thus

$$T_{T^{-1}U^2q} = T_{R^2q} \quad , \quad \hat{T}_{T^{-1}U^2q} = \hat{T}_{R^2q} \; . \tag{2.39}$$

The automorphism T takes $(x_q, y_q)$ to $(\omega x_q, \omega^{-1} y_q)$, leaving $t_q$ unchanged. This is equivalent to replacing the suffixes 0, n in (2.13) by 1, n+1 (and multiplying $\mu_q$ by $\omega$, which has no effect on $T_q$). It follows that

$$T_{Tq} = \left[ \frac{\xi_{Tq}(y_p - \omega\, x_q)(y_{p'} - \omega^{-1} y_q)}{\xi_q \mu_p \mu_{p'} (\omega\, x_p - y_q)(x_{p'} - x_q)} \right]^L X^{-1} T_q \; , \tag{2.40}$$

where

$$\begin{aligned} \xi_q &= W_{pq}(0)\, \overline{W}_{p'q}(0) \; , \\ \hat{\xi}_q &= W_{p'q}(0)\, \overline{W}_{pq}(0) \; , \end{aligned} \tag{2.41}$$

and X is the spin-shift operator:

$$X_{\sigma\sigma'} = \prod_{J=1}^{L} \delta(\sigma_J, \sigma'_J + 1) \; , \tag{2.42}$$

taking $\delta(a, b) = 1$ if $a = b$, modulo N ; else $\delta(a, b) = 0$. A corresponding relation for $\hat{T}$ can be obtained from (2.40) by interchanging p with p′ and replacing T, $\xi$ by $\hat{T}$, $\hat{\xi}$. One also has the elementary commutation relations

$$X T_q = T_q X \quad , \quad X \hat{T}_q = \hat{T}_q X \; . \tag{2.43}$$

**The Function $f_{pq}$**

The function $f_{pq}$ enters the commutation relations (2.32) and plays an important role in the following equations. It has been evaluated [16], using the identity

$$\begin{aligned} \det{}_N[\overline{W}_{pq}(i-j)] &= [\overline{W}_{pq}(0)]^N N^{N/2} e^{i\pi(N-1)(N-2)/12} \\ &\times \prod_{j=1}^{N-1} \frac{(t_p - \omega^j t_q)^j}{(x_p - \omega^j x_q)^j (y_p - \omega^j y_q)^j} \; . \end{aligned} \tag{2.44}$$

It is convenient to define

$$w_{pq} = \overline{W}_{pq}(0)/W_{pq}(0) . \tag{2.45}$$

(In this paper we make no specific normalization of the weight functions, so $W_{pq}(0)$, $\overline{W}_{pq}(0)$ are arbitrary functions of $p$ and $q$.) From (2.6) - (2.8) we can deduce the elementary identities:

$$k'(t_p^N - t_q^N) = \mu_q^N (x_p^N - x_q^N)(y_p^N - x_q^N) = \mu_q^{-N}(x_p^N - y_q^N)(y_p^N - y_q^N) . \tag{2.46}$$

Using these, from (2.6) - (2.13), (2.20) and (2.44) one can obtain the following relations satisfied by $f_{pq}$:

$$f_{p,Rq} f_{qp} = N\, k'^{(1-N)/N} w_{p,Rq} w_{qp} , \tag{2.47}$$

$$f_{pq} f_{qp} = N\, w_{pq} w_{qp} \frac{(x_p - x_q)(y_p - y_q)(t_p^N - t_q^N)}{(x_p^N - x_q^N)(y_p^N - y_q^N)(t_p - t_q)} , \tag{2.48}$$

$$\frac{f_{p,\bar{q}k\ell}}{f_{pq}} = \frac{w_{p,\bar{q}k\ell}\, k'^{(j-N)/N} \mu_q^{k-\ell+1} (x_p,x_q)_{\ell-1,N-1} (y_p,y_q)_{k,N-1}}{w_{pq} (t_p,t_q)_{j-1,N-1} (x_p,y_q)_{-1,k-1} (y_p,x_q)_{0,\ell-1}} , \tag{2.49}$$

where $j = k+\ell$. These are true for all rapidities $p$, $q$ and all integers $k$, $\ell$. Note that $k'$ is still the real parameter entering (2.4) - (2.8), but $k$ has changed its meaning.

The power of $1/N$ in (2.20) means that (2.47) - (2.49) are only necessarily true to within factors that are signs or integer powers of $\omega$. In (2.49) these powers may depend on $k$ and $\ell$, but in all three equations they must be independent (being integers) of the continuous variables $p$ and $q$. Since $f_{pq}$ only enters (2.32) via the ratio $f_{pq}/f_{p'q}$, such factors cancel out of this ratio and we can ignore them.

Taking ratios of (2.49) with $(k,\ell) = (0,0), (1,0), (0,1)$, we obtain

$$\frac{f_{p,Uq}}{f_{pq}} = \frac{w_{p,Uq}\, k'^{1/N}\, \mu_q^{-1}\,(t_p - t_q)}{w_{pq}\,(x_p - x_q)\,(y_p - \omega x_q)} \tag{2.50a}$$

$$\frac{f_{p,R^2q}}{f_{p,Uq}} = \frac{w_{p,R^2q}\, k'^{1/N}\, \mu_q\,(t_p - \omega t_q)}{w_{p,Uq}(x_p - y_q)\,(y_p - y_q)} \tag{2.50b}$$

## 3. FIRST HIERARCHY OF FUNCTIONAL RELATIONS

### Simplest case

Here we deliberately choose $r$ so that $W_{qr}(n)$ vanishes for some $n$. The simplest case is $r = Rq = \bar{q}01$. Then from (2.35), (2.12) and (2.24)

$$\frac{W_{q,Rq}(n)}{W_{q,Rq}(0)} = \begin{bmatrix} a_q d_q \\ b_q c_q \end{bmatrix}^n \frac{(\omega,1)_{0,n}}{(1,1)_{0,n}}$$

$$\frac{\overline{W}^{(f)}_{q,Rq}(n)}{\overline{W}^{(f)}_{q,Rq}(0)} = \begin{bmatrix} a_q c_q \\ b_q d_q \end{bmatrix}^n \frac{(\omega,1)_{0,n}}{(1,1)_{0,n}} . \tag{3.1}$$

The RHS of both equations is zero for $n = 1,\ldots,N-1$, so

$$\frac{W_{q,Rq}(n)}{W_{q,Rq}(0)} = \frac{\overline{W}^{(f)}_{q,Rq}(n)}{\overline{W}^{(f)}_{q,Rq}(0)} = \delta_{n,0} , \tag{3.2}$$

modulo $N$. Let us choose $W_{q,Rq}(0)$, $\overline{W}^{(f)}_{q,Rq}(0)$ to be finite and non-zero.

Now consider the relation (2.26), with $r = Rq$. Taking $a = c$ and $n = 0$ we observe that $R_{pqr}$ is finite and non-zero (the RHS of (2.19) is indeterminate). Taking $n \neq 0$, or $a \neq c \pmod N$, we deduce that

$$V_{p,q,Rq}(a,c;\, n) = 0 \text{ if } a = c \text{ and } n \neq 0 ,$$

$$V_{p,Rq,q}(a,c;\, n) = 0 \text{ if } n = 0 \text{ and } a \neq c . \tag{3.3}$$

Now use these results, true for all $p$, $q$, in (2.29). If $a = d$ the only non-zero term in the sum is $n = 0$, and even this vanishes if $b \neq c$, so

$$U_{q,Rq}(a,b,c,d) = 0 \text{ if } a = d \text{ and } b \neq c \ .$$

Thus if $\sigma_J = \sigma''_J$, for any $J$, the RHS of (2.27) vanishes unless $\sigma_{J+1} = \sigma''_{J+1}$, $\sigma_{J+2} = \sigma''_{J+2}$, etc. It follows that

$$T_q \hat{T}_{Rq} = \mathcal{T}^{(0,1)}_q + \bar{\mathcal{T}}^{(0,1)}_q \ , \tag{3.4}$$

where

$$\begin{aligned} &[\mathcal{T}^{(0,1)}_q]_{\sigma\sigma''} = 0 \text{ unless } \sigma_J = \sigma''_J \ , \forall J \ , \\ &[\bar{\mathcal{T}}^{(0,1)}_q]_{\sigma\sigma''} = 0 \text{ unless } \sigma_J \neq \sigma''_J \ , \forall J \ . \end{aligned} \tag{3.5}$$

From (2.29),

$$U_{q,Rq}(a,b,b,a) = N\, V_{p,q,Rq}(a,a;\,0)\, V_{p',Rq,q}(-b,-b;\,0) \ , \tag{3.6}$$

while from (2.12b) and (2.12a)

$$W_{pq}(n)\, \overline{W}_{p,Rq}(-n) = W_{pq}(0)\, \overline{W}_{p,Rq}(0) \ , \tag{3.7}$$

$$\begin{aligned} \frac{W_{p,Rq}(-n)\ \overline{W}_{pq}(n)}{W_{p,Rq}(0)\ \overline{W}_{pq}(0)} &= \frac{(c_p b_q,\ \omega^{-1} b_p c_q)_{0,n}}{(a_p d_q,\ d_p a_q)_{0,n}} \times \frac{(\omega a_p d_q,\ d_p a_q)_{0,n}}{(c_p b_q,\ b_p c_q)_{0,n}} \\ &= \frac{(y_q - y_p)(x_p - x_q)\omega^n}{(y_q - \omega^n y_p)(x_p - \omega^n x_q)} \\ &= \frac{(x_p - x_q)(y_p - y_q)}{x_p y_p - x_q y_q} \left\{ \frac{x_p}{x_p - \omega^n x_q} - \frac{y_q}{y_q - \omega^n y_p} \right\} . \end{aligned} \tag{3.8}$$

From (2.25), it follows that $V_{p,q,Rq}(a,a;\,0) = W_{pq}(0)\, \overline{W}_{p,Rq}(0)$,

$$V_{p,Rq,q}(-b,-b;0) = W_{p,Rq}(0)\, \overline{W}_{pq}(0) \frac{(x_p - x_q)(y_p - y_q)}{x_p y_p - x_q y_q}$$

(3.9)

$$\times \left\{ \frac{x_p^N}{x_p^N - x_q^N} - \frac{y_q^N}{y_q^N - y_p^N} \right\} .$$

Thus

$$U_{q,Rq}(a,b,b,a) = S_{p,p',q} , \tag{3.10}$$

where $S_{p,p',q}$ is independent of a and b:

$$S_{p,p',q} = N\, W_{pq}(0)\, \overline{W}_{p,Rq}(0)\, W_{p',Rq}(0)\, \overline{W}_{p'q}(0) \tag{3.11}$$

$$\times \frac{(x_{p'}-x_q)(y_{p'}-y_q)(t_{p'}^N-t_q^N)}{(x_{p'}^N-x_q^N)(y_{p'}^N-y_q^N)(t_{p'}-t_q)} .$$

The elements of $\mathcal{T}_q^{(0,1)}$ are given by (2.27), with the restrictions $\sigma_J = \sigma_J''$, $\forall J$. From (3.10) it follows at once that

$$\mathcal{T}_q^{(0,1)} = S_{p,p',q}^L\, 1 , \tag{3.12}$$

1 being the $N^L$ by $N^L$ identity matrix. We shall not discuss $\bar{\mathcal{T}}_q^{(0,1)}$, but move on to discuss the general situation.

**General case**

Now we consider the product $T_q \hat{T}_r$ when

$$r = \bar{q}(k,\ell) \; , \; 1 \le k + \ell \le N . \tag{3.13}$$

The argument is a generalization of (3.1) - (3.12). From (2.37), (2.12) and (2.24),

(3.14)

$$\frac{W_{qr}(n)}{W_{qr}(0)} = \left[\frac{a_q d_q}{b_q c_q}\right]^n \frac{(\omega^\ell,1)_{0,n}}{(1,\omega^k)_{0,n}} ,$$

$$\frac{\overline{W}_{qr}^{(f)}(n)}{\overline{W}_{qr}^{(f)}(0)} = \left[\frac{a_q c_q}{b_q d_q}\right]^n \frac{(\omega^\ell,1)_{0,n}}{(1,\omega^k)_{0,n}} .$$

Let $\zeta_{k\ell}$ be the set of consecutive integers $\{-k, 1-k, 2-k, \ldots, \ell-1\}$, with $k+\ell$ elements. We say that $n \in \zeta_{k\ell}$ if $n$ is equal, modulo $N$, to one of the elements of $\zeta_{k\ell}$. If $0 \in \zeta_{k\ell}$, it follows at once from (3.14) that we can normalize $W_{qr}(n)$, $\overline{W}^{(f)}_{qr}(n)$ so that they are finite and non-zero for $n \in \zeta_{k,\ell}$, zero otherwise. Thus

$$W_{qr}(n) = \overline{W}^{(f)}_{qr}(n) = 0 \ , \ n \notin \zeta_{k\ell} \ . \tag{3.15}$$

This is still true even if $0 \notin \zeta_{k\ell}$, but then we should first express the general formulae (2.12), (2.24) in terms of ratios with some element other than $n = 0$, say with $n = -k$, before taking $r = \bar{q}(k, \ell)$.

From (2.26),

$$\begin{aligned} &V_{pqr}(a,c;\, n) = 0 \ \text{ if } \ a-c \in \zeta_{k\ell} \ \text{ and } \ n \notin \zeta_{k\ell} \ , \\ &V_{prq}(a,c;\, n) = 0 \ \text{ if } \ n \in \zeta_{k\ell} \ \text{ and } \ a-c \notin \zeta_{k\ell} \ . \end{aligned} \tag{3.16}$$

Substituting into (2.29):

$$U_{qr}(a,b,c,d) = 0 \ \text{ if } \ a-d \in \zeta_{k\ell} \ \text{ and } \ b-c \notin \zeta_{k\ell} \ . \tag{3.17}$$

Thus the RHS of (2.27) vanishes unless either all of $\sigma_J - \sigma''_J \in \zeta_{k\ell}$ (modulo $N$), or none of them do so. An "all or nothing" property.

It follows that

$$T_q \hat{T}_r = \mathcal{T}^{(k,\ell)}_q + \overline{\mathcal{T}}^{(k,\ell)}_q \ , \tag{3.18}$$

where

$$\begin{aligned} &[\mathcal{T}^{(k,\ell)}_q]_{\sigma\sigma''} = 0 \ \text{ unless } \ \sigma_J - \sigma''_J \in \zeta_{k\ell} \ , \forall J \ , \\ &[\overline{\mathcal{T}}^{(k,\ell)}_q]_{\sigma\sigma''} = 0 \ \text{ unless } \ \sigma_J - \sigma''_J \notin \zeta_{k\ell} \ , \forall J \ . \end{aligned} \tag{3.19}$$

**The star-weight $U_{qr}(a,b,c,d)$**

We still have to evaluate the elements of $\mathcal{T}_q^{(k,\ell)}$, $\overline{\mathcal{T}}_q^{(k,\ell)}$. Both have elements given by the RHS of (2.27), where the star-weight $U_{qr}(a,b,c,d)$ is given by (2.29). The difference is that for $\mathcal{T}_q^{(k,\ell)}$ the elements are zero unless

(3.20a) $$a-d\,,\ n\,,\ b-c \in \zeta_{k\ell}\ ,$$

while for $\overline{\mathcal{T}}_q^{(k,\ell)}$ they are zero unless

(3.20b) $$a-d\,,\ n\,,\ b-c \notin \zeta_{k\ell}\ .$$

To simplify the notation we shall introduce integers $j$, $\alpha$, $m$, $\beta$, related to $k$, $\ell$, $a$, $b$, $c$, $d$, $n$ by

(3.21) $$j = k+\ell,\ \alpha = a-d+k,\ m = n+k,\ \beta = b-c+k\ .$$

thus

(3.22) $$1 \le j \le N\ ,$$

and the elements of $\mathcal{T}_q^{(k,\ell)}$ are non-zero when

(3.23) $$0 \le \alpha, m, \beta < j\ .$$

Let us define

(3.24)
$$\Omega_{pq}^{k\ell} = \frac{y_p^{j-1}(\mu_q/\mu_p)^k\, W_{pq}(0)\, \overline{W}_{pr}(0)}{(y_p,x_q)_{0,\ell-1}\,(y_q,x_p)_{-k,0}}$$

$$\overline{\Omega}_{pq}^{k\ell} = \frac{\omega^{-jk}\,(\mu_p\mu_q)^{\ell}\, y_p^{N-1-j}\, W_{pr}(0)\, \overline{W}_{pq}(0)}{(y_p,\, y_q)_{k,N-1}\,(x_q,x_p)_{0,-\ell}}\ .$$

Then, from (2.37) and (2.10) - (2.12),

$$W_{pq}(a-e)\,\overline{W}_{pr}(e-d) = \Omega^{k\ell}_{pq}\,\omega^{k(e-d)} \left[\frac{d_p b_q}{b_p d_q}\right]^{\alpha}$$

(3.25a)

$$\times\,(1,\,\omega^{a-e}x_q/y_p)_{0,j-\alpha-1}\,(1,\,\omega^{d-e-k}x_p/y_q)_{0,\alpha}\,,$$

$$W_{pr}(e-c)\,\overline{W}_{pq}(b-e) = \frac{\overline{\Omega}^{k\ell}_{pq}\,(-b_p c_q)^j\,y_q^{N-1}\,(a_p d_q/c_p b_q)^{\beta}}{\omega^{j(1-j)/2}\,(1-x_q^N/x_p^N)\,(a_p d_q)^j y_p^{N-1}}$$

(3.25b)

$$\times\,\omega^{b+kc-(k+1)e}\,(1,\,\omega^{b-e}y_p/y_q)_{0,N-\beta-1}\,(1,\,\omega^{c-e+\ell-1}x_q/x_p)_{0,N+\beta-j}\,.$$

We have cast these equations in a form that is particularly suitable for the case (3.23), but they are in fact true for all integers a, b, c, d, n .

For any non-negative integers $m, m'$, $i$, with $0 \le i \le m + m'$, we can define a function $\Phi(x,y)_i^{m,m'}$ by the polynomial identity

(3.26)
$$(1\,,\,xu)_{0,m}\,(1\,,\,yu)_{0,m'} = \sum_{i=0}^{m+m'} u^i\,\Phi(x,y)_i^{m,m'}\,,$$

true for all complex numbers u, x, y . (The LHS is a polynomial of degree $m + m'$ in u , the $\Phi(x,y)_i^{m,m'}$ are its coefficients.) These functions $\Phi$ are special cases of the basic hypergeometric function [29] and are discussed in Appendix A. For the moment we need only note that they are homogeneous in x, y , of degree i, i.e.

(3.27)
$$\Phi(vx,vy)_i^{m,m'} = v^i\Phi(x,y)_i^{m,m'}$$

for all $v \in \mathbb{C}$ .

When (3.23) is satisfied, the indices $j - \alpha - 1$, $\alpha$, $N - \beta - 1$, $N + \beta - j$ in (3.25) are all non-negative (and less than N) , so we can expand each RHS in powers of $\omega^{-e}$ by using (3.26). For (3.25a), the RHS is a Laurent polynomial in $\omega^{-e}$, being a linear combination of the j terms

$$\omega^{ke}\,,\,\omega^{(k-1)e}\,,\ldots,\,\omega^{-(\ell-1)e}\,,$$

so from (2.25) we see that $V_{pqr}(a,d;\, n)$ vanishes unless $n \in \zeta_{k\ell}$, in agreement with (3.16), otherwise it is the coefficient of $\omega^{-ne}$ in the expansion.

The RHS of (3.25b) is a linear combination of $\omega^{-(k+1)e}$, $\omega^{-(k+2)e}$, ..., $\omega^{-(2N-\ell)e}$. Provided (3.23) is satisfied, it follows from (2.25), with e negated, that $V_{prq}(-c,-b;n)$ is the coefficient of the $\omega^{(n-N)e}$ term only.

Thus we obtain, provided the restrictions (3.23) are satisfied,

$$V_{pqr}(a,d;\, n) = \Omega_{pq}^{k\ell} \left[ \frac{d_p b_q}{b_p d_q} \right]^{\alpha} \omega^{nd-mk} \times (y_p y_q)^{-m}\, \Phi(t_p,\, \omega^{\alpha} t_q)_m^{\alpha, j-\alpha-1} \,, \tag{3.28a}$$

$$V_{prq}(-c,-b;\, n) = \frac{\overline{\Omega}_{pq}^{k\ell}\, (-b_p c_q)^j\, (a_p d_q / c_p b_q)^{\beta}}{\omega^{j(1-j)/2}\, (x_p^N - x_q^N)\, (a_p d_q)^j\, y_p^{N-1}} \times \omega^{kc-mb}\, x_p^{m+1}\, y_q^m\, \Phi(t_p,\, \omega^{j-\beta-1} t_q)_{N-m-1}^{N-\beta-1,\, N+\beta-j} \,, \tag{3.28b}$$

using the definition (2.6) of $t_p$.

To obtain the elements of $\mathcal{T}_q^{(k,\ell)}$, we must consider the case when instead of (3.23) we have

$$j \le \alpha,\, m,\, \beta < N \,, \tag{3.29}$$

which imply that $a-d$, $n$, $b-c \notin \zeta_{k\ell}$. In this case we replace the factor $(1,\, \omega^{a-e} x_q/y_p)_{0,j-\alpha-1}$ in (3.25a) by the equivalent expression

$$(1,\, \omega^{a-e} x_q/y_p)_{o,N+j-\alpha-1} / (1 - x_q^N / y_p^N) \,. \tag{3.30a}$$

Similarly, we replace the last factor in (3.25b) by

$$(1,\, \omega^{c-e+\ell-1} x_q/x_p)_{o,\beta-j}\, (1 - x_q^N / x_p^N) \,. \tag{3.30b}$$

The indices $N + j - \alpha - 1$, $\alpha$, $N - \beta - 1$, $\beta - j$ that then occur are non-negative (and less than $N$), so we can use the identity (3.26). The RHS of (3.25a) becomes a Laurent polynomial in $\omega^{-e}$, with terms $\omega^{ke}, \omega^{(k-1)e}, \ldots, \omega^{-(N+\ell-1)e}$, while the RHS of (3.25b) has terms $\omega^{-(k+1)e}, \omega^{-(k+2)e}, \ldots, \omega^{-(N-\ell)e}$. From (2.25), $V_{pqr}(a,d;n)$ is therefore the coefficient of $\omega^{-ne}$ in (3.25a). Also, negating $e$ in (2.25), $V_{prq}(-c,-b;n)$ is the coefficient of $\omega^{(n-N)e}$ in (3.25b): it vanishes if we violate (3.29) and take $n \in \zeta_{k\ell}$, in agreement with the second of the relations (3.16).

Hence when the restrictions (3.29) are satisfied

(3.31a)
$$V_{pqr}(a,d;n) = \frac{\Omega^{k\ell}_{pq}}{1-x_q^N/y_p^N} \left[\frac{d_p b_q}{b_p d_q}\right]^{\alpha} \omega^{nd-mk}$$
$$\times\ (y_p y_q)^{-m}\, \Phi(t_p, \omega^{\alpha} t_q)_m^{\alpha,N+j-\alpha-1} ,$$

(3.31b)
$$V_{prq}(-c,-b;n) = \overline{\Omega}^{k\ell}_{pq} \left[\frac{a_p d_q}{c_p b_q}\right]^{\beta} \omega^{kc-mb+j(j-1)/2}$$
$$\times \left[\frac{-b_p c_q}{a_p d_q}\right]^{j} t_p^{1-N} (x_p y_q)^m\, \Phi(t_p, \omega^{j-\beta-1} t_q)_{N-m-1}^{N-\beta-1,\beta-j} .$$

**Alternative forms for $V_{pqr}$, $V_{prq}$**

If we take $j = 1$, $\alpha = m = \beta = 0$, satisfying (3.23), the function $\Phi$ in (3.28a) is simply unity, but in (3.28b) is $\Phi(t_p,t_q)_{N-1}^{N-1,N-1}$. This is a polynomial of degree $N-1$ in $t_p$ or $t_q$, but a very simple one:

(3.32)
$$\Phi(t_p,t_q)_{N-1}^{N-1,N-1} = (t_p^N - t_q^N)/(t_p - t_q) .$$

Similar simplifications occur for $j = 2, 3$, etc. They in fact follow (as does so much of this paper) from the star-triangle relation (2.26). Substituting into this the expressions (3.14), (3.28) for $W_{qr}$, $\overline{W}^{(f)}_{qr}$, $V_{pqr}$, $V_{prq}$, after some rearrangement we obtain

$$\Phi(t_p, \omega^{j-\beta-1} t_q)^{N-\beta-1,\, N+\beta-j}_{N-m-1} = \gamma^{(j)}_{pq}\, t_p^{-\beta-m}\, t_q^{\beta-m}\, \omega^{j(\beta-m)} \tag{3.33}$$

$$\times \frac{(1,1)_{\beta,\beta+N-j}}{(1,1)_{m,m+N-j}}\, \Phi(t_p, \omega^{\beta} t_q)^{\beta, j-\beta-1}_{m} ,$$

for all integers m, $\beta$, j satisfying (3.22) and (3.23). The factor $\gamma^{(j)}_{pq}$ includes the $R_{pqr}$ in (2.26), and is independent of $\beta$, m.

We can evaluate $\gamma^{(j)}_{pq}$ by taking $m = \beta = 0$, giving

$$\begin{aligned} \gamma^{(j)}_{pq} &= \Phi(t_p, \omega^{j-1} t_q)^{N-1,N-j}_{N-1} \\ &= t_p^{j-1} (t_p, t_q)_{j-1,N-1} , \end{aligned} \tag{3.34}$$

using the formula (A.5) of Appendix A.

We can use the identity (3.33) to simplify (3.28b) and (3.31a). Define the following functions:

$$\bar{h}^{(j)}_{pq} = \frac{\omega^{j(j-1)/2}\, t_p^{j-N}\, (t_p, t_q)_{j-1,N-1}}{(1-x_q^N/x_p^N)\ (-a_p d_q/b_p c_q)^j} , \tag{3.35}$$

$$h^{(j)}_{pq} = \frac{\omega^{j(j+1)/2}\, y_p^{-j}\, (t_p,\ t_q)_{-1,\ j-1}}{(1-x_q^N/y_p^N)\ (-b_p d_q/d_p c_q)^j} , \tag{3.36}$$

$$\eta_{q,j,\alpha} = \omega^{j\alpha}\, t_q^{\alpha}\, (1,1)_{\alpha,\alpha+N-j} , \tag{3.37}$$

$$F_{pq}(j,\alpha,m) = \left[\frac{d_p}{b_p}\right]^{\alpha} \left[\frac{c_p}{b_p}\right]^{m} \Phi(t_p, \omega^{\alpha} t_q)^{\alpha, j-\alpha-1}_{m} \tag{3.38}$$

for $0 \le \alpha,\ m < j \le N$ . Also apply to both equations (3.31) the elementary identity (A.7). Then (3.28) and (3.31) become:

$$V_{pqr}(a,d;\ n) = \Omega^{k\ell}_{pq}\, \omega^{nd-mk}\, (b_q/d_q)^{\alpha}\, y_q^{-m}\, F_{pq}(j,\alpha,m) , \tag{3.39a}$$

(3.39b) $V_{prq}(-c,-b;\, n) = \overline{\Omega}^{k\ell}_{pq}\, \overline{h}^{(j)}_{pq}\, \omega^{kc-mb} \left[\frac{d_q}{b_q}\right]^{\beta} \frac{\eta_{q,j,\beta}}{\eta_{q,j,m}}\, y_q^m\, F_{pq}(j,\beta,m)$ ,

(3.40a)
$$V_{pqr}(a,d;\, n) = \Omega^{k\ell}_{pq}\, h^{(j)}_{pq}\, \omega^{nd+\ell m-j^2} \left[\frac{c_q}{a_q}\right]^{\alpha-j} x_q^{m-j} \times \frac{\eta_{r,N-j,\alpha-j}}{\eta_{r,N-j,m-j}}\, F_{pr}(N-j,\, \alpha-j,\, m-j) \;,$$

(3.40b) $V_{prq}(-c,\, -b;\, n) = \overline{\Omega}^{k\ell}_{pq}\, \omega^{j\beta-mb+kc} \left[\frac{a_q}{c_q}\right]^{\beta-j} x_q^{j-m}\, F_{pr}(N-j,\, \beta-j,\, m-j)$ .

The equations (3.39) apply when the restrictions (3.23) are satisfied; (3.40) when (3.29) are satisfied. The functions $\eta$ , F in (3.40) are defined with q replaced by $r = \bar{q}(k,\ell)$: the effect of this is to replace $t_q$ in (3.37) and (3.38) by $t_r = \omega^j t_q$ .

**The Matrices $\mathcal{T}^{(k,\ell)}_q$ , $\overline{\mathcal{T}}^{(k,\ell)}_q$**

The non-zero elements of $\mathcal{T}^{(k,\ell)}_q$ are equal to the RHS of (2.27), where $U_{qr}$ is given by (2.29), the restrictions (3.23) being satisfied. Thus we can use the above results (3.39) for $V_{pqr}$ , $V_{prq}$ . The factors explicitly involving $\omega^{-kd}$ , $\omega^{kc}$ , $b_q/d_q$ and $y_q^m$ cancel out of the product in (2.27). Define the following scalar functions:

(3.41) $$\lambda^{(k,\ell)}_q = [N\, \Omega^{k\ell}_{pq}\, \overline{\Omega}^{k}_{p'q}{}^{\ell}]^{-L} \;,$$

(3.42) $$H^{(j)}_{pq} = [h^{(j)}_{pq}]^L \;, \quad \overline{H}^{(j)}_{pq} = [\overline{h}^{(j)}_{pq}]^L \;.$$

Then we obtain

(3.43) $$\lambda^{(k,\ell)}_q\, \mathcal{T}^{(k,\ell)}_q = \overline{H}^{(j)}_{p'q}\, \tau^{(j)}_{k,q} \;,$$

where the matrix $\tau^{(j)}_{k,q}$ has elements

(3.44a)
$$[\tau^{(j)}_{k,q}]_{\sigma\sigma''} = \sum_{m_1,\ldots,m_L} \prod_{J=1}^{L} [\omega^{-m_J(\sigma_{J+1}-\sigma''_J+k)} \frac{\eta_{q,j,\sigma_J-\sigma''_J+k}}{\eta_{q,j,m_J}} \times F_{pq}(j,\, \sigma_J-\sigma''_J+k,\, m_J)\, F_{p'q}(j,\, \sigma_{J+1}-\sigma''_{J+1}+k,\, m_J)] \;,$$

provided $-k \le \sigma_J - \sigma''_J < j-k$ for $J = 1,\ldots,L$; the sum being over all values $0,\ldots,j-1$ of $m_1,\ldots,m_L$. All other elements are defined by

$$[\tau^{(j)}_{k,q}]_{\sigma\sigma''} = 0 \text{ if } j-k \le \sigma_J - \sigma''_J < N-k \text{ for any } J \;, \tag{3.44b}$$

together with the $Z_N$ invariance that $[\tau^{(j)}_{k,q}]_{\sigma\sigma''}$ is unchanged by incrementing any spin $\sigma_1,\ldots,\sigma''_L$ by $N$.

This $\tau^{(j)}_{k,q}$ is in fact the transfer matrix of a "j by N" model, namely a model defined on the lattice of Fig. 4, with N-state chiral Potts spins $\sigma_i$ on open circles, j-state spins $m_i$ on filled squares, and one-, two-, and three-spin interactions (on sites, edges, and triangles) corresponding to the various factors in (3.44).

Similarly, using (3.40), the second matrix $\bar{\mathcal{T}}^{(k,\ell)}_q$ is given by

$$\lambda^{(k,\ell)}_q \bar{\mathcal{T}}^{(k,\ell)}_q = H^{(j)}_{pq}\, \tau^{(N-j)}_{k-j,r} \;. \tag{3.45}$$

From (3.18), we have therefore obtained the hierarchy of matrix relations

$$\lambda^{(k,\ell)}_q\, T_q\, \hat{T}_{\bar{q}k\ell} = \bar{H}^{(j)}_{p'q}\, \tau^{(j)}_{k,q} + H^{(j)}_{pq}\, \tau^{(N-j)}_{k-j,\bar{q}k\ell} \;, \tag{3.46}$$

true for any rapidity $q$ and all integers $k, \ell, j = k+\ell$ satisfying (3.22), i.e. $1 \le j \le N$. The case $j = N$ is special in that the second term on the RHS of (3.46) does not occur, which is equivalent to defining

$$\tau^{(0)}_{k,q} = 0 \;. \tag{3.47a}$$

With this convention, the hierarchy (3.46) can immediately be extended to include the case $j = 0$.

Another special case is $j = 1$. Then $\tau^{(j)}_{k,q}$ is just a spin k-shift operator:

$$\tau^{(1)}_{k,q} = X^{-k} \;, \tag{3.47b}$$

and for $k = 0$ we can regain (3.12) from (3.43). The first non-trivial $\tau$-matrix occurs for $j = 2$ ; then in (3.42a) one only need the following values (or ratios of values) of the functions $\eta$ and $F$ :

$$\eta_{q,2,1}/\eta_{q,2,0} = -\omega t_q ,$$

$$\text{(3.48)} \qquad F_{pq}(2,0,0) = 1 , \; F_{pq}(2,0,1) = -\omega c_p t_q/b_p ,$$

$$F_{pq}(2,1,0) = d_p/b_p , \; F_{pq}(2,1,1) = -\omega a_p/b_p .$$

**Commutation Properties**

We can apply the staggered similarity transformation (2.34) to diagonalize $T_q$, $\hat{T}_{\bar{q}k\ell}$ in (3.46). We want to show that this diagonalizes the two terms on the RHS individually, so

$$\text{(3.49)} \qquad P_W^{-1} \tau_{k,q}^{(j)} P_W = \text{diagonal} ,$$

for all j, k, q .

We can do this by a simple trick. The functions $\eta$ and F in (3.37) - (3.38) depend on q only via $t_q$. Hence so do all the matrices $\tau_{k,q}^{(j)}$. (In fact they are Laurent polynomials in $t_q$.)

Now replace q in (3.46) by the conjugate value $\bar{q}(0,0) = UR^{-1}q$. From (2.35) - (2.38), this leaves $t_q$ unchanged, but interchanges $x_q$ with $y_q$. The two $\tau$-matrices on the RHS of (3.46) are therefore unchanged. On the other hand, because $x_q^N$ occurs explicitly in (3.35) and (3.36), the ratio of $\overline{H}_{p'q}^{(j)}$ to $H_{pq}^{(j)}$ is changed. Thus the two $\tau$-matrices now occur in a different linear combination. Both these linear combinations C are diagonalized by the q-independent similarity transformation $C \to P_W^{-1} C P_W$, so the individual matrices must be. This proves (3.49). In particular, the matrices satisfy the commutation relations

$$\text{(3.50)} \qquad (\tau_{k,q}^{(j)}, \tau_{k',q'}^{(j')}) = (\tau_{k,q}^{(j)}, T_{q'}\hat{T}_{q''}) = 0$$

for all rapidities q, q′, q″ and all integers j, k, j′, k′ .

**Symmetries**

The parameter k enters the definition (3.44) only in combinations of the form $\sigma_I - \sigma''_J + k$ . Using the definition (2.42) of the spin shift operator X , it follows that

$$X\,\tau^{(j)}_{k,q} = \tau^{(j)}_{k,q}\,X = \tau^{(j)}_{k-1,q} \quad . \tag{3.51}$$

Multiplying (3.46) by X , using (2.43), we can deduce the simple relation

$$\hat{T}_{\bar{q}(k,\ell)} = [\lambda^{(k-1,\ell+1)}_q / \lambda^{(k,\ell)}_q]\, X^{-1}\, \hat{T}_{\bar{q}(k-1,\ell+1)} \quad . \tag{3.52}$$

Indeed, this is a direct consequence of the spin-shift relation (2.40), or rather its analogue with T , $\xi$ replaced by $\hat{T}$ , $\hat{\xi}$ , and p interchanged with p′ .

There is therefore a redundancy in our notation: without loss of generality, we could have restricted our attention to a fixed value of k . For N odd and j even, a natural choice is to take k = (j-N-1)/2 : then (3.44a) applies provided all $\sigma_J - \sigma''_J$ lie between (N+1-j)/2 and (N-1+j)/2 . This range is symmetric, in that if $\sigma_J = \sigma''_J + s$ is an allowed value (mod N), then so is $\sigma_J = \sigma''_J - s$ . This is the choice used (for j=2) in ref. 24.

However, for N and j even, there is no natural symmetric choice of k . Rather than use an artificial convention, we have preferred here to allow k to be arbitrary.

The relations (3.46) possess another symmetry. Replace k, ℓ, j therein by -ℓ, N-k, N-j . Then replace q by $\bar{q}k\ell$ (and hence $\bar{q}k\ell$ by q) . Then use the commutation relation (2.32a). We regain the original form of (3.46) (with the two terms on the RHS interchanged) provided that

$$\left[\frac{f_{pq}\, f_{p'r}}{f_{pr}\, f_{p'q}}\right]^L \frac{\lambda^{(-\ell,N-k)}_r}{\lambda^{(k,\ell)}_q} = \frac{H^{(N-j)}_{pr}}{\overline{H}^{(j)}_{p'q}} = \frac{\overline{H}^{(N-j)}_{p'r}}{H^{(j)}_{pq}} \tag{3.53}$$

for all integers k, $\ell$, j = k + $\ell$, and all rapidities p, p′, q, r satisfying $r = \bar{q}(k,\ell)$.

From (2.49) and the definitions (3.24), (3.35), (3.36), we can establish the relations

$$\Omega_{pr}^{(-\ell,N-k)} = \omega^{k^2+k\ell+\ell^2}\,\bar{\Omega}_{pq}^{k\ell}$$
(3.54)
$$\bar{\Omega}_{pr}^{(-\ell,N-k)} = \omega^{-k^2-k\ell-\ell^2}\,\Omega_{pq}^{k\ell} ,$$

$$h_{pr}^{(N-j)} = \bar{h}_{pq}^{(j)} , \quad \bar{h}_{pr}^{(N-j)} = h_{pq}^{(j)} ,$$
(3.55)
$$h_{pq}^{(j)}\,\bar{h}_{pq}^{(j)} = \omega^{j^2}\mu_q^{N-2j}/k' ,$$

(3.56) $$f_{pr}/f_{pq} = k'^{(j-N)/N}\,\omega^{-k^2}\,\mu_q^{1-j}\,\Omega_{pq}^{k\ell}\,/\,[\bar{\Omega}_{pq}^{k\ell}\,\bar{h}_{pq}^{(j)}] .$$

We can then readily verify (3.53), using (3.41) and (3.42).

## 4. OTHER FUNCTIONAL RELATIONS

### The $\tau^{(2)}$, T relation

Here we first establish a relation between $\tau^{(2)}$ and T, analogous to the T, Q relation of the eight-vertex model (eq. 87 of ref. 13, eq. 10.5.32 of ref.15). Our method is basically that of eqs. (5.1) - (5.26) of ref. 24.

We start in a similar way to the eight-vertex derivation, attempting to parallel (9.8.2) - (9.8.7) of ref. 15. Let Q be an $N^L$-dimensional vector with entries

(4.1) $$Q(\sigma_1,..,\sigma_L) = g_1(\sigma_1)\,g_2(\sigma_2)...g_L(\sigma_L) ,$$

the $g_J(a)$ being some functions, as yet arbitrary, satisfying the $Z_N$ periodicity condition

(4.2) $$g_J(a+N) = g_J(a) .$$

Then from (3.42)

$$[\tau^{(2)}_{k,q}\, Q]_\sigma = \mathrm{Tr}[G_1(\sigma_1)\, G_2(\sigma_2) \quad G_L(\sigma_L)] \;, \tag{4.3}$$

where each $G_J(a)$, for $a \in \mathbb{Z}_N$, is a two-by-two matrix, whose entry $(m, m')$, for $m, m' = 0, 1$, is

$$\begin{aligned}[G_J(a)]_{m,m'} &= \sum_d \omega^{-ma+m'(d-k)} \frac{\eta_{q,2,a-d+k}}{\eta_{q,2,m'}} F_{pq}(2, a-d+k, m') \\ &\times F_{p'q}(2,a-d+k,m)\, g_J(d) \;,\end{aligned} \tag{4.4}$$

the sum being over the allowed values $a+k-1$, $a+k$ of $d$. (To obtain this we replaced $J$ by $J-1$ in the two factors in (3.42a) obtaining $\sigma_{J+1}$, and replaced $m_{J-1}$, $m_J$, $\sigma_J, \sigma_J''$ by $m, m', a, d$.)

Now look for a set of two-by-two invertible matrices $P_1,\dots,P_L$ such that, for $J=1,\dots,L$,

$$G_J(a) = P_J\, H_J(a)\, P^{-1}_{J+1} \;, \tag{4.5}$$

where $H_J(a)$ is triangular:

$$H_J(a) = \begin{bmatrix} g_J'(a) & g_J'''(a) \\ 0 & g_J''(a) \end{bmatrix} . \tag{4.6}$$

If we can do this, then it follows from (4.3) that

$$[\tau^{(2)}_{k,q}\, Q]_\sigma = g_1'(\sigma_1)\dots g_L'(\sigma_L) + g_1''(\sigma_1)\dots g_L''(\sigma_L) \;. \tag{4.7}$$

Let $p_J$ be the first column of $P_J$, then the requirement that $H_J(a)$ be triangular is equivalent, from (4.5) and (4.6), to requiring that

$$G_J(a)\, p_{J+1} = g_J'(a)\, p_J \;. \tag{4.8}$$

Without loss of generality, we can take $p_J$ (strictly $p_J^T$) to be the vector $(1,-r_J)$. Then (4.8) is

(4.9) $$\sum_{m'=0}^{1} [G_J(a)]_{m,m'} (-r_{J+1})^{m'} = g_J'(a)\,(-r_J)^m$$

for a=0,...,N-1 ; m=0,1 . This is the analogue of the "propagation-through-a-vertex" equations of the six- and eight-vertex models (eqs. 9.8.9-9.8.11 and 10.5.1 of ref. 15).

Using (4.4), it is natural to perform the $m'$ summation before the d summation, and to replace d by $a+k-\nu$, where $\nu = 0,1$, thereby breaking (4.9) into two equations:

(4.10a) $$\sum_{m'=0}^{1} \omega^{m'(a-\nu)} (-\omega t_q)^{\nu-m'} F_{pq}(2,\nu,m')\,(-r_{J+1})^{m'} = s_J(a,\nu) ,$$

(4.10b) $$\omega^{-ma} \sum_{\nu=0}^{1} F_{p'q}(2,\nu,m)\, s_J(a,\nu)\, g_J(a+k-\nu) = g_J'(a)(-r_J)^m ,$$

where $s_J(a,\nu)$ is an intermediate variable and we have used the first of the relations (3.46). We can naturally associate $r_{J+1}$ with the $m'$ site in Fig.4; $r_J$, $g_J$, $g'_J$ with the sites m, d, a ; and $s_J(a,\nu)$ with the edge (a,d) .

From (4.10a) and (3.48):

(4.11a) $$\frac{s_J(a,1)}{s_J(a,0)} = -\frac{\omega\, d_p t_q - \omega^a a_p r_{J+1}}{b_p - \omega^a c_p r_{J+1}} ,$$

while from (4.10b), eliminating $g_J'(a)$ between the m = 0,1 equations,

(4.11b) $$\frac{g_J(a+k)}{g_J(a+k-1)} = -\frac{s_J(a,1)}{s_J(a,0)}\,\frac{a_{p'} - \omega^{a-1} d_{p'} r_J}{c_{p'} t_q - \omega^{a-1} b_{p'} r_J} .$$

Multiplying these equations gives an expression for the ratio $g_J(a+k)/g_J(a+k-1)$, for a=0,...,N-1 . The only constraint on this result is that the periodicity condition (4.2) be satisfied, i.e. the product over a=0,...,N-1 of the ratios be unity. This gives at once

$$\frac{a_{p'}^N - d_{p'}^N r_J^N}{c_{p'}^N t_q^N - b_{p'}^N r_J^N} \cdot \frac{d_p^N t_q^N - a_p^N r_{J+1}^N}{b_p^N - c_p^N r_{J+1}^N} = 1 , \tag{4.12}$$

which is a bilinear relation between $r_{J+1}^N$ and $r_J^N$.

Given $r_1^N$, (4.12) determines successively $r_2^N, r_3^N$, etc. Because of the toroidal boundary conditions, we must have $r_{L+1}^N = r_1^N$, so $r_1^N$ is a fixed point of the Lth composition of this mapping. This Lth composition is not (for general p, p′;q) the identity, from which it follows that $r_1^N$ must be a fixed point of the mapping itself, i.e. $r_1^N = r_2^N = ... = r_L^N$.

Setting $r_{J+1}^N = r_J^N = r^N$ in (4.12), we obtain a quadratic equation for $r^N$. Its solutions are

$$r^N = a_q^N / d_q^N \quad \text{or} \quad b_q^N / c_q^N . \tag{4.13}$$

(One can verify those by inspection, using (2.6) and the identities (2.14).)

Choosing the first solution, it follows that we can take (as in (5.10) of ref. 24)

$$r_J = \omega^{1-\beta_{J-1}} a_q / d_q , \tag{4.14}$$

for J = 1,...,L, the $\beta_1,...,\beta_L$ being arbitrary integers. Then the equations (4.11) give

$$\frac{g_J(a+k)}{g_J(a+k-1)} = \frac{d_p b_q - \omega^{a-\beta_J} a_p c_q}{b_p d_q - \omega^{a+1-\beta_J} c_p a_q} \cdot \frac{\omega a_{p'} d_q - \omega^{a+1-\beta_{J-1}} d_{p'} a_q}{c_{p'} b_q - \omega^{a-\beta_{J-1}} b_{p'} c_q} . \tag{4.15}$$

We can now solve for (4.10a) for $s_i(a,\nu)$, and (4.10b) for $g_i'(a)$, obtaining

$$g_J'(a) = \frac{a_{p'}}{b_p}\left[1 - \frac{t_q}{t_{p'}}\right] g_J(a+k) \frac{b_p d_q - \omega^{a-1-\beta_J} c_p a_q}{a_{p'} d_q - \omega^{a-\beta_{J-1}} d_{p'} a_q} . \tag{4.16}$$

From (4.4) and (3.46),

(4.17) $$\det G_J(a) = \frac{\omega a_p a_{p'}}{b_p b_{p'}} g_J(a+k-1)\, g_J(a+k) \left[1 - \frac{t_q}{t_{p'}}\right] \left[1 - \frac{\omega t_q}{t_p}\right] ,$$

so from (4.5) and (4.6), we can choose the $P_J$ so that

(4.18) $$g''_J(a) = \frac{\omega a_p}{b_{p'}} \left[1 - \frac{\omega t_q}{t_p}\right] g_J(a+k-1) \frac{a_{p'} d_q - \omega^{a-\beta_{J-1}} d_{p'} a_q}{b_p d_q - \omega^{a+1-\beta_J} c_p a_q} .$$

Comparing (4.15) with (2.12), we observe a remarkable property: $g_J(a+k)$ is (to within a normalization) just a product of two $W, \overline{W}$ functions. So indeed are $g'_J(a)$ and $g''_J(a)$. We can normalize so that

(4.19a) $$g_J(a+k) = y_p y_{p'} \overline{W}_{p',U_q}(a-\beta_{J-1})\, W_{p,Uq}(a-\beta_J)/\xi_{u_q} ,$$

(4.19b) $$g'_J(a) = \frac{(y_p - \omega x_q)(t_{p'} - t_q)}{x_{p'} - x_q} \overline{W}_{p'q}(a-\beta_{J-1})\, W_{pq}(a-\beta_J)/\xi_q ,$$

(4.19c) $$g''_J(a) = \frac{(y_{p'} - y_q)(t_p - \omega t_q)}{x_p - y_q} \overline{W}_{p'q''}(a-\beta_{J-1}) W_{pq''}(a-\beta_J)/\xi_{q''} ,$$

where $q'' = U^2 T^{-1} q$ .

Substituting (4.19a) into (4.1), and comparing the result with (2.15a), we see that $Q$ is proportional to a column vector of the transfer matrix $T_{Uq}$. Similarly, the vectors whose elements are the terms on the RHS of (4.7) are proportional to column vectors of $T_q$ and $T_{q''}$. Allowing $\{\beta_1,\ldots,\beta_L\}$ to take all possible $N^L$ values, we span over all columns of the matrices, and (4.7) becomes a matrix identity:

(4.20) $$(y_p y_{p'}/\xi_{Uq})^L \tau^{(2)}_{k,q} X^k T_{Uq} = \left[\frac{(y_p - \omega x_q)(t_{p'} - t_q)}{(x_{p'} - x_q)\xi_q}\right]^L T_q + \left[\frac{(y_{p'} - y_q)(t_p - \omega t_q)}{(x_p - y_q)\xi_{R^2 q}}\right]^L T_{R^2 q} .$$

Here we have used the invariance (2.39).

This is the "T,Q" relation (5.14) of ref. 24. It can be put into diagonal form by using the staggered similarity transformation (2.34), (3.49). (From (2.43), X commutes with all the matrices of this paper, including $P_W$ and $P_B$.

One can obtain a second relation from (4.20) by post-multiplying by $\hat{T}_r$ (taking r to be any rapidity such that $\hat{T}_r$ is invertible), using the commutation relations (2.32a) and (3.50) so as to bring $T_r$ to the left side of each matrix product, then pre-multiplying by $T_r^{-1}$. Using (2.50), we obtain

$$(y_p y_{p'} / \hat{\xi}_{Uq})^L \hat{T}_{Uq} \tau^{(2)}_{k,q} X^k = \left[\frac{(y_{p'}-\omega x_q)(t_p-t_q)}{(x_p-x_q)\hat{\xi}_q}\right]^L \hat{T}_q + \left[\frac{(y_p-y_q)(t_{p'}-\omega t_q)}{(x_{p'}-y_q)\hat{\xi}_{R^2q}}\right] \hat{T}_{R^2q} \quad . \tag{4.21}$$

**The $\tau^{(2)},\tau^{(j)}$ relations**

We have obtained the hierarchy of relations (3.46), and the relations (4.20), (4.21) between $\tau^{(2)}$, T and $\hat{T}$. From these we can readily deduce another hierarchy of relations, this time involving only the $\tau^{(j)}$.

First replace k in (4.21) by m (m any integer). Then replace q by $\bar{q}(k-1,\ell)$, so $Uq = \bar{q}(k,\ell)$ and $R^2q = \bar{q}(k,\ell+1)$. Then pre-multiply (4.21) by $T_q$. Each term now contains a matrix product of the type $T_q\hat{T}_{\bar{q}k\ell}$, so we can use (3.46) to express these in terms of $\tau^{(j)}$ matrices.

The resulting scalar coefficients can be simplified by using the identities

$$\left[\frac{\hat{\xi}_{\bar{q}k\ell}(y_{p'}-\omega^k y_q)(t_p-\omega^{j-1}t_q)}{\hat{\xi}_{\bar{q}(k-1,\ell)}y_p y_{p'}(x_p-\omega^{k-1}y_q)}\right]^L \frac{\lambda_q^{(k,\ell)}}{\lambda_q^{(k-1,\ell)}} = \frac{H^{(j)}_{pq}}{H^{(j-1)}_{pq}} = \frac{z(\omega^{j-1}t_q)\bar{H}^{(j)}_{p'q}}{\bar{H}^{(j-1)}_{p'q}} \quad , \tag{4.22a}$$

$$\left[\frac{\hat{\xi}_{\bar{q}k\ell}(y_p-\omega^\ell x_q)(t_{p'}-\omega^j t_q)}{\hat{\xi}_{\bar{q}(k,\ell+1)}y_p y_{p'}(x_{p'}-\omega^\ell x_q)}\right]^L \frac{\lambda_q^{(k,\ell)}}{\lambda_q^{(k,\ell+1)}} = \frac{\bar{H}^{(j)}_{p'q}}{\bar{H}^{(j+1)}_{p'q}} = \frac{z(\omega^j t_q)H^{(j)}_{pq}}{H^{(j+1)}_{pq}} \quad , \tag{4.22b}$$

where the function $z(t_q)$ is defined by

$$z(t_q) = \left[\omega\mu_p\mu_{p'}(t_p - t_q)(t_{p'} - t_q) \,/\, (y_p y_{p'})^2\right]^L \quad , \tag{4.23}$$

regarding $\mu_p, \mu_{p'}, t_p, t_{p'}, y_p, y_{p'}$ as fixed constants.

Then we obtain

$$\bar{H}^{(j)}_{p'q} A^{(k,\ell)}_q + H^{(j)}_{pq} \bar{A}^{(k,\ell)}_q = 0 \quad , \tag{4.24}$$

where

$$A^{(k,\ell)}_q = \tau^{(j)}_{kq}\tau^{(2)}_{m,\bar{q}(k-1,\ell)} X^m - z(\omega^{j-1}t_q)\,\tau^{(j-1)}_{k-1,q} - \tau^{(j+1)}_{k,q}$$

(4.25)

$$\bar{A}^{(k,\ell)}_q = \tau^{(N-j)}_{-\ell,\bar{q}k\ell}\,\tau^{(2)}_{m,\bar{q}(k-1,\ell)} X^m - \tau^{(N-j+1)}_{-\ell,\bar{q}(k-1,\ell)} - z(\omega^j t_q)\,\tau^{(N-j-1)}_{-\ell-1,\bar{q}(k,\ell+1)} \quad .$$

We want to show that each term in (4.24) must separately vanish: we can do so by an argument similar to that used to obtain (3.50). The matrices $\tau^{(j)}_{kq}$, and the coefficients $g^{(j)}_q$, depend on q only as single-valued functions of $t_q$. Thus replacing q by $\bar{q}(0,0)$ leaves $t_q$, $A^{(k,\ell)}_q$ and $\bar{A}^{(k,\ell)}_q$ unchanged. On the other hand, it interchanges $x_q$ with $y_q$. Because of the explicit $x^N_q$ dependence in (3.35) and (3.36), this changes the ratio of $\bar{H}^{(j)}_{p'q}$ to $H^{(j)}_{pq}$. Hence we get a different linear combination of A and $\bar{A}$, so (4.24) implies

$$A^{(k,\ell)}_{pq} = \bar{A}^{(k,\ell)}_q = 0 \quad . \tag{4.26}$$

The matrix $\tau^{(j)}_{k,q}$ depends on q only via $t_q$, so we can write it as $\tau^j_k(t_q)$. Replacing $k,\ell,j$, $\omega^{j-1}t_q$ in $\bar{A}^{(k,\ell)}_q$ by $N-\ell$, $-k$, $N-j$, $t_q$, we can write the two equations (4.26) explicitly as

$$\tau^j_k(t)\tau^2_m(\omega^{j-1}t)X^m = z(\omega^{j-1}t)\tau^{j-1}_{k-1}(t) + \tau^{j+1}_k(t) \quad , \tag{4.27a}$$

$$\tau_k^j(\omega t)\tau_m^2(t)X^m = z(\omega t)\tau_{k-1}^{j-1}(\omega^2 t) + \tau_k^{j+1}(t) \quad , \tag{4.27b}$$

((writing $t_q$ simply as $t$ ).

These are our desired relations between the $\tau^{(j)}$ matrices. They are true for $j = 1,\ldots,N-1$ and all integers $k,m$. For $j = 1$ they are trivially satisfied because of (3.47) and (3.51). For $j = 2,3,4,\ldots$ they can be regarded as successively giving $\tau^{(3)}$, $\tau^{(4)}$, $\tau^{(5)}$,... as functions of $\tau^{(2)}$. Although the individual equations (4.27a) differ from (4.27b), the two sets are equivalent in that they both yield the same results for $\tau^{(3)}$, $\tau^{(4)}$, $\tau^{(5)}$,.... (We verify this below.)

We can take $j = N$ in the above working, but then the last term in (4.21) leads to the expression $T_q \hat{T}_{\bar{q}(k,\ell+1)}$, with $k+\ell+1 = N+1$. We can evaluate this from (3.46), but first have to use $\bar{q}(k,\ell+1) = \bar{q}(k,\ell+1-N)$ to reduce $j$ in (3.46) from $N+1$ to $1$. This alters the last terms in (4.25) and we no longer have the simple decomposition (4.26), but must use (4.24) as it stands. The end result is that we can extend both (4.27a) and (4.27b) to $j = N$, provided we take

$$\tau_k^{N+1}(t) = z(t)\,\tau_{k-1}^{N-1}(\omega t) + (\alpha_q+\bar{\alpha}_q)\tau_k^1(t) \quad , \tag{4.27c}$$

where

$$\alpha_q = \left[\mu_q^N(y_p^N-x_q^N)(y_{p'}^N-x_q^N) \,/\, k' y_p^N y_{p'}^N\right]^L \tag{4.28}$$

$$\bar{\alpha}_q = \left[\mu_q^{-N}(y_p^N-y_q^N)(y_{p'}^N-y_q^N) \,/\, k' y_p^N y_{p'}^N\right]^L \quad . \tag{4.29}$$

Using (2.46), we can deduce the relation

$$\alpha_q\,\bar{\alpha}_q = z(t_q)\, z(\omega t_q) \cdots z(\omega^{N-1} t_q) \quad . \tag{4.30}$$

The equations (4.27), for $j = 2,\ldots,N$, now form a closed set of functional relations for $\tau^{(2)},\ldots,\tau^{(N)}$. We shall discuss them, for special values of $N$, in the next section.

The $\tau^{(j)}$ in terms of the $T_q$

Using (2.39) and (2.40), we can write the functional relation (4.20) as

(4.31) $$\tau^{(2)}_{k,q} X^k T_{Uq} = \varphi_q T_q + \bar{\varphi}_{Uq} X T_{U^2q}$$

where the scalar coefficients $\varphi_q$, $\bar{\varphi}_q$ are given by

(4.32) $$\varphi_q = \left[\frac{\xi_{Uq}[y_p - \omega x_q)[t_{p'} - t_q)}{\xi_q y_p \, y_{p'}(x_{p'} - x_q)}\right]^L$$

(4.33) $$\bar{\varphi}_q = z(t_q) / \varphi_q \ .$$

We can regard (4.31) as expressing $\tau^{(2)}$ in terms of $T_q$, $T_{Uq}$, $T_{U^2q}$. Substituting into (4.27a), we can successively solve for $\tau^{(3)}$, $\tau^{(4)}$,..., obtaining

(4.34) $$\tau^{(j)}_{k,q} = \tau^j_k(t_q) = \sum_{m=0}^{j-1} \varphi_q \varphi_{Uq} \cdots \varphi_{U^{m-1}q} \times$$

$$\times \bar{\varphi}_{U^{m+1}q} \bar{\varphi}_{U^{m+2}q} \cdots \varphi_{U^{j-1}q} T_q \left[T_{U^m q}\right]^{-1} T_{U^j q} \left[T_{U^{m+1}q}\right]^{-1} X^{j-m-k-1} \ .$$

One can verify that these expressions, for $j = 1,...,N+1$, satisfy not only (4.27a), but also the alternative relations (4.27b) and the boundary condition (4.27c). One needs to use the commutation relations

(4.35) $$(T_q T_r^{-1}, T_{q'} T_{r'}^{-1}) = 0 \ , \ T_q T_r^{-1} T_{q'} = T_{q'} T_r^{-1} T_q \ ,$$

which are true for all rapidities $q,r,q',r'$ and are readily derivable from (2.33). (Replace each $T_q$ in (4.35) by $T_q \hat{T}_o$, where $o$ is some rapidity such that $\hat{T}_o$ is invertible. This leaves (4.35) unchanged and one can immediately use (2.33) to verify it.) One also needs (4.33) and the relations

$$\varphi_q \varphi_{Uq} \varphi_{U^2q} \cdots \varphi_{U^{N-1}q} = \alpha_q \ ,$$

(4.36)

$$\bar{\varphi}_q \, \bar{\varphi}_{Uq} \, \bar{\varphi}_{U^2 q} \cdots \bar{\varphi}_{U^{N-1} q} = \bar{\alpha}_q \, .$$

**$\hat{T}_q$ in terms of $T_q$**

Two other identities that we shall need are:

$$\frac{H^{(j)}_{pq}}{\bar{H}^{(j)}_{p'q}} = \frac{\bar{\varphi}_q \, \bar{\varphi}_{Uq} \cdots \bar{\varphi}_{U^{j-1} q}}{\varphi_{U^j q} \, \varphi_{U^{j+1} q} \cdots \varphi_{U^{N-1} q}} \, , \tag{4.37}$$

$$\lambda_q^{(0,j)} \, \varphi_{U^j q} \, \varphi_{U^{j+1} q} \cdots \varphi_{U^{N-1} q} \, / \, \bar{H}^{(j)}_{p'q} = \beta_{U^j q} \, , \tag{4.38}$$

where

$$\beta_q = \left[ \frac{(y_p y_{p'})^{1-N} (y_p^N - x_q^N)(y_{p'}^N - y_q^N)}{N \xi_q \, \hat{\xi}_{\bar{q}} \, (y_p - x_q)(y_{p'} - y_q)} \right]^L \, . \tag{4.39}$$

We have written $\bar{q}$ for the "conjugate" $\bar{q}(0,0)$ of $q$; from (2.38) going from $q$ to $\bar{q}$ is equivalent to interchanging $x_q$ with $y_q$, and inverting $\mu_q$. Replacing $q$ by $U^j q$ in (4.39) is simply equivalent to replacing $\xi_q$, $\hat{\xi}_{\bar{q}}$, $x_q$, $y_q$ by $\xi_{q(j,0)}$, $\hat{\xi}_{\bar{q}(0,j)}$, $\omega^j x_q$, $y_q$. Thus the RHS of (4.38) depends on $j$ and $q$ only via these last four scalar quantities.

Since $\tau^{(j)}_{k,q}$ depends on $q$ only as a single-valued function of $t_q$, (3.46) is unchanged by replacing $\tau^{(N-j)}_{k-j,\bar{q}k\ell}$ therein by $\tau^{(N-j)}_{k-j,qj0}$, where $qj0 = q(j,0) = U^j_q$. Doing this, then substituting the expression (4.34) for each matrix $\tau$, one obtains a functional relation satisfied by the matrices $T$, $\hat{T}$. At first sight there are many such relations, for different values of $k$ and $\ell$, but it follows immediately from (3.51) and (3.52) that any two such relations are identical if they have the same value of $j = k+\ell$. Further, taking $k = 0$, $\ell = j$ and using (4.37) and (4.38), we find that all the relations reduce to a single matrix functional relation, namely

$$\hat{T}_{\bar{q}} = \sum_{m=0}^{N-1} C_{m,q} T^{-1}_{U^m q} T_q T^{-1}_{U^{m+1} q} X^{-m-1} , \tag{4.40}$$

where the coefficients $C_{m,q}$ are

$$C_{m,q} = \varphi_q \varphi_{Uq} \cdots \varphi_{U^{m-1} q} \bar{\varphi}_{U^{m+1} q} \bar{\varphi}_{U^{m+2} q} \cdots \bar{\varphi}_{U^{N-1} q} / \beta_q , \tag{4.41}$$

and, as mentioned above,

$$\bar{q} = \bar{q}(0,0) , \quad x_{\bar{q}} = y_q , \quad y_{\bar{q}} = x_q . \tag{4.42}$$

Yet other functional relations for $T_q$ can be obtained from (4.34) or (4.31) by using the fact that $\tau^{(j)}_{k,q}$ depends on q only via $t_q$, so

$$\tau^{(j)}_{k,q} = \tau^{(j)}_{k,\bar{q}} . \tag{4.43}$$

The relation (3.46) is particularly simple when $j = N$, since then the last term vanishes due to (3.47a). Without loss of generality we can take $k = 0$ and hence $\ell = N$. Then from (4.38) the relation reduces to

$$\beta_q T_q \hat{T}_{\bar{q}} = \tau^{(N)}_{0,q} . \tag{4.44}$$

To summarize thus far: we have obtained four sets of functional relations: the hierarchies (3.46) and (4.27); the single "$\tau$,T" relation (4.20), or equivalently (4.21) or (4.31); and the "$\hat{T}$,T" relation (4.40). To each of these there is a corresponding relation obtained by interchanging T with $\hat{T}$, $\tau$ with $\hat{\tau}$, $\xi$ with $\hat{\xi}$, and p with p′. This introduces matrices $\hat{\tau}$ that we have not yet discussed: these act on "black" rows of the lattice and from (4.20), (4.21) and their corresponding equations

$$\tau^{(2)}_{k,q} = T^{-1}_{Uq} \tau^{(2)}_{k,q} T_{Uq} = \hat{T}_{Uq} \tau^{(2)}_{k,q} (\hat{T}_{Uq})^{-1} . \tag{4.45}$$

**Diagonalization**

The spin-translation operator X commutes with the $T$, $\hat{T}$, $\tau$, $\hat{\tau}$ matrices, and all these matrices can be diagonalized by the transformations (2.34) and (3.49) (together with a corresponding transformation for $\hat{\tau}$, involving $P_B$). The functional relations then become scalar functional relations for eigenvalues: if these can be solved, then the partition function Z can be calculated from (2.16) or (2.17). The diagonal forms of $T_q$ and $\hat{T}_q$ are simply related, since from (2.32)

$$(\hat{T}_q)_{diag} = (T_q)_{diag}(f_{pq} / f_{p'q})^L D \quad , \tag{4.46}$$

where $(T_q)_{diag}$, $(\hat{T}_q)_{diag}$ are the right-hand-sides of (2.34) and D is a diagonal matrix, independent of q. Similarly, from (4.44)

$$(\hat{\tau}^{(2)}_{k,q})_{diag} = (\tau^{(2)}_{k,q})_{diag} \quad . \tag{4.47}$$

**Inhomogeneous case**

All our functional relations can be readily generalized to the inhomogeneous case when the L column-rapidities $p_1,...,p_{2L}$ are distinct. In the scalar coefficients one simply separates out factors dependent on p from those dependent on p′. For the former, any Lth power is replaced by a product over $p_1,p_3,p_5,...,p_{2L-1}$; for the latter by a product over $p_2,p_4,...,p_{2L}$. (Note from fig.1 that the checkerboard case we have been considering corresponds to $p_{2i-1} = p$, $p_{2i} = p'$, $\forall i$.) Thus for instance (3.41) becomes

$$\lambda_q^{(k,\ell)} = \prod_{i=1}^{L} \left[ N \, \Omega^{k\ell}_{p_{2i-1},q} \, \bar{\Omega}^{k\ell}_{p_{2i},q} \right]^{-1} . \tag{4.48}$$

**Homogeneous case**

Conversely, one can immediately specialize to the fully homogeneous lattice model by taking $p' = p$ in all the above equations. Then one has an extra symmetry:

(4.49) $$\hat{T}_q = T_q S_R = S_R T_q \ ,$$

where $S_R$ is the spatial translation operator, with eigenvalues $2\pi\ell_R/L$ ($\ell_R = 0,\ldots,L-1$) used in Ref. 18. This $S_R$ then commutes with all our matrices, so $\ell_R$ plays the role of a quantum number.

## 5. PARTICULAR VALUES OF N

### N = 2

When $N = 2$ the solvable chiral Potts model reduces to the Ising model; solved by Onsager in 1944 [14]. Eliminating $\tau^{(3)}$ between (4.27a) (with $j = 2$) and (4.27c), we obtain a functional relation for $\tau^{(2)}$:

(5.1) $$\tau_o^2(t)\,\tau_o^2(-t) = [z(t) + z(-t)]\,X + (\alpha_q + \bar{\alpha}_q)\,\mathbf{I} \ .$$

(We have taken $m = k = 0$ in (4.27a); because of (3.51), this involves no loss of generality.)

Also, (4.40) reduces to

(5.2) $$T_{Uq}\,\hat{T}_{\bar{q}} = \left[\frac{2\,\xi_{Uq}\,\hat{\xi}_{\bar{q}}\,(x_{p'} + x_q)}{y_{p'} + y_q}\right]^L \left\{\left[\frac{t_{p'} - t_q}{x_{p'}^2 - x_q^2}\right]^L + X\left[-\mu_p\mu_{p'}\frac{t_p + t_q}{y_p^2 - x_q^2}\right]^L\right\}$$

and (4.44) becomes

(5.3) $$T_q\,\hat{T}_{\bar{q}} = \left[\frac{2\,\hat{\xi}_q\,\xi_{\bar{q}}\,y_p\,y_{p'}}{(y_p + x_q)\,(y_{p'} + y_q)}\right]^L \tau_o^2(t) \ .$$

The right-hand sides of (5.1) and (5.2) are known functions. In principle it is straightforward to solve (5.1) for the polynomial $\tau_o^2(t)$ by sharing out the zeros of the

RHS between $\tau_0^2(t)$ and $\tau_0^2(-t)$. Using (4.45), it should also be possible to solve (5.2), or (5.3), for $T_q$ and $\hat{T}_q$, so as to obtain the known result (e.g. (7.9.7) of Ref. 15).

**N = 3**

Eliminating $\tau^{(3)}$, $\tau^{(4)}$ between (4.27a) (with j = 2,3) and (4.27c), we obtain a functional relation for $\tau_k^2(t)$. It is convenient to choose (without loss of generality) k = 1, giving

$$\tau_{-1}^2(t)\,\tau_{-1}^2(\omega t)\,\tau_{-1}^2(\omega^2 t) = z(t)\,\tau_{-1}^2(\omega t) + z(\omega t)\,\tau_{-1}^2(\omega^2 t) + z(\omega^2 t)\,\tau_{-1}^2(t) + (\alpha_q + \bar{\alpha}_q)\,I \;. \tag{5.4}$$

In fact this relation can be derived quite easily from (4.31). Replacing q therein by Ua and $U^2q$, we obtain a total of three equations, homogeneous and linear in $X^{-1}T_{Uq}$, $X\,T_{U^2q}$, $T_q$. The determinant of coefficients must vanish. Taking k = -1, we obtain

$$\begin{vmatrix} \tau^{(2)}_{-1,q} & -\bar{\varphi}_{Uq} & -\varphi_q \\ -\varphi_{Uq} & \tau^{(2)}_{-1,Uq} & -\bar{\varphi}_{U^2q} \\ -\bar{\varphi}_q & -\varphi_{U^2q} & \tau^{(2)}_{-1,U^2q} \end{vmatrix} = 0 \;. \tag{5.5}$$

Expanding and using (4.33) and (4.36), we obtain (5.4).

This procedure generalizes to arbitrary integers N. In general the functional relation explicitly involves the spin-shift operator X, as in (5.1). However, for N odd we can eliminate X by choosing $k = -(N-1)/2$.

Similarly, we can eliminate X from the functional relation (4.40) by writing $Uq$ as $U^4q$ and noting from (2.39) and (2.40) that

$$T_{U^2q} = \left[\frac{\xi_{U^2q}}{\omega\xi_{R^2q}} \frac{(y_p-\omega^2 x_q)(y_{p'}-y_q)}{\mu_p \mu_{p'} (x_p-y_q)(x_{p'}-\omega x_q)}\right]^L X^{-1} T_{R^2q} . \tag{5.6}$$

We can also use the simple identity $\hat{T}_{T\bar{q}} = \hat{T}_{R^3q}$ , and it is convenient to define a function A by

$$A_p(x,y) = \left[N^{\frac{1}{2}} \frac{x_p y_p - xy}{(x_p-x)(y_p-y)}\right]^L . \tag{5.7}$$

Then (4.40) can be written as

$$\begin{aligned} \left[\xi_{R^{-2}q}\, \hat{\xi}_{R^3q}\, \xi_{R^2q}\right]^{-L} & T_{R^{-2}q}\, \hat{T}_{R^3q}\, T_{R^2q} \\ & = \xi_q^{-L} A_{p'}(x_q,y_q)\, A_p(\omega^2 y_q,x_q)\, T_q \\ & + \xi_{R^2q}^{-L}\, A_p(y_q,\omega x_q)\, A_p(\omega^2 y_q,x_q)\, T_{R^2q} \\ & + \xi_{R^{-2}q}^{-L}\, A_{p'}(x_q,y_q)\, A_{p'}(\omega^2 x_q,\omega^2 y_q)\, T_{R^{-2}q} . \end{aligned} \tag{5.8}$$

For the homogeneous case, when $\hat{T}_q$ is related to $T_q$ by (4.49), this is the functional relation postulated by Albertini, McCoy and Perk [17].

As yet the relation (5.8) has only been solved in the "superintegrable" case $p' = \bar{p}$ (see next section). Rather than attempting to solve it directly for $T_q$ and $\hat{T}_q$, it may be easier to first solve (5.4) for the polynomial $\tau^2_{-1}(t)$ , then calculate $\tau^3_1(t)$ from (4.27a):

$$\tau^3_1(t) = \tau^2_{-1}(t)\, \tau^2_{-1}(\omega t) - z(\omega t)\, I \tag{5.9}$$

and then use (4.44), which can be put into the form

$$(5.10) \qquad T_q \hat{T}_{R^3 q} = \left[\xi_q \hat{\xi}_{R^3 q}\right]^L A_p(y_q,\omega x_q)\, A_{p'}(\omega^2 x_q,\omega^2 y_q)$$

$$\times\, \tau_1^3(t_q) \,/\, z(\omega t_q) \quad .$$

## 6. SUPERINTEGRABLE CASE

In Ref. 18 and 21 it was noted that the eigenvalue spectrum of $T_q$ simplifies for the case $p' = p = \bar{p}$, i.e. $x_p = x_{p'} = y_p = y_{p'} = \eta^{-1}$, where $\eta = [(1+k')\,/\,(1-k')]^{1/2N}$. This case was further considered in Ref. 22, and extended to the row-inhomogeneous situation where both q and $k'$ vary arbitrarily from row to row.

For a given row, let us write the parameters $k'$, $x$, $y$, $\mu$ of Refs. 21 and 22 as $k'_B$, $\alpha$, $\beta$, $\nu$. Then the weight functions $W$, $\overline{W}$ for the superintegrable case are

$$W(n) \,/\, W(o) = \nu^n\, (1,\beta)_{o,n} \,/\, (1,\alpha)_{o,n}$$

$$(6.1)$$

$$\overline{W}(n) \,/\, \overline{W}(o) = \nu^{-n}\, (\omega,\alpha)_{o,n} \,/\, (1,\beta)_{o,n}$$

where

$$\nu^N = (\alpha^N - 1) \,/\, (\beta^{N-1}) \quad .$$

Now return to the general solvable checkerboard-type model of the previous sections, with $p'$ different from p. Recall that we have two types of row: "white-black" rows with transfer matrix $T_q$ given by (2.15a); and "black-white" rows with transfer matrix $\hat{T}_q$ given by (2.15b). For the former, $T_q$ involves only the weight functions $W_{pq}$ and $\overline{W}_{p'q}$. From (2.13), sufficient conditions for them to take the superintegrable form (6.1) are that

$$\alpha = x_q/y_p = x_q/x_{p'} \,,$$
(6.2) $$\beta = x_p/y_q = y_{p'}/y_q \,,$$
$$\nu = \mu_p y_q / \mu_q y_p = y_q/\mu_{p'}\mu_q x_{p'} \,.$$

The q-parameters cancel out of these restrictions, leaving

(6.3) $$x_{p'} \,,\, y_{p'} \,,\, \mu_{p'} = y_p \,,\, x_p \,,\, \mu_p^{-1} \,.$$

For black-white rows we obtain the same conditions, with p interchanged with $p'$. This leaves (6.3) unchanged, so we see that a sufficient condition for the checkerboard model (more strictly, the "alternating vertical rapidities" model) to be superintegrable is that

(6.4) $$p' = \bar{p}(o,o) = \bar{p} \,,$$

where $\bar{p}$ is the rapidity conjugate to p. The horizontal rapidities $q_1, q_2, \ldots, q_M$ can still vary arbitrarily from row to row, but here we shall take the $k$, $k'$ of eqns. (2.4) - (2.8) to be fixed, the same for all edges of $\mathcal{L}$. Note that this $k'$ is no longer the same as $k'_B$; in fact for a given white-black row

(6.5) $$\begin{aligned} k'_B &= (\alpha^N - 1)(\beta^N - 1) / (\beta^N - \alpha^N) \\ &= (x_q^N - y_p^N)(x_p^N - y_q^N) / (t_p^N - t_q^N) \\ &= k'(x_q^N - y_p^N) / [\mu_p^N(x_q^N - x_p^N)] \,, \end{aligned}$$

using (2.46). (For the isotropic case, $p' = p$, $y_p = x_p$, $\mu_p = 1$ and we regain $k'_B = k'$.)

In this section we shall consider this "alternating superintegrable" model, with M+2 free parameters $k'$, p, $q_1, \ldots, q_M$. It is intermediate in generality between the homogeneous superintegrable model (only two free parameters: $k'$ and q) and the inhomogeneous model of Ref. 22 (2M free parameters $\alpha_1, \ldots, \alpha_M$ and $\beta_1, \ldots, \beta_M$).

Consider the lattice $\mathcal{L}$ of Fig. 1, with periodic boundary conditions from right to left, as in the previous sections; but now impose fixed-spin boundary conditions at top and bottom:

$$\sigma_i = a_B \quad \text{if} \quad i \in \text{row } 1$$
$$= 0 \quad \text{if} \quad i \in \text{row } M+1 \ , \tag{6.6}$$

where $a_B$ is a given integer (the same for all columns) between 0 and N-1 . Let $q_i$ be the rapidity for the edges between rows i and i+1 : thus $q_1, q_3, q_5, \ldots$ correspond to white-black rows ; $q_2, q_4, q_6 \ldots$ to black-white rows. Define

$$p_i = p \quad \text{if } i \text{ odd} \ ,$$
$$= p' = \bar{p} \quad \text{if } i \text{ even} \ , \tag{6.7}$$

$$g(p,q) = N\, W_{pq}(0)\, \overline{W}_{\bar{p}q}(o)\, x_q^{N-1} \frac{y_p - x_q}{y_p^N - x_q^N} \ , \tag{6.8}$$

$$\ell_{pq} = (x_q^N + y_p^N) \,/\, (x_q^N - y_p^N) \ , \tag{6.9}$$

$$C = g(p_1,q_1)\, g(p_2,q_2) \cdots g(p_M,q_M) \ , \tag{6.10}$$

$$Y = y_{p_1} y_{p_2} \cdots y_{p_M} \,/\, (x_{q_1} x_{q_2} \cdots x_{q_M}) \ , \tag{6.11}$$

$$B(p,q,\theta) = \frac{1}{1 + \ell_{pq}} \begin{bmatrix} \ell_{pq} - \cos\theta & -\sin\theta \\ -\sin\theta & \ell_{qp} + \cos\theta \end{bmatrix} \tag{6.12}$$

$$\mathfrak{B}(\theta) = B(p_1,q_1,\theta)\, B(p_2,q_2,\theta) \cdots B(p_M,q_M,\theta) \ , \tag{6.13}$$

$$D(\cos\theta) = (1,0)\, \mathfrak{B}(\theta) \begin{bmatrix} 1 \\ 0 \end{bmatrix} \ . \tag{6.14}$$

Thus C , Y , $D(\cos\theta)$ are scalars; B , $\mathfrak{B}$ are two-by-two matrices. Also, let

$$P(z^N) = z^{-Q} \sum_{n=0}^{N-1} \omega^{Qn} [(1-z^N)/(1-\omega^{-n} z)]^L \ . \tag{6.15}$$

This is a polynomial in $z^N$ , of degree

(6.16) $$m = \text{integer part of } [(NL-L-Q)/N] \quad .$$

Let its zeros be $z_1^N, z_2^N, \ldots, z_m^N$, and let

(6.17) $$\cos\theta_j = (1+z_j^N)/(1-z_j^N) , \quad j = 1,\ldots,m \quad .$$

(The $z_j^N$ are negative real, so the $\theta_j$ are real.)

Then from eqns. (3.2) - (3.9), (7.1) and (7.2) of Ref. 22, the partition function of this superintegrable chiral Potts model, for all $L$ and $M$, is given by

(6.18) $$Z_{a_B} = N^{-1} \sum_{Q=0}^{N-1} \omega^{-Qa_B} \tilde{Z}_Q \quad ,$$

where $\tilde{Z}_Q$ is independent of $a_B$ and is given by

(6.19) $$\tilde{Z}_Q = C^L Y^Q D(\cos\theta_1) D(\cos\theta_2) \cdots D(\cos\theta_m) \quad .$$

For the general inhomogeneous model of Ref. 22 we can take $p_1,\ldots,p_m$ and $q_1,\ldots,q_M$ to be arbitrary (related to $\alpha_1,\ldots,\alpha_M$ and $\beta_1,\ldots,\beta_M$ by (6.2)). Here we are specializing to the case when $p_1,\ldots,p_M$ are given by (6.7). For this case we can simplify (6.13) and (6.14). Define

(6.20) $$s_p = (y_p/x_p)^{N/2} \mu_p^{-N} \quad ,$$

$$S(p) = \begin{bmatrix} 1 & 0 \\ 0 & s_p \end{bmatrix} \quad .$$

Because $x_{\bar{p}}, y_{\bar{p}}, \mu_{\bar{p}} = y_p, x_p, \mu_p^{-1}$, we have $s_{\bar{p}} = s_p^{-1}$ and

$$S(p_i) S(p_{i+1}) = \mathbf{I} \quad \forall i \quad .$$

Hence (6.14) is unchanged if we replace $B(p,q,\theta)$ in (6.13) by

(6.21) $$\tilde{B}(p,q,\theta) = S(p) B(p,q,\theta) S(p) \quad .$$

From (2.7) and (2.8) we can establish the identities

$$\ell_{pq} - 1 - s_p^2(\ell_{qp}+1) = 2k\,\mu_p^{-N}\,y_p^N\,/\,k' \quad ,$$

$$\ell_{pq} + 1 - s_p^2(\ell_{qp}-1) = 2k\,\mu_p^{-N}\,x_p^{-N}\,/\,k' \quad ,$$

Define

$$h_{pq\theta} = \frac{1}{2}\left[\ell_{pq} + s_p^2\,\ell_{qp} - (1-s_p^2)\cos\theta\right] \quad ,$$

$$\gamma_{p,\theta} = k\left[t_p^{N/2}\cot(\theta/2) + t_p^{-N/2}\tan(\theta/2)\right]\,/\,2k' \quad ,$$

where $t_p = x_p y_p$ is defined in (2.6). Then it follows that

$$\tilde{B}(p,q,\theta) = \frac{1}{1+\ell_{pq}}\left[h_{pq\theta}\,\mathbf{I} + s_p \sin\theta\, J_{p,\theta}\right] \quad , \tag{6.22}$$

where $J_{p,\theta}$ is the two-by-two matrix

$$J_{p,\theta} = \begin{bmatrix} \gamma_{p,\theta} & -1 \\ -1 & -\gamma_{p,\theta} \end{bmatrix} .$$

Note that $J_{p,\theta}$ is independent of $q$ and is unchanged by replacing $p$ by $\bar{p}$ (because $t_{\bar{p}} = t_p$). It follows that the matrices $\tilde{B}(p_i,q_i,\theta)$ all commute (for $i = 1,\ldots,M$). Hence (6.14) reduces to

$$D(\cos\theta) = D_+(\cos\theta) + D_-(\cos\theta) \tag{6.23}$$

where

$$D_\pm(\cos\theta) = \frac{1}{2}\left[1 \pm \gamma_{p,\theta}\,/\,(1+\gamma_{p,\theta}^2)^{1/2}\right] \times \prod_{i=1}^{M} \frac{h_{p_iq_i\theta} \pm s_{p_i}\sin\theta\,(1+\gamma_{p,\theta}^2)^{1/2}}{1+\ell_{p_iq_i}} . \tag{6.24}$$

Substituting this result into (6.19), we can deduce those eigenvalues of $T_q$ and $\hat{T}_q$ that contribute to the partition function. (Because of the fixed-spin boundary conditions, only some of the eigenvalues so contribute: this point is discussed in [21] and [22].) The eigenvalues of $T_q \hat{T}_r$, for arbitrary $q$ and $r$, are $\Lambda_{pq} \Lambda_{\bar{p}r}$, where

$$\Lambda_{pq} = g(p,q)^L (y_p/x_q)^Q \times \prod_{j=1}^{m} \frac{h_{pq\theta} \pm s_p \sin\theta \, (1+\gamma_{p\theta}^2)^{1/2}}{1 + \ell_{pq}} . \tag{6.25}$$

The sign choice can be made independently for each term in the product, so there are $2^m$ such eigenvalues for each value of $Q$, where $Q = 0,1,\ldots,N-1$. From (6.8) - (6.23), (2.48) and (2.49), we can establish that

$$\begin{aligned} g(\bar{p},q) / g(p,q) &= \mu_p^{N-1} f_{pq} / f_{\bar{p}q} \\ \tilde{B}(\bar{p},q,\theta) &= \tilde{B}(\bar{p},\bar{p},\theta)\, \tilde{B}(p,q,\theta) . \end{aligned} \tag{6.26}$$

From these relations we can explicitly verify that the diagonal forms of $T_q$ and $\hat{T}_q$ satisfy the constraint (4.46).

**Alternative form for Z**

The function $D(\cos\theta)$ is a polynomial in $\cos\theta$ of degree

$$r = \text{integer part of } [(M+1)/2] . \tag{6.27}$$

From (6.19), to within simple factors $\tilde{Z}_Q$ is the product of $D(c)$ over the $m$ zeros of the polynomial $(c+1)^m \, P[(c-1)/(c+1)]$, i.e. it is the resultant [29] of these two polynomials of degree $r$, $m$, respectively.

Let $c_1,\ldots,c_r$ be the zeros of $D(c)$, and define

$$\zeta_i = [(c_i - 1) / (c_i + 1)]^{1/N} . \tag{6.28}$$

Since $D(1) = Y^N$, and $D(-1) = 1$, we can choose $\zeta_1 \ldots \zeta_r$ so that

$$\zeta_1 \cdots \zeta_r = Y \quad . \tag{6.29}$$

Expanding each $D(\cos\theta)$ in (6.19) as a product of its r factors, we obtain

$$\tilde{Z}_Q = C^L Y^Q \prod_{j=1}^{r} \left[ P(\zeta_j^N) / P(1) \right] \quad , \tag{6.30}$$

From (6.15), $P(1) = N^L$ ; it follows that (6.18) can be written as

$$Z_{a_B} = C^L \sum \left[ \lambda_{1,n_1} \lambda_{2,n_2} \cdots \lambda_{r,n_r} \right]^L \tag{6.31}$$

where

$$\lambda_{i,n} = N^{-1} (1 - \zeta_i^N) / (1 - \zeta_i \omega^{-n}) \tag{6.32}$$

and the sum is over all integers $n_1,\ldots,n_r$ such that $0 \le n_i < N$ and

$$n_1 + n_2 + \cdots + n_r = a_B \quad , \bmod N \quad . \tag{6.33}$$

We can regard the lattice of Figure 1 as made up of L columns, each containing two vertical rapidity lines ($p$ and $\bar{p}$). Let $T_{col}(a_B)$ be the column-to-column transfer matrix that adds such a column. Then it is of dimension $N^{r-1}$, and from (6.31) all its eigenvalues are given by

$$\Lambda_{col}(\mathbf{n}) = C \lambda_{1,n_1} \lambda_{2,n_2} \cdots \lambda_{r,n_r} \tag{6.34}$$

where $n = \{n_1,\ldots,n_r\}$ is any one of the $N^{r-1}$ sets of integers $n_1,\ldots,n_r$ that satisfy (6.33).

**Homogeneous case: finite-size corrections at criticality**

When $\bar{p} = p$, $x_p = y_p = \eta^{-1}$, we regain the homogeneous model. The thermodynamics of this model were discussed in Ref. [22], and it was noted that the usual predictions of conformal invariance fail for $N > 2$. The model is intrinsically anisotropic, with correlation length exponents

$$\nu_{hor} = 2/N \ , \quad \nu_{vert} = 1 \tag{6.35}$$

in the horizontal and vertical directions. This suggests that at criticality the finite=size corrections should depend on the width $L$ and the height $M$ of the lattice via the ratio $L^{N/2} / M$.

In fact we find that this is the case, and

$$\tilde{Z}_Q = e^{-LMf-4Lf_s} Z_Q^{(c)}(L^{N/2}/M) \ . \tag{6.36}$$

Here $f$, $f_s$ are the bulk and surface free energies. They are given in section 6 of Ref. [22], in terms of a parameter

$$\begin{aligned} G &= (\alpha^N\beta^N - 1) / (\beta^N - \alpha^N) \\ &= k(x_q^N - y_q^N) / [1 + k' - (1-k')x_q^N y_q^N] \ . \end{aligned} \tag{6.37}$$

The behaviour of the finite-size correction function $Z_Q^{(c)}(x)$ can be obtained from eq. (6.39) of [22] for the limits $x \to 0$ and $x \to \infty$. In fact the techniques therein can be used to obtain the function for all $x$, and here we wish to report the result. Define

$$\begin{aligned} x &= L^{N/2} / M \ , \\ q &= \exp[-(\pi G x/2)^{2/N}] \ , \\ \delta &= \frac{2 \sin(\pi/N)}{N\pi} \int_0^\infty \ln(1 + e^{-2\pi v})\, v^{(2-N)/N}\, dv \ . \end{aligned} \tag{6.38}$$

Then

$$Z_Q^{(c)}(x) = q^{-\delta} \prod_{m=1}^{\infty} \left\{1 + \sum_{n=1}^{N-1} \omega^{Qn}\, q^{(1-\omega^{-n})(m-\frac{1}{2})^{2/N}}\right\} \ . \tag{6.39}$$

This infinite product converges rapidly when $|q|$ is small, i.e. for large $x$. For small $x$ it is convenient to use an alternative (and equivalent) expression that follows from the "resultant" transformation from (6.14) to (6.30). Define

$$q' = \exp\left\{- 4[\pi/\sin(\pi/N)]^{N/2} / (Gx)\right\} , \tag{6.40}$$

$$F(z) = 1 + \sum_{n=1}^{N-1} \omega^{Qn} \exp[\pi z(\omega^{-n}-1) / \sin(\pi/N)] , \tag{6.41}$$

$$\delta'_Q = \frac{N}{4\pi} \int_0^\infty z^{(N-2)/2} \ln F(z)\, dz . \tag{6.42}$$

The function $F(z)$ has zeros at

$$z = \omega^{1/2} t_1 , \ \omega^{1/2} t_2 , \ldots \tag{6.43}$$

where $t_1 , t_2 , \ldots$ are real positive and can be ordered so that

$$0 < t_1 < t_2 < t_3 < \cdots . \tag{6.44}$$

Then the alternative form is

$$Z_Q^{(c)}(x) = 2^{Q/N} q'^{-\delta'} \prod_{k=1}^{\infty} \left[1 + (q')^{t_k^{N/2}}\right] . \tag{6.45}$$

For the $N = 2$ case we regain the Ising model and the above results become

$$\delta = \frac{1}{24} , \ \delta'_0 = \frac{1}{48} , \ \delta'_1 = -\frac{1}{24} , \tag{6.46}$$

$$Z_Q^{(c)} = q^{-1/24} \prod_{m=1}^{\infty} \left[1 + (-1)^Q q^{2m-1}\right] ,$$

$$Z_0^{(c)} = q'^{-1/48} \prod_{k=1}^{\infty} (1 + q'^{k-\frac{1}{2}}) , \tag{6.47}$$

$$Z_1^{(c)} = 2^{\frac{1}{2}} q'^{1/24} \prod_{k=1}^{\infty} (1 + q'^{k}) ,$$

and $q$, $q'$ are related by

$$\ln q \;\; \ln q' = 2\pi^2 \; . \tag{6.48}$$

Remembering that we have fixed-spin boundary conditions at the top and bottom of the lattice, periodic at the sides, these $N = 2$ results are of the form predicted by conformal invariance [25,26]. It would be interesting to investigate whether some more general anisotropic theory would lead to the expressions (6.39) and (6.45).

**Rotated superintegrable model**

In sections 2 through 5 we have considered the general solvable chiral Potts model, with arbitrary horizontal rapidities $q_1,\ldots,q_M$ and rapidities $p$ and $p'$. At the end of section 4 we indicated how these results could be immediately extended to arbitrary vertical rapidities $p_1,\ldots,p_{2L}$.

So far in this section we have regarded the superintegrable case as a special case of the alternating vertical rapidity model, when $p' = \bar{p}$. However, we can obviously rotate the superintegrable model through 90°, so that the spins fixed at the value $a_B$ are on the right, as in Fig. 5. Using the relations [8]

$$\frac{W_{pq}(n)}{W_{pq}(0)} = \frac{\overline{W}_{R^{-1}q,p}(n)}{\overline{W}_{R^{-1}q,p}(0)} \; ,$$

(6.49)

$$\frac{\overline{W}_{pq}(n)}{\overline{W}_{pq}(0)} = \frac{W_{R^{-1}q,p}(-n)}{W_{R^{-1}q,p}(0)} \; ,$$

the model becomes a special case of the general model, with $M$ vertical rapidities $p_1',\ldots,p_M'$ and $2L$ alternating horizontal rapidities $q'$, $q''$, $q'$,...,$q''$, where

$$p_i' = R^{-1} q_{M+1-i} \; ,$$

(6.50)

$$q' = \bar{p} \, , \quad q'' = p \; .$$

Thus now the vertical rapidities are arbitrary, but the horizontal rapidities are restricted. Let us write $T_{col}(a_B)$ simply as $\mathcal{T}_{q'}$ : with this orientation it becomes the double-row transfer matrix

$$\mathcal{T}_{q'} = T_{q'} \hat{T}_{q''} , \tag{6.51}$$

where $q'' = \overline{q'}$. From (3.18), (3.43), (4.38) and (4.44), this is the matrix $\mathcal{T}^{(0,N)}_{q'} = \tau^{(N)}_{0,q'} / \beta_{q'}$ in full generality, except that now we are imposing fixed spin boundary conditions at right and left, instead of the periodic boundary conditions used in sections 2 through 5. We know all its eigenvalues: they are given by (6.34).

As in [22], take M to be odd, and define single-edge transfer matrices $E^{ij}_{pq}$ and $F^{i}_{pq}$, of dimension $N^{r-1}$, by

$$\begin{aligned} \left[E^{ij}_{pq}\right]_{\sigma,\sigma'} &= W_{pq}(\sigma_i - \sigma_j) \prod_{j=1}^{r-1} \delta(\sigma_j, \sigma'_j) \\ \left[F^{i}_{pq}\right]_{\sigma,\sigma'} &= W_{pq}(\sigma_i - \sigma'_i) \prod_{\substack{j=1 \\ \neq i}}^{r-1} \delta(\sigma_j, \sigma'_j) . \end{aligned} \tag{6.52}$$

Similarly, define $\bar{E}^{i,i+1}_{pq}$, $\bar{F}^{i}_{pq}$ with $W_{pq}$ replaced by $\overline{W}_{pq}$. Drop the primes on $p'_1, \ldots, p'_M, q'$. Then $q'' = \bar{q}$ and

$$\begin{aligned} T_q &= E^{r-1,r}_{p_M,q} \bar{F}^{r-1}_{p_M-1,q} \cdots \bar{F}^{1}_{p_2,q} E^{01}_{p_1,q} \\ \hat{T}_q &= \bar{E}^{10}_{p_1q} F^{1}_{p_2q} \cdots F^{r-1}_{p_{M-1},q} \bar{E}^{r,r-1}_{p_M,q} . \end{aligned} \tag{6.53}$$

Thus with these boundary conditions $T_q$ and $\hat{T}_q$ are products of single-edge transfer matrices. They can readily be inverted, using the identities (3.7), (3.3) and (3.9), which imply

(6.54)
$$(E^{ij}_{pq})^{-1} = \bar{E}^{ji}_{p,Rq} \,/\, [W_{pq}(0)\, \overline{W}_{p,Rq}(0)]$$
$$(\bar{F}^{i}_{pq})^{-1} = F^{i}_{p,Rq} \,/\, [\overline{W}_{pq}(0)\, W_{p,Rq}(0) v_{pq}] \quad ,$$

where

(6.55)
$$v_{pq} = \frac{N(x_p - x_q)(y_p - y_q)(t_p^N - t_q^N)}{(t_p - t_q)(x_p^N - x_q^N)(y_p^N - y_q^N)} \quad .$$

It follows that

(6.56)
$$T_q^{-1} = \rho_q \hat{T}_{Rq} \quad ,$$

where the scalar coefficient $\rho_q$ is given by

(6.57)
$$\rho_q^{-1} = \prod_{j\,\mathrm{odd}} \left[\overline{W}_{p_j q}(0)\, \overline{W}_{p_j,Rq}(0)\right]$$
$$\times \prod_{j\,\mathrm{even}} \left[\overline{W}_{p_j q}(0)\, W_{p_j,Rq}(0)\, v_{p_j q}\right] \quad .$$

The first product is over $j = 1,3,5,\ldots,M$ and the second over $j = 2,4,6,\ldots,M-1$ .

**Functional relations for $\tau^{(2)},\ldots,\tau^{(N)}$**

For brevity, write

(6.58)
$$\rho_q^{(j)} = \rho_q\, \rho_{Uq}\, \rho_{U^2 q} \cdots \rho_{U^{j-1} q} \quad ,$$
$$\tau_q^{(j)} = \tau_{Uq}^{-1}\, \tau_{U^2 q}^{-1} \cdots \tau_{U^{j-1} q}^{-1} \quad .$$

Then from (6.51) and (6.56), using the definitions (2.35) - (2.38), we can deduce that

(6.59)
$$T_q\, \hat{T}_{\bar{q}(0,j)} = \tau_q^{(j)} \,/\, \rho_q^{(j)} \quad ,$$

(6.60a)
$$\tau_q^{(j)}\, \tau^{(2)}_{U^{j-1} q} = \tau_q^{(j+1)} \quad ,$$

(6.60b) $$\mathcal{T}_q^{(2)} \mathcal{T}_{Uq}^{(j)} = \mathcal{T}_q^{(j+1)} ,$$

(6.60c) $$\mathcal{T}_q^{(N+1)} = \rho_q^{(N)} \, I .$$

For $j = 1,\ldots,N$, we can use (3.46) to express the LHS of (6.59) in terms of $\tau$-matrices. However, now the left-hand spins ($\sigma_1$ and $\sigma_1''$ in (2.27)) are fixed to zero. Hence $\sigma_1 - \sigma_1''$ belongs to the set $\zeta_{0j}$ defined before (3.15), and only the first term on the RHS of (3.46) occurs. Comparing (3.46) and (6.60), we see that

(6.61) $$\mathcal{T}_q^{(j)} \propto \tau_{0,q}^{(j)} ;$$

i.e. the $\mathcal{T}_q^{(j)}$ are (to within scalar factors) the $\tau_{0,q}^{(j)}$, with fixed-spin boundary conditions.

With this identification, we see that (6.60a-c) correspond to the functional relations (4.27)a-c).

The fixed-spin boundary conditions mean that we no longer have the commutation relation (2.32b). Nevertheless, one can use the method of Ref. [24] to show that $\mathcal{T}_q^{(2)}$, $\mathcal{T}_r^{(2)}$ commute, for all rapidities q and r (this is certainly true for M even: presumably there is some appropriate extension to M odd). Thus the relations (6.60) must be satisfied by the individual eigenvalues of $\mathcal{T}_q^{(j)}$. Remembering that the $\tau_{0q}^{(j)}$ depend on q only as polynomials in $t_q$ (again, this is true for M even: one needs an appropriate extension to M odd), it should be possible to derive the eigenvalue result (6.34) directly from the relations (6.60). (Note that $p$, $\bar{p}$, $q_j$ in (6.7)-(6.14) have now been replaced, according to (6.50), by $\bar{q}$, $q$, $Rp_{M+1-j}$.)

Such a re-derivation of (6.34) would be useful, since the method given in Ref. [22] is not completely rigorous: it assumes that each of the $2^m$ eigenvalues (6.25) occurs just once, and that a certain similarity transformation is independent of $k_B'$ (see the remarks after eq. (2.20) of Ref. [22]).

The result (6.34) shows that the diagonal form of $\mathcal{T}_q$ is basically a direct product of N-dimensional matrices. An intriguing corollary is that the eigenvalues of a corresponding Hamiltonian have a very simple direct-sum structure [23]. The form is very reminiscent of the Clifford algebra results for the corresponding Ising model Hamiltonian. It is natural to ask whether the result could have been obtained by an appropriate extension of the Clifford algebra method, using operators that "$\omega$-commute", i.e. $AB = \omega BA$ [30].

## 7. SUMMARY

In section 2 we have defined the N-state chiral Potts model, and its transfer matrices $T_q$, $\hat{T}_q$ for the two types of row, using toroidal (periodic) boundary conditions. In (3.44) we defined an associated tau-matrix $\tau_{k,q}^{(j)}$ (also written as $\tau_k^j(t_q)$), and noted that it is the transfer matrix of the model of Fig. 4, with N-state spins $\sigma_i$ on open circles, j-state spins $m_i$ on filled squares. By drawing vertical (horizontal) lines through circles (squares), this can be expressed as the "j by N" vertex model of Ref. [24].

These tau-matrices depend on the horizontal rapidity q only as polynomials in the single variable $t_q$. In this respect they are much simpler than $T_q$ and $\hat{T}_q$, which are rational functions of $x_q$ and $y_q$, these variables being related by the high-genus algebraic constraint (2.8). Two tau-matrices with the same j and $t_q$, but different k, are simply related by (3.51), which implies

$$\tau_{k,q}^{(j)} = X^{-k}\, \tau_{0,q}^{(j)} \quad . \tag{7.1}$$

In section 3 we show that the $T_q$, $\hat{T}_q$ matrices are related to the $\tau_{k,q}^{(j)}$ by the hierarchy of functional relations (3.46). The matrices satisfy the commutation relations (2.32), (3.50), and can be simultaneously diagonalized by the transformations (2.34), (3.49), where $P_B$ and $P_W$ depend on the modulus $k'$ and the vertical rapidities $p, p'$ (or $p_1,\ldots,p_{2L}$), but **are independent of q**.

In section 4 we establish another functional relation (4.20). This is (in changed notation) the result (5.14), (5.20) of Ref. [24]. It is also written here in the forms (4.21), (4.31). We go on to deduce a third functional relation: the hierarchy (4.27), which is a closed set of relations for the tau-matrices. Finally we obtain (4.40), which is a closed relation (when combined with (2.32) and (4.46)) for the matrices $T_q$, $\hat{T}_q$.

These are the main results of this paper. From them it should be possible to calculate all the eigenvalues of the various transfer matrices. Note from (3.46) (with $j = N$) that the zeros of $T_q$ are also zeros of $\tau_{k,q}^{(N)}$.

In section 5 we have written down the relations explicitly for $N = 2$ (the Ising case) and for $N = 3$. For the latter, we obtain the $T_q$, $\hat{T}_q$ relation postulated in [17].

In section 6 we discuss the "superintegrable case" of the solvable chiral Potts model, using the generalization to a row-inhomogeneous model given in [22]. We show that the model is superintegrable provided only that the vertical rapidities are alternately $p, \bar{p}, p, \ldots, \bar{p}$, where $\bar{p} = \bar{p}(0,0)$ is the "conjugate" of $p$, as defined by (2.37), (2.38), (6.4). The horizontal rapidities $q_1, \ldots, q_M$ are not restricted. For fixed spin boundary conditions at top and bottom of the lattice (periodic from right to left), the partition function $Z$ can be calculated exactly. The result is given in (6.7) - (6.19), and in an equivalent alternative form in (6.31). For the homogeneous case (when $\bar{p} = p$ and $q_1 = \ldots = q_M = q$) at criticality (when $k = 0$), the finite-size corrections are given in (6.38) - (6.48). Because of the intrinsic anisotropy of the model, they do not have the form predicted by conformal invariance. It is an interesting question whether the arguments of conformal invariance can be extended to include this model.

Finally, in the last half of section 6 we consider the superintegrable model after a 90° rotation. Now the vertical rapidities are arbitrary, but the horizontal ones alternate: $q, \bar{q}, q, \bar{q}, \ldots$. Thus the two-row transfer matrix is $T_q \hat{T}_{\bar{q}}$, which from (3.46) is proportional to $\tau_{0,q}^{(N)}$. From (6.46) its inverse (now with fixed-spin boundary conditions at right and left) is proportional to $T_{U^{-1}q} \hat{T}_{Rq}$, i.e. from (3.46) to

$\tau^{(2)}_{0,U^{-1}q}$ . Since $\tau^{(2)}$ is a simpler and much sparser matrix than $\tau^{(N)}$, this explains the observation in [22] that the inverse of $T_{col}$ therein is simpler than $T_{col}$ itself. Also, we see that the "inverse superintegrable" model defined in [22], with transfer matrix $T^{-1}_{col}$, is just the "two-state by N-state" model of [24] and this paper.

With fixed-spin boundary conditions, it follows that the eigenvalues of $\tau^{(N)}$ are proportional to those obtained in [22] for $T_{col}$, and given here in (6.34). In particular, they have a simple direct product structure. Further, the relations (4.27) simplify to (6.60), and it should be possible to solve these directly and verify (6.34).

The corresponding Hamiltonian has an intriguingly simple direct-sum structure, as is discussed in [22] and [23].

## APPENDIX A : THE FUNCTIONS Φ

The functions $\Phi(x,y)_i^{m,n}$ are defined by (3.26) and are non-zero for $0 \le i \le m+n$. Define the Gaussian polynomial [31]

$$\begin{bmatrix} n \\ r \end{bmatrix} = \frac{(1-\omega^{n-r+1})(1-\omega^{n-r+2}) \cdots (1-\omega^{n})}{(1-\omega)(1-\omega^2) \cdots (1-\omega^r)} \tag{A1}$$

for $0 \le r \le n$, and zero otherwise. Then for all $x,y \in \mathbb{C}$ and $n \ge 0$

$$(x,y)_{0,n} = \sum_{r=0}^{n} \omega^{r(r+1)/2} \begin{bmatrix} n \\ r \end{bmatrix} x^{n-r}(-y)^r \quad . \tag{A2}$$

It follows that

$$\Phi(x,y)_i^{m,n} = (-1)^i \sum_{r=0}^{i} \omega^{[r(r+1)+(i-r)(i-r+1)]/2} \times \begin{bmatrix} m \\ r \end{bmatrix} \begin{bmatrix} n \\ i-r \end{bmatrix} x^r y^{i-r} \quad . \tag{A3}$$

To establish the result (3.34), note from (2.10) that

$$(1,xu)_{0,N-1} = (1-x^N u^N) / (1-xu) \quad . \tag{A4}$$

Hence $\Phi(x,y)_i^{N-1,n}$ is the coefficient of $u^i$ in the Taylor expansion of

$$(1-x^N u^N)\,(1,yu)_{0,n} / (1-xu) \quad .$$

For $i < N$ this is the same as the coefficient of $(1,yu)_{0,n} / (1-xu)$, but for any Taylor expandable function $f(x)$ the coefficient of $u^i$ in $f(yu) / (1-xu)$ is the polynomial part of $x^i f(y/x)$. Hence for $0 \le n \le i < N$

$$\begin{aligned} \Phi(x,y)_i^{N-1,n} &= x^i (1,y/x)_{0,n} \\ &= x^{i-n}(x,y)_{0,n} \quad . \end{aligned} \tag{A5}$$

The result (3.34) now follows.

Another identity follows from the simple relation

(A6) $$(1,x^{-1})_{0,m} = (-x)^{-m}\,\omega^{m(m+1)/2}\,(1,\omega^{-m-1}x)_{0,m} \quad .$$

Using this in (3.26), we get

(A7) $$\Phi(x,y)_i^{m,n} = (-1)^{m+n}\,\omega^{-(m+n)(m+n+1)/2}\,\omega^{m^2+(n+1)i}$$
$$x^{i-n}\,y^{i-m}\,\Phi(x,\omega^{n-m}y)_{m+n-i}^{n,m}$$

for all $x,y \in \mathbb{C}$ and all non-negative integers $i,m,n$ satisfying $i \le m+n$ .

## REFERENCES


1. H. Au-Yang, B. M. McCoy, J. H. H. Perk, S. Tang and M. L. Yan, Phys. Lett. **123A** (1987) 219-223.

2. B. M. McCoy, J. H. H. Perk, S. Tang and C. H. Sah, Phys. Lett. **125A** (1987) 9-14.

3. J. H. H. Perk, Preprint ITP-SB-87-57, to be published in Proc. 1987 Summer Institute on Theta functions (Am. Math. Soc.)

4. H. Au-Yang, B. M. McCoy, J. H. H. Perk and S. Tang in Vol. 1 of "Algebraic Analysis", eds. M. Kashiwara and T. Kawai (1988) Academic.

5. L. Dolan and M. Grady, Phys. Rev. **D25** (1982) 1587-1604.

6. S. Howes, L. P. Kadanoff and M. den Nijs, Nucl. Phys. **B215** [FS7] (1983) 169-208.

7. G. von Gehlen and V. Rittenberg, Nucl. Phys. **B257** [FS14] 351-370.

8. R. J. Baxter, J. H. H. Perk and H. Au-Yang, Phys. Lett. **A128** (1988) 138-142.

9. H. Au-Yang and J. H. H. Perk, Proc. of Taniguchi Symposium, Vol 19 (1989) Kinokumiya/Academic.

10. E. H. Lieb, Phys. Rev. **162** (1967) 162-172.

11. E. H. Lieb, Phys. Rev. Lett. **18** (1967) 1046-1048.

12. E. H. Lieb, Phys. Rev. Lett. **19** (1967) 108-110.

13. R. J. Baxter, Ann. Physics (N.Y.) **76** (1973) 1-24, 25-47, 48-71.

14. L. Onsager, Phys. Rev. **65** (1944) 117-149.

15. R. J. Baxter, "Exactly Solved Models in Statistical Mechanics" (1982) Academic.

16. R. J. Baxter, J. Stat. Phys. **52** (1988) 639-667.

17. G. Albertini, B. M. McCoy and J. H. H. Perk, Phys. Lett. **135A** (1989) 159-166.

18. G. Albertini, B. M. McCoy, J. H. H. Perk and S. Tang, Nucl. Phys. **B314** (1989) 741-763.

19. G. Albertini, B. M. McCoy and J. H. H. Perk, Phys. Lett. 139A (1989) 204-212.

20. H. Itoyama, B. M. McCoy and J. H. H. Perk, Preprint ITP-SB-89-46, "Level crossing, spontaneous parity violation, high-$T_c$ superconductivity mechanisms and the chiral Potts chain."

21. R. J. Baxter, Phys. Lett. **133A** (1988) 185-189.

22. R. J. Baxter, Preprint CMA-R12-89, "Superintegrable chiral Potts model: thermodynamic properties, an "inverse" model, and a simple associated Hamiltonian", to appear in J. Stat. Phys.

23. R. J. Baxter, "A simple solvable $Z_N$ Hamiltonian", to appear in Phys. Lett. A.

24. V. V. Bazhanov and Yu. G. Stroganov, Preprint CMA-R34-89, "Chiral Potts model as a descendant of the six-vertex model", submitted to J. Stat. Phys.

25. H. W. J. Blöte, J. L. Cardy and M. P. Nightingale, Phys. Rev. Lett. **56** (1986) 742-745.

26. I. Affleck, Phys. Rev. Lett **56** (1986) 746-748.

27. V.B. Matveev and A.O. Smirnov, "Some comments on the solvable chiral Potts model", pre-print.

28. S.C. Milne, Adv. in Mathematics **72**(1988), 59-131.

29. P.J. Davis, "Circulant Matrices", §3.2, John Wiley, New York (1979).

30. T.T. Truong, *in* "Clifford Algebras and their application in Mathematical Physics", eds. J.S.R. Chisholm and A.K. Common (D. Reidel, Dordrecht, Holland, 1986) 541.

31. G.E. Andrews, "The Theory of Partitions" p.35 (Addison-Wesley, Reading, Mass., 1976).

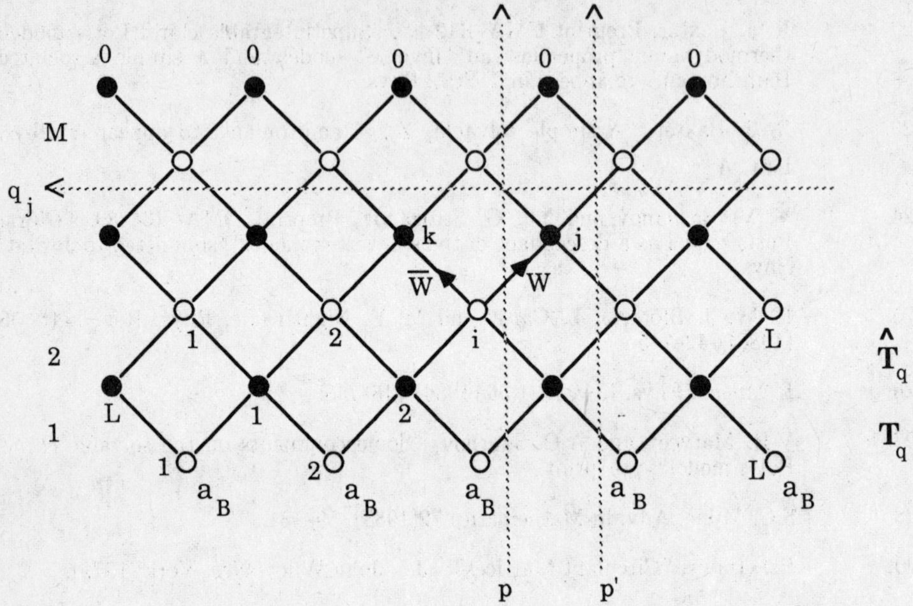


Figure 1. The square lattice $\mathcal{L}$, showing two vertical and one horizontal rapidity lines. Throughout this paper it has periodic (cylindrical) boundary conditions from right to left. In sections 2 through 5 it also has periodic (toroidal) boundary conditions from top to bottom. In section 6 fixed-spin boundary conditions are imposed at top and bottom, with values 0 and $a_B$, as shown.

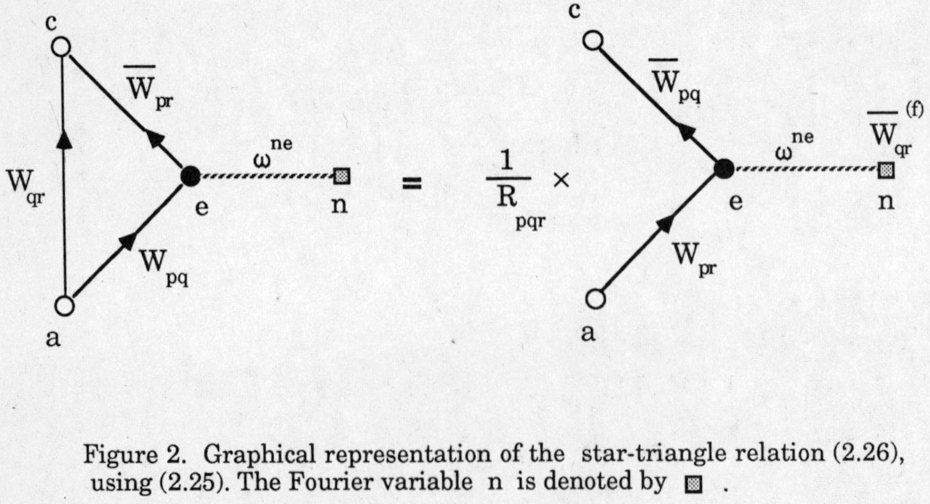


Figure 2. Graphical representation of the star-triangle relation (2.26), using (2.25). The Fourier variable n is denoted by ▩ .

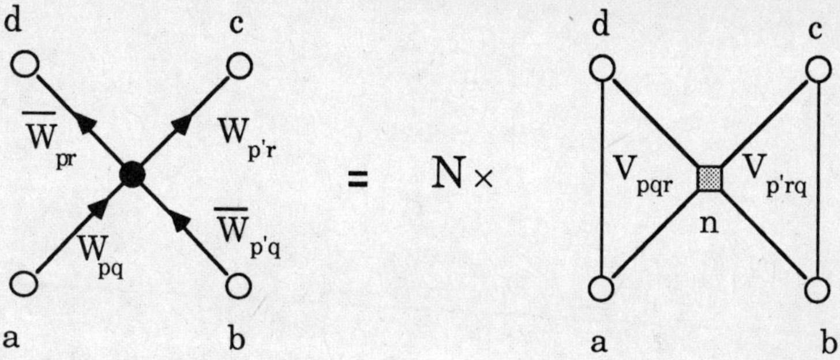


Figure 3: Graphical representation of the star weight $U_{qr}$(a,b,c,d), as given by eqns. (2.28)-(2.29).

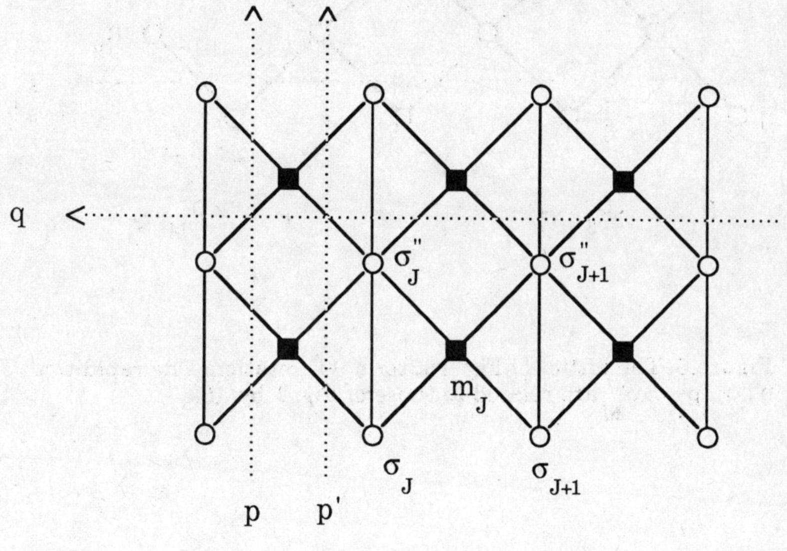


Figure 4. The lattice of the " j by N " model with row transfer matrix $\tau^{(j)}_{k,q}$ . The $m_J$ take values 0,1,...,j-1 ; the $\sigma_J$ take values 0,1,...,N-1.

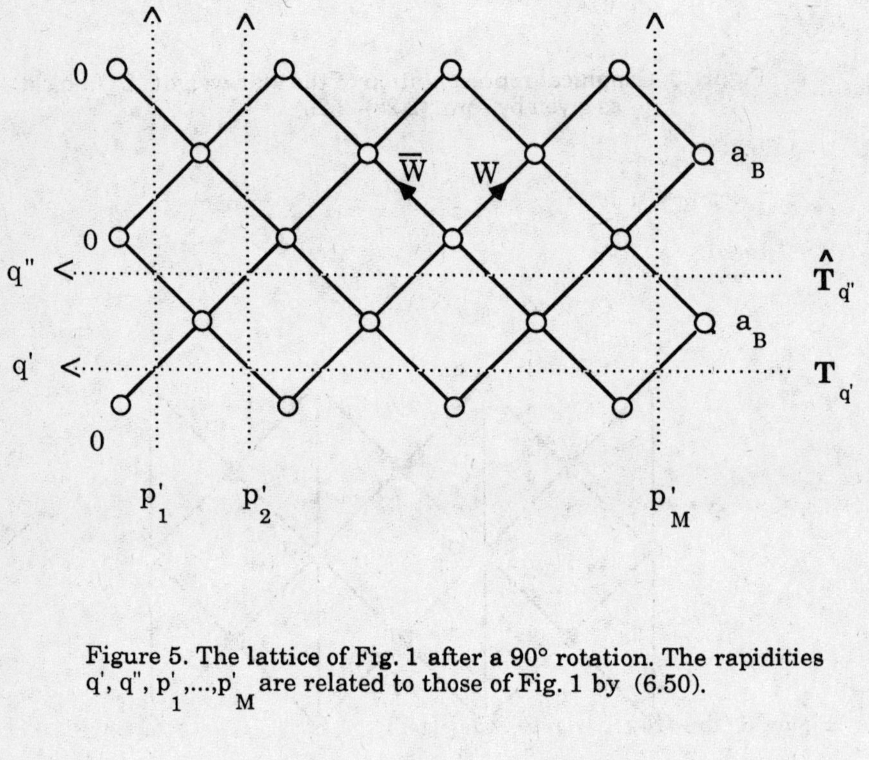


Figure 5. The lattice of Fig. 1 after a 90° rotation. The rapidities $q'$, $q''$, $p'_1,\ldots,p'_M$ are related to those of Fig. 1 by (6.50).

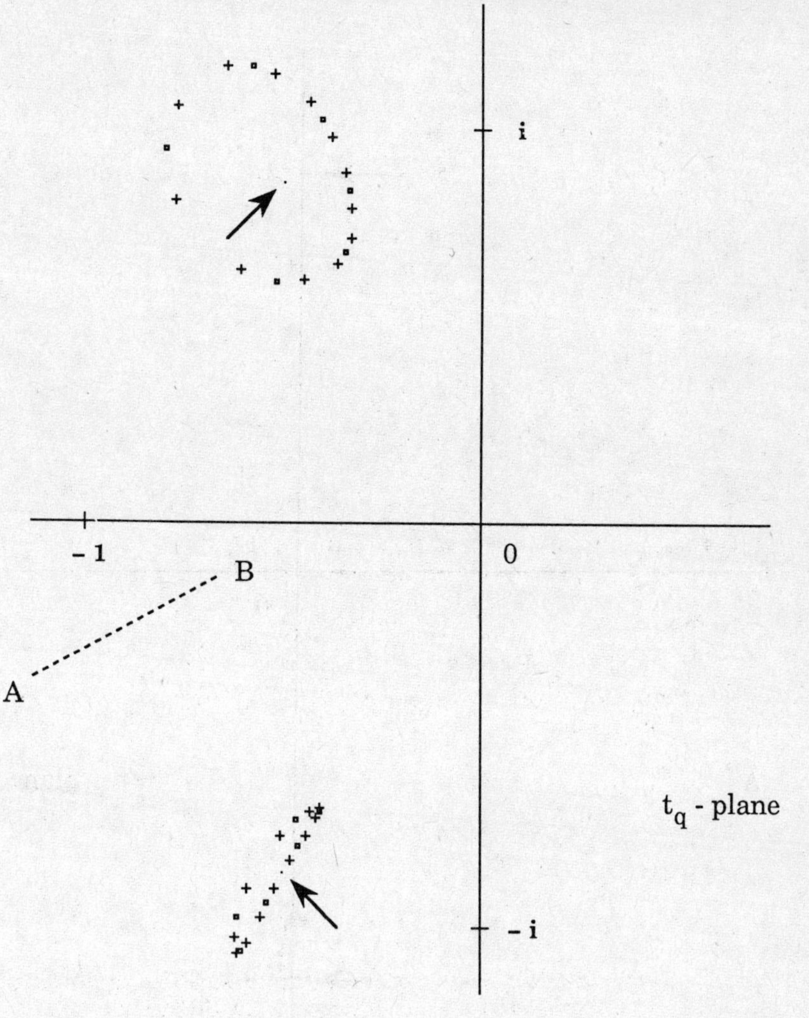


Figure 6. The (N-1)L zeros of the eigenvalue of $\tau_0^N(t_q)$ corresponding to the maximum eigenvalue of $T_q$. They are determined from (4.27), with $k' = 0.3$, $p' = p$, $t_p = \exp(-1.2\,\pi i/N)$, $x_p = \exp(-3.75498i/N)$, $y_p = \exp(-0.01493i/N)$, $\exp(i\pi/N)\,\mu_p^2\,x_p/y_p = (0.0259360)^{1/N}$ = real and positive; and with $N = 3$. Points marked ▪ are for $L = 6$ ; + for $L = 12$. From (4.44), these are also the zeros of the maximum eigenvalue of $T_q$, those below the line AB being zeros on the Riemann sheet with $|\mu_q| < 1$, those above on the sheet with $|\mu_q| > 1$. The arrows point to the Nth roots of unity, other than one.

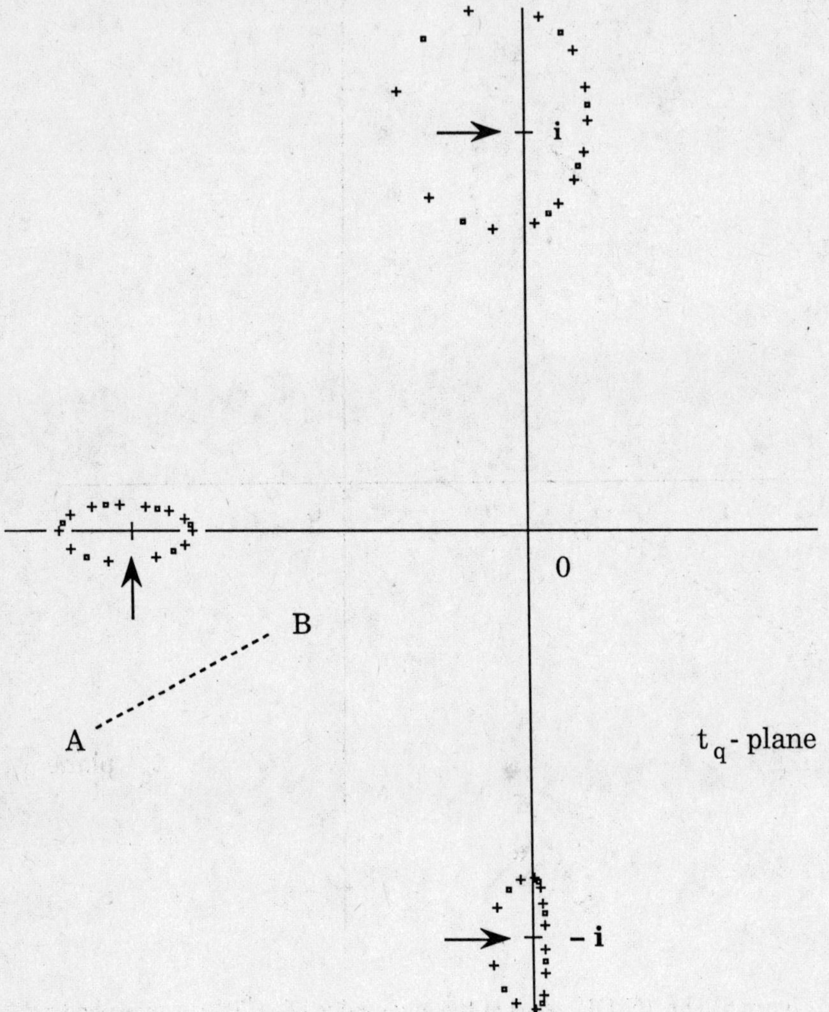


Figure 7. The same as Figure 6, except with N = 4 .

International Journal of Modern Physics B, Vol. 4, No. 5 (1990) 871–893
© World Scientific Publishing Company

# INVESTIGATION OF EXCITATION SPECTRA OF EXACTLY SOLVABLE MODELS USING INVERSION RELATIONS

A. KLÜMPER

*Institut für Theoretische Physik, Universität zu Köln, Zülpicher Str. 77, D-5000 Köln 41, W.-Germany*

We study the excitations of two-dimensional classical vertex models and related one-dimensional quantum Hamiltonians. The main mathematical means we use are functional equations which are set up and solved for the excitation functions. The method is first applied to the eight-vertex model as a prototype and compared with the traditional solution procedure of the Bethe Ansatz. In the last part we investigate some $q$-state models, notably the biquadratic spin-1 quantum antiferromagnet and its generalizations, which recently attracted interest due to Haldane's conjecture.

## 1. Introduction

We investigate exactly solvable two-dimensional vertex models and related one-dimensional quantum chains. These models are exactly solvable in the sense that they satisfy the Star-Triangle/Yang-Baxter equation [1] and hence possess a continuous family of commuting transfer matrices. Instead of solving the models by means of the Bethe Ansatz [2,1,3] we determine the spectra of the corresponding transfer matrices by applying functional/inversion relations for the eigenvalues. Since Stroganov's solution of some 3-state vertex models [4] inversion relations have been used many times to calculate partition functions of two-dimensional classical models [1].

In Sect. 2 we describe a method to determine also the excitations taking the eight-vertex model as an example. In Sect. 2.1 the definition of the eight-vertex model and the Bethe Ansatz equations of Baxter's solution [5,1] are presented. In Sect. 2.2 two functional relations for the transfer matrix and consequently for all eigenvalues are derived. In Sect. 2.3 Stroganov's method for calculating the partition function is shortly reviewed. In Sect. 2.4 the next-largest eigenvalues of the transfer matrix are determined by appropriate functional equations. This method is contrasted with the traditional procedure of the Bethe Ansatz. We also give some remarks on the classification of the excited states and the asymptotics of the 2-point correlations with horizontal, vertical and diagonal separation. In Sect. 2.5 the relation of the spin-1/2 $XYZ$-chain and the eight-vertex model transfer matrix is pointed out. The spectrum of the Hamiltonian for instance can be calculated directly from the eigenvalues of the transfer matrix.

In Sect. 3 we study the antiferromagnetic spin-1 chain as well as some $q$-state vertex models ($q > 2$) and associated quantum spin chains. We follow the route outlined in Sect. 2. In Sect. 3.1 the vertex models are defined. In Sect. 3.2 two functional relations for the transfer matrices are derived which are solved in Sect. 3.3 for the partition function and in Sect. 3.4 for the excitations. In Sect. 3.5 the related quantum spin chains are investigated. Renewed interest in quantum spin chains is due to Haldane's conjecture that integral spin Heisenberg chains are non-critical in contrast to half-integer spin Heisenberg chains [6,7]. These ideas initiated an intensive study [8-16] of the phase diagram of the bilinear-biquadratic spin-1 Hamiltonian which is the most general $SU(2)$ invariant 3-state model with nearest neighbour interaction

$$H_{\text{bil.biqu.}} = \sum_{j=1}^{N} \left[\cos\Theta\ \vec{S}_j \cdot \vec{S}_{j+1} + \sin\Theta\left(\vec{S}_j \cdot \vec{S}_{j+1}\right)^2\right].$$

In the sector $\frac{\pi}{2} \le \Theta \le \frac{5}{4}\pi$ the model is ferromagnetic and the ground state trivial. The Uimin [17] $\Theta = \frac{\pi}{4}$ and Takhtajan-Babujian [18,19] $\Theta = -\frac{\pi}{4}$ points are exactly integrable and critical. For the model characterized by $\tan\Theta = 1/3$ the ground state is a valence bond solid [13,14] and the existence of a non-vanishing mass-gap is known as well as the 2-spin correlation function which decays exponentially. The first family of quantum chains investigated in Sect. 3.5 is of particular interest in this connection. Each of these spin-$s$ chains ($s = 1, 3/2, 2, ...$) is defined basically as the sum of the projectors onto the singlets of neighbouring spins. The symmetry group of this family is $SU(q)$ ($q = 2s + 1$). For $s = 1$ the corresponding Hamiltonian is the biquadratic antiferromagnetic spin-1 chain respectively the $\Theta = -\pi/2$ point of the general bilinear-biquadratic model. The exact solution of this model [20-22] shows that it is non-critical and possesses a small but non-vanishing mass gap $\Delta_{bq} = 0.1731788...$ .This was obtained also independently by Barber and Batchelor [23] (building on [24]) and proves a conjecture by Affleck [25]. The large correlation length $\xi = 21.0728505...$ explains the problems faced in numerical treatments of finite chains [11,15,16].

Finally we like to remark that the method of inversion relations is also applicable to IRF models [1], e.g. to the self dual Potts model [26-28], and to models of itinerant particles, notably to the Hubbard model [29].

## 2. The Eight-Vertex Model

### 2.1 *Definition and Bethe Ansatz Equations*

The zero-field eight-vertex model is defined on a square lattice of $M$ rows and $N$ columns with cyclic boundary conditions ($N$ even). To each bond a spin variable is attached which can take two values and is visualized by an arrow pointing either up (right) or down (left). Only eight vertex configurations are allowed which are shown

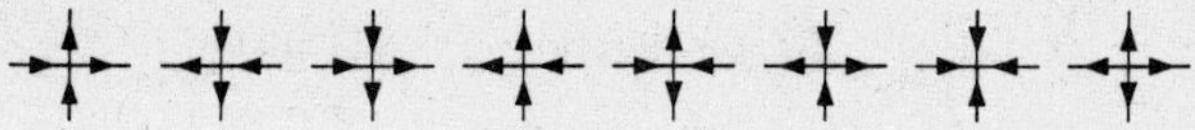


Figure 1. Representation of the eight vertex configurations and the assignment of the Boltzmann weights, $a, ..., d$.

in Fig.1. There are four Boltzmann weights associated with the vertices which in standard notation are denoted by $a, b, c$ and $d$.

The partition function of this model is

$$Z = \sum_{\text{Conf.}} a^{n_1} b^{n_2} c^{n_3} d^{n_4}, \tag{1}$$

where the sum is over all allowed arrow configurations and $n_j$ $(j = 1, \cdots, 4)$ is the number of vertices with the corresponding Boltzmann weight. The most convenient way to calculate thermodynamic properties, i.e. partition function and correlation functions, is to use the transfer matrix method. The transfer matrix $T$ describes the contribution of a single row with fixed lower and upper spin configuration to the partition function. Therefore the matrix elements are polynomials in the Boltzmann weights $(a, b, c, d)$ and the matrix dimension is $2^N \times 2^N$. As is well-known the largest eigenvalue of the transfer matrix determines the partition function and the free energy, whereas the other eigenvalues and also the corresponding eigenstates are needed for the calculation of the correlation functions.

A solution method employed quite generally is to look for a family of commuting transfer matrices especially by means of the Yang/Baxter or Star-Triangle equation [1]. The commutativity of these matrices implies the existence of an infinite number of conservation laws which seems to entail exact solvability. For the eight-vertex model Baxter [1,5] has shown that two matrices $T$ and $T'$ with weights $a$, $b$, $c$, $d$ and $a'$, $b'$, $c'$, $d'$, respectively, commute if $\Delta = \Delta'$ and $\Gamma = \Gamma'$ , where these quantities are defined by

$$\begin{aligned} \Delta &:= \frac{a^2 + b^2 - c^2 - d^2}{2(ab + cd)}, \\ \Gamma &:= \frac{ab - cd}{ab + cd}. \end{aligned} \tag{2}$$

Because of various symmetry properties [30,1] it is sufficient to consider the antiferroelectric region

$$c > a + b + d, \qquad \text{i.e.} \quad \Delta < -1, \quad |\Gamma| < 1. \tag{3}$$

The Boltzmann weights are suitably parametrized in terms of theta functions such that $\Delta$ and $\Gamma$ are independent of the so-called spectral variable $v$ :

$$a = -i\rho\Theta(i\lambda)H\Big[\frac{1}{2}i(\lambda - v)\Big]\Theta\Big[\frac{1}{2}i(\lambda + v)\Big],$$

$$\begin{aligned} b &= -i\rho\Theta(i\lambda)\Theta\Big[\frac{1}{2}i(\lambda-v)\Big]H\Big[\frac{1}{2}i(\lambda+v)\Big], \\ c &= -i\rho H(i\lambda)\Theta\Big[\frac{1}{2}i(\lambda-v)\Big]\Theta\Big[\frac{1}{2}i(\lambda+v)\Big], \\ d &= i\rho H(i\lambda)H\Big[\frac{1}{2}i(\lambda-v)\Big]H\Big[\frac{1}{2}i(\lambda+v)\Big]. \end{aligned} \tag{4}$$

Instead of the variables $a$, $b$, $c$, $d$ one has the new variables $\rho$, $k$, $\lambda$, $v$, where $\rho$ is just the scale, $\lambda$ and $v$ appear as arguments of the two theta functions $H$, $\Theta$, and the modulus $k$ which determines the quarter-periods $K$, $K'$ is not explicitly written down in this notation. The definitions of the theta functions and the related elliptic functions and relations among them can be found in [31,1,32].

The restriction to the antiferroelectric region implies for the new variables

$$\rho > 0, \quad 0 < k < 1, \quad 0 < \lambda < K', \quad -\lambda < v < \lambda. \tag{5}$$

From now on $\rho$, $k$, $\lambda$ are considered as real fixed parameters, whereas the variable $v$ may take all complex values beyond the physical regime (5).

Baxter derived for each eigenvalue $\Lambda(v)$ of the transfer matrix $T(v)$ the equation

$$\Lambda(v)q(v) = s\,[\Phi(v-\lambda)q(v+2\lambda) + \Phi(v+\lambda)q(v-2\lambda)], \tag{6}$$

where $s = \pm 1$, and $\Phi(v)$ is defined by

$$\Phi(v) := [\rho h(v/2)]^N, \tag{7}$$

$$h(v) := -i\Theta(0)H(iv)\Theta(iv). \tag{8}$$

$q(v)$ can be represented explicitly by

$$q(v) = \exp(\tau v)\prod_{j=1}^{N/2} h\Big(\frac{v-v_j}{2}\Big), \tag{9}$$

with $N/2$ unknown zeros $v_j$ in the period rectangle (width $2K'$, height $4K$) of the complex plane. These "moments" $v_j$ which lie symmetric to the imaginary axis satisfy a sum-rule [1] from which $s = \pm 1$ and the constant $\tau$ can be calculated. (6) shows that $\Lambda(v)$ is the quotient of two analytic functions, but $\Lambda(v)$ itself must be analytic. Therefore

$$\Phi(v_j-\lambda)q(v_j+2\lambda) + \Phi(v_j+\lambda)q(v_j-2\lambda) = 0 \tag{10}$$

has to hold for all $j = 1, \cdots, N/2$. With the definition of

$$p(v) := \frac{\Phi(v-\lambda)q(v+2\lambda)}{\Phi(v+\lambda)q(v-2\lambda)}, \tag{11}$$

the Bethe Ansatz (BA) equations (10) can be written as

$$p(v_j) = -1, \quad j = 1, \cdots, N/2. \tag{12}$$

It is often convenient to use two functional properties of $q(v)$:

$$\begin{aligned} q(v+4iK) &= s\,q(v), \\ q(v+2K') &= rs\,q^{-\frac{N}{4}}\exp(N\pi v/4K)q(v), \end{aligned} \tag{13}$$

where $r = \pm 1$ has its origin in the existence of two ground states and $q$ (without argument) denotes the nome of the theta functions. The foregoing equations, in particular (6)-(12) are the fundamental equations to determine all eigenvalues $\Lambda(v)$ of the transfer matrix. We point out that for finite, but large $N$ this is a very complicated system of coupled non-linear equations for which generally no standard method of solution exists.

### *2.2 Functional Relations for the Transfer Matrix*

Before passing to the calculation of the eigenvalues we look a little bit closer at the BA equations. For $v = -\lambda$ the second summand in (6) has a zero of order $N$. Therefore in a neighbourhood of $v = -\lambda$ the first summand dominates with order $O(e^N)$ over the second summand. Near $v = +\lambda$ this relation is reversed. Hence in a neighbourhood of $v = -\lambda$

$$\begin{aligned} \Lambda(v)\,q(v) &= s\,\Phi(v-\lambda)\,q(v+2\lambda)\cdot[1+O(e^{-N})], \\ \Lambda(v+2\lambda)\,q(v+2\lambda) &= s\,\Phi(v+3\lambda)\,q(v)\cdot[O(e^{-N})+1], \end{aligned} \tag{14}$$

which implies

$$\Lambda(v)\Lambda(v+2\lambda) = \Phi(v-\lambda)\Phi(v+3\lambda)\cdot[1+O(e^{-N})] = \phi(v)^N\cdot[1+O(e^{-N})] \tag{15}$$

where

$$\phi(v) := \rho^2\,h\left(\frac{v-\lambda}{2}\right)h\left(\frac{v+3\lambda}{2}\right). \tag{16}$$

This is a remarkable relation reducing a product of the unknown function $\Lambda(v)$ to a known function. The correction term vanishes exponentially in the thermodynamic limit. We want to remark that with some effort the validity of (15) can be proven for the whole strip $-\min(2\lambda, K') < \mathrm{Re}(v) < 0$ [32].

A fundamental relation like (15) should have a deeper reason than some rather technical properties of the BA equations. Indeed a vertex identity like that shown in Fig.2 is the key stone of an inversion relation for the transfer matrix $T(v)$

$$T(v)\cdot T(v+2\lambda) = \phi(v)^N\cdot[I_N + O(e^{-N})] \tag{17}$$

where $I_N$ is the identity matrix and $O(e^{-N})$ denotes a correction term [20] vanishing in the thermodynamic limit[1]. The inversion relation is exact even for finite systems at $v = -\lambda$ since $T(\pm\lambda)$ are left- and right-shift operators. (15) obviously is a direct consequence of (17).

[1]This relation was discovered independently in [33] and [34] but must also have been known to the authors of [4] and [35].

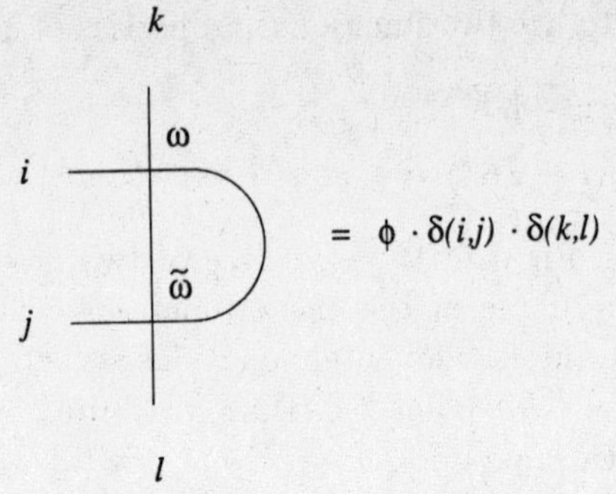


Figure 2. Graphical representation of a vertex identity which generally is at the bottom of a matrix inversion relation like (17). $\omega$ is an arbitrary set of weights, $\tilde{\omega}$ and $\phi$ are functions of $\omega$. On the left hand side a summation must be performed over the spin variables of the inner bonds.

Another quite simple symmetry relation is derived from $T^T(a,b,c,d) = T(b,a,c,d)$. The replacement $v \to -v$ interchanges the weights $a$ and $b$ hence we have $T^T(v) = T(-v)$ or

$$T^+(v) = T(-v^*). \tag{18}$$

This can be written immediately in terms of the eigenvalues $\Lambda(v)$

$$\Lambda^*(v) = \Lambda(-v^*). \tag{19}$$

The relations (15,19) will serve in the next sections to simplify the calculation of the largest and next-largest eigenvalues of the transfer matrix $T(v)$ in the thermodynamic limit.

### *2.3 Largest Eigenvalue of the Transfer Matrix and Partition Function*

In this section we shortly describe the calculation of the eigenvalue $\Lambda_0(v)$ which is largest in the physical region $-\lambda \le \mathrm{Re}(v) \le \lambda$. It is convenient to define a limiting function

$$\psi(v) := \lim_{N\to\infty} \Lambda_0^{1/N}(v), \tag{20}$$

which has the meaning of the partition function per site.

The standard method within the Bethe Ansatz is to take the logarithm of (12) and to rewrite it as a linear integral equation for the density $\rho_0(v)$ of moments which are distributed densely on the imaginary axis in the thermodynamic limit. This integral equation can be solved via Fourier transform leading finally to an explicit expression for $\psi(v)$.

As is well-known there is an alternative method employing functional relations providing a shortcut for calculating $\Lambda_0(v)$ [4,1]. The required properties of $\psi(v)$ are

- analyticity in the physical region $-\lambda \le \mathrm{Re}(v) \le \lambda$,[2]
  no zeros therein,

[2] by this, analyticity in some open region containing the strip $-\lambda \le \mathrm{Re}(v) \le \lambda$ is understood

- periodicity $\psi(v + 4iK) = \psi(v)$,
- inversion relation, $\psi(v)\psi(v+2\lambda) = \phi(v)$. (21)

[Of course these properties can be proven by means of the BA equations but they are quite obvious.]

The first two properties of $\psi(v)$ enable one to write $\ln \psi(v)$ as a Fourier series. The Fourier coefficients then can be read off from the inversion relation leading directly and unambiguously to the result

$$\ln[\psi(v)] := \ln(\rho\gamma/x) - \sum_{m=1}^{\infty} \frac{(x^{3m} + q^m x^{-m})(z^m + z^{-m})}{m(1-q^m)(1+x^{2m})}, \tag{22}$$

where

$$\begin{aligned} x &:= exp(-\pi\lambda/2K), \\ z &:= exp(-\pi v/2K), \\ \gamma &:= q^{1/4}\Theta(0)\prod_{m=1}^{\infty}(1-q^{2m})^2. \end{aligned} \tag{23}$$

This is the shortest and most elegant solution for the partition function of the eight-vertex model. It avoids completely the involved Bethe Ansatz and is based mainly on the inversion relation (21). [Note that the symmetry (19) applied to $\Lambda_0(v)$ yields $\psi(v) = \psi(-v)$. This relation was not needed to derive $\psi(v)$ but of course is satisfied by the result].

## *2.4 Next-largest Eigenvalues and Correlation Length*

Here we are interested in the "next-largest" eigenvalues $\Lambda(v)$ which are defined as all those eigenvalues for which the ratio $\Lambda/\Lambda_0$ has a finite (non-zero) thermodynamic limit. It is convenient to define

$$l(v) := \lim_{N\to\infty} \frac{\Lambda(v)}{\Lambda_0(v)} \tag{24}$$

which we call excitation function.

To compute this function within the Bethe Ansatz it is obvious to look for a moment distribution which is similar to that in the "ground state" $\Lambda_0(v)$.

- a finite number $m$ of moments may have a non-zero real part ("complex moments"), while
- all the other moments, $\frac{N}{2} - m$, should lie on the imaginary axis ("imaginary moments").

The complex moments usually are grouped in so-called strings [36,37] in order to satisfy the BA equations. This concept was first criticized in [38-40] for related models. We will comment on the "string hypothesis" for the eight-vertex model at the end of this section. The distribution of the imaginary moments can be expressed again by an integral equation for some density function $\rho(v)$. But here the solution is more involved and complicated than for $\Lambda_0(v)$ [37]. For instance it is impossible to foresee that there is only one kind of excitations despite the possibility of different kinds of strings (e.g. 1-strings, 2-strings).

We now describe a procedure using functional relations to derive the excitation function directly without being bothered with strings etc. $l(v)$ must satisfy a simple inversion relation obtained by dividing equations (15) applied to $\Lambda(v)$ and $\Lambda_0(v)$: $l(v)\,l(v+1) = 1$. In the same way (19) gives rise to the symmetry relation $l^*(v) = l(-v^*)$. Following [20,32,28] the excitation function can be determined. The required properties of $l(v)$ are

- analyticity in $-\lambda \leq \text{Re}(v) \leq \lambda$,
  zeros are allowed,
- periodicity, $l(v+4iK) = l(v)$,
- inversion relation, $l(v)\,l(v+2\lambda) = 1$, (25)
- symmetry relation, $l^*(v) = l(-v^*)$.

The first property (analyticity) is an assumption made only for some region containing $-\lambda \leq \text{Re}(v) \leq \lambda$. Using the inversion relation, however, the excitation function can be continued onto the whole complex plane. For simplicity this continuation is denoted also by $l(v)$. Since $l(v)$ has zeros in the physical region the continuation is a meromorphic function. Applying the inversion relation twice we obtain

$$l(v+4\lambda) = \frac{1}{l(v+2\lambda)} = l(v). \tag{26}$$

Therefore $l(v)$ has two independent periods, the new real period $4\lambda$ and the old "trivial" period $4iK$. A meromorphic function defined on the whole complex plane having two independent periods is an elliptic function. As is well-known it is determined up to a constant factor by its zeros and poles. We obtain

$$l(v) = C \cdot \prod_{j=1}^{\nu} \sqrt{k_1}\, snh\Big[\frac{K_1}{2K}(v - \Theta_j)\Big] \tag{27}$$

where $snh$ is the elliptic $snh$-function of modulus $k_1 \in (0,1)$ which is defined by requiring that the corresponding quarter-periods $K_1$, $K_1'$ satisfy

$$\frac{K_1'}{K_1} = \frac{\lambda}{K}. \tag{28}$$

$\Theta_j$ are the zeros of $l(v)$ in the physical region whose number is denoted by $\nu$. [The identity (27) is proven by dividing the left hand side by the product on the right hand side. The ratio is an elliptic function without any zeros or poles. According to Liouville's theorem this function has to be constant.]

A simple argument shows that $\nu$ must be even. [Each $snh$-factor on the right hand side of (27) is $4iK$-antiperiodic, but $l(v)$ is periodic.] Inserting (27) into the inversion relation one obtains $1 = l(v)\,l(v+1) = C^2 \cdot \left(\prod \cdots\right) = C^2 \cdot 1$ whereby $C$ is determined: $C = \pm 1$. The final result for the excitation function $l(v)$ is

$$l(v) = \pm \prod_{j=1}^{\nu} \sqrt{k_1}\, snh\Big[\frac{K_1}{2K}(v - \Theta_j)\Big]. \tag{29}$$

For $\nu = 2$, this is basically the result of [37] which surprisingly turned out to be independent of the distribution of the complex moments. In our derivation, however, the reason for this fact is obvious.

We now address to the distribution of the zeros $\Theta_j$ whose number is $\nu = 2, 4, \ldots$ . They play the role of excitation parameters/rapidities. The location is restricted by the symmetry relation $l^*(v) = l(-v^*)$. It implies that for each $\Theta_j$ there must be an $\Theta_i$ such that $\Theta_j = -\Theta_i^*$. In the simplest case ($i = j$) this is satisfied if $\mathrm{Re}(\Theta_j) = 0$. Then there are no further restrictions on $\Theta_j$ and the excitation function is a superposition of free states. This is actually the generic type of excitations for the eight-vertex model. In terms of a Bethe Ansatz like solution the existence of free particles can be traced back to 1- and 2-strings, i.e. strings consisting of 1 or 2 moments, in the repulsive parameter range $\lambda < \frac{K'}{2}$ or to so-called type II-strings in the attractive parameter range $\lambda > \frac{K'}{2}$ [32 ,41]. Really conjugate complex pairs of zeros ($i \neq j$) are only allowed in the attractive range [32,41] and are due to so-called type I-strings. The corresponding states can be interpreted physically as bound states.

The last considerations concerning the occurence of bound states could be made in [32,41] unfortunately only by employing the BA equations again. However, the existence of bound states seems to be related to some kind of "competition" between the "new real period" $4\lambda$ and the "old real period" $2K'$. In the repulsive regime we have $4\lambda < 2K'$ in contrast to the attractive regime where $4\lambda > 2K'$!

Classification of excited states

Before passing to the calculation of the correlation length we want to make some remarks about the distribution of complex moments in the BA solution of the eight-vertex model. We restrict ourselves to the simpler repulsive regime $\lambda < K'/2$ (the general case was investigated in [41]) where only 1- and 2-strings are allowed [41] in contrast to the "string hypothesis" [36]. The positions of the strings are described conveniently by some new variables $\chi$. In [41] a set of equations was derived relating the $\chi$ variables to the rapidities $\Theta$

$$\prod_{\Theta \in \mathcal{R}} \frac{\hat{H}(\chi - \Theta + \lambda)}{\hat{H}(\chi - \Theta - \lambda)} = -\exp\Big(\pi n \frac{\lambda}{K}\Big) \prod_{\tilde{\chi} \in \mathcal{S}} \frac{\hat{H}(\chi - \tilde{\chi} + 2\lambda)}{\hat{H}(\chi - \tilde{\chi} - 2\lambda)}. \tag{30}$$

where $n$ is some integer which can be determined from a sum-rule [41] and

$$\begin{aligned} \mathcal{R} &:= \text{set of all imaginary rapidities,} \\ \mathcal{S} &:= \text{set of all string variables,} \end{aligned} \tag{31}$$

$$\hat{H}(v) \;:=\; H\left(\frac{K_2}{2iK}\cdot v\right).$$

$H$ is a theta function of a new modulus $k_2$ with quarter-periods $K_2$, $K_2'$ which must satisfy

$$\frac{K_2'}{K_2} = \frac{K'-\lambda}{2K}. \tag{32}$$

The solutions of (30) describe 1- or 2-strings which do not necessarily lie on the axis $\mathrm{Re}(v) = K'$ or symmetrically about the axis $\mathrm{Im}(v) = 0$, respectively. The only restriction is that the overall distribution of moments is symmetric about the axis $\mathrm{Im}(v) = 0$. This again is in contrast to the string hypothesis.

The set of equations (30) allows to classify the excitations. Although the eigenvalues of the transfer matrix are specified by a set of rapidities $\Theta_1, ..., \Theta_\nu$ the spectrum is not completely known. The missing degeneracies of the eigenvalues (in principal) can be calculated from (30) by determining all $\chi$-distributions for a given set of $\Theta$-parameters. For $\nu = 2$ and 4 this was carried out in [41] yielding a degeneracy of 4 and 16. This suggests that a general $\nu$-particle excitation is $2^\nu$-fold degenerate. It is also supported by the similarity of (30) ("higher level BA equations") to the BA equations of an inhomogeneous eight-vertex model with chain length $\nu$. Since this matrix has $2^\nu$ eigenvalues and the BA equations are assumed to be complete, we also expect $2^\nu$ different solutions for (30).

To count the states of all $\nu$-particle bands one has to take into account an overall degeneracy of 2 due to the doubly degenerate ground state (or the discrete quantum number $r = \pm 1$), the number $2^\nu$ of all $\nu$-particle bands as well as the number of states per band which is in leading order $\binom{N/2}{\nu}$: $2 \cdot 2^\nu \cdot \binom{N/2}{\nu} \simeq 2 \cdot \binom{N}{\nu}$. Furthermore we conjecture that the total number of $\nu$-particle states is exactly $2 \cdot \binom{N}{\nu}$ which implies the completeness of the spectrum constructed from the original BA equations because of

$$\sum_{\substack{\nu=0 \\ \nu\,\mathrm{even}}}^{N} 2 \cdot \binom{N}{\nu} = 2^N. \tag{33}$$

Correlation length

From the knowledge of the excitations of the transfer matrix the exponential decay of the 2-spin correlation functions can be calculated. Particularly simple is the case of purely horizontally or purely vertically separated spins. Due to the symmetry property these correlation functions are related and the corresponding correlation lengths are equal. The correlation of two spins separated by $R$ rows is

$$C_{(0,R)} = \frac{\mathrm{Tr}\,\sigma\, T^R \sigma T^{M-R}}{\mathrm{Tr}\, T^M} \tag{34}$$

where $\sigma$ is a spin operator, $M$ the number of all rows and $T$ the transfer matrix. Following the treatment in [37] the asymptotic behaviour ($R$ large) of $C_{(0,R)}$ can be obtained in the thermodynamic limit ($N, M \to \infty$) from the band of lowest excitations $l(v; \Theta_1, \Theta_2)$

$$C_{(0,R)} - C_{(0,\infty)} = \int_{-2iK}^{2iK} d\Theta_1 \int_{-2iK}^{2iK} d\Theta_2\, g(\Theta_1, \Theta_2)\, l(v; \Theta_1, \Theta_2)^R, \tag{35}$$

where $g(\Theta_1, \Theta_2)$ is an unknown weight function and the integration is extended over all 2-particle excitations. Assuming some analyticity and periodicity properties of $g$ a saddle point integration can be performed yielding

$$C_{(0,R)} - C_{(0,\infty)} \simeq (\pm 1)^R k_1^R. \tag{36}$$

This shows that the correlation length is given by

$$\xi = -\frac{1}{\ln k_1}. \tag{37}$$

Similarly the correlation length in other directions can be calculated. This is achieved without too much effort for diagonally separated spins $C_{(R,\pm R)}$. Since the spins are now separated also by $R$ columns one has to introduce $R$ shift operators $T(\pm\lambda)$ in a formula analogous to (34)

$$C_{(R,\pm R)} = \frac{\mathrm{Tr}\,\sigma\, T(v)^R \{T(\pm\lambda)^R \sigma\, T(\pm\lambda)^{-R}\}\, T^{M-R}}{\mathrm{Tr}\, T^M}. \tag{38}$$

Following the standard procedure we obtain

$$C_{(R,\pm R)} - C_{(\infty,\pm\infty)} \simeq \left( \frac{k_1}{dn\left[\frac{K_1}{4K}(v \mp \lambda), k_1'\right]^2} \right)^{2R} \tag{39}$$

where $dn$ is the elliptic $dn$-function of modulus $k_1'$ which is conjugate to $k_1$, $k_1^2 + k_1'^2 = 1$. From (39) the correlation length can be read off immediately. Note that (39) depends on the spectral parameter in contrast to (36). This dependence is particularly obvious if one specializes $v = \mp\lambda$ where the corresponding correlation length is infinite $C_{(R,\pm R)} - C_{(\infty,\pm\infty)} \simeq 1$. This of course is no critical behaviour but just a consequence of the reduction of the transfer matrix to a simple shift operator entailing strict alignment of the spins on the diagonal. Critical behaviour is observed only if $k_1 = 1$ where all correlation lengths become infinite.

### *2.5 The XYZ-Chain*

Usually there is a one-dimensional quantum Hamiltonian $H$ which is related to a two-dimensional classical model. The Hamiltonian corresponding to the eight-vertex model is the one-dimensional anisotropic spin-1/2 Heisenberg model

$$H_{XYZ} = -\frac{1}{2} \sum_{j=1}^{N} (J_X \sigma_j^X \sigma_{j+1}^X + J_Y \sigma_j^Y \sigma_{j+1}^Y + J_Z \sigma_j^Z \sigma_{j+1}^Z), \tag{40}$$

where $\sigma_{N+1} = \sigma_1$ and $\sigma^{X,Y,Z}$ are the Pauli spin matrices. It was shown by Sutherland [42] and Baxter [5] that the logarithmic derivative of $T(v)$ at $v = -\lambda$ leads to (40) with $J_Y/J_X = \Gamma, J_Z/J_X = \Delta$:

$$H_{XYZ} = \text{const.} - \alpha\,(\ln T)'(-\lambda), \tag{41}$$

where $\alpha := 2J_X snh\,\lambda/[1 + k\,snh^2\,\lambda] > 0$. The momentum operator $P$ [which is the generator of translations: $e^{iP}$ = shift operator = $T(-\lambda)$] is given by

$$P = -i \ln T(-\lambda). \tag{42}$$

The complete solution of the eight-vertex model therefore yields the total spectrum of the quantum-mechanical $XYZ$-chain. For instance the ground state energy can be calculated from the largest eigenvalue $\Lambda_0(v)$. The energy-momentum excitations are determined by the excitation function $l(v)$

$$\begin{aligned} E - E_0 &= -\alpha(\ln l)'(-\lambda), \\ P - P_0 &= -i \ln l(-\lambda). \end{aligned} \tag{43}$$

Since $l(v)$ is a product of independent factors the excitations turn out to be sums of independent elementary momenta and energies, $p_i$ and $\epsilon(p_i)$

$$\begin{aligned} E - E_0 &= \sum_{i=1}^{\nu} \epsilon(p_i), \\ P - P_0 &= \sum_{i=1}^{\nu} p_i, \end{aligned} \tag{44}$$

where the dispersion $\epsilon(p)$ is given by

$$\epsilon(p) = \alpha \frac{K_1}{2K} \sqrt{(1 - k_1)^2 + 4\,k_1 \sin^2 p}. \tag{45}$$

This is the dispersion of free particles. We want to close this section by remarking that for the attractive regime $\lambda > \frac{K'}{2}$ in addition bound states appear with different dispersions.

## 3. $q$-State Vertex Models

We study some exactly solvable $q$-state vertex models ($q > 2$) and associated quantum spin chains. The largest as well as the next-largest eigenvalues of the transfer matrices are derived following the way outlined in Sect. 2 for the eight-vertex model. The main difference is that everything is based now on inversion relations and the required analyticity properties are more or less assumed. (For the eight-vertex model these assumptions could be checked against the BA equations.) From the excitation functions the asymptotics of the 2-point correlations are calculated as well as the spectra of the related quantum spin chains.

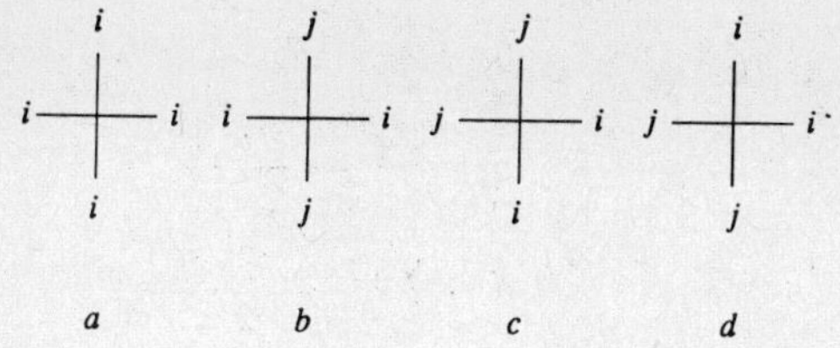


Figure 3. Representation of the four vertex types and the assignment of the Boltzmann weights $a, ..., d$. The spin variables are $i \neq j = 1, ..., q$.

### *3.1 Definition of the Vertex Models*

We consider vertex models on a square lattice of $M$ rows and $N$ columns where $M$ and $N$ are assumed to be even and periodic boundary conditions are imposed in both directions. To each bond a spin variable is attached which can take $q$ ($> 2$) values $1, \cdots, q$. Here we confine ourselves to a certain generalization of the eight-vertex model. Only four types of vertex configurations are allowed which are depicted in Fig.3. To all vertices of each type one Boltzmann weight $a$, $b$, $c$ or $d$ is assigned. We investigate three solutions of the star-triangle equation

1)

$$\begin{aligned} a &= 1, \qquad & c &= \frac{\omega z + 1 - \omega}{z+1}, \\ b &= 0, \qquad & d &= 1 - c = \frac{(1-\omega)z + \omega}{z+1}, \end{aligned} \tag{46}$$

where $\omega := \frac{1}{2}\left(1 + \sqrt{\frac{q+2}{q-2}}\right)$ is a constant and $z$ is a (complex) variable. In what follows it will be useful sometimes to parametrize $z$ as

$$z = \exp(\ln \alpha \cdot v), \tag{47}$$

where

$$\alpha := \left(\frac{\omega}{\omega - 1}\right)^2 = (q-2)^2 \omega^4 \tag{48}$$

is a constant and $v$ is the so-called spectral variable.

2)

$$\begin{aligned} a &= 1 - \tfrac{q-2}{2}\left(\tfrac{1}{4} - v^2\right), \qquad & c &= \tfrac{1}{2} + v, \\ b &= \pm\tfrac{q-2}{2}\left(\tfrac{1}{4} - v^2\right), \qquad & d &= \tfrac{1}{2} - v, \end{aligned} \tag{49}$$

where $v$ is the spectral variable.

3)

$$a = 1, \qquad c = \frac{\sqrt{q-1}\,z - 1}{q-2},$$
$$b = 0, \qquad d = \frac{\sqrt{q-1}\,z^{-1} - 1}{q-2}, \tag{50}$$

where $z$ is a variable which will be parametrized in the following as

$$z = \exp(\ln(q-1) \cdot v). \tag{51}$$

We remark that the maximal (continuous) symmetry of the row-to-row transfer matrices of cases 1) and $2^-$), i.e. case 2) with minus-sign, is $SO(q)$. For cases $2^+$) and 3) there is only the trivial symmetry $\{id\}$.

Models 1) and 3) for $q = 3$ were known to Stroganov [4] who derived the partition function in these cases by solving inversion relations. For the general cases 1) – 3) Perk and Schultz similarly calculated the partition function [43,44]. In [44] some more exactly solvable models can be found within a classification of all solutions of the star-triangle equation for a more general class of vertex models than studied here. In [45,46] the partition function was calculated again for cases 1) and 3) by mapping "non-intersecting string models" onto the self dual Potts model. Family 2) and its partition function were found first in [47], the Bethe ansatz equations were derived by an analytic ansatz in [48] and the algebraic Bethe ansatz was performed in [49].

### *3.2 Functional Relations for the Transfer Matrix*

In this section we describe some properties of the row-to-row transfer matrices of the models defined above.

A well-known consequence of the star-triangle equation is the analyticity or meromorphicity of the eigenvalues $\Lambda(v)$ of the matrix $T(v)$ if the matrix elements are analytic or meromorphic, respectively.

- For families 2) and 3) each $\Lambda(v)$ is analytic.

  For family 1) $\Lambda(v)$ is analytic up to poles of maximal order $N$ at $v = (2n+1)\frac{\pi i}{\ln \alpha}$, $n$ integer (corresponding to $z = -1$).

Furthermore some periodicity properties exhibited by all entries of $T(v)$ have to be shared by each eigenvalue $\Lambda(v)$.

- For families 1) and 3) each $\Lambda(v)$ has a period $\frac{2\pi i}{\ln \alpha}$ or $\frac{2\pi i}{\ln(q-1)}$, respectively.

  For family 2) one has the asymptotic behaviour $\Lambda(v) \simeq (v^2)^N$, $v \to \pm i\infty$.

A simple symmetry relation is derived from $T^T(a,b,c,d) = T(a,b,d,c)$. Since the weights $c$ and $d$ are interchanged for all families by the substitution $v \longrightarrow -v$ we have $T^T(v) = T(-v)$ or

$$T^+(v) = T(-v^*). \tag{52}$$

From this one immediately derives the symmetry relation

$$\Lambda^*(v) = \Lambda(-v^*). \tag{53}$$

At the special points $v = \mp 1/2$ the Boltzmann weights $(a,b,c,d)$ are equal to (1,0,0,1) and (1,0,1,0). The corresponding transfer matrix reduces to a left- or right-shift operator, respectively. $T(-1/2)$ and $T(1/2)$ are inverse. A similar inversion relation is still valid in some neighbourhood of $v_0 = -1/2$

$$T(v) \cdot T(v+1) = \phi(v)^N \cdot I_N + O(e^{-N}), \tag{54}$$

where $I_N$ is the identity matrix and $O(e^{-N})$ is a correction vanishing exponentially in the thermodynamic limit. $\phi$ is defined differently for the three cases

$$\begin{aligned}
&1) \qquad \phi(v) = \frac{\omega^2 \alpha z}{(z+1)(\alpha z+1)} \cdot (1-\alpha^{-1/2}z)(1-\alpha^{-3/2}z^{-1}),\\
&2) \qquad \phi(v) = \Big[1-(1/2+v)^2\Big]\Big[1-(\tfrac{q-2}{2})^2(1/2+v)^2\Big],\\
&3) \qquad \phi(v) = (\tfrac{q-1}{q-2})^2 \cdot \Big[1-(q-1)^{-1/2}z\Big]\Big[1-(q-1)^{-3/2}z^{-1}\Big],
\end{aligned} \tag{55}$$

where $z$ is parametrized by $v$ according to (47) or (51). The proof of (54) is based again on a vertex identity shown in Fig.2 [20,22]. (54) directly implies for the eigenvalues $\Lambda(v)$

$$\Lambda(v) \cdot \Lambda(v+1) = \phi(v)^N + O(e^{-N}). \tag{56}$$

### *3.3 Largest Eigenvalue of the Transfer Matrix and Partition Function*

The eigenvalue $\Lambda_0(v)$ which is largest in the physical region $-1/2 \le \mathrm{Re}(v) \le 1/2$ or equivalently the partition function per site

$$\psi(v) := \lim_{N\to\infty} \Lambda_0^{1/N}(v), \tag{57}$$

is calculated from the inversion relation. The required properties of $\psi(v)$ are basically

- analyticity in the physical region $-1/2 \le \mathrm{Re}(v) \le 1/2$,
  no zeros therein,
- periodicity/asymptotic behaviour,
- inversion relation, $\psi(v)\psi(v+1) = \phi(v)$. (58)

A suitable ansatz (together with an uniqueness argument) [20,22] yields the following result

1)

$$\psi(v) = \omega\sqrt[4]{\alpha}\frac{\sqrt{z}}{z+1}F(z)\,F\Big(\frac{1}{z}\Big), \tag{59}$$

where

$$F(z) = \prod_{n=0}^{\infty}\frac{1-z/\alpha^{2n+3/2}}{1-z/\alpha^{2n+5/2}}. \tag{60}$$

2)

$$\begin{aligned}\psi(v) &= F(v)\,F(-v)\\ F(v) &= \sqrt{2(q-2)}\cdot\frac{\Gamma\Big(\frac{v}{2}+\frac{5}{4}\Big)}{\Gamma\Big(\frac{v}{2}+\frac{3}{4}\Big)}\cdot\frac{\Gamma\Big(\frac{v}{2}+\frac{3}{4}+\frac{1}{q-2}\Big)}{\Gamma\Big(\frac{v}{2}+\frac{1}{4}+\frac{1}{q-2}\Big)}.\end{aligned} \tag{61}$$

3)

$$\begin{aligned}\psi(v) &= \frac{q-1}{q-2}F(z)\,F\Big(\frac{1}{z}\Big),\\ F(z) &= \prod_{n=0}^{\infty}\frac{1-z/(q-1)^{2n+3/2}}{1-z/(q-1)^{2n+5/2}}.\end{aligned} \tag{62}$$

In Sect. 3.5 the ground state energy of the related quantum spin chain is calculated from $\psi(v)$.

### 3.4 Next-Largest Eigenvalues and Correlation Length

Here all eigenvalues $\Lambda(v)$ of the transfer matrix are determined for which

$$l(v) := \lim_{N\to\infty}\frac{\Lambda(v)}{\Lambda_0(v)} \tag{63}$$

is finite. The excitation function $l(v)$ must satisfy a simple inversion relation obtained by dividing equations (56) applied to $\Lambda(v)$ and $\Lambda_0(v)$: $l(v)\,l(v+1) = 1$. In the same way (53) gives rise to the symmetry relation $l^*(v) = l(-v^*)$. Following the reasoning of Sect. 2.4 the excitation function can be determined [20,28,32]. The required properties of $l(v)$ are

- analyticity in $-1/2 \leq \mathrm{Re}(v) \leq 1/2$, zeros are allowed,
- periodicity/asymptotic behaviour,
- inversion relation, $l(v)\,l(v+1) = 1$, (64)
- symmetry relation, $l^*(v) = l(-v^*)$.

1) The solution is

$$l(v) = \pm \prod_{j=1}^{\nu} \sqrt{k}\, snh[K'(v - \Theta_j)], \tag{65}$$

where $snh$ is the elliptic $snh$-function of modulus $k \in (0,1)$ which is defined by requiring that the corresponding quarter-periods $K$, $K'$ satisfy

$$\frac{K'}{K} = \frac{\ln \alpha}{\pi}. \tag{66}$$

$\Theta_j$ are the zeros of $l(v)$ in the physical region whose number $\nu$ is even. All $\Theta_j$ are independent (imaginary) excitation parameters/rapidities.

2) The reasoning here is basically the same as in Sect. 2.4. The only change is that one has to take into account the absence of an imaginary period. The asymptotic behaviour $l(v) \simeq$ const., $v \to \pm i\infty$, gives rise to the occurence of tan-factors in place of $snh$-functions. The result is

$$l(v) = \pm \prod_{j=1}^{\nu} \tan\left[\frac{\pi}{2}(v - \Theta_j)\right], \tag{67}$$

where $\nu = 2, 4, \cdots$ and the $\Theta_j$ are imaginary parameters.

3) This case is very similar to case 1) where

$$l(v) = \pm \prod_{j=1}^{\nu} \sqrt{k}\, snh[K'(v - \Theta_j)], \tag{68}$$

but here the modulus $k$ is defined by

$$\frac{K'}{K} = \frac{\ln(q-1)}{\pi}. \tag{69}$$

Knowing the next-largest eigenvalues of the transfer matrices it is possible to derive the correlation length[3]. For case 2) $\xi$ is always infinite. For cases 1) and 3) we have the following asymptotics of the 2-spin correlations

$$C_{(0,R)} - C_{(0,\infty)} \simeq (\pm 1)^R k^R, \tag{70}$$

$$C_{(R,\pm R)} - C_{(\infty,\pm\infty)} \simeq \left(\frac{k}{dn\left[\frac{K'}{2}(v \mp 1/2), k'\right]^2}\right)^{2R}, \tag{71}$$

where $k'$ is the conjugate of the modulus $k$, $k^2 + k'^2 = 1$. In the next section the excitation functions are used for caculating the energy-momentum excitations of the related quantum spin chains.

[3] Some IRF models [1] can be treated similarly employing inversion relations and yielding asymptotics of the correlations like (70,71). The self dual Potts model can be solved in this way after mapping it onto a particular IRF model involving a rotation of the lattice by an angle of $\pi/4$ [28]. Therefore the vertical correlation in the Potts model is given by a formula like the right hand side of (71) and the diagonal correlation by the right hand side of (70).

### 3.5 Quantum Spin Chains

Usually there is a one-dimensional quantum Hamiltonian $H$ which is related to a two-dimensional classical model. For many exactly solvable models the Hamiltonian limit of the transfer matrix is performed at a special point $v_0$ of the spectral variable where $T(v_0)$ is a simple shift-operator [1,3,5,42]. This guarantees that the derivative $(\ln T)'(v_0)$ is a sum of local interactions. Therefore the logarithmic derivative of $T(v)$ qualifies as a Hamiltonian

$$H = -(\ln T)'(v_0) = -\sum_{j=1}^{N} h_j, \tag{72}$$

where $v_0$ for our models is always equal to $-1/2$ and $h_j$ denotes a 2-spin operator $h$ acting on sites $j$ and $j+1$. The matrix elements of $h$ are

1) $$h^{(\beta_1,\beta_2)}_{(\alpha_1,\alpha_2)} = \frac{2\omega-1}{q+2}\cdot\ln\alpha\cdot\left[\delta(\alpha_1,\alpha_2)\cdot\delta(\beta_1,\beta_2)-\delta(\alpha_1,\beta_1)\cdot\delta(\alpha_2,\beta_2)\right],$$

2$^-$) for the minus-sign in (49) we have

$$h^{(\beta_1,\beta_2)}_{(\alpha_1,\alpha_2)} = -\frac{q-2}{2}\cdot\delta(\alpha_1,\beta_2)\cdot\delta(\alpha_2,\beta_1)+\delta(\alpha_1,\alpha_2)\cdot\delta(\beta_1,\beta_2)-\delta(\alpha_1,\beta_1)\cdot\delta(\alpha_2,\beta_2),$$

2$^+$) and for the plus-sign

$$\begin{aligned} h^{(\beta_1,\beta_2)}_{(\alpha_1,\alpha_2)} &= -(q-2)\cdot\delta(\ alpha_1,\alpha_2,\beta_1,\beta_2) \\ &+\frac{q-2}{2}\cdot\delta(\alpha_1,\beta_2)\cdot\delta(\alpha_2,\beta_1)+\delta(\alpha_1,\alpha_2)\cdot\delta(\beta_1,\beta_2)-\delta(\alpha_1,\beta_1)\cdot\delta(\alpha_2,\beta_2), \end{aligned}$$

3) $$\begin{aligned} h^{(\beta_1,\beta_2)}_{(\alpha_1,\alpha_2)} &= \frac{\ln(q-1)}{q-2}\cdot\Big[(q-2)\cdot\delta(\alpha_1,\alpha_2,\beta_1,\beta_2) \\ &+\delta(\alpha_1,\alpha_2)\cdot\delta(\beta_1,\beta_2)-(q-1)\cdot\delta(\alpha_1,\beta_1)\cdot\delta(\alpha_2,\beta_2)\Big]. \end{aligned} \tag{73}$$

The $\delta$-symbol with four arguments is equal to 1 if all entries are identical, 0 otherwise.

We want to add that in (72) periodic boundary conditions are used and that the momentum operator $P$ is given by

$$P = -i\ln T(v_0). \tag{74}$$

We briefly describe the symmetries of the Hamiltonians. The corresponding symmetry groups have to cover the symmetries of the transfer matrices but partly turn out to be bigger. The maximal symmetry of case 1) is $Sl(q)$ (comprising $SU(q)$ and $SO(q)$) yet with different representations on even and odd sites. For case 2$^-$) the maximal symmetry is $SO(q)$, for 2$^+$) it is the trivial group $\{id\}$. Case 3) has a $q$-dimensional abelian symmetry but with different representations on even and odd sites. Apparently the first two Hamiltonians can be written more nicely. Let $\vec{S}$ be the $q(q-1)/2$-vector of the generators of the group $SO(q)$ which coincides for $q=3$ with the usual $SU(2)$ spin-1 operator. We then have

$$\begin{aligned} 1) \qquad H &= -\frac{2\omega-1}{q^2-4}\ln\alpha\cdot\sum_{j=1}^{N}\left[\left(\vec{S}_j\cdot\vec{S}_{j+1}\right)^2-(q-1)\right], \\ 2^-) \qquad H &= \frac{1}{2(q-2)}\cdot\sum_{j=1}^{N}\left[(q-2)^2\left(\vec{S}_j\cdot\vec{S}_{j+1}\right)+(q-4)\left(\vec{S}_j\cdot\vec{S}_{j+1}\right)^2+q\right]. \end{aligned} \tag{75}$$

The first Hamiltonian can be formulated furthermore in a simple way in terms of $SU(2)$ spin-$s$ operators $\vec{s}_j$ (with $2s+1=q$)

$$1) \qquad H = -\frac{2\omega-1}{q+2}\ln\alpha\cdot\sum_{j=1}^{N}P\left[\left(\vec{s}_j+\vec{s}_{j+1}\right)^2\right], \tag{76}$$

where the polynomial $P[\cdots]$ basically describes a projection onto the singlet state of the neighbouring spins

$$P[x] = -1+q\cdot\prod_{j=1}^{2s}\left[1-\frac{x}{j(j+1)}\right]. \tag{77}$$

By (72,74) the Hamiltonian $H$ and the momentum operator $P$ are related to $T(v)$. Therefore all energy-momentum eigenvalues can be derived from the spectrum of $T(v)$. The ground state energy $\epsilon_0 := \lim_{N\to\infty} E_0/N$ is determined by the partition function (per site)

$$\epsilon_0 = -(\ln\psi)'(v_0). \tag{78}$$

More interesting are the (low-lying) energy-momentum excitations which are calculated from the excitation function $l\,(v)$

$$\begin{aligned} E-E_0 &= -(\ln l)'(v_0), \\ P-P_0 &= -i\ln l\,(v_0). \end{aligned} \tag{79}$$

Since $l\,(v)$ is a product of independent factors the excitations turn out to be sums of independent elementary momenta and energies, $p_i$ and $\epsilon(p_i)$

$$\begin{aligned} E-E_0 &= \sum_{i=1}^{\nu}\epsilon(p_i), \\ P-P_0 &= \sum_{i=1}^{\nu}p_i, \end{aligned} \tag{80}$$

where the dispersion $\epsilon(p)$ is given by

$$\begin{aligned} 1),3) \qquad \epsilon(p) &= K'\sqrt{(1-k)^2+4k\sin^2 p}, \\ 2) \qquad \epsilon(p) &= \pi|\sin p|. \end{aligned} \tag{81}$$

($k$, $K'$ are defined differently for cases 1) and 3)!)

Obviously the Hamiltonian of case 1) has a gap $\epsilon(0)$ multiplied by the minimal value of $\nu$ which is 2

$$1) \qquad \Delta = 2\epsilon(0) = 2K'(1-k) = \ln(\alpha)\cdot \prod_{n=1}^{\infty}\left[\frac{1-(1/\alpha)^{n/2}}{1+(1/\alpha)^{n/2}}\right]^2, \tag{82}$$

where $\alpha$ is defined in (48). The Hamiltonians of case 2) are gapless. In case 3) the gap is

$$3) \qquad \Delta = \ln(q-1)\cdot \prod_{n=1}^{\infty}\left[\frac{1-(q-1)^{-n/2}}{1+(q-1)^{-n/2}}\right]^2. \tag{83}$$

These results are consistent with the findings for the correlation length $\xi$ which is identical with the horizontal correlation length of the corresponding vertex model, see (70). Therefore cases 1) and 3) are non-critical and case 2) is critical for all $q > 2$.

We discuss now the Hamiltonians 1) and $2^-$) for the lowest value $q = 3$. For this value both Hamiltonians 1) and $2^-$) turn out to be special cases of the bilinear-biquadratic spin-1 Hamiltonian, see (75). Model 1) essentially is the biquadratic Hamiltonian, $\tilde{H} = -\sum\left(\vec{S}_j\cdot\vec{S}_{j+1}\right)^2$. From (59,78) the ground state energy can be calculated numerically

$$\tilde{e}_0 = -2.796863\ldots. \tag{84}$$

The biquadratic Hamiltonian is non-critical since it has a non-zero energy gap $\tilde{\Delta}$ and a finite correlation length $\xi$ which proves the conjecture of [25]. $\tilde{\Delta}$ can be calculated from (82) by taking into account the scale factor for $\tilde{H}$

$$\tilde{\Delta} = \sqrt{5}\prod_{n=1}^{\infty}\left[\frac{1-(1/\alpha)^{n/2}}{1+(1/\alpha)^{n/2}}\right]^2 = 0.1731788\ldots \tag{85}$$

where $\alpha = \left[\frac{1}{2}(1+\sqrt{5})\right]^4$ was inserted. From (70) we obtain the correlation length

$$\xi = -1/\ln k = 21.0728505\ldots \tag{86}$$

where $k$ was calculated from requirement (66) and from standard formulas for elliptic quantities, see [1,31,32]. These results, notably (84,85), have been found independently also by Barber and Batchelor using a quite different method [23] thereby clarifying and extending the observations made by Parkinson [24]. The rather large value of the correlation length (86) may explain the difficulties in drawing definite conclusions from the numerical treatment of finite chains. The largest chain considered in [16] had length 26! The degeneracy of the ground state which could not be determined within our approach is expected on physical grounds to be twofold due to dimerization.

The second Hamiltonian $2^-$) coincides for $q = 3$ with the (spin-1) Takhtajan-Babujian model [18,19]. The ground state energy and the dispersion (81) confirm known results [19].

Very recently the work in [23] was extended to higher spin chains [50,51] which are identical with family 1) for arbitrary $q$. In [50,51] these $q$-state Hamiltonians could be mapped to the quantum Hamiltonian limit of the $q^2$-state Potts model for finite chains with free ends and finally to Bethe Ansatz solvable spin-1/2 $XXZ$ models. The conclusions of [50,51] are in accordance with our findings [21,22].

## Acknowledgments

My attendance at the conference was made possible by a travel grant from the Deutsche Forschungsgemeinschaft. I am grateful to the Centre for Mathematical Analysis for supporting my participation at the workshop.

## References

1. Baxter, R.J.: *Exactly Solved Models in Statistical Mechanics.* London, Academic Press 1982
2. Bethe, H.: *Z. Phys.* **71**, 205 (1931)
3. Takhtajan, L.A.: in *Lecture Notes in Physics* **242**, 175-219, Berlin Heidelberg: Springer Verlag 1985
4. Stroganov, Yu.G.: *Phys. Lett.* **74A**, 116 (1979)
5. Baxter, R.J.: *Ann. Physics* **70**, 193 (1972); **80**, 323 (1972)
6. Haldane, F.D.M.: *Phys. Lett.* **93 A** , 464 (1983); *Phys. Rev. Lett.* **50**, 1153 (1983); *J. Appl. Phys.* **57** , 3359 (1985)
7. Affleck, I.: *J. Phys.: Cond. Matter* **1**, 3047 (1989)
8. Parkinson, J.B., Bonner, J.C., Müller, G., Nightingale, M.P., Blöte, H.W.J.: *J. Appl. Phys.* **57**, 3319 (1985)
9. Blöte, H.W.J., Capel, H.: *Physica* **139 A**, 387 (1986); Nightingale, M.P., Blöte, H.W.J.: *Phys. Rev. B* **33**, 659 (1986)
10. Schulz, H.J., Ziman, T.: *Phys. Rev. B* **33**, 6545 (1986)
11. Sólyom, J.: *Phys. Rev. B* **36**, 8642 (1987)
12. Papanicolaou, N.: *Nucl. Phys.* **B 305** [FS 23], 367 (1988)
13. Affleck, I., Kennedy, T., Lieb, E.H., Tasaki, H.: *Commun. Math. Phys.* **115** , 477 (1988)
14. Knabe, S.: *J. Stat. Phys.* **52** , 627 (1988)

15. Bonner, J.C., Parkinson, J.B., Oitmaa, J., Blöte, H.W.J.: *J. Appl. Phys.* **61**, 4432 (1987)

16. Chang, K., Affleck, I., Hayden, G.W., Soos, Z.G.: J. Phys.: *Condens. Matter* **1**, 153 (1989)

17. Uimin, G.V.: *JETP Lett.* **12**, 225 (1970); Sutherland, B.: *Phys. Rev. B* **12**, 3795 (1975)

18. Takhtajan, L.A.: *Phys. Lett.* **87 A**, 479 (1982)

19. Babujian, H.M.: *Phys. Lett.* **90 A**, 479 (1982)

20. Klümper, A.: Dissertation (1988), Universität zu Köln (unpublished)

21. Klümper, A.: *Europhys. Lett.* **9**, 815 (1989)

22. Klümper, A.: submitted to *J. Phys. A*

23. Barber, M.N., Batchelor, M.T.: *Phys. Rev. B* **40**, 4621 (1989)

24. Parkinson, J.B.: *J. Phys. C: Solid State Phys.* **20**, L1029 (1987); **21**, 3793 (1988); *J. Physique* **C8**, 1413 (1988)

25. Affleck, I.: *Nucl. Phys.* **B 265**[FS15], 409 (1986)

26. Maillard, J.M., Rammal, R.: *J. Phys.* **A16**, 353 (1983)

27. Jaeckel, M.T., Maillard, J.M.: J. Phys. **A15**, 2241 (1982)

28. Klümper, A., Schadschneider, A., Zittartz, J.: *Z. Phys.* **B 76**, 247 (1989)

29. Klümper, A., Schadschneider, A., Zittartz, J.: *Z. Phys.* **B** (in press)

30. Fan, C., Wu, F.Y.: *Phys.Rev.* **B 2**, 723 (1970)

31. Abramowitz, M., Stegun, I.A.: Handbook of Mathematical Functions. Washington, U.S. National Bureau of Standards 1964; New York, Dover 1965

32. Klümper, A., Zittartz, J.: *Z. Phys.* **B 71**, 495 (1988)

33. Klümper, A.: Diplom-Arbeit (1987), Universität zu Köln, unpublished

34. Pearce, P.A.: *Phys.Rev.Lett.* **58**, 1502 (1987)

35. Tetel'man, M.G.: *Sov.Phys. JETP* **55**(2), 306 (1982)

36. Takahashi, M., Suzuki, M.: *Prog. Theor. Phys.* **48**, 2187 (1972)

37. Johnson, J.D., Krinsky, S., McCoy, B.M.: *Phys. Rev. A* **8**, 2526 (1973)

38. Woynarovich, F.: *J. Phys. A* **15**, 2985 (1982)

39. Destri, C., Lowenstein, J.H.: *Nucl. Phys.* **B 205 [FS5]**, 369 (1982)

40. Babelon, O., de Vega, H.J., Viallet, C.M.: *Nucl. Phys.* **B 220 [FS8]**, 13 (1983)

41. Klümper, A., Zittartz, J.: *Z. Phys.* **B 75**, 371 (1989)

42. Sutherland, B.: *J. Math. Phys.* **11**, 3183 (1970)

43. Schultz, C.L.: *Phys. Rev. Lett.* **46**, 629 (1981)

44. Perk, J.H.H., Schultz, C.L.: in *Non-Linear Integrable Systems – Classical Theory and Quantum Theory*, Proc. RIMS Symposium, M. Jimbo and T. Miwa, eds. (World Scientific, Singapore, 1983)

45. Perk, J.H.H., Wu, F.Y.: *Physica* **138A**, 100 (1986); *J. Stat. Phys.* **42**, 727 (1986)

46. Owczarek, A.L., Baxter, R.J.: *J. Stat. Phys.* **49**, 1093 (1987)

47. Zamolodchikov, A.B. and Al.B.: *JETP Lett.* **26**, 457 (1977); *Nucl. Phys.* **B133**, 525 (1978)

48. Reshetikhin, N.Yu.: *Sov. Phys. JETP* **57**, 691 (1983)

49. de Vega, H.J., Karowski, M.: *Nucl. Phys.* **B 280** [FS18], 225 (1987)

50. Affleck, I.: *J. Phys. C: Solid State Phys.* (in press)

51. Batchelor, M.T., Barber, M.N.: preprint, Australian National University, CMA-R35-89

International Journal of Modern Physics B, Vol. 4, No. 5 (1990) 895 – 905
© World Scientific Publishing Company

# THE CORNER TRANSFER MATRIX OF SOME FREE-FERMION SYSTEMS AND MEIXNER'S POLYNOMIALS

T.T. TRUONG and I. PESCHEL

*Fachbereich Physik, Freie Universität Berlin, Arnimallee 14, D-1000 Berlin 33, West-Germany.*

Corner transfer matrices of some free-fermion vertex systems on a finite lattice, are exactly diagonalised in the Hamiltonian limit with the help of a special class of orthogonal polynomials: the Meixner polynomials. We present the derivation, discuss the asymptotic behavior for a large lattice and compare the results with numerical computation.

## 1. Introduction

In this work we are concerned with the finite quantum spin chain described by the Hamiltonian:

$$\mathcal{H}_N = -\sum_{n=1}^{N-1} n\mathcal{H}_{n,n+1} \tag{1}$$

where

$$\mathcal{H}_{n,n+1} = \sigma_n^x\sigma_{n+1}^x + \lambda\sigma_n^y\sigma_{n+1}^y + h(\sigma_n^z + \sigma_{n+1}^z) \tag{2}$$

and $\sigma_n^\alpha(\alpha = x, y, z)$ are Pauli matrices. This is an inhomogeneous $XY$ model with a magnetic field, where all parameters increase linearly along the chain. Such Hamiltonians arise naturally in the study of Baxter's corner tranfer matrix (CTM) for two-dimensional statistical models. For $\lambda = 0, \exp(-u\mathcal{H}_N), u$ being a constant, is essentially the CTM of an Ising model, while for $h = 0$ it corresponds to the CTM of a doubled Ising model [1,2]. Previous studies have been done mostly on infinite systems, but recently systems of finite size also have been treated [2]. They are of particular interest at the critical point where conformal invariance predicts the stucture of the low-lying eigenvalues of $\mathcal{H}_N$ [3]. In the two cases mentioned, we found that $\mathcal{H}_N$ can be diagonalised with the help of special polynomials, namely the Carlitz polynomials. The eigenvalues of $\mathcal{H}_N$ are then related to the zeros of these polynomials. The characteristic feature is that the low-lying levels are equidistant for large systems (i.e. $N \to \infty$), confirming a theorem established long ago by Baxter [4].

In the present paper we take both $\lambda$ and $h$ to be non-zero. The Hamiltonian then still is of free-fermion type and thus completely soluble, but it exhibits some new features. It can be related to the CTM of an eight-vertex model with an electric field proportional to $h$. This connection is outlined in the appendix.

Here we shall treat two special cases. The first is the isotropic chain ($\lambda = 1$) which is critical for $|h| \leq 1$ and shows a band filling and a Fermi velocity depending on $h$ in the homogeneous case. We will find the analogous phenomena here. The second is the so-called *disorder line* for $\lambda = h^2$. This line arises from a particular interplay between the anisotropy parameter $\lambda$ and the field $h$. In the homogeneous chain, it separates the oscillatory and non-oscillatory regimes of the spatial spin correlation functions [5]. Moreover, on this line the solution simplifies remarkably and we shall see that this feature carries over to the inhomogeneous chain. The diagonalisation of $\mathcal{H}_N$ is relatively easy in both cases and requires no use of elliptic functions. We find that it leads to orthogonal polynomials of Meixner's type namely Gottlieb, Laguerre and Pollaczek polynomials. The resulting eigenvalues of $\mathcal{H}_N$ are discussed and illustrated by numerical calculations. On the critical line, the predictions of conformal invariance are verified.

## 2. The Isotropic Chain

In this section we perform the diagonalisation of $\mathcal{H}_N$ for $\lambda = 1$ and arbitrary $h$, thereby generalising a previous work [6]. The method is to rewrite $\mathcal{H}_N$ in terms of fermion operators $c_n^+, c_n$ which gives:

$$\begin{aligned}\mathcal{H}_{n,n+1} &= 2\{(1+\lambda)[c_n^+ c_{n+1} + c_{n+1}^+ c_n] + (1-\lambda)[c_n^+ c_{n+1}^+ + c_{n+1} c_n] \\ &\qquad + h[c_n^+ c_n + c_{n+1}^+ c_{n+1} - 1]\}\end{aligned} \tag{3}$$

In the following we consider:

$$\widehat{\mathcal{H}}_N = \frac{1}{2}[\mathcal{H}_N - hN(N-1)] \tag{4}$$

and denote its single-fermion eigenvalues by $\varepsilon$. Without loss of generality we will assume $h \geq 0$.

For $\lambda = 1$ only the $c^+ c$-terms remain and one has just to diagonalise the $N \times N$ tridiagonal matrix $\mathbf{A}$ related to $\widehat{\mathcal{H}}_N$:

$$\begin{pmatrix} h & 1 & 0 & \cdots & 0 & 0 \\ 1 & 3h & 2 & \cdots & 0 & 0 \\ 0 & 2 & 5h & \cdots & 0 & 0 \\ \vdots & \vdots & \vdots & \ddots & \vdots & \vdots \\ 0 & \cdot & \cdot & \cdots & (2N-3)h & (N-1) \\ 0 & \cdot & \cdot & \cdots & (N-1) & (N-1)h \end{pmatrix} \tag{5}$$

It is useful to view the eigenvalue equation

$$(n-1)\psi_{n-1} + n\psi_{n+1} + h(2n-1)\psi_n = \varepsilon\psi_n \tag{6}$$

as a three-way recursion relation for the amplitude $\psi_n(\varepsilon, h)$, subject to the boundary condition:

$$\psi_{N+1} + h\psi_N = 0 \tag{7}$$

This condition follows from the last of the equations and selects the eigenvalues $\varepsilon = \varepsilon_l, l = 1, 2, \ldots, N$.

A recursion of this type suggests that $\psi_n$ is essentially an orthogonal polynomial [7] and this is indeed the case. Let

$$\psi_{n+1}(\varepsilon, h) = (-1)^n Q_n(\varepsilon, h), \qquad n \geq 0 \tag{8}$$

Depending on the value of $h$, we have the following cases:

a) $h \geq 1$

Put $h = \cosh\mu, \varepsilon = (1 + 2x)\sinh\mu, Q_n(\varepsilon, h) = e^{n\mu} l_n(x, \mu)$. Then $l_n$ fulfills:

$$(n+1)l_{n+1} = [(n+1)e^{-2\mu} + n + (e^{-2\mu} - 1)]l_n - ne^{-2\mu}l_{n-1}. \tag{9}$$

This is the recursion relation of the Gottlieb polynomials [8].

b) $h = 1$

With $x = \varepsilon, Q_n(\varepsilon, h) = L_n(x)$ one finds:

$$(n+1)L_{n+1} = (2n + 1 - x)L_n - nL_{n-1} \tag{10}$$

which defines the Laguerre polynomials [9].

c) $h \leq 1$

With $h = \cos\theta, Q_n(\varepsilon, h) = P_n(x, \theta)$ one has

$$(n+1)P_{n+1} = [(2n-1)\cos\theta + 2x\sin\theta]P_n - nP_{n-1} \tag{11}$$

which is the definition of the Pollaczek polynomials [10].

All three types of polynomials belong to Meixner's class [11] and have been studied in detail in the cited literature. What we need here are the expressions for large $n$ which allow us to determine the eigenvalues $\varepsilon_l$ for a large but finite system. This gives the following results:

a) $h \geq 1$

The Gottlieb polynomials have the asymptotic form, valid for bounded $x$ [8]

$$l_N(x, \mu) \simeq (-1)^N \frac{\Gamma(1+x)}{(1 + e^{2\mu})^{(1+x)} N! \Gamma(1 + x - N)} \tag{12}$$

Thus $l_n$ has zeros at the integer values $x = x_l = l \geq 0$. At these points also (7) is fulfilled, so the low-lying eigenvalues are:

$$\varepsilon_l = (2l+1)\sqrt{h^2 - 1}, \qquad l = 0, 1, \ldots \tag{13}$$

The $\varepsilon_l$ are therefore equidistant with a spacing dependent on $h$. The values in (13) can be easily checked by diagonalising numerically the matrix $\mathbf{A}$. Results for a chain

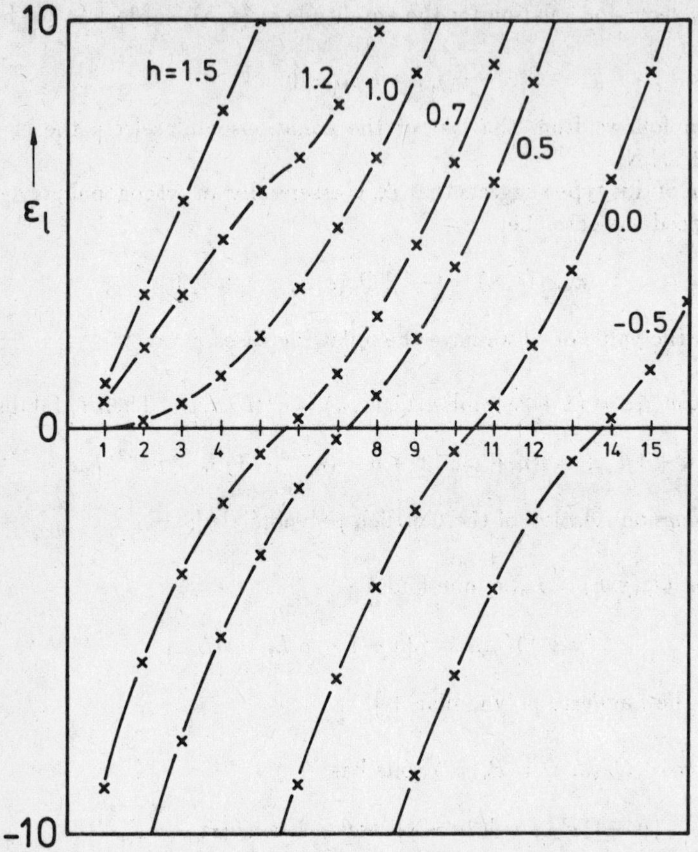


Figure 1. Smallest eigenvalues $\varepsilon_l$ of $\mathcal{H}_N$ in the *isotropic case*, $\lambda = 1$, for a chain of 20 sites. The eigenvalues are numbered according to their magnitude. The curves are guides for the eye.

of $N = 20$ sites are shown in Fig. 1. Note that at higher values of $\varepsilon_l$ there is a break in the curves coming from the reduced field (i.e.$(N-1)h$ instead of $(2N-1)h$) at the right end of the chain.

b)$h = 1$

Using the properties of the Laguerre polynomials [9], equation (7) can be transformed into the equivalent condition:

$$\varepsilon L^{(1)}_{N-1}(\varepsilon) = 0 \tag{14}$$

where $L^{(\alpha)}_n$ is a generalised Laguerre polynomial [9]. One solution is the zero mode $\varepsilon_o = 0$ with the amplitude $\psi_{n+1} = (-1)^n$. The other solutions may obtained from Fejer's formula for $L^{(\alpha)}_N(x)$, (see p.233 of [12])

$$L^{(\alpha)}_N(x) \simeq \sqrt{\pi} e^{x/2} x^{-\alpha/2-1/4} \cos[2\sqrt{Nx} - \frac{\alpha\pi}{2} - \frac{\pi}{4}] \tag{15}$$

The corresponding low-lying eigenvalues are given by:

$$\varepsilon_l = \frac{\pi^2}{4(N-1)}(l+\frac{1}{4})^2 \qquad l = 1, 2, \ldots \tag{16}$$

They have an unusual quadratic dependence on $l$. This is also seen in numerical calculations (see Fig. 1).

c) $h \leq 1$

To obtain the necessary asymptotic expression we reexpress $\psi_{n+1}$ in terms of another class of Pollaczek polynomials: the $\hat{P}_n(\cos\theta; a, b)$ [13] for which a large order ($n \to \infty$) expansion has been readily worked out by G. Szegö [14]. Note however that his work [12] uses the original parameters $a$ and $b$ of Pollaczek [13] but (1.7) of his appendix in [12] fails to be consistent with the rest of the text. In any case we have

$$\hat{P}_n(\cos\theta; a, b) \simeq \frac{A(\theta)}{\sqrt{N}} \cos[N\theta + \frac{a\cos\theta + b}{2\sin\theta} \ln N + B(\theta)] \tag{17}$$

where $A(\theta)$ and $B(\theta)$ are known functions of $\theta$ independent of $N$. The passage to the $\hat{P}_n(\cos\theta; a, b)$ is given by (4.4) in the appendix of [12]:

$$\psi_{n+1} = (-1)^n P_n(-\frac{\varepsilon}{2\sin\theta}, \theta) = (-1)^n \hat{P}_n(\cos\theta; a, b) \tag{18}$$

with $\varepsilon = -(a\cos\theta + b)$. Inspection shows that one may choose $a \geq |b|$ for $\varepsilon$ as required by the definition of $\hat{P}_n$. Then the boundary condition (7) is asymptotically equivalent to:

$$\sin[(N-1)\theta + \frac{\varepsilon}{2\sin\theta} \ln N + B(\theta)] = 0 \tag{19}$$

The low-lying levels are then:

$$\varepsilon_l = \frac{2\sin\theta}{lnN}[l\pi - (N-1)\theta - B(\theta)] \qquad l = 1, 2, \ldots. \tag{20}$$

This formula contains two features which are also apparent in the curves of Fig. 1. Firstly, the whole spectrum is shifted (via the term $(N-1)\theta$) if one changes $h = \cos\theta$. This is analogous to the shift of the cosine band of eigenvalues $\varepsilon(q) = 4\cos(q) + 2h$ in the homogeneous chain. Secondly, the level spacing for small $\varepsilon$ is $\Delta\varepsilon = \frac{2\pi}{\ln N}\sqrt{1-h^2}$, it changes also with the field. The factor $\sqrt{1-h^2}$ is precisely proportional to the Fermi velocity in the homogeneous case. Dividing it out, one obtains from (20) the conformal result [3]. One can also check this by numerical calculations but one has to use much larger systems or modified chains with some sites on the left end cut-off [15].

A last remark on this case concerns the filling of the negative energy levels in the construction of the ground state and the excited states. Depending on the strength of the field $h$, the magnitude of the first excited state may be less, equal or larger than the $\Delta\varepsilon$ spacing. Thus by manipulating $h$ one may bring the excited states to the zero level each time generating a double degeneracy of the spectrum.

## 3. The Disorder Line

The special role of the line $\lambda = h^2$ was first noted by E. Barouch and B. McCoy [5] in their study of the homogeneous spin chain. They found that for $\lambda \geq h^2$ the correlation functions oscillate while for $\lambda \leq h^2$ they are monotonic. At $\lambda = h^2$ they become constant, for example $< \sigma_n^x \sigma_m^x >= (1 - h^2)$. Such a phenomenon occurs also in the triangular Ising model and the corresponding line was called by J. Stephenson [16] a *disorder line.* Both problems are esentially the same and there is also a connection to Glauber's kinetic Ising model [17]. Disorder lines have been found and studied in various systems (see [18] for the latest works on disorder lines.).

A typical feature of these lines is the simple structure of the ground state. In our case, it was noted previously [17] that the product states:

$$|\Phi_\pm\rangle = \prod_{j=n}^{n+1} [\cos\frac{\theta}{2}|+\rangle_j \pm \sin\frac{\theta}{2}|-\rangle_j] \tag{21}$$

are degenerate eigenstates of $\mathcal{H}_{n,n+1}$ for $\lambda = h^2 \leq 1$ belonging to the largest eigenvalue $(1 + h^2)$. Here $h = \cos\theta$ and $|\pm\rangle_j$ are the eigenfunctions of $\sigma_j^z$. The spins in (21) are parallel and form an angle $\theta$ with the $z$-axis in the $xz$-plane. Extending the product over all sites, one obtains two states which are the ground states of $\mathcal{H}_N$ (and also for its homogeneous counterpart). The resulting ground state energy is:

$$E_0 = -\frac{N(N-1)}{2}(1 + h^2) \tag{22}$$

A simple ground state does not imply that the whole problem is simple, but here it is the case. If one seeks to diagonalise $\mathcal{H}_N$ by the standard procedure [19] one has to find the eigenvalues $\varepsilon_l^2$ of an $N \times N$ matrix $(\mathbf{A} - \mathbf{B})(\mathbf{A} + \mathbf{B})$. In general this matrix is a pentadiagonal matrix and quite complicated. However on the line $\lambda = h^2$, it turns out to be the square of the following tridiagonal matrix $\mathbf{M}$:

$$\begin{pmatrix} h^2 & h & 0 & \cdots & 0 & 0 \\ h & (1+2h^2) & 2h & \cdots & 0 & 0 \\ 0 & 2h & (2+3h^2) & \cdots & 0 & 0 \\ \vdots & \vdots & \vdots & \ddots & \vdots & \vdots \\ 0 & 0 & 0 & \cdots & (N-2)+(N-1)h^2 & (N-1)h \\ 0 & 0 & 0 & \cdots & (N-1)h & (N-1) \end{pmatrix} \tag{23}$$

We remark that the same feature exists in the homogeneous chain. The associated recursion relation for the components $\psi_n$ of the eigenvectors is:

$$(n-1)h\psi_{n-1} + nh\psi_{n+1} + [(n-1) + h^2 n]\psi_n = \varepsilon\psi_n \tag{24}$$

with the boundary condition:

$$\psi_{N+1} + h\psi_N = 0 \tag{25}$$

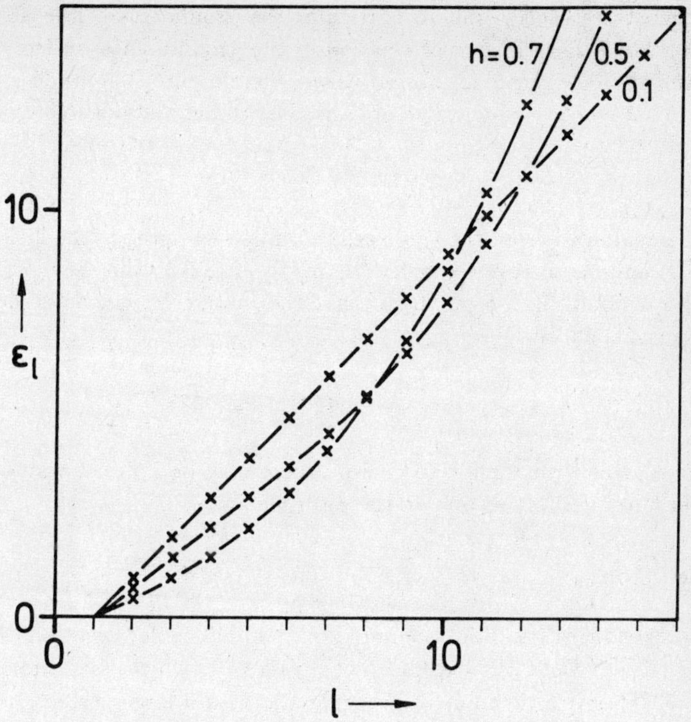


Figure 2. Smallest eigenvalues $\varepsilon_l$ of $\mathcal{H}_N$ on the *disorder line*, $\lambda = h^2$ for a chain of 20 sites. The curves are guides for the eye.

Since we can map the case $h \geq 1$ onto the case $h \leq 1$ by factorising out $h$ in $\mathcal{H}_N$ and exchanging the rôles of $\sigma_n^x$ with $\sigma_n^y$ we can choose $h = \exp(-\mu)$ with $\mu \geq 0$ and set

$$\psi_{n+1} = (-1)^n e^{n\mu} l_n(x, \mu) \tag{26}$$

Thus $\psi_n$ is given by the Gottlieb polynomials of Sect. 2 with the relation:

$$\varepsilon = x(1 - h^2) \tag{27}$$

The eigenvalues $\varepsilon_l$ are determined by the equation

$$l_N(x, \mu) = e^{2\mu} l_{N-1}(x, \mu) \tag{28}$$

which for large $N$ using (12) yields the spectrum:

$$\varepsilon_l = l(1 - h^2) \qquad l = 0, 1, 2, \ldots \tag{29}$$

This can be easily seen in numerical calculations. In Fig. 2 we have illustrated the situation for a chain of 20 sites. One observes that for fixed $N$, the linear region of the spectrum becomes smaller with increasing $h$. Equation (29) shows that one eigenvalue $\varepsilon_0$ is zero. This is exactly true for all $N$. The eigenvector in this case is given by $\psi_{n+1} = (-h)^n$, i.e. its corresponds to a localised state. Because of this fact, all eigenstates of $\mathcal{H}_N$ and in particular the ground state, are at least doubly degenerate. Thus there is a *level crossing* in the ground state on the disorder line. We also see that all eigenvalues are positive. Therefore in the ground state all single particle levels are occupied and one obtains the ground state energy as

$$E_0 = -\sum_l \varepsilon_l \tag{30}$$

This can be evaluated from the trace of the matrix $\mathbf{M}$ without the knowledge of the individual $\varepsilon_l$ and one recovers exactly the result of (22) in this way.

Finally we would like to point out that the matrix $\mathbf{M}$ can be associated with a Hamiltonian of the form:

$$\mathcal{H}_N = \sum_{n=1}^{N-1} n[\sigma_n^x \sigma_{n+1}^x + \sigma_n^y \sigma_{n+1}^y + h\sigma_n^z + h^{-1}\sigma_{n+1}^z] \tag{31}$$

corresponding to an isotropic spin chain. There may be a direct way of relating the original Hamiltonian to this one on the disorder line.

## 4. Conclusions

We have diagonalised the Hamiltonian (1) in two particular cases and found its low-lying eigenvalues. With the exception of $\lambda = h = 1$, which is at the end point of a critical line, the levels turn out to be equidistant for a large system, as found previously in other cases. The solutions were obtained by special orthogonal polynomials. In Fig. 3 we have collected all the cases solved so far and indicated the corresponding polynomials. It is also possible to treat the problem for arbitrary $\lambda$ and $h$, but this leads to elliptic functions and will be deferred to a future work. There are two other interesting open points. One is the possible connection of $\mathcal{H}_N$ to the CTM of a triangular Ising model. This is strongly suggested by the appearance of the disorder line in $\mathcal{H}_N$. The other is the calculation of the ground state energy $E_0$. We have seen that it is easy to obtain it on the disorder line, but an analytical calculation on the critical line would be more relevant in view of the conformal predictions.

## Appendix: $\mathcal{H}_N$ and the Corner Transfer Matrix

The standard configurations of an 8-vertex model round a vertex (see Fig. 4) may be described conveniently by the operator $\mathcal{R}_{n,n+1}$ [20]:

$$\mathcal{R}_{n,n+1} = J_x\sigma_n^x\sigma_{n+1}^x + J_y\sigma_n^y\sigma_{n+1}^y + J_z\sigma_n^z\sigma_{n+1}^z + H(\sigma_n^z + \sigma_{n+1}^z) + C.$$

Its matrix elements are labelled by the two sets of variables $\sigma_n, \sigma_{n+1}$ and $\sigma_n', \sigma_{n+1}'$ as shown in Fig. 4. The coefficients $J_x, J_y, J_z, H$ and $C$ are related to the set of configuration weights $\omega_j, j = 1, \ldots, 8$ where $\omega_3 = \omega_4, \omega_5 = \omega_6, \omega_7 = \omega_8$, as follows:

$$J_x = \frac{1}{2}(\omega_3 + \omega_7), \quad J_y = \frac{1}{2}(\omega_3 - \omega_7), \quad J_z = \frac{1}{2}(\omega_1 + \omega_2 - 2\omega_5),$$

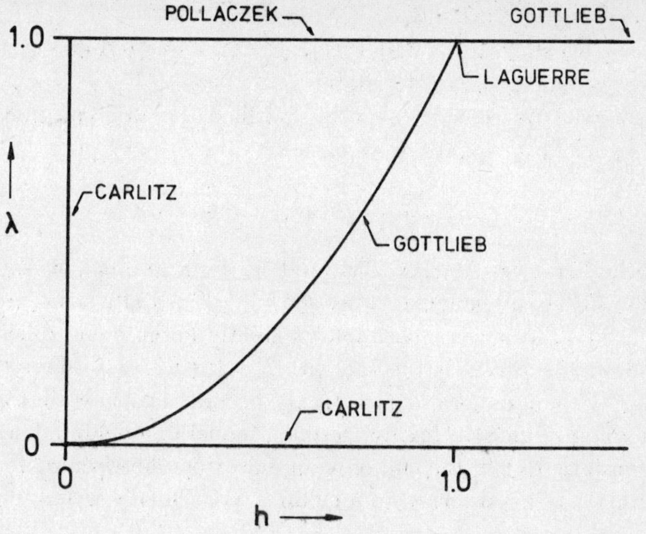


Figure 3. Parameter space of the Hamiltonian $\mathcal{H}_N$. Indicated are the polynomials which occur in the solution on various special lines.

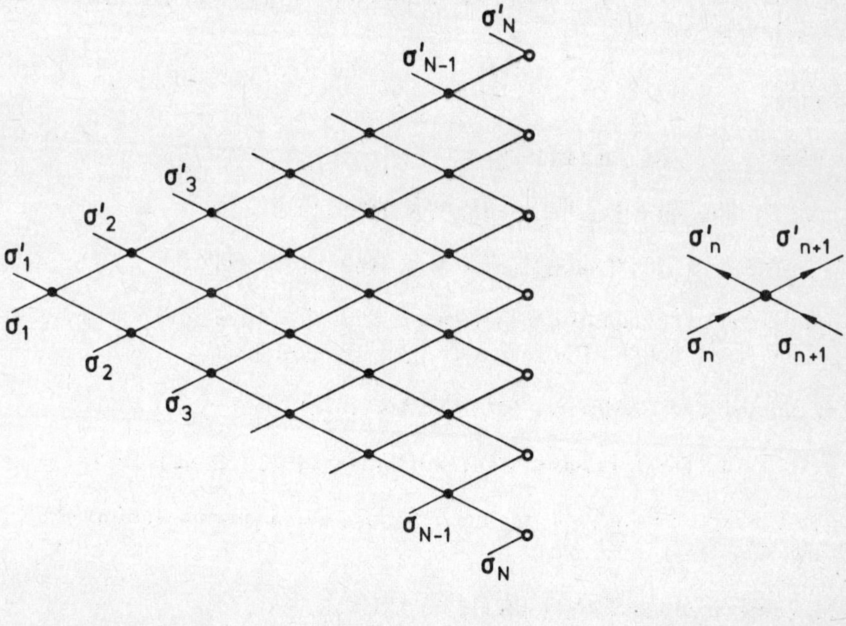


Figure 4. The wedge-shaped portion of an eight-vertex model which gives rise to the Corner Transfer Matrix connected with the Hamiltonian $\mathcal{H}_N$. Nearby an individual vextex is also shown.

$$H = \frac{1}{2}(\omega_1 - \omega_2), \qquad C = \frac{1}{4}(\omega_1 + \omega_2 + 2\omega_5)$$

For simplicity we shall argue first with a 6-vertex free fermion system in a field defined by the canonical parametrisation: $\omega_1 = \cos u e^{2uh}, \omega_2 = \cos u e^{-2uh}, \omega_3 = \sin u, \omega_5 = 1, \omega_7 = 0$. We consider now the so-called Hamiltonian limit which corresponds here to a small $u$ expansion; we obtain:

$$\mathcal{R}_{n,n+1} = \mathcal{I} + u\mathcal{H}_{n,n+1} + \ldots$$

Now a Corner Transfer Matrix $\mathbf{V}$ is built up by a product of $\mathcal{R}_{n,n+1}$ operators anchored at the sites of a triangular array shown in Fig. 4. Its structure is thus quite complicated, however at small $u$ it simplifies greatly because one collects only terms linear in $u$ and obtains precisely the $\mathcal{H}_N$ with $\lambda = 1$, (it is easy to see that the factor $n$ in $\mathcal{H}_N$ is simply the number of times $\mathcal{H}_{n,n+1}$ occurs"vertically" in this limit).

Finally in going to an 8-vertex free fermion model in a field one needs an elliptic parametrisation [21]. But in the end only an anisotropy between $\sigma_n^x \sigma_{n+1}^x$ and $\sigma_n^y \sigma_{n+1}^y$ emerges and this may be absorbed in a factor $\lambda$ as originally writen down.

## Acknowledgments

One of the author (T.T.) would like to thank B. Davies and the participants of the Workshop on Conformal Invariance, the Yang-Baxter equations and Integrability in Canberra for discussions.

## References

1. B. Davies, *Physica* **A154** (1988) 1.
2. T.T. Truong and I. Peschel, *Zeit.f.Phys.* **B75** (1989) 119.
3. I. Peschel and T.T. Truong, *Zeit.f.Phys.* **B69** (1987) 385.
4. R.J. Baxter in *"Fundamental Problems in Statistical Mechanics V"* Ed. by E.G.D. Cohen, North-holland Publishing Co., Amsterdam (1980)
5. E. Barouch and B. Mc Coy, *Phys.Rev* **A3** (1971) 786.
6. T.T. Truong and I. Peschel, *J.Phys.:Math.Gen.* **A21** (1988) L 1029.
7. T.S. Chihara, *"An Introduction to Orthogonal Polynomials"*, Gordon and Breach, New York (1978)
8. M. Gottlieb, *Amer.J.Math.* **60** (1938) 453.

9. E.D. Rainville, *"Special Functions"* chap.12, The Macmillan Company, New york (1960)

10. F. Pollaczek, *Compt.Rend.Acad.Sci.Paris* **230** (1950) 1563.

11. J. Meixner, *J.London Math.Soc.* **9** (1934) 6.

12. G. Szegö, *"Orthogonal Polynomials"*, Amer.Math.Soc. Colloq. Publications **23** (1959) pp.233.

13. F. Pollaczek, *Compt.Rend.Acad.Sci.Paris* **228** (1949) 1363.

14. G. Szegö, *Proc.Amer.Math.Soc.* **1** (1950) 731.

15. I. Peschel, *J.Phys.:Math.Gen* **A21** (1988) L185.

16. J. Stephenson, *J.Math,Phys.* **11** (1970) 420.

17. I. Peschel and V. Emery, *Zeit.f.Phys.* **B43** (1981) 241.

18. M.T. Batchelor and J.M.J. van Leeuwen, *Physica A* **154** (1989) 365.

19. E.H. Lieb, T. Schultz and D. Mattis, *Ann.Phys.(N.Y.)* **16** (1961) 407.

20. I. Peschel and F. Rys, *Phys.Lett.* **91A** (1982) 187.

21. B. Felderhof, *Physica* **66** (1973) 279.

International Journal of Modern Physics B, Vol. 4, No. 5 (1990) 907 – 912
© World Scientific Publishing Company

# CONFORMAL PROPERTIES OF CRITICAL ISING MODEL CORNER TRANSFER MATRICES

Brian DAVIES

*Mathematics Department, The Faculties, The Australian National University, Canberra, ACT 2601, Australia*

and

Paul A. PEARCE

*Centre for Mathematical Analysis, Australian National University, Canberra, ACT 2601, Australia*

*and*

*Mathematics Department, University of Melbourne, Parkville Vic. 3052, Australia*

The scaling spectra of finite-size Ising model corner transfer matrices (CTMs) are studied at criticality, using the fermion algebra. The low-lying eigenvalues collapse like $1/\log N$ for large $N$ as predicted by conformal invariance. The shift in the largest eigenvalue is evaluated analytically using a generalized Euler-Maclaurin summation formula giving $\pi c/6\log N$ with central charge $c = 1/2$. The spectrum generating functions, for both fixed and free boundary conditions, are expressed simply in terms of the $c = 1/2$ Virasoro characters $\chi_\Delta(q)$ with modular parameter $q = \exp(-\pi/\log N)$ and conformal dimensions $\Delta = 0, 1/2, 1/16$.

## 1. Introduction

Recently much progress has been made in the theory of critical phenomena by generalizing the notion of scale invariance of critical systems to include invariance under conformal and modular transformations [1,2,3,4]. In particular, the universality classes of critical two-dimensional lattice models and conformal field theories can be classified according to the value of the central charge $c$ of the corresponding conformal Virasoro algebra. For $c < 1$, a complete classification has been obtained [5] yielding the unitary series with $c = 1 - 6/m(m+1)$ and $m = 3, 4, 5, \ldots$ . In this case the critical exponents are quantized and only a finite number of scaling operators enter the theory. These theories are realized within statistical mechanics by the $A$–$D$–$E$ lattice models [6,7,8,9,10]. The additional assumption of modular invariance on a torus allows the complete operator content of these theories to be elucidated [2]. More precisely, the finite-size scaling partition function at criticality is given as a modular invariant combination of the characters of the associated Virasoro algebra [11,12,13].

The study of conformal invariance in other geometries and the effect of boundary conditions has been initiated by Cardy [14]. It now appears that the operator content of the theory on an annulus plays a privileged role [15] and that this is intimately connected with the very special properties of CTMs introduced by Baxter [16]. Perhaps the most striking evidence for this is the mysterious occurrence of Virasoro characters in the spectra of CTMs of lattice models off-criticality [17,18,19,20]. This naturally leads to the possibility of extending the technical machinery of Virasoro algebras to integrable non-critical models [21,22,23,24].

Following Peschel and Truong [25,26,27,28] we pursue here the study of the spectra of lattice model CTMs at criticality. In this case the occurrence of the Virasoro algebra is expected as a consequence of the conformal invariance of the critical system. Indeed, on the basis of conformal invariance, two important facts have been established [29,25]. First, for a corner of large but finite size $N$ and opening angle $\theta$, the low-lying eigenvalues are equally spaced with level splitting given by $\epsilon = \pi\theta/\log N$. Second, the universal finite-size correction to the largest eigenvalue or ground-state energy is given by $\pi c/24 \log N$ where $c$ is the central charge.

This paper is a brief report of investigations for the Ising model. We extend results of Peschel and Truong in two directions. First, we are able calculate the universal finite-size corrections for the CTM generators analytically. This is achieved using fermion techniques developed by Davies [30,31] and a generalized Euler-Maclaurin summation formula [32]. We find the well known Ising result $c = 1/2$. Second, we explore the dependence of the critical CTM spectra on boundary conditions. Specifically, we obtain the low-lying eigenvalues for both fixed and free boundary conditions. The generating functions of these eigenvalues are expressed simply in terms of the $c = 1/2$ Virasoro characters $\chi_\Delta(q)$ with modular parameter $q = \exp(-\pi/\log N)$ and conformal dimensions $\Delta = 0, 1/2, 1/16$. We treat only the Ising model here for the technical convenience that there is a fermion representation. Nevertheless, the clear general picture that emerges should apply to lattice models in general and to the $A$–$D$–$E$ models in particular. Technical details are ommitted since these may be found in a complete account published elsewhere [33].

## 2. Ising model Corner Transfer Matrices and their Generators

The basic CTM for a lattice with given centre spin $s_1$ is the product of two-spin operators, one for each interaction. In this paper we are interested in the critical Ising model with interaction strengths $J_i$ (normalised by $kT$) given as $\sinh 2\hat{J}_1 = \sinh 2J_2 = \tan 2u$. Here we have introduced dual variables $\hat{J}_i$ and $u$ is the spectral parameter. Rotation of the lattice through $90^\circ$ is achieved by $u \to \lambda - u$, where $\lambda = \pi/4$ is the crossing parameter; the physical region is $0 < u < \lambda$. For more details consult [33].

Now we make some general observations which are independent of the boundary conditions, but apply only to the Ising model. There are $2^N$ eigenvalues $\Lambda$ of the CTM and the free-fermion property means that they have direct product form:

$$\Lambda = \Lambda_{max}\lambda_1^{n_1}\dots\lambda_N^{n_N} \tag{1}$$

where the occupation numbers $n_j$ take the values 0 or 1. The collection of all eigenvalues may be represented by the generating functions

$$\chi_\pm(\lambda_1\dots\lambda_N) = \Lambda_{max}\prod_{j=1}^{N}(1\pm\lambda_j) \tag{2}$$

By this we mean that the $2^N$ terms in this product are in one-to-one correspondence with the eigenvalues of the CTM. This formula is exact: we shall show that in the large $N$ limit it leads to the Virasoro characters. One more exact result is needed. The maximum eigenvalue $\Lambda_{max}$ may be determined from the individual $\lambda_j$ by an inversion identity [30] via

$$(2\cot 2u)^{N_1} = \Lambda_{max}^2\prod_{j=1}^{N}\lambda_j \tag{3}$$

where $N_1$ is the number of bonds of type $J_1$.

For a non-critical system, and in the thermodynamic limit ($N\to\infty$), Baxter [16] has shown that there is an operator $H$ called the generator such that $A_n(u) = \exp[-uH]$. At criticality the situation is more delicate because the spectrum collapses. However we may still define a generator by expanding $A(u)$ about zero: $A_n(u)\approx I-uH$. Expressed in fermion operators we find that for free boundaries

$$H_0 = \sum_{j=1}^{N}4(j-1)(f_j^\dagger f_{j+1}-1/2)-\sum_{j=1}^{N-1}(2j-1)(f_j^\dagger f_{j+1}+f_{j+1}^\dagger f_j+f_j^\dagger f_{j+1}^\dagger-f_jf_{j+1}) \tag{4}$$

which is the critical limit of the high-temperature case [30], and for fixed boundary conditions

$$H_\pm = \sum_{j=1}^{N}2(2j-1)(f_j^\dagger f_{j+1}-1/2)-\sum_{j=1}^{N-1}2j(f_j^\dagger f_{j+1}+f_{j+1}^\dagger f_j+f_j^\dagger f_{j+1}^\dagger-f_jf_{j+1}) \tag{5}$$

which is the critical limit of the low-temperature case. The eigenvalues $\epsilon_j$ of the generators are related to the $\lambda_j$ of (1) by $\lambda_j=\exp(u\epsilon_j)$. They are given by an eigenvalue problem for a real symmetric tridiagonal matrix $\mathcal{H}_0$ (resp $\mathcal{H}_\pm$) of dimension $2N$[30]. The matrices $\mathcal{H}_0$ (resp $\mathcal{H}_\pm$) are bipartite and have the following structure: $(\mathcal{H}_0)_{ij}=\frac{1}{2}(i-1)\delta_{i,j+1}+\frac{1}{2}(j-1)\delta_{i+1,j}$, $(\mathcal{H}_\pm)_{ij}=\frac{1}{2}i\delta_{i,j+1}+\frac{1}{2}j\delta_{i,j+1}$. Compared with $\mathcal{H}_\pm$, $\mathcal{H}_0$ is shifted one row down and one column to the right, which eliminates the last row and column of $\mathcal{H}_\pm$ and introduces a border of zeros as the first row and column of $\mathcal{H}_0$. The border of zeros gives a zero eigenvalue with eigenvector (1,0,0,0,....): this informs us that $H_0$ falls into two identical block, one for each value of the centre spin. The remainder of the spectrum, in both cases, is found by diagonalising the same matrix: for free boundary conditions it has $2N-1$ rows and columns and for fixed boundary conditions $2N$. Because the matrices are symmetric and bipartite, the eigenvalues occur in real pairs of opposite sign.

## 3. Asymptotic Analysis

The methods given for diagonalisation do not lead to closed form expressions for the spectra. For the purpose of studying the conformal properties of $A(u)$, we want to know the structure of the low-lying states, including the universal finite size corrections to $\Lambda_{max}(u)$ as a function of $N$. In general, the spectrum of $\mathcal{H}$ is the only thing we can handle analytically. This will give the first term in a Taylor expansion of $A(u)$. Away from criticality, this is sufficient because of the exponential relation between $A(u)$ and $\mathcal{H}$. For the low-lying states, this exponential property survives even at criticality [33].

CTMs are themselves partition functions of a quadrant. According to the theory of conformal invariance, partition functions for a large but finite critical system may be factored into the product of a non-universal part, which contains the information about bulk free energy, etc., and a universal part which is manifested in the finite size corrections. Truong and Peschel [26] and Davies [30] have shown that the low-lying eigenvalues of Ising and six-vertex CTMs are found from the zeros of Meixner polynomials [34], and that these zeros collapse like $1/\log N$ for large $N$. The eigenvalues of $\mathcal{H}_0$ and $\mathcal{H}_\pm$ are given by an implicit equation which can be approximated for large $N$ using steepest descents [33]. The result is that the low-lying part of the spectrum collapses for large $N$. Asymptotically, the excitation energies are $\epsilon_j \approx \pi j/\log N$ for $\mathcal{H}_0$ and $\epsilon_j \approx \pi(j-1/2)/\log N$ for $\mathcal{H}_\pm$.

We must also calculate the maximal eigenvalue $\Lambda_{max}(u)$, at least for small $u$, from the inversion identity (3). Using a generalised Euler-Maclaurin formula [32] to estimate the sum and its finite size corrections, we get for free boundaries [33]

$$\log \Lambda_{max} \approx A_0 N^2 + B_0 N + C_0 + \pi/24 \log N \tag{6}$$

The coefficients $A_0$, $B_0$, $C_0$, will only be correct for infinitesimal $u$, but the next term is universal as it depends only on the low-lying states. For fixed boundaries we get

$$\log \Lambda_{max} \approx A_\pm N^2 + B_\pm N + C_\pm - \pi/48 \log N \tag{7}$$

We shall write $\Lambda_0$ and $\Lambda_\pm$for the universal parts of $\Lambda_{max}$ and discard the non-universal parts. Now we may recover the modular invariant part as Virasoro characters. First, for free boundary conditions the CTM has two identical blocks, labelled by the value of the centre spin. The asymptotic values which we need are $\Lambda_0 \approx q^{1/24}, \lambda_j \approx q^j$ with $q = \exp(-\pi/\log N)$. Substituting this into (2) we find that asymptotically the generator of the eigenvalue spectrum is the Virasoro character

$$\chi_{1/16}(q) = q^{1/24} \prod_{j=1}^{\infty} (1 + q^j) \tag{8}$$

For fixed boundary conditions, there are the two blocks of the CTM for which the centre spin is aligned (resp, opposite) to the boundary spins. The relevant asymptotic values are $\Lambda_\pm \approx q^{-1/48}, \lambda_j \approx q^{j-1/2}$. Thus the asymptotic limit of the generators are expressed as the linear combinations $\chi_0(q) \pm \chi_{1/2}(q)$ of two Virasoro characters

$$\chi_0(q) \pm \chi_{1/2}(q) = 2q^{-1/48} \prod_{j=1}^{\infty}(1 \pm q^{j-1/2}) \tag{9}$$

## 4. Conclusion

There is increasing evidence of a close relation between CTMs and the Virasoro algebra. The Virasoro characters $\chi_\Delta(q)$ occur in the spectrum of CTMs in two distinct instances. First, for solvable models such as the *A–D–E* models associated with classical Lie algebras, they appear in the exact expressions obtained for the noncritical local state probabilities in the thermodynamic limit. In this case the modular parameter $q$ is related to the deviation from the critical temperature. Second, the Virasoro characters appear in the spectrum of CTMs at criticality. This has been established for the Ising model in this paper and is expected to hold more generally for the *A–D–E* models. In this instance the modular parameter $q$ is related to the finite size geometry. From the point of view of scaling, however, there should be just one scale in the theory away from the critical point of the bulk system, given by the correlation length $\xi$.

Indeed, as pointed out by Saleur and Bauer [20], both cases are correctly described by the single formula $q = \exp(-\pi\theta/\log\xi)$, where the opening or effective angle $\theta$ is $2\pi$ for a plane. For the sake of definiteness, let us consider the $L$-state ABF models so that $L = 3$ is the Ising model. Then in the ordered regime III of Andrews et. al. [6] the modular parameter of the Virasoro characters is given by $q = \exp[-4\pi\theta/(L+1)\epsilon]$, where $p = \exp(-\epsilon)$ is the temperature variable. The effective angle $\theta$ of a corner is related [35,36] to the spectral parameter $u$ of the corner by $\theta = (L+1)u$. If we use the result $\xi \approx p^{-\nu}$, $\nu = (L+1)/4$ to eliminate $\epsilon$ in favour of $\xi$ we see that this modular parameter agrees with Saleur and Bauer. Similarly, returning to the result for a finite system at criticality, and using the finite-size scaling hypothesis $\xi \approx N$, $N$ large, also gives agreement for a value of the angle $\theta$ equal to one radian.

## Acknowledgments

We would like to thank M. N. Barber for helplful discussions. PAP also acknowledges the support of Australian Research Council grant No. A68716350.

## References

1. Belavin A A Polyakov A M and Zamolodchikov A B, *Nucl. Phys. B* **241** (1984), 333.
2. Cardy J L, *Nucl. Phys. B* **270** (1986), 186.
3. Cardy J L, 1987, *Conformal Invariance, in Phase Transitions and Critical Phenomena vol 11, eds. Domb C and Lebowitz J L* (Academic Press)

4. Cardy J L, 1988, *Les Houches Lectures, Session XLIV (1988) Fields, Strings and Critical Phenomena, eds. Brzin E and Zinn-Justin J* (1989)
5. Friedan D Qiu Z and Shenker S, *Vertex operators in Mathematics and Physics eds. Lapowski J Mandelstam S and Singer I M* (Springer 1984)
6. Andrews G E Baxter R J and Forrester P J, *J. Stat. Phys.* **35** (1984), 193.
7. Huse D, *Phys. Rev. B* **30** (1984), 3908.
8. Pasquier V, *Nucl. Phys. B* [FSS] **285** (1986), 162.
9. Pasquier V, *J. Phys. A* **20** (1987), 5707.
10. Pasquier V, *J. Phys. A* **20** (1987), L1229.
11. Itzykson C and Zuber J-B, *Nucl. Phys. B* [FSS] **17** (1986), 580.
12. Capelli A Itzykson C and Zuber J-B, *Nucl. Phys. B* **280** (1987), 445.
13. Capelli A Itzykson C and Zuber J-B, *Comm. Math. Phys.* **113** (1987), 1.
14. Cardy J L, *Nucl. Phys. B* **275** (1986), 200.
15. Cardy J L, 1989, Santa Barbara preprint UCSBTH-89-06
16. Baxter R J, 1982, *Exactly Solved Models in Statistical Mechanics* (London: Academic Press).
17. Date E Jimbo M Miwa T and Okado M, *Phys. Rev. B* **35** (1987), 2105.
18. Date E Jimbo M Miwa T and Okado M, 1989, *Proceedings of AMS Summer Institute on Theta Functions* (RIMS-590)
19. Kuniba A and Yajima T, *J. Stat. Phys.* **52** (1987), 829.
20. Saleur H and Bauer M, *Nucl. Phys. B* **320** (1989), 591.
21. Thacker H B, *Physica D* **18** (1986), 348.
22. Itoyama H and Thacker H B, *Phys. Rev. Lett.* **58** (1986), 1395.
23. Itoyama H and Thacker H B, 1988, Fermilab-Pub-88/49-T
24. Itoyama H and Thacker H B, 1988, Fermilab-Conf-88/142-T
25. Peschel I and Truong T T, *Z. Phys. B* **69** (1987), 385.
26. Truong T T and Peschel I, *J. Phys. A* **21** (1988), L1029.
27. Truong T T and Peschel I, *Int. J. Mod. Phys.* (1990), this issue.
28. Peschel I, *J. Phys. A* **21** (1988), L185.
29. Cardy J L and Peschel I, *Nucl. Phys. B* **300** (1988), 377.
30. Davies B, *Physica A* **154** (1988), 1.
31. Davies B, *Physica A* **156** (1989), 171.
32. Lyness J N and Ninham B W, *Mathematics of Computation* **20** (1966), 162.
33. Davies B and Pearce P A, *J. Phys. A* (1990), to appear.
34. Chihara T S, *An introduction to Orthogonal Polynomials* (Gordon and Breach, New York, 1978)
35. Barber M N Peschel I and Pearce P A, *J. Stat. Phys.* **37** (1984), 497.
36. Kim D and Pearce P A, *J. Phys. A* **20** (1987), L451.

International Journal of Modern Physics B, Vol. 4, No. 5 (1990) 913–927
© World Scientific Publishing Company

# EXCHANGE ALGEBRA AND EXOTIC SUPERSYMMETRY IN THE CHIRAL POTTS MODEL

**D. Bernard and V. Pasquier**
*Service de Physique Théorique*†
de Saclay
F-91191 Gif-sur-Yvette Cedex

**Abstract**

We obtain an exchange algebra for the Chiral Potts model, the elements of which are linear in the parameters defining the rapidity curve. This enables us to connect the Chiral Potts model to a $U_q(GL(2))$ algebra. On the other hand, looking at the model from the $S$-matrix point of view relates it to a $\mathbb{Z}_N$ generalisation of the supersymmetric algebra.

† Laboratoire de l'Institut de Recherche Fondamentale du Commissariat à l'Energie Atomique

## I. Introduction

The Chiral Potts model is an intriging $2D$ lattice statistical model which generalizes the Ising and Fatteev-Zamolodchikov models. See e.g. Ref.[1] for an introduction to the model. The Boltzmann weights satisfy the star-triangle relation (or Yang-Baxter equation). One of the particularity of the model resides in the fact that contrary to the usual case the spectral parameters, or the rapidities, live on a non-trivial algebraic curve. In particular translation invariance in the spectral parameter is broken.

In this note we describe algebraic structures related to the Chiral Potts model. On one hand inspired by the remarks made by Bazhanov and Stroganov[2] on the connection between the Chiral Potts model and the 6-vertex model, we construct an exchange algebra for the Chiral Potts model. In a remarkable way the objects belonging to the algebra turn out to be linear in the parameters defining the rapidity curve. Moreover commutation relations of these objects are described by the 6-vertex $R$–matrix. On the other hand, inspired by recent works made by Zamolodchikov on the perturbation of conformal field theories[3], we look at $\mathbb{Z}_N$-related operators commuting with the transfert matrices. The construction of these operators is also inspired by a $RSOS$ - like differential calculus on quantum groups. In a not-less remarkable way, these operators turn out to be generators of a $\mathbb{Z}_N$ - generalization of the supersymmetric algebra. Moreover, despite the different motivations underlying the two constructions (the 6-vertex-like and $\mathbb{Z}_N$–susy one), the latters are equivalent.

One of the obvious consequence of the following is the relation between the Fatteev-Zamolodchikov model and the quantum algebra $\mathcal{U}_q(GL(2))$ with $q$ a root of unity. But, contrary to the $RSOS$ minimal models it is the cyclic representations of $\mathcal{U}_q(GL(2))$, $(q^N = 1)$ which play a role.

The paper is organized as follows; below we fix the notations and give the Boltzmann weights of the Chiral Potts model. In section 2, we present the 6-vertex like exchange algebra whereas the section 3 deals with $\mathbb{Z}_N$-exotic supersymmetry.

The Chiral Potts model[4,5,8] can be formulated as an IRF model on a square lattice (rotated by 45°). The spins $\alpha$ located on the vertices of the lattice either take their values in $\mathbb{Z}_N$ or are equal to a neutral element which we denote by a star; $\alpha = k \in \mathbb{Z}_N$ or $\alpha = *$. The admissible configurations for adjacent spins are $(* - - - k)$ or $(k - - - *)$. Each link of the lattice carries a rapidity. The rapidity lines are orthogonal to the edges of the lattice. The rapidities $p$ leave on an algebraic curve which can be parametrized as follows :

$$p = (a_p; b_p; c_p; d_p)$$

with

$$\begin{aligned} a_p^N + k'b_p^N &= kd_p^N \\ k'a_p^N + b_p^N &= kc_p^N \end{aligned} \tag{1.1}$$

where $k^2 + k'^2 = 1$.

The $N = 2$ case is the Ising case. There are two families of Boltzmann weights :

$W_{pq}(n-m) =$ 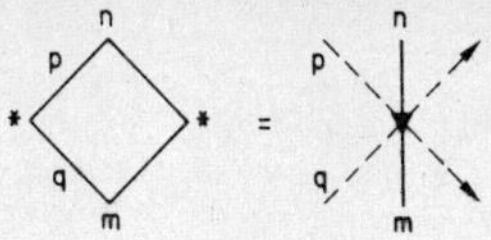

and

$\bar{W}_{pq}(n-m) =$ 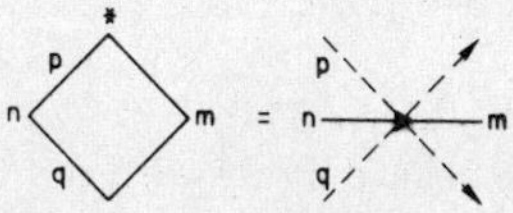

Their expression are

$$\frac{W_{pq}(n)}{W_{pq}(0)} = \prod_{j=1}^{n} \left( \frac{d_p b_q - a_p c_q \omega^j}{b_p d_q - c_p a_q \omega^j} \right) \tag{1.2a}$$

$$\frac{\bar{W}_{pq}(n)}{\bar{W}_{pq}(0)} = \prod_{j=1}^{n} \left( \frac{\omega a_p d_q - d_p a_q \omega^j}{c_p b_q - b_p c_q \omega^j} \right) \tag{1.2b}$$

with $\omega = \exp(i2\pi/N)$. The conditions (1.1) on the rapidities ensure the periodicity of the Boltzmann weights.

$$W_{pq}(n+N) = W_{pq}(n) \quad ; \quad \bar{W}_{pq}(n+N) = \bar{W}_{pq}(n) \tag{1.3}$$

They satisfy the recursion relations

$$W_{pq}(n+1)\left(b_p d_q - c_p a_q \omega^{n+1}\right) = W_{pq}(n)\left(d_p b_q - a_p c_q \omega^{n+1}\right) \tag{1.4a}$$

$$\bar{W}_{pq}(n+1)\left(c_p b_q - b_p c_q \omega^{n+1}\right) = \bar{W}_{pq}(n)\left(a_p d_q - d_p a_q \omega^{n}\right)\omega \tag{1.4b}$$

Remarkably the Boltzmann weigths of the Chiral Potts model satisfy the star-triangle (or Yang-Baxter) equation

$$\begin{aligned} \sum_{\ell=1}^{N} & \bar{W}_{qr}(m-\ell) W_{pr}(n-\ell) \bar{W}_{pq}(\ell-k) \\ & = R_{pqr} W_{pq}(n-m) \bar{W}_{pr}(m-k) W_{qr}(n-k) \end{aligned} \tag{1.5}$$

for some function $R_{pqr}$.

There are various automorphisms which preserve the relations (1.1). We will be interested in two of them, called $R$ and $S$ :

$$R(a;b;c;d) = (b;\omega a;d;c) \tag{1.6a}$$

$$S(a;b;c;d) = (\omega^{-1/2}c;d;a;\omega^{-1/2}b) \tag{1.6b}$$

They satisfy $R^{2N} = S^2 = (RS)^2 = 1$.

The Boltzmann weigths behave nicely with respect to these automorphisms. First they are crossing symmetric

$$W_{pR(q)}(n) = \bar{W}_{qp}(n) \quad ; \quad \bar{W}_{pR(q)}(n) = W_{qp}(N-n) \tag{1.7}$$

Secondly they satisfy the following inversion relation :

$$W_{S(p)S(q)}(n) = W_{qp}(n) \quad ; \quad \bar{W}_{S(p)S(q)} = \bar{W}_{qp}(N-n) \tag{1.8}$$

## II. Exchange algebra

A convenient way to obtain solutions of the Yang-Baxter equation is to solve an equation[6]

$$(L_p \otimes L_q)\, R_{pq} = R_{pq}\, (L_q \otimes L_p) \tag{2.1}$$

where $L_p, L_q$ are matrices, the elements $L_{\alpha\beta}$ of which obey quadratic relations determined by (2.1). We shall here define analogous of $L$ operators with $\otimes$ replaced by another operation.

We define $X$ and $Y$ by

$$\begin{aligned} X|n\rangle &= \left|n+\frac{1}{2}\right\rangle \\ Y|n\rangle &= \omega^n|n\rangle \end{aligned} \tag{2.2}$$

$\omega = e^{2\pi i/N}$, $n \in \mathbb{Z}_{N/2}$ obeying the relation $YX = \omega^{1/2}XY$.

To the Boltzmann weights $\bar{W}_{pq}(n)$ we associate the operator

$$\bar{W}_{pq} = \sum_n \bar{W}_{pq}(n) X^{2n} \tag{2.3}$$

We define an operation $*$ by

$$Y^k X^p * Y^q X^\ell = \delta_{p,\ell} Y^{k+q} X^p \tag{2.4}$$

$\bar{W}_{pq}$ is then completely determined by the exchange relation:

$$\left(L_{p,+} * L'_{q,-}\right) \bar{W}_{pq} = \bar{W}_{pq} \left(L_{q,+} * L'_{p,-}\right)$$

with

$$\begin{aligned} L_{p,+} &= a_p Y^{-1} X - c_p X^{-1} \\ L'_{q,-} &= d_q X - \omega^{-1} b_q Y^{-1} X^{-1} \end{aligned} \tag{2.5}$$

A graphical representation of $L_+, L_-$ where $*$ acquires a natural interpretation is

$(L_+)_{nn'} =$ 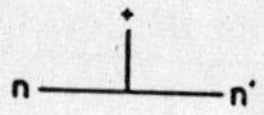

$(L_-)_{nn'} =$ 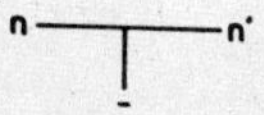 (2.6)

A second exchange relation with $Y^{-1}$ replaced by $Y$ is:

$$\left(L_{p,-} * L'_{q,+}\right) \bar{W}_{pq} = \bar{W}_{pq} \left(L_{q,-} * L'_{p,+}\right)$$

$L_- =$  $= dX + bYX^{-1}$

$L'_+ =$ 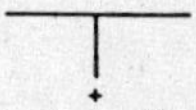 $= aYX - cX^{-1}$ (2.7)

We observe that (2.5), (2.7) can be put in the more general form:

$$\left(L_{p,\alpha} * L'_{q,\beta}\right) \bar{W}_{pq} = \bar{W}_{pq} \left(L_{q,\alpha} * L'_{p,\beta}\right) \tag{2.8}$$

$$\alpha, \beta = \pm$$

because no new relation occurs when $\alpha = \beta$. The graphical form of (2.8) is:

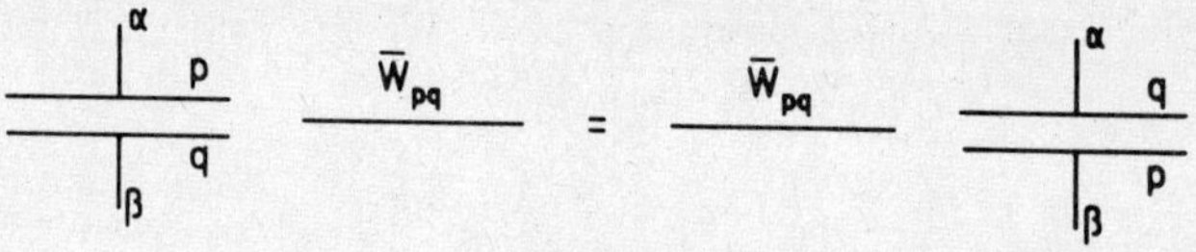

(2.9)

To obtain the exchange relation determining the weights $W_{pq}(n)$ we consider the operator

$$W_{pq} = \sum_n \hat{W}_{pq}(n) \left(Y_1 Y_2^{-1}\right)^n \tag{2.10}$$

$Y_1$ and $Y_2$ act in two different spaces. The coefficients $\hat{W}_{pq}(n)$ are determined by the relation:

$$\sum_{\alpha=\pm} \left(L'_{p,\alpha} \otimes L_{q,\alpha}\right) W_{pq} = W_{pq} \sum_{\alpha=\pm} \left(L'_{q,\alpha} \otimes L_{p,\alpha}\right) \tag{2.11}$$

or graphically

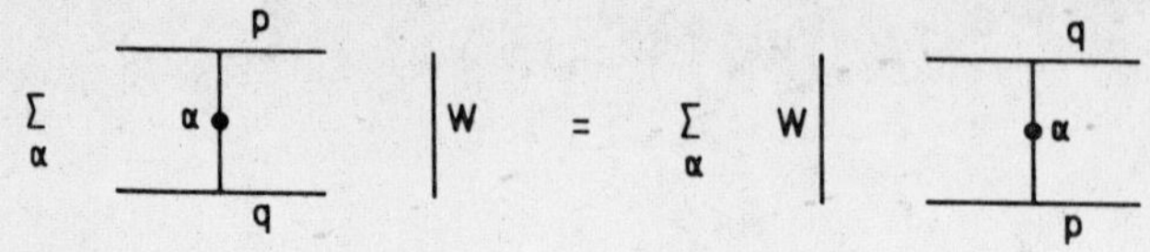


(2.12)

It gives

$$\hat{W}_{pq}(m) = \sum_{n\in\mathbb{Z}_N} \omega^{-mn} W_{pq}(n) \tag{2.13}$$

A plausibility argument for the Yang-Baxter equation (1.5) is now:

$$\begin{aligned}&\sum_\alpha \left(L'_{p,\alpha} \otimes L_{q,\alpha} * L'_{r,\beta}\right) \bar{W}_{qr}(X_2)\, W_{pr}\left(Y_1Y_2^{-1}\right) \bar{W}_{pq}(X_2)\\ &\quad= \sum_\alpha \bar{W}_{qr}(X_2)\, W_{pr}\left(Y_1Y_2^{-1}\right) \bar{W}_{pq}(X_2) \left(L'_{r,\alpha} \otimes L_{q,\alpha} * L_{p,\beta}\right)\end{aligned} \tag{2.14}$$

The same relation being satisfied with $\bar{W}(X_2)\, W\left(Y_1Y_2^{-1}\right) \bar{W}(X_2)$ replaced by $W\left(Y_1Y_2^{-1}\right)$ $\bar{W}(X_2)\, W\left(Y_1Y_2^{-1}\right)$. Let us put $L, L'$ into a matrix form:

$$\begin{aligned}L &= \begin{pmatrix} -cX^{-1} & aY^{-1}X \\ bYX^{-1} & dX \end{pmatrix}\\ L' &= \begin{pmatrix} -cX^{-1} & \omega^{-1}bY^{-1}X^{-1} \\ aYX & dX \end{pmatrix}\end{aligned} \tag{2.15}$$

Now (2.8, 2.11) read

$$\begin{aligned}\left(L_p \dot{*} L'_q\right)_{\alpha\beta} \bar{W}_{pq} &= \bar{W}_{pq} \left(L_q \dot{*} L'_p\right)_{\alpha\beta}\\ \left(L'_p \dot{\otimes} L_q\right)_{\alpha\beta} W_{pq} &= W_{pq} \left(L'_q \dot{\otimes} L_p\right)_{\alpha\beta}\end{aligned} \tag{2.16}$$

where . means that the ordinary matrix product is taken. We now define a $T$ matrix by

$$T_{\alpha\beta} = (\overset{1}{L} \dot{*} \overset{2}{L'} \dot{\otimes} \overset{3}{L} \cdots \dot{*} \overset{n}{L'})_{\alpha\beta} \tag{2.17}$$

The connection with a six vertex model is made clear if we consider the matrix

$$\begin{aligned}\overset{pq}{L}_{\alpha\beta} &= \begin{pmatrix} \overset{p}{L} \dot{*} \overset{q}{L'} \end{pmatrix}_{\alpha\beta}\\ &= \begin{pmatrix} a_pa_qX + c_pc_qX^{-1} & \omega^{-1/2}\left(a_pd_qX - c_pb_qX^{-1}\right) Y^{-1} \\ Y\left(d_pa_qX - b_pc_qX^{-1}\right) & d_pd_qX - \omega^{-1}b_pb_qX^{-1} \end{pmatrix}\end{aligned} \tag{2.18}$$

In analogy with a six vertex model we find two decompositions of $\overset{pq}{L}$ into triangular matrices

$$\overset{pq}{L} = L_+ + L_- = \hat{L}_+ + \hat{L}_-$$

with

$$\begin{aligned}
L_+ &= \begin{pmatrix} -c_pX^{-1} & a_pY^{-1}X \\ b_pYX^{-1} & 0 \end{pmatrix} \dot{*} \begin{pmatrix} 0 & \omega^{-1}b_qY^{-1}X^{-1} \\ a_qYX & d_qX \end{pmatrix} \\
L_- &= \begin{pmatrix} -c_pX^{-1} & 0 \\ b_pYX^{-1} & d_pX \end{pmatrix} \dot{*} \begin{pmatrix} -c_qX^{-1} & \omega^{-1}b_qY^{-1}X^{-1} \\ 0 & d_qX \end{pmatrix} \\
\hat{L}_+ &= \begin{pmatrix} -c_pX^{-1} & a_pY^{-1}X \\ 0 & d_pX \end{pmatrix} \dot{*} \begin{pmatrix} -c_qX^{-1} & \omega^{-1}b_qY^{-1}X^{-1} \\ 0 & d_qX \end{pmatrix} \\
\hat{L}_- &= \begin{pmatrix} 0 & a_pY^{-1}X \\ b_pYX^{-1} & d_pX \end{pmatrix} \dot{*} \begin{pmatrix} -c_qX^{-1} & \omega^{-1}b_qY^{-1}X^{-1} \\ a_qYX & 0 \end{pmatrix}
\end{aligned} \tag{2.19}$$

each $L$, $\hat{L}$ is written $L * L'$ and it is easy to check that $L_+$, $L'_+$; $L_-L'_-$;... obey the exchange realtions (2.16) on their own. The operators $(T_+)_{12}$, $(T_-)_{21}$, $(\hat{T}_+)_{12}$, $(\hat{T}_-)_{21}$ obtained from (2.17) with $L_+$; $L'_+$,... are the analogous of Jimbo's $e$, $f$, $\hat{e}$, $\hat{f}$ [7]. Their commutation relations derive from the six vertex exchange relation :

$$R(\lambda_1/\lambda_2)\ \overset{pq}{L}(\lambda_1) \otimes \overset{pq}{L}(\lambda_2) = \overset{pq}{L}(\lambda_2) \otimes \overset{pq}{L}(\lambda_1) R(\lambda_1/\lambda_2)$$

with

$$R(\lambda) = \begin{pmatrix} \alpha & & & \\ & \beta & \gamma & \\ & \bar{\gamma} & \beta & \\ & & & \alpha \end{pmatrix}$$

$$\begin{aligned}
\alpha = \omega^{1/2}\lambda - \omega^{-1/2}\lambda^{-1}\ &,\ \gamma = \lambda\left(\omega^{1/2} - \omega^{-1/2}\right) \\
\beta = \lambda - \lambda^{-1}\ &,\ \bar{\gamma} = \lambda^{-1}\left(\omega^{1/2} - \omega^{-1/2}\right) \\
L(\lambda) = \lambda L_+ + \lambda^{-1}L_-&
\end{aligned} \tag{2.20}$$

We shall not write them explicitly here because they will be studied more extensively in the next section.

## III. Exotic supersymmetry

In the previous section we defined an exchange algebra for the Chiral Potts model which was related to the 6-vertex model. We will now explain another way to find the same operators. This new approach generalizes a construction recently used by Zamolodchikov in his study of the perturbations of the tricritical Ising and tricritical Potts conformal field theories. In this framework the operators we define generate an algebra which is a $\mathbb{Z}_N$ generalization of the usual supersymmetric algebra.

This section is divided in three parts. In the first one we characterize the operators we are looking at, and we explain how the commutation relations reduce to the $\mathbb{Z}_N$ invariance and to the recursion relations satisfied by the Boltzmann weights. In the second part we describe the commutation relations between these operators and their relations with the 6-vertex model. Finally we take the Fatteev-Zamolodchikov limit (self dual genus zero case) in which a field theoretical interpretation is possible and we exhibit how the operators act as generators of an exotic supersymmetry

We use from the beginning the field theoretical (F.T.) language (even if in the genus $> 1$ cases a F.T. interpretation is not justified) in order to easely make contact with F.T. in the genus zero limiting case.

*a) The conserved charges*

The space of states on which the operators act are made of chains of spins. The spin $\alpha$ is either an element of $\mathbb{Z}_N$ or a neutral element which we denote by a star; $\alpha = k$ or $*, k \in \mathbb{Z}_N$ (two $\mathbb{Z}_N$-spins or two stars cannot be adjacent). Moreover each link between two adjacent spins carries a rapidity $p$ which is a point on the rapidity surface (1.1) defined above. In QFT the link between two adjacent spins $\alpha$ and $\beta$ is interpreted as a kink with rapidity $p$. We denote them by $K_{\alpha\beta}(p)$.

The Boltzmann weights, $W_{pq}(n)$ and $\bar{W}_{pq}(n)$, define a factorizable $S-$matrix acting on the kinks. On 2-kink states the action of $S$ is

$$< K_{n*}(p)K_{*m}(q) \mid S =< K_{n*}(q)K_{*m}(p) \mid W_{pq}(n-m) \tag{3.1a}$$

$$< K_{*n}(p)K_{n*}(q) \mid S = \sum_m \bar{W}_{pq}(n-m) < K_{*m}(q)K_{m*}(p) \mid \tag{3.1b}$$

The action of $S$ propagates on N-link states, $\langle K_{\alpha_0\alpha_1}(p_1)K_{\alpha_1\alpha_2}(p_2)...K_{\alpha_{N-1}\alpha_N}(p_N)\rvert$, like a factorizable $S-$matrix.

A graphical representation of $S$ is

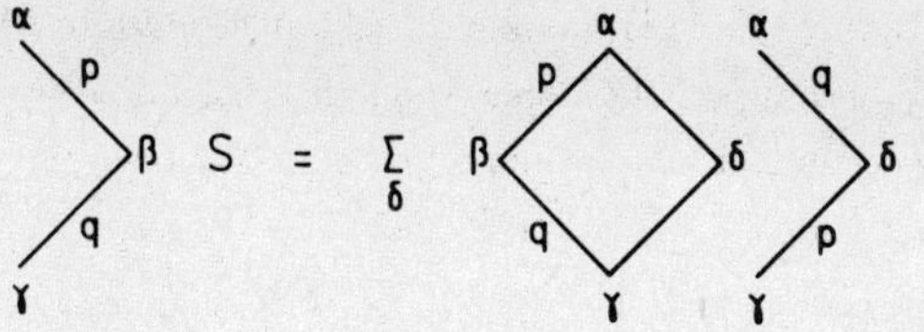


(3.2)

By construction the $S$−matrix is $\mathbb{Z}_N$-invariant. This $\mathbb{Z}_N$ invariance can be written as a Yang-Baxter (Y.B.) equation. To the generator $\sigma$ of $\mathbb{Z}_N$; $\sigma(*) = *$, $\sigma(k) = k+1$, we associate an operator $\Xi$ which acts on the kinks by translation of the spins. Namely

$$\langle K_{\alpha\beta}(p) \mid \; \Xi = \theta(p) < K_{\sigma(\alpha)\sigma(\beta)}(p) \mid \tag{3.3a}$$

$$\left\langle K_{\alpha_0\alpha_1}(p_1)...K_{\alpha_{N-1}\alpha_N}(p_N) \right| = \left(\prod_{j=1}^{N} \theta(p_j)\right) \left\langle K_{\sigma(\alpha_0)\sigma(\alpha_1)}(p_1)...K_{\sigma(\alpha_{N-1})\sigma(\alpha_N)}(p_N) \right| \tag{3.3b}$$

where $\theta$ is an arbitrary function.

Graphically, $\Xi$ can be represented by a square-box

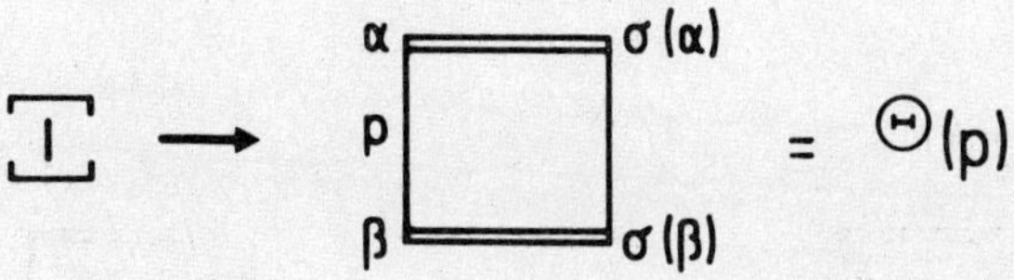


(3.4)

The "comultiplication" defining the action on N-link states consists in the juxtaposition of the boxes. The $\mathbb{Z}_N$ invariance of the $S$-matrix is then equivalent to the following Y.B. equation

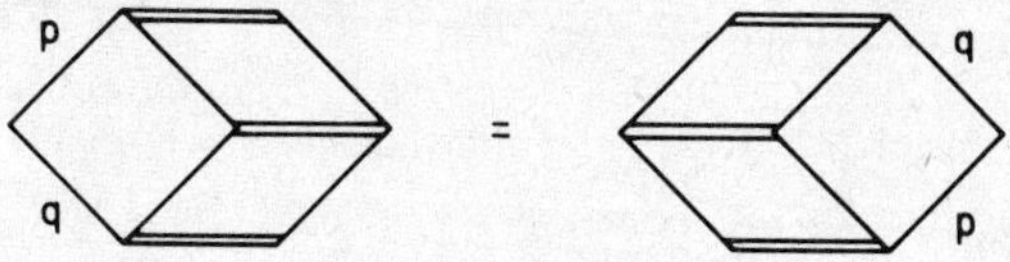


(3.5)

The operators, denoted by $Q$, we are looking at should commute with the $S-$matrix. In the QFT framework they become on-shell conserved charges. Morever we want them to be associated to the $\mathbb{Z}_N$-invariance of the $S-$matrix. As a starting point we demand that the operators act on the 1-link states as follows :

$$< K_{\alpha\beta}(p) \mid Q = \Lambda^{\alpha}_{\beta}(p) < K_{\sigma(\alpha)\beta}(p) \mid \tag{3.6}$$

for some functions $\Lambda^{\alpha}_{\beta}(p)$. Graphically

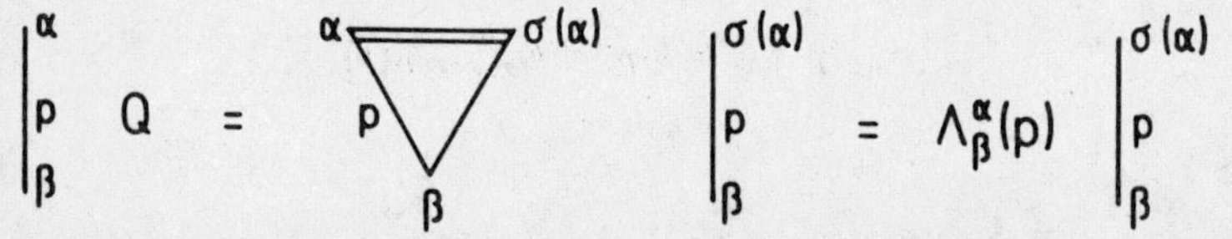


(3.7)

By analogy with the differential calculus on a quantum group we define the action of $Q$ on N-link states by adding boxes on the top of the triangles. Explicitly

$$\begin{aligned}&\langle K_{\alpha\alpha_1}(p_1)...K_{\alpha_{N-1}\alpha_N}(p_N)\big|\ Q \\ &= \sum_{k=1}^{N}\left(\prod_{j=1}^{k-1}\Theta(p_j)\right)\Lambda^{\alpha_{k-1}}_{\alpha_k}(p_k)\ \langle K_{\sigma(\alpha_0)\sigma(\alpha_1)}(p_1)...K_{\sigma(\alpha_{k-1})\alpha_k}(p_k)...K_{\alpha_{N-1}\alpha_N}(p_N)\big|\end{aligned} \tag{3.8}$$

Graphically $Q$ is represented by

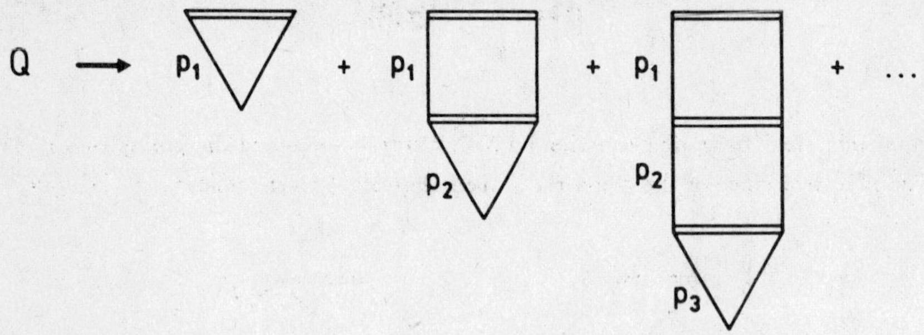


(3.9)

The function $\Theta_{(p)}$ and $\Lambda^{\alpha}_{\beta}(p)$ have to be determined by requiring that $Q$ commutes with the $S-$matrix.

The Y.B. equation (3.5) (or the $\mathbb{Z}_N$ invariance of the $S-$matrix) ensures that $Q$ action on N-kink states will commute with $S$ provided that $Q$ acting on 2-kink states does. The latter statement is equivalent to the following recursion equations for the Boltzmann weights

$$\left(\Lambda^k_*(q) + \theta(q)\Lambda^*_\ell(p)\right) W_{pq}(k-\ell) = W_{pq}(k-\ell+1)\left(\Lambda^k_*(p) + \theta(p)\Lambda^*_\ell(q)\right) \tag{3.10}$$

$$\left(\Lambda^*_k(p) - \Lambda^*_\ell(q)\right) \bar{W}_{pq}(k-\ell) = \bar{W}_{pq}(k-\ell+1)\left(\theta(q)\Lambda^{\ell-1}_*(p) - \theta(p)\Lambda^k_*(q)\right) \tag{3.11}$$

Solutions to these equations exist because of the recursion relations (1.4) defining the Boltzmann weights of the Chiral Potts models. One of the solution is given by

$\theta(p) =$ 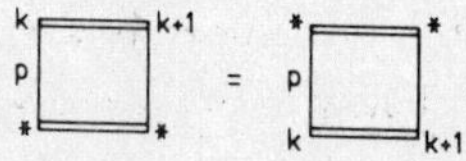 $= \left(\dfrac{c_p}{d_p}\right)$ (3.12a)

$\Lambda^*_k(p) =$ 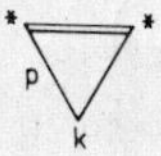 $= \left(\dfrac{a_d}{d_p}\right)\omega^{-k}$ (3.12b)

$\Lambda^k_*(p) =$ 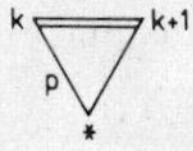 $= -\left(\dfrac{b_p}{d_p}\right)\omega^{-k-1}$ (3.12c)

Let us denote by $Q_+$ the operator defined by the above boxes and triangles.

The other solutions to the commuting equations can be generated from $Q_+$ by using the automorphisms of the rapidity surface. The automorphisms $R$ and $S$ act on the N-kink states by acting simultaneously on all the kink rapidities. Also let us denote by $C$ the conjugation charge which flips all the spins : $C(k) = N-k$, $C(*) = *$. Using these notations, from $Q_+$ we define the operators $Q_-$ and $\bar{Q}_+$ and $\bar{Q}_-$ by

$$Q_- = C^{-1}R^{-1}Q_+RC \tag{3.13a}$$

$$\bar{Q}_- = S^{-1}Q^t_+S \tag{3.13b}$$

$$\bar{Q}_- = C^{-1}R^{-1}\bar{Q}_+RC \tag{3.13c}$$

We will not generate new conserved charges by conjugating again by the automorphisms. The operators we produce in that way are gauge equivalent to these ones. Using Eqs. (1.7) and (1.8), it is an easy check to verify that $Q_\pm$ and $\bar{Q}_\pm$ commute with the $S-$matrix. Moreover they all have a structure analogous to $Q_+$ with their associated boxes and triangles.

The notation has been introduced for the following reasons : *i)* the lower indice ($\pm$) counts the amount by which the spins of the kinks are translated under the action of the $Q's$ (the $Q'_+s$ are associated to the generator $\sigma$ and the $Q'_-s$ to $\sigma^{-1}$); *ii)* the presence or the absence of a bar on the top of the $Q's$ arises from the Q.F.T. interpretation. In this framework the $Q_\epsilon$ (or the $\bar{Q}_\epsilon$) will be interpreted as the left (or the right) generators of a generalized supersymmetry.

*b) Commutation relations and relationship with the 6-vertex model.*

One of the crucial properties of the $Q's$ is their local character i.e. the boxes and the triangles only depends on the kink rapidity on which they act. By explicit computation it is easy to see that the following $\omega$-commutators

$$\bar{Q}_\epsilon Q_{\epsilon'} - \omega^{\epsilon\epsilon'} Q_\epsilon \bar{Q}_{\epsilon'} = T_{\epsilon\epsilon'} \tag{3.14}$$

preserve this property. In other words all the cross-terms cancel in these $\omega-$commutators such that the $T'_{\epsilon\epsilon'}s$ are also local operators. The operators $T_{\epsilon\epsilon'}$ admit a description in terms of boxes and triangles whose expressions follow from those associated to the operators $Q$. But instead of writting cumbersome expressions let us concentrate on chains with an even number of kinks. It is on such even-chains that the connection with the 6-vertex construction can be made. There are two kinds of even chain : either the chain begins with a star, or it begins with a $\mathbb{Z}_N-$spin. We call the first ones the star-chains and the second ones the spin-chains.

The spins-chains are of the form $\left\langle \prod_{j=1}^{N-1} K_{k_j *}(p_j)K_{*k_{j+1}}(p'_j)\right|$ and the commutators (3.14) act diagonally on such chains. More precisely, up to overall constants, we have

$$\left\langle \prod_{j=1}^{N-1} K_{k_j *}(p_j)K_{*k_{j+1}}(p'_j)\right| T_{\pm\pm} = 0 \tag{3.15}$$

$$\left\langle \prod_{j=1}^{N-1} K_{k_j *}(p_j)K_{*k_{j+1}}(p'_j) \mid T_{\pm\mp} = \left(\omega^{\pm 2k_1} - \left(\prod_{j=1}^{N-1} \rho^{\pm}(j)\right)\omega^{\pm 2k_N}\right)\left\langle \prod_{j=1}^{N-1} K_{k_j *}(p_j)K_{*k_{j+1}}(p'_j)\right|$$

with $\rho(j) = \bar{\omega}\frac{(b_j b'_j)(d_j d'_j)}{(a_j a'_j)(c_j c'_j)}$.

Note the almost topological nature of these commutators. In the field theoretical limit (the Fateev-Zamolodchikov case) all the $\rho(j)'s$ become equal to one and the commutators aquire a true topological nature.

The star chains are of the form $\left\langle \prod_{j=1}^{N} k_j\,(p_j;p'_j)\right|$ where we have defined $k(p;q) = K_{*k}(p)K_{k*}(q)$. On these chains the actions of the $T'_{\epsilon\epsilon'}s$ are (up to overall constants)

$$\left\langle \prod_{j=1}^{N} k_j\,(p_j;p'_j)\right| T_{\pm\mp} = 0 \tag{3.16}$$

$$\left\langle \prod_{j=1}^{N} K_{k_j}(p_j;p_j') \right| T_{\pm\pm} = \left\langle \prod_{j=1}^{N} k_j\,(p_j;p_j') \right| - \left( \prod_{j=1}^{N} \eta^{\pm 1}(j) \right) \left\langle \prod_{j=1}^{N} (k_j \pm 2)\,(p_j;p_j') \right|$$

with $\eta(j) = \bar{\omega}\frac{(b_j b_j')(c_j c_j')}{(a_j a_j')(d_j d_j')}$.

The spins $k(p;q)$ are the spin-variables of the 6-vertex like construction described in the first section. Acting on star-chains the Q-operators are not new operators : they can be expressed in terms of the elements of the transfert matrices defined in section 1. More precisely we have (up to overall constants)

$$\begin{aligned} Q_+ &= \hat{L}_{12}^{(+)} \cdot \left(\hat{L}_{22}^{(+)}\right)^{-1} \quad ; \quad Q_- = L_{21}^{(-)} \cdot \left(L_{11}^{(-)}\right)^{-1} \\ \bar{Q}_+ &= \hat{L}_{21}^{(-)} \cdot \left(\hat{L}_{11}^{(-)}\right)^{-1} \quad ; \quad \bar{Q}_- = L_{12}^{(+)} \cdot \left(L_{22}^{(+)}\right)^{-1} \end{aligned} \tag{3.17}$$

Thus in the case of star chains the commutation relations (3.16) turn out to be nothing but the commutation relations of $\mathcal{U}_q(GL(2))$.

We find quite surprising that the two constructions (sect. 2 and 3) reveal themselves to be equivalent because one was found by looking at a 6-vertex like exchange algebra and the other one by looking at operators which are RSOS-analogue of differential operators on quantum groups. Both constructions have their own advantages : the first one reveals the relation with the 6-vertex models whereas the second one reveals the connection with exotic supersymmetry and can be defined on any chains (not only on even chains).

*c) The field theoretical limit*

The limit we are interested in the self-dual genus zero limit in which the Chiral Potts model reduces to the Fateev-Zamolodchikov $\mathbb{Z}_N$-invariant model. In this case the Boltzmann weights can be interpreted as matrix elements of the factorizable $S-$matrix of a relativistic $(1+1)$ quantum field theory. In this limit

$$k' = 1\;; \quad c = d\;; \quad b = \omega^{1/2} a \tag{3.18}$$

on the Boltzmann weights reduce to

$$\frac{W_{pq}(n)}{W_{pq}(0)} = \prod_{j=1}^{n} \left( \frac{1 - a_{pq}\omega^{j-1/2}}{a_{pq} - \omega^{j-1/2}} \right) \; ; \; \frac{\bar{W}_{pq}(n)}{\bar{W}_{pq}(0)} = \omega^{n/2} \prod_{j=1}^{n} \left( \frac{a_{pq} - \omega^{j-1}}{1 - a_{pq}\omega^{j}} \right) \tag{3.19}$$

where $a_{pq} = a_p/a_q$.

Contact with field theory is made by writting $a = \exp\left(\frac{\theta}{N}\right)$ with $\theta$ the rapidity of the kink $K_{\alpha\beta}(\theta)$ (i.e. : the left and the right momenta of the kink $K_{\alpha\beta}(\theta)$ are respective $p = m\,\exp(\theta)$ and $\bar{p} = m\,\exp(-\theta)$. The relations (1.7) then express the crossing-symmetry properties of the $S-$matrix.

In this limiting case, the conserved charges $Q_\pm$ and $\bar{Q}_\pm$ act as follows

$$\langle K_{\alpha_0\alpha_1}(\theta_1)...K_{\alpha_{N-1}\alpha_N}\,|\,Q_\pm = \sum_{j=1}^{N} \mathrm{m}e^{\theta_j/N}\lambda_{\alpha_j}^{\alpha_{j-1}}(\pm)\,\langle K_{\sigma^{\pm1}(\alpha_0)\sigma^{\pm1}(\alpha_1)}(\theta_1)... ...K_{\sigma^{\pm1}(\alpha_{j-1})\alpha_j}(\theta_j)...K_{\alpha_{N-1}\alpha_N}(\theta_N)| \tag{3.20}$$

$$\langle K_{\alpha_0\alpha_1}(\theta_1)...K_{\alpha_{N-1}\alpha_N}(\theta_N)\,|\,\bar{Q}_\pm = \sum_{j=1}^{N} \mathrm{m}e^{-\theta_j/N}\bar{\lambda}_{\alpha_j}^{\alpha_{j-1}}(\pm)\,\langle K_{\sigma^{\pm1}(\alpha_0)\sigma^{\pm1}(\alpha_1)}(\theta_1)... ...K_{\sigma^{\pm1}(\alpha_{j-1})\alpha_j}(\theta_j)...K_{\alpha_{N-1}\alpha_N}(\theta_N)|$$

with $\lambda_k^*(\pm) = \omega^{\mp k}$ and $\lambda_*^k(\pm) = -\omega^{\mp k-1/2}$.

In the $N = 2$ and $N = 3$ these operators are equivalent to those recently introduced by Zamolodchikov in his study of the perturbation of the tricritical Ising and tricritical Potts conformal field theories.

The commutation relation are as above

$$\bar{Q}_\epsilon Q_{\epsilon'} - \omega^{\epsilon\epsilon'} Q_{\epsilon'}\bar{Q}_\epsilon = T_{\epsilon\epsilon'} \tag{3.21}$$

But now $T_{\epsilon\epsilon'}$ has a topological nature because in the Fatteev-Zamolodchikov $\eta(j) = \rho(j) = 1$. Moreover cyclicity property of $\mathbb{Z}_N$ implies that

$$Q_\pm^N = P \quad , \quad \bar{Q}_\pm^N = \bar{P} \tag{3.22}$$

where $P$ and $\bar{P}$ are the left and right momentum operators.

Equations (3.21) and (3.22) show that $Q_\pm$ and $\bar{Q}_\pm$ generate a $\mathbb{Z}_N$ generalization of the supersymmetric algebra.

It is a challenge to find an action for which the operators $Q$ and $T$ are the conserved and topological charges. Certainly the solution of this problem has to be found in the differential geometry on quantum spaces generated by objects satisfying $\omega-$statistics.

Acknowledgements

V. Pasquier thanks the participants to the Workshop held at the Australian National University for fruitful discussions, particularly R.J. Baxter, B. Mac Coy, V.E. Korepin, J. Perk for introducing him to the Chiral Potts model. The authors also thank M. Gaudin for numerous helpful comments and discussions.

**References**

[1] H. Au-Yang and J.H.H Perk, Proc. of Taniguchi Symposium, Vol. 19 (1989) Kinokumiya/academic

[2] V.V. Bazhanov and Yu-G-Stroganov, Preprint CMA-R34-89, "Chiral Potts model as a descendent of the six vertex model", submitted to J. Stat. Phys.

[3] A.B. Zamolodchikov, Preprint

[4] H. Au Yang, B. Mc Coy, J.H.H. Perk, S. Tang and M.L. Yan, Phys. Lett. 123A (1987) 219-223

[5] R.J. Baxter, J.H.H. Perk and H. Au Yang, Phys. Lett. A128 (1988) 138-142

[6] L. Faddeev, Les Houches Lectures (1982), Amsterdam : Elsevier (1984)

[7] M. Jimbo, A q-difference Analogue of U(g) and the Yang-Baxter equation, Letters in Mathematical Physics 10 (1985) 63-69

[8] M. Gaudin, rapport interne du CEA, Février 1989.

International Journal of Modern Physics B, Vol. 4, No. 5 (1990) 929–942
© World Scientific Publishing Company

# CRITICAL SPIN-1 VERTEX MODELS AND $O(n)$ MODELS

Bernard NIENHUIS
*Instituut voor Theoretische Fysica, Universiteit van Amsterdam*
*Valckenierstraat 65, 1018 XE Amsterdam, Netherlands.*

A method is presented by which critical and multicritical points of spin-1 (three-state) vertex models and classical $O(n)$ models on two-dimensional lattices are determined. It is a straightforward generalization of the ideas that earlier led to the determination of critical points and critical exponents of a honeycomb $O(n)$ model. On the square lattice the methods leads to tricritical as well as critical loci. For $n = 2$ a larger critical manifold is found than for other values of $n$. At the critical and multicritical points thus produced the models turn out to be soluble. The method is applicable to $O(n)$ models and spin-1 vertex models on any planar lattice.

## 1. Introduction

The $O(n)$ model in two dimensions as the $n$-component generalization of the classical Heisenberg model has received considerable attention. The physical applications are numerous, among which are phase transitions of systems with one- and two-component order parameters as well as polymer systems. The calculation of the exact value of the critical coupling constant and the critical exponents [1] of an $O(n)$ model on the honeycomb lattice started a new direction of research to the further properties of this specific model. This eventually led to the exact value of the free energy [2] at the critical point and the direct computation of the central charge and the critical exponents [3] from asymptotic finite size corrections of the free energy of systems infinite in one direction by means of the Bethe Anstatz.

In this paper we present a method by which critical and multicritical points of a wider class of $O(n)$ models can be determined. The method is applied to a square lattice, which results in several classes of critical points, for $O(n)$ models with general $n$, and some only for the $O(2)$ or $XY$ model.

A central theme is the existence of an intimate relation between the $O(n)$ model, the Potts model, vertex models and the Coulomb Gas. The relation of various lattice spin models to the Coulomb Gas has been utilized widely, and a recent review [4] will provide some details which are not repeated here.

The basis of the method rests in the following observations. The Potts model can be mapped onto a Coulomb Gas with electric charges of integer magnitude. At the critical point there are no charges of magnitude one, but only two and larger. Thermal deviations from criticality of of the Potts model, are associated with the appearance of excitations with unit electric charges *of one sign only.* Charge neutrality is provided for by the presence of the double charges of both signs which abound in the model. Thus the Potts model away from the critical temperature, maps onto a Coulomb Gas in which the elementary charges come in one sign only.

A consequence of this is that unit charges of the opposite sign constitute a critical fluctuation in the equivalent Coulomb Gas, even when the Potts model itself is not critical. These fluctuations cannot be translated into Potts-like variables. This picture implies that the locus of the Potts model in the larger, embedding phase diagram of the Coulomb Gas, forms a critical subspace. Therefore, when another model, being mapped onto the Coulomb Gas, intersects with the domain spanned by a Potts model, this first model will be critical at the intersection. It is this argument that allowed us earlier [1] to compute the critical points of honeycomb $O(n)$ models.

While the qualitative argument is based on observations concerning Potts models with only neighbor interaction on planar lattices, we assume that it is universally valid for Potts models with more general interactions, provided they have the same symmetry of the Hamiltonian, in terms of the order parameter. We use the word Potts model for lattice models in which the dynamical variables assume one out of $q$ discrete values, with interactions that are invariant under all permutations of these $q$ values.

In this paper we proceed to compute intersection points between Potts and $O(n)$ models on a square lattice by transforming the former models into the latter, via a sequence of intermediary models, which include spin-1 (i.e. three-state) vertex models. Based on the arguments above, we claim that such points are critical points of the $O(n)$, in the following sense. The free energy of the model is singular at the intersection point, as a function of the $O(n)$ model parameters. At the intersection point itself there are correlations in the $O(n)$ which decay in space as a power of the distance. We follow closely the method of ref. [1], reviewed in [4]. In the transformations only the bulk properties are considered. The questions concerning boundary conditions are not difficult, but would interrupt the flow of the argument. They are discussed qualitatively at the end of this paper.

## 2. From the Potts Model to a Dense Loop Model

It is central to find a coincidence between a Potts model and an $O(n)$ model, as they are reexpressed in terms of a vertex model. As in ref. [5] the Potts model is mapped on a spin-$\frac{1}{2}$ (i.e. two-state) vertex model. By merging the lattice edges of the vertex model one obtains a spin-1 vertex model, and this in its turn can be mapped on an $O(n)$ model. Each step in this process is possible only provided the original Potts model parameters satisfy some restrictions. In the interest of clarity these restrictions are anticipated in our notation.

We first turn to the $q$-state Potts model on a square lattice, with interactions on the faces of the lattice. The Boltzmann weight $\mathcal{W}_p$ is given as the product over the faces of the lattice of functions $f$ of the spin variables on the four corners of the face:

$$\mathcal{W}_p = \prod_{\substack{l \\ m \quad k \\ j}} f(s_j, s_k, s_l, s_m). \tag{1}$$

Here the variables $s$ assume one of $q$ different values, and the function $f$ is invariant under simultaneous permutation of these $q$ values in all four arguments. Therefore $\mathcal{W}_p$ has the global permutation symmetry $S(q)$ of the Potts model. For reasons of computational convenience we write

$$f(a,b,c,d) = f_4(a,b,c,d)\ f_2(a,b)\ f_2(b,c)\ f_2(c,d)\ f_2(d,a), \tag{2}$$

in which

$$\begin{aligned} f_4(a,b,c,d) &= u + v\delta_{a,c} + w\delta_{b,d} \\ [f_2(a,b)]^2 &= 1 + x\delta_{a,b}. \end{aligned} \tag{3}$$

We define the square of $f_2$ because each pair of nearest-neighbor spins is coupled by two factors $f_2$ arising from the two adjacent faces.

Generalizing the approach of ref. [5] we introduce the surrounding lattice $\mathcal{L}$ of the original square lattice $\mathcal{S}$, in which the interactions in the Potts model are represented by vertices. The two types of interaction, $f_2$ and $f_4$ , are represented by two types of vertices located on the edges and faces respectively of $\mathcal{S}$. We refer to the vertices of $\mathcal{L}$ by the symbol for the interaction (3) that centers on it. Each vertex of type $f_4$ is connected to the four adjacent vertices of type $f_2$ *by two links each.* The resulting lattice $\mathcal{L}$, drawn diagonally, is shown in Fig. 1.

We develop the partition sum in powers of $u$ , $v$ , $w$ and $x$, by expanding the products (1) and (2). The terms in this expansion may be represented by graphs $\mathcal{G}$ on $\mathcal{S}$, in which the Kronecker-$\delta$'s between spins are denoted by bonds between the corresponding sites. The graphs $\mathcal{G}$ then have vertical, horizontal and diagonal bonds, with weight $v$ , $w$ and $x$ , respectively. The partition sum thus expanded is given by

$$Z = \sum_{\mathcal{G}} q^C\ x^{N_x}\ v^{N_v}\ w^{N_w}\ u^{N-N_v-N_w}, \tag{4}$$

where $C$ is the number of connected components in $\mathcal{G}$ ; $N_x$, $N_v$ and $N_w$ are the number of diagonal, vertical and horizontal bonds respectively; and $N$ is the total number of vertices of type $f_4$.

The graphs $\mathcal{G}$ on $\mathcal{S}$ can be mapped into a polygon decomposition $\mathcal{P}$ of $\mathcal{L}$ as follows. The legs of each vertex of $\mathcal{L}$ are joined into non-intersecting pairs, such that each pair of (octagonal) faces adjacent to the vertex are joined *if and only if* they are connected directly by a bond of $\mathcal{G}$. The resulting local polygon configurations at each vertex are shown in Fig. 2. Since this yields a one-to-one correspondence between graphs and polygon decompositions, we may write the partition sum (4) as a sum on

Figure 1. The surrounding lattice $\mathcal{L}$. The sites of the original lattice $\mathcal{S}$ are shown by dots. The vertices with four and eight legs are referred to as type $f_2$ and $f_4$, respectively.

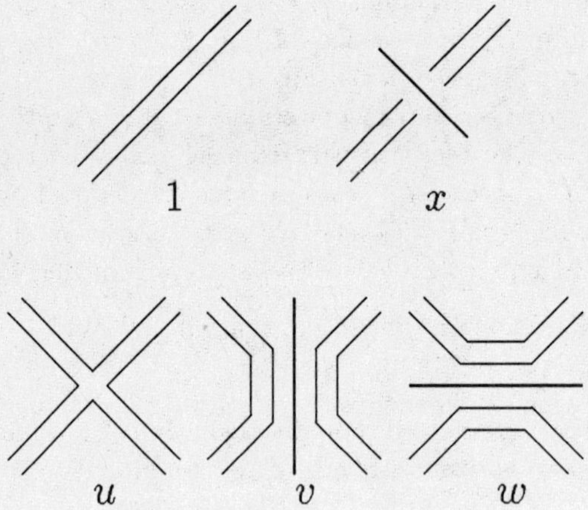


Figure 2. The ways to decompose the vertices of type $f_4$ (weights $u$, $v$ and $w$) and type $f_2$ (weights 1 and $x$). The heavy lines represent bonds of $\mathcal{G}$ and the thin lines the edges of the polygon decomposition of $\mathcal{L}$.

$\mathcal{P}$ rather than $\mathcal{G}$. The polygons of $\mathcal{P}$ come in two types, (i) those corresponding with the outer perimeter of the components of $\mathcal{G}$ and (ii) those inscribing the circuits of $\mathcal{G}$. Therefore the number of polygons is given by $p = C + S$ , where $S$ is the number of circuits of $\mathcal{G}$. With Euler's relation $C = N - N_x - N_v - N_w + S$ the variable $C$ may be eliminated: $2C = N - N_x - N_v - N_w + p$ , and the partition sum rewritten

$$Z = q^{N/2} \sum_{\mathcal{P}} q^{p/2} \left(x\, q^{-1/2}\right)^{N_x} \left(v\, q^{-1/2}\right)^{N_v} \left(w\, q^{-1/2}\right)^{N_w} u^{N_4 - N_v - N_w}. \qquad (5)$$

In this form the partition sum is essentially that of a loop model. By that we mean a sum of configurations of loops or closed polygons on the edges of the lattice. The loops themselves are non-local objects as there is no limit to there size, but the interaction is the product of factors, each of which is determined by only a few adjacent segments of the entire configuration. The factor $q^{1/2}$ per polygon makes the model essentially different from a two-state (occupied and non-occupied) vertex model, because it cannot be locally assessed. In this case the polygons together cover all the edges of the lattice. Below we see another variety of loop models in which the loops are free to cover part of the lattice.

## 3. From the Loop Model to a Spin-$\frac{1}{2}$ Vertex Model

We now proceed to turn the model into a spin-$\frac{1}{2}$ vertex model. Since this process is entirely analogous to [5], we review it only briefly. Each polygon is given an orientation by placing arrows on all of its edges. The weight $q^{1/2}$ per polygon is now rewritten as the sum

$$q^{1/2} = e^{-i\phi} + e^{i\phi}. \qquad (6)$$

The two terms are associated with, respectively, the clockwise and anticlockwise orientation of the polygon. Furthermore, the phase factors $e^{\pm i\phi}$ are distributed over the entire polygon by associating a phase with each turn that an observer makes when he follows the polygon in the direction of the arrows. The phases corresponding with the possible turns are shown in Fig. 3. Thus the partition sum (5) is unchanged if the factor $q^{p/2}$ is replaced by the product over the phase factors from summed over all polygon orientations. Like in the interactions (3) a distinction between the vertical and the horizontal directions is allowed in these phase factors.

The total of all the arrows of all the polygons in a decomposition $\mathcal{P}$ forms an ice type arrow configuration $\mathcal{A}$. Therefore the partition sum (5) may be written as

$$Z = q^{N/2} \sum_{\mathcal{P}} \sum_{\mathcal{A}|\mathcal{P}} weights. \qquad (7)$$

in which the *weights* are the product of the powers of $u$, $v$, $w$ and $x$ from (4) and the phase factors from Fig. 3. The second sum is over all ice configurations that induce a consistent orientation of the polygons of $\mathcal{P}$. By the simple change of order of the summations the model is turned into a spin-$\frac{1}{2}$ vertex model, on the lattice $\mathcal{L}$.

$$Z = q^{N/2} \sum_{\mathcal{A}} \sum_{\mathcal{P}|\mathcal{A}} weights, \qquad (8)$$

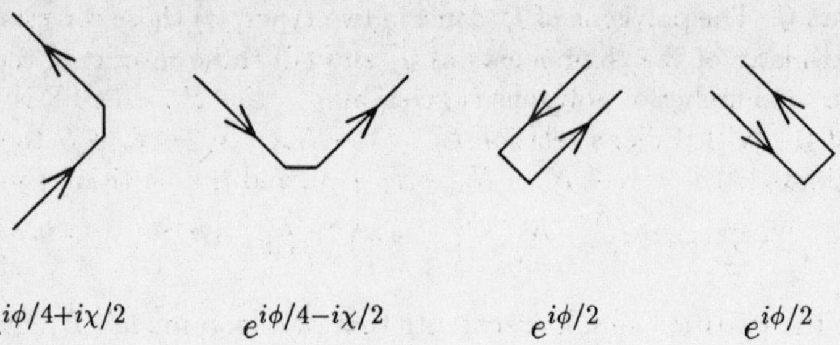


Figure 3. The phase factors associated with the turning of arrows along the edges of a polygon. Figures obtained from these by rotation over $\pi$ have the same phase factor, and by inversion of all the arrows the opposite phase.

where the first sum is over all ice confgurations $\mathcal{A}$, and the second over all polygon decompositions $\mathcal{P}$ consistent with $\mathcal{A}$, i.e. all those $\mathcal{P}$ in which connect the arrows head-to-tail. The full description of this vertex model is very tedious, it has vertices of four and of eight legs. The latter admit $\binom{8}{4} = 70$ different ice type arrow configurations, of which, however, only 32 have non-zero weight. Fig. 4 gives an example of the calculation of the weight of one such arrow configuration.

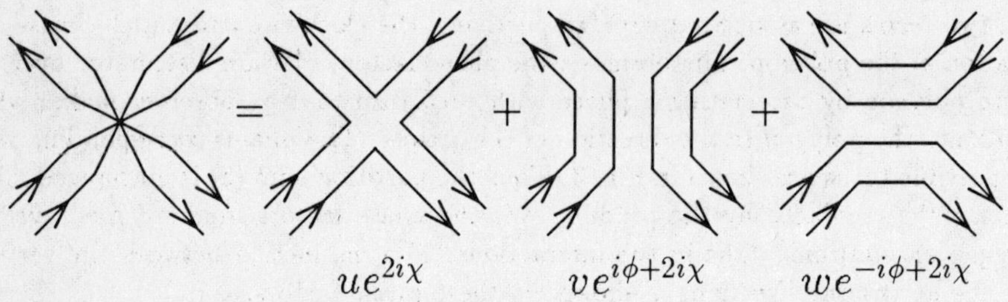


Figure 4. The calculation of the vertex weight. The vertex is decomposed, according to Fig. 2 and phases are given according to Fig. 3. Only those decompositions are taken into account, which connect the different arrows head-to-tail.

## 4. From a Spin-$\frac{1}{2}$ to a Spin-1 Vertex Model

We intend to transform this complicated spin-$\frac{1}{2}$ vertex model into a much simpler spin-1 model, by combining the edges that connect the same two vertices (i.e. the leaf-shaped objects of Fig. 1) into a single edge. These new edges can be in three states, up, down or null corresponding to the 'sum' of the arrows of the component edges. Whenever a new edge is in the null state it has an internal degree of freedom, corresponding to the choice between the up-down and the down-up order of its composite edges. The intended transformation from a spin-$\frac{1}{2}$ to a spin-1 model, amounts to the summation over these internal degrees of freedom. We demand therefore, that the internal variables decouple from each other at the vertices of type $f_2$, so that the summation involves only four such internal variables, arranged as a clover leaf about the vertices of type $f_4$ (see Fig. 1).

The weights of the six possible arrow configuration of the vertices of type $f_2$ are shown in Fig. 5. The internal degrees of freedom of the two composite edges joining in this vertex, decouple when the weights factorize into separate functions of the two leaves joining in the vertex. This is indeed the case when $x = -1$. The resulting factorized weights for the lower leaf are also indicated in Fig. 5. In terms of the original Potts model (3) this implies that all nearest-neighbors pairs of spins must be in different states.

Once the vertices of type $f_2$ are thus factorized, it is convenient to combine the separate factors each with the weight of the vertex of type $f_4$ nearest to it. After the summation of the internal degrees of freedom, there remains a 19-vertex model, with vertex configurations as shown in Fig. 6. The resulting weights are (after multiplication with their common denominator $2\cos\phi$)

$$\begin{aligned}
\omega_1 &= v + w - u(1 - 2\cos 2\phi) \\
\omega_2 = \omega_4 &= e^{i\phi/2+i\chi}\left(ue^{-2i\phi} + v\right) \\
\omega_3 = \omega_5 &= e^{-i\phi/2-i\chi}\left(ue^{2i\phi} + v\right) \\
\omega_6 = \omega_8 &= e^{i\phi/2-i\chi}\left(ue^{-2i\phi} + w\right) \\
\omega_7 = \omega_9 &= e^{-i\phi/2+i\chi}\left(ue^{2i\phi} + w\right) \\
\omega_{10} = \omega_{11} = \omega_{12} = \omega_{13} &= -u \\
\omega_{14} = \omega_{15} &= v \\
\omega_{16} = \omega_{17} &= w \\
\omega_{18} &= e^{2i\chi}\left(2u\cos\phi + ve^{i\phi} + we^{-i\phi}\right) \\
\omega_{19} &= e^{-2i\chi}\left(2u\cos\phi + ve^{-i\phi} + we^{i\phi}\right)
\end{aligned} \tag{9}$$

We claim that these weights represent a critical manifold within the larger class of more general 19-vertex models. This, as argued in the introduction, is a consequence of the fact that the model (9) is equivalent with the Potts model (3). This claim is the central point of this paper. In the following sections we give the applications to the special cases when this 19-vertex model reduces to $O(n)$ or loop models. It will be seen that in these special cases the model is soluble, and its criticality may then be demonstrated from its exact solution as well.

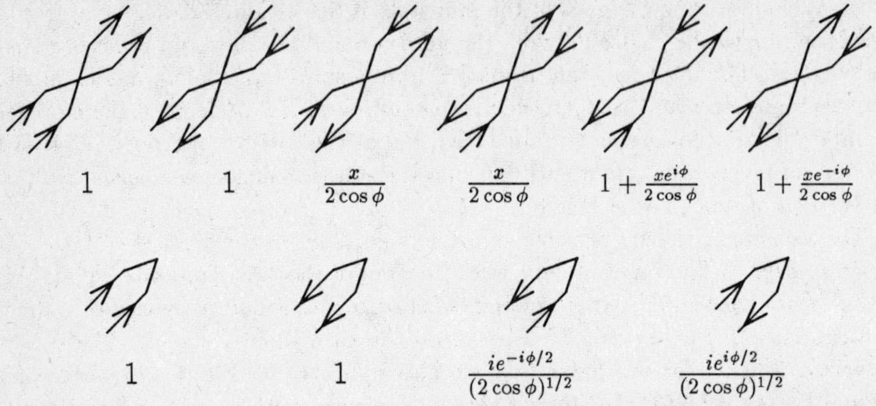


Figure 5. Decoupling of internal variables. The $f_2$ vertex states with their weights are shown above. When $x=-1$ the weights factorize, and the factors of the lower half of the vertex are shown below. The same weights apply to the upper half, for the states obtained by rotation over $\pi$.

Figure 6. The states of the 19-vertex model

## 5. From the 19-Vertex Model to a Loop Model

Under some conditions the 19-vertex (9) is equivalent to a loop model on the dual lattice of $\mathcal{S}$. The loops consist of those edges that carry an arrow, and the arrows on the same loop give it a consistent orientation, clockwise or anticlockwise. The possible local configurations at a single vertex of the loop model are shown in Fig. 7, and the weights of these configurations we denote by $\rho_j$.

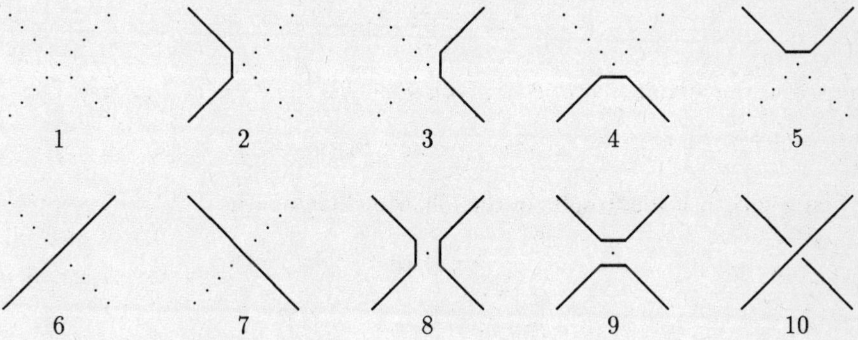


Figure 7. The vertex states of the loop model. In state 10 two separate polygon segments cross.

The correspondence between vertex and loop configurations is not one-to-one. As in the case of (7) and (8) a model of oriented polygons serves as an intermediary. When in this model the orientation of the loops is summed out, the remaining variables specify a loop configuration, and when the polygon decompositions are summed over, the remaining arrows yield a configuraton of the 19-vertex model. Each of the loop configurations of Fig. 7 corresponds to those vertex states which, on all occupied edges carry arrows that are connected head-to-tail by the bonds.

We consider two possibilities to make the connection.

(i) The loops are not allowed to cross each other, $\rho_{10} = 0$. The fugacity may take any value between 2 and $-2$, resulting from local phase factors associated with the corners of each polygon, and summed over the two possible orientations, entirely analogous to the polygon decompositions described above.

(ii) The vertex weights are invariant under inversion of all arrows and the fugacity is two. In this case there is no need to suppress intersections, $\rho_{10} \neq 0$.

First consider the case that the fugacity arises from phase factors associated with the turns in the polygon. Then it is essential that the polygons do not intersect themselves, i.e. $\rho_{10} = 0$. Otherwise the fugacity of a loop would depend on the topology of its self-intersections. The vertex weights $\omega_2, .., \omega_9$ specify not only the loop weights $\rho_2, .., \rho_5$ but also the phase factors that determine the fugacity. Once the phase factors are fixed, the weights $\omega_{14}, .., \omega_{19}$ (four independent values) give only $\rho_8$ and $\rho_9$, which are therefore overdetermined. Consequently, this scheme only works if the 19-vertex weights (9) satisfy the additional constraint that the values of $\rho_8$ and $\rho_9$ are consistent with the phases of $\omega_2, .., \omega_9$:

$$\begin{aligned} \omega_{18} &= \omega_{14}\frac{\omega_4}{\omega_5} + \omega_{16}\frac{\omega_9}{\omega_8}, \\ \omega_{19} &= \omega_{15}\frac{\omega_5}{\omega_4} + \omega_{17}\frac{\omega_8}{\omega_9}. \end{aligned} \tag{10}$$

Espressed in the original parameters, both yield the same simple equation

$$u^3 + u^2 v + u^2 w + uvw(2\cos 2\phi - 1) = 0. \tag{11}$$

The first solution, $u = 0$, results in the following loop model

$$\begin{aligned} \rho_1 &= v + w \\ \rho_2 = \rho_3 &= v \\ \rho_4 = \rho_5 &= w \\ \rho_6 = \rho_7 &= 0 \\ \rho_8 &= v \\ \rho_9 &= w \\ \rho_{10} &= 0 \\ n &= 2\cos 2\phi, \end{aligned} \tag{12}$$

where $n$ is the fugacity of the loops. In [6] it is shown that this model satisfies the Yang-Baxter equations, and in [7] that it is equivalent with a $q$-state Potts model with $q^{1/2} = 1 + 2\cos 2\phi$. This is the more remarkable because this model originated from a Potts model with $q^{1/2} = 2\cos\phi$. The equivalence proves that the model is critical when $n \leq 1$, consistent with numerical analysis [7].

The second solution of (11)

$$w = \frac{u^2 + uv}{v(1 - 2\cos 2\phi) - u} \tag{13}$$

is conveniently parametrised by

$$\begin{aligned} u &= -\sin 2\psi - \sin 3\phi \\ v &= \sin(2\psi - 2\phi) + \sin\phi \\ w &= \sin(2\psi + 2\phi) + \sin\phi. \end{aligned} \tag{14}$$

The vertex model resulting from the substitution of this $u$, $v$ and $w$ into (9) turns out to be one proposed by Izergin and Korepin[8,9]. The model satisfies the Yang-Baxter equations[9,6], with $\psi$ and $\chi$ as spectral parameters, and is therefore soluble. It is critical, for real values of $\phi$ and $\psi$, which has been demonstrated by numerical [7] and analytical [10] studies.

The loop version of the model is given by

$$\begin{aligned} \rho_1 &= \sin\phi + \sin 3\phi - \sin 5\phi + \sin 2\psi \\ \rho_2 = \rho_3 &= 2\sin(2\phi)\cos(3\phi/2 - \psi) \\ \rho_4 = \rho_5 &= 2\sin(2\phi)\sin(3\phi/2 + \psi) \\ \rho_6 = \rho_7 &= \sin 3\phi + \sin 2\psi \\ \rho_8 &= \sin\phi - \sin(2\phi - 2\psi) \\ \rho_9 &= \sin\phi + \sin(2\phi + 2\psi) \\ \rho_{10} &= 0 \\ n &= -2\cos 4\phi. \end{aligned} \tag{15}$$

Here only the variable $\psi$ plays the role of a spectral parameter. For each value of $n$ there are four essentially different values of the coupling constants $\rho_j$. These several critical points play different roles in the phase diagram[6]. When the spectral parameter $\psi = \phi/2$, the value $\rho_8 = 0$, so that the loops may not pass each other vertically in a single vertex. In that case each vertex (with four legs) may as well be drawn as two vertices with three legs, connected by one vertical, as indicated in Fig. 8. The topology of the lattice is now turned into that of a honeycomb lattice, and indeed the loop parameters (15) reflect the hexagonal symmetry, and correspond to the known critical values [1].

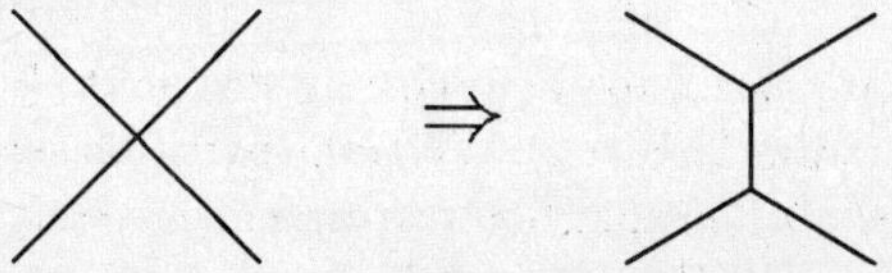

Figure 8. Interpretation of a vertex of the square lattice as two vertices of a honeycomb lattice.

Finally we consider the case when the 19-vertex weights (9) are invariant under inversion of the arrows. This symmetry is obeyed when

$$\begin{aligned} u &= \cos\phi - \cos 2\chi, \\ v &= -\cos\phi + \cos(2\phi - 2\chi), \\ w &= -\cos\phi + \cos(2\phi + 2\chi). \end{aligned} \tag{16}$$

When these values are substituted into (9) the result is the 19-vertex model proposed by Zamolodchikov and Fateev[11]. They showed that it satisfies the Yang-Baxter equations (with the spectral parameter $\chi$), and thus is soluble. Later numerical and analytical analysis[12] demonstrated that for real values of the parameters the model is critical. The loop version, other than the cases above, has a fixed fugacity $n = 2$, and a non-zero crossing term

$$\begin{aligned}
\rho_1 &= \cos 3\phi - 2\cos\phi + \cos 2\chi \\
\rho_2 = \rho_3 &= -2\sin(2\phi)\sin(\phi/2-\chi) \\
\rho_4 = \rho_5 &= -2\sin(2\phi)\sin(\phi/2+\chi) \\
\rho_6 = \rho_7 &= \cos 2\chi - \cos\phi \\
\rho_8 &= \sin(2\phi)\sin(2\chi) - \sin(2\phi)\sin(\phi) \\
\rho_9 &= -\sin(2\phi)\sin(2\chi) - \sin(2\phi)\sin(\phi) \\
\rho_{10} &= \cos(2\phi)\cos(2\chi) - \cos(2\phi)\cos(\phi) \\
n &= 2
\end{aligned} \tag{17}$$

## 6. From the Loop Model to the $O(n)$ Model

The configurations of loop models (12),(15) and (17) may be interpreted as the terms in the high temperature expansion of $O(n)$ models. The dynamical variables are $n$-component vectors, and live on the edges of the square lattice (the dual of $\mathcal{S}$). The Boltzmann weight is the product over the vertices of an interaction factor, depending on the four spins around it.

$$\begin{aligned}
\mathcal{W}_o &= \prod_{\substack{m\ l \\ j\ k}} \rho_1 + \rho_2\,(\vec{s}_j\cdot\vec{s}_m) + \rho_3\,(\vec{s}_k\cdot\vec{s}_l) + \rho_4\,(\vec{s}_j\cdot\vec{s}_k) + \rho_5\,(\vec{s}_l\cdot\vec{s}_m) \\
&+ \rho_6\,(\vec{s}_j\cdot\vec{s}_l) + \rho_7\,(\vec{s}_k\cdot\vec{s}_m) + \rho_8\,(\vec{s}_j\cdot\vec{s}_m)\,(\vec{s}_k\cdot\vec{s}_l) \\
&+ \rho_9\,(\vec{s}_j\cdot\vec{s}_k)\,(\vec{s}_l\cdot\vec{s}_m) + \rho_{10}\,(\vec{s}_j\cdot\vec{s}_l)\,(\vec{s}_k\cdot\vec{s}_m)
\end{aligned} \tag{18}$$

The spin variables $\vec{s}$ are taken to be uniformly distributed over the surface of the hypersphere of square radius $(\vec{s}\cdot\vec{s}) = n$. When the partition sum is expanded in powers of the $\rho_j$, the diagrams of the expansion correspond precisely with the configuration of polygons on the lattice.[13,4] The number of spin-components $n$ enters the weights of the diagrams as the fugacity of the loops. According to the arguments above, the parametrizations of the $\rho$, (12), (15) and (17) constitute critical subspaces of this model. The specific role of these critical points, has been discussed in separate publications[7,10,6].

## 7. Discussion

The results of this paper are two-fold. In the first place a scheme is proposed by which critical or multicrital loci of $O(n)$ and loop models in two dimensions may be found. Secondly, the scheme is applied to the square lattice, and several critical manifolds are thus located. It turns out that the model is exactly soluble in these critical manifolds, so that its criticality could be ascertained by independent means (refs. [12,10,6]).

The scheme itself is not restricted to the square lattice. The generalization to other planar lattices is immediate, and has been applied successfully to the honeycomb, triangular and Kagome lattices[14]. The results suggest that critical points of higher rank (in the sequence critical, tricritical, tetracritical etc.) are found for lattices of higher coordination number. For the honeycomb lattice the method produces only a critical point, for the square lattice besides a critical point also a tricritical point [7,6], and for the triangular lattice the solutions include yet a third class of critical behavior[14].

When the method is viewed as a scheme to construct critical spin-1 vertex models, a second generalization suggests itself, namely to vertex models with more than three states per edge. The author has generalized[14] the transformation from a Potts model to a vertex model, (3) to (9), and constructed spin-$\frac{3}{2}$ and spin-2 models. Very likely also models of higher spin may thus be produced. This gives a procedure only to construct critical vertex models. In a sense this is akin to the fusion procedure[15] by which higher spin vertex models are constructed from two-state models. However in the case discussed in this paper, it is only after the further restriction of (9) to a loop or $O(n)$ model, that the model is soluble. As we can not fathom why this is the case, we also do not know as yet how to generalize this process to higher spin.

The long string of model transformations as presented in this paper, is applicable to bulk properties. The mappings are not intrinsically faithful in the sense that they equate finite lattice partition sums or entire spectra of the transfer matrix. However, the transformations can be made accurate also in this respect, if the appropriate boundary conditions are applied, which ascertain that loops which wind round handles of a finite geometry or hit a boundary are weighed appropriately. It should be noted that for this reason the values of the central charge is not automatically carried over by the transformations in this paper. Consequently when boundary conditions are not chosen appropriately, two systems, of which we argue that they are equivalent, may have different values of the central charge.

This work was supported in part by the foundation for Fundamental research of Matter (FOM) and the Royal Netherlands Academy of Sciences (KNAW). Hospitality of the Center of Mathematical Analysis of the Australian National University in Canberra, where this work was presented, is gratefully acknowledged.

## References

1. B. Nienhuis, *Phys. Rev. Lett.* **49** (1982), 1062

2. R.J. Baxter, *J. Phys. A* **19** (1986), 2821 and *J. Phys. A* **20** (1987), 5241

3. M.T. Batchelor and H.W.J. Blöte, *Phys. Rev. Lett.* **61** (1988), 138 and *Phys. Rev. B* **39** (1989), 2391

4. B. Nienhuis, Coulomb Gas representations of phase transitions in two dimensions, in Phase Transitions and Critical Phenomena, eds. C. Domb and J.L. Lebowitz (Academic, London, 1987)

5. R.J. Baxter, S.B. Kelland and F.Y. Wu, *J. Phys. A* **9** (1976), 397

6. B. Nienhuis, proceedings of the IUPAP conference STATPHYS 17, Rio de Janeiro 1989, (North-Holland 1990)

7. H.W.J. Blöte and B. Nienhuis, *J. Phys. A* **22** (1989), 1415

8. A.G. Izergin and V.E. Korepin, *Comm. Math. Phys.* **79** (1981), 303

9. N.Yu. Reshetikhin, *Zh. Eksp. Teor. Fiz.* **84** (1983), 1190 [*Sov. Phys. JETP* **57** (1983), 691]; V.I. Vichirko and N.Yu. Reshetikhin, *Teor. Mat. Fiz.* **56** (1983), 260 [*Theor. Math. Phys.* **56** (1983), 805]

10. M.T. Batchelor, B. Nienhuis and S.O. Warnaar, *Phys. Rev. Lett.* **62** (1989), 2425

11. A.B. Zamolodchikov and V.A. Fateev, *Yad. Fiz. (USSR)* **32** (1980), 581, [translation: *Sov. J. Nucl. Phys.* **32** (1980), 298]

12. F.C. Alcaraz and M.J. Martins, *J. Phys.* **A21** (1988), 4397

13. E. Domany, D. Mukamel, B. Nienhuis and A. Schwimmer, *Nucl. Phys. B* **190** [FS3] (1981), 279

14. B. Nienhuis, in progress

15. P.P. Kulish, N.Yu. Reshetikhin and E.K. Sklyanin, *Lett. Math. Phys.* **5** (1981), 393

International Journal of Modern Physics B, Vol. 4, No. 5 (1990) 943–952
© World Scientific Publishing Company

# CONFORMAL INVARIANCE IN ONE-DIMENSION AND A TWO-COMPONENT LOG-GAS

P.J. FORRESTER
*Department of Mathematics, La Trobe University, Bundoora, Victoria 3083, Australia.*

and

M.L. ROSINBERG
*Université Pierre et Marie Curie (Paris VI), Structure et Reactivite aux Interfaces, Bâtiment F(74)4, Place Jussieu 75230, Paris Cedex 05, France.*

Conformal invariance in one-dimension implies the correlation functions must be constant. It is demonstrated by an exact solution that all the correlations between like species in a two-component lattice gas with the logarithmic potential have the conformal invariance property at the metal- insulator transition. A further exactly solvable isotherm of the model system is studied and the corresponding density of zeros for the grand partition function is obtained explicitly. A phase transition along this isotherm can be induced by appropriate choice of the parameters.

## 1. Introduction

As is well known, the conformal invariance of correlations at a critical point is a property which can be exploited to derive quantitative information regarding critical exponents and correlations in two dimensions (see e.g. [1]). In three and higher dimensions the conformal group has only a finite number of parameters in contrast to the infinite number of parameters of the two-dimensional conformal group. This fact is responsible for the inability of conformal invariance to provide useful results in three or higher dimensions.

What of the situation in one-dimension? There, any general one-to-one coordinate transformation is conformal (the lattice will simply translate, stretch or compress). As noted by Cardy [1], this requirement is extremely restrictive in that it implies the correlation functions must be constant at the critical point. In general this is not the case (see e.g. [2] where the correlation of the $1/r^2$ Ising model is calculated by the renormalisation method), and thus the correlations of a one-dimensional system at criticality are not generally conformally invariant.

How does this fact fit into the general assumptions for conformal invariance? According to Cardy [1], these requirements may be summarized as:

(1) Scale invariance

(2) Translation invariance

(3) Rotation invariance

(4) Short-range interactions

The first three of these assumptions will typically apply to a one- dimensional system. However, the fourth requirement will not. A well known theorem says that one-dimensional systems with short-range interactions will not exhibit a phase transition at non-zero temperature.

As well as being sufficient conditions for conformal invariance, it has also been stated that they are necessary [1, pp.64-65]. Hence, for phase transitions in one-dimension, since the interactions must be long-ranged, the correlations cannot be conformally invariant i.e. they cannot all be constant.

In this paper, the correlations at criticality of a two-component one-dimensional lattice gas with a long range log-potential will be calculated exactly. These results "almost" provide a counter example to the necessity of short-range interactions for conformal invariance by exhibiting a conformal invariance. In particular, it will be shown that the correlations between like species in the lattice gas vanish identically at criticality. The behaviour of the system is thus analogous to a perfect gas, and indeed it will be shown that the partition function at the critical coupling is precisely that of a perfect gas on a lattice.

We should mention that this result concerning the behaviour of the correlations at criticality has been previously reported [3,4]. However, the model as formulated in this paper has an extra parameter — the distance of shortest approach between the two species — which wasn't introduced in the previous studies. Furthermore, the method of calculations adopted here is very different from that used in [4]. Here techniques introduced in [3] and [5] are modified and consequently a procedure which highlights the underlying mathematical structure of the exact solution is obtained.

The format of the paper is as follows. In Sec. 2 the precise definition of the model is given and it will be shown that at two special values of the coupling the grand partition function and correlations can be written as determinants. In Sec. 3, the determinants are evaluated and the exact results for the partition function and correlations given. The final section discusses the location of the zeros of the polynomial representing the grand partition function.

## 2. Determinant Identities for the Grand Partition Function and Correlations

### *2.1 Definition of the model*

The model system to be studied is a two-component, one-dimensional lattice gas with positive and negative charges of equal magnitude interacting via the log-potential.

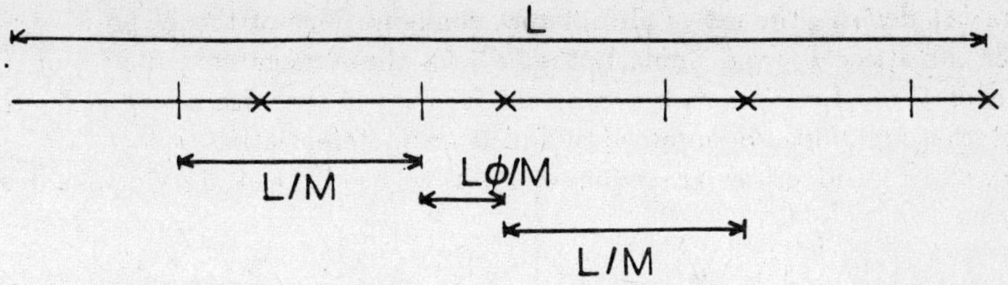


Figure 1. The lattice geometry and spacings. The sublattices $L_1$ and $L_2$ have sites denoted by crosses and dashes respectively.

More precisely, divide a line of length $L$ into $M$ intervals so that there are sites at the points $nL/M$, $n = 1, 2, \ldots, M$. Introduce an interlacing lattice at the points $(n - \phi)L/M$, $n = 1, \ldots, M$, $0 < \phi < 1$ (see Fig. 1). Denote these lattices $L_1$ and $L_2$ respectively. Allow $N$ ($\le M$) positive charges to occupy $L_1$ and $N$ negative charges to occupy $L_2$. The charges are two-dimensional so that they interact via the logarithmic potential. In periodic boundary conditions, the potential is

$$V(\theta_1, \theta_2) = -q_1 q_2 \log \left[ |e^{2\pi i\theta_1/L} - e^{2\pi i\theta_2/L}| (L/2\pi) \right] \tag{2.1}$$

Denote the coordinates of the positive charges by $m_k L/M$ and the coordinates of the negative charges by $(\ell_k - \phi)/M$, where $m_k, \ell_k = 1, \ldots, M$.

The grand partition function is a polynomial in the fugacity $\zeta^2$. As explained in [4, Sec. 4.1], the polynomial can be made reciprocal by defining a scaled fugacity $\xi$. Computation of the ground state energy shows that the appropriate choice is

$$\xi = \left( \frac{\pi M}{L \sin \pi\phi} \right)^{\Gamma} \zeta^2 \tag{2.2}$$

The coupling constant is

$$\Gamma = q^2 / k_B T \tag{2.3}$$

Thus the model involves three dimensionless parameters: $\Gamma, \xi$ and the cutoff $\phi$.

### 2.2 *The grand partition function at $\Gamma = 2$*

As noted in [3] and [4], it follows immediately from the Cauchy double alternate determinant formula that the Boltzmann factor at $\Gamma = 2$, for each configuration in the above model, can be written as

$$W_{N2} = \left( \frac{\pi}{L} \right)^{2N} \left( \det \left[ \sin^{-1} (m_j - (\ell_k - \phi))/M \right]_{j,k=1,\ldots,N} \right)^2 \tag{2.4}$$

The grand partition function with variable site fugacities $\zeta(1 + a(j))$ and $\zeta(1 + b(j))$ for $L_1$ and $L_2$ respectively is given by

$$\Xi_2(a; b) = \sum_{N=0}^{M} (\xi \sin^2 \pi\phi/M)^N \sum_{\{\ell\},\{m\}} W_{N2} \prod_{j=1}^{N} (1 + a(\ell_j))(1 + b(\ell_j)) \tag{2.5}$$

where $\{\ell\}$ denotes the set of all different configurations of the $N$ positive particles on the sublattice $L_1$, and similarly the $\{m\}$ for the $N$ negative particles on $L_2$. The grand partition function with a constant fugacity is the case $a = b = 0$, while the correlation functions are obtained by functional differentiation.

Two further identities are required. First, we need the straightforward result

$$\left(\det\Big[f(x_j - y_k)\Big]_{j,k=1,\dots,N}\right)^2 = \det\begin{bmatrix} \mathbf{0}_N & f(x_j - y_k) \\ f(y_j - x_k) & \mathbf{0}_N \end{bmatrix} \tag{2.6}$$

where $\mathbf{0}_N$ denotes the zero matrix of order $N$ and $f(x_j - y_k)$ is the element in the $jth$ row and $kth$ column of the $N \times N$ upper right hand block. Similarly the meaning of $f(y_j - x_k)$ for the lower left hand block. Second, it follows from an expansion in minors that for any $f(x)$ odd and set of numbers $r_1, r_2, \dots, r_M$ and $s_1, s_2, \dots, s_M$

$$\sum_{N=0}^{M} \zeta^N \sum_{x\in\{r\},y\in\{s\}} \det\begin{bmatrix} \mathbf{0}_N & f(x_j - y_k) \\ f(y_j - x_k) & \mathbf{0}_N \end{bmatrix} = \det(\mathbf{1}_{2M} + \zeta^{\frac{1}{2}}\mathbf{K}), \tag{2.7}$$

where the sum is over the set of all combinations of $r_1, \dots, r_M$ taken $N$ at a time for $\{x_1, \dots, x_N\}$, and all combinations of $s_1, \dots, s_M$ taken $N$ at a time for $\{y_1, \dots, y_N\}$. The matrix $\mathbf{K}$ is given by

$$\mathbf{K} = \begin{bmatrix} \mathbf{0}_M & f(r_j - s_k) \\ f(s_j - r_k) & \mathbf{0}_M \end{bmatrix} \tag{2.8}$$

and $\mathbf{1}_{2M}$ denotes the $2M \times 2M$ identity matrix.

Application of (2.6) and (2.7) to (2.5) gives the determinant identity

$$\Xi_2(a;b) = \det[\mathbf{1}_{2M} + \{\xi^{\frac{1}{2}}(\sin\pi\phi)/M\}\mathbf{J}] \tag{2.9}$$

where

$$\mathbf{J} = \begin{bmatrix} \mathbf{0}_M & (1+a(j))\sin^{-1}\pi(j-k+\phi)/M \\ (1+b(j))\sin^{-1}\pi(j-\phi-k)/M & \mathbf{0}_M \end{bmatrix} \tag{2.10}$$

### *2.3 The grand partition function at $\Gamma = 4$*

To derive a determinant formula for the grand partition function at $\Gamma = 4$, we begin by deriving a determinant formula for the Boltzmann factor. Our technique is a slight modification of that used in [4] to derive the formula

$$\left(\frac{\prod_{1\le j<k\le N}(x_k - x_j)(y_k - y_j)}{\prod_{j=1}^{N}\prod_{k=1}^{N}(x_k - y_j)}\right)^4 = \det\begin{bmatrix} (x_j - y_k)^{-1} & (x_j - y_k)^{-2} \\ (x_j - y_k)^{-2} & 2(x_j - y_k)^{-3} \end{bmatrix}_{j,k=1,\dots,N} \tag{2.11}$$

where the determinant consists of $2 \times 2$ blocks as specified.

Consider the Cauchy doubly alternant in the form

$$D_{2N} = \det\Big[\sin^{-1}(\theta_j - \phi_k)\Big]_{j,k=1,\dots 2N} \tag{2.12}$$

where

$$D_{2N} := \frac{\prod_{1\le j<k\le 2N} \sin(\theta_k - \theta_j)\sin(\phi_k - \phi_j)}{\prod_{j=1}^{2N}\prod_{k=1}^{2N} \sin(\theta_k - \phi_k)} \tag{2.13}$$

By taking the limits $\theta_{j+N} \to \theta_j$ and $\phi_{j+N} \to \phi_j$ with manipulations as detailed in [4], we derive the analogue of (2.11)

$$(D_N)^4 = \det \begin{bmatrix} \sin^{-1}(\theta_j - \phi_k) & \frac{\partial}{\partial\phi_k}\sin^{-1}(\theta_j - \phi_k) \\ \frac{\partial}{\partial\phi_k}\sin^{-1}(\theta_j - \phi_k) & \frac{\partial^2}{\partial(\phi_k)^2}\sin^{-1}(\theta_j - \phi_k) \end{bmatrix} \tag{2.14}$$

Use of (2.14) gives that the Boltzmann factor at $\Gamma = 4$, $W_{N4}$, can be expressed as the block determinant

$$W_{N4} = (\pi/L)^{4N} \det[\mathbf{A}(m_j, \ell_k)]_{j,k=1,\dots,N} \tag{2.15}$$

where $\mathbf{A}(m_j, \ell_k)$ is the $2 \times 2$ block

$$\mathbf{A}(m_j, \ell_k) = \begin{bmatrix} \sin^{-1}\frac{\pi(m_j-\ell_k+\phi)}{M} & -\frac{M}{\pi}\frac{\partial}{\partial\phi}\sin^{-1}\frac{\pi(m_j-\ell_k+\phi)}{M} \\ -\frac{M}{\pi}\frac{\partial}{\partial\phi}\sin^{-1}\frac{\pi(m_j-\ell_k+\phi)}{M} & \left(\frac{M}{\pi}\right)^2\frac{\partial^2}{\partial\phi^2}\sin^{-1}\frac{\pi(m_j-\ell_k+\phi)}{M} \end{bmatrix} \tag{2.16}$$

To express the grand partition function as a determinant, the following result of Rosinberg [5] is required.

Let $f_1(x)$ and $f_2(x)$ be even functions, $g_1(x)$ and $g_2(x)$ odd functions and $\{r_1, \dots, r_M\}$, $\{s_1, \dots, s_M\}$ be sets of numbers. Then

$$\left(\sum_{N=0}^{M} \zeta^N \sum_{x\in\{s\},y\in\{r\}} \det\Big[\mathbf{A}_1(x_j - y_k)\Big]_{2N\times 2N}\right)^2 = \det\left(\mathbf{1}_{4M} + \zeta^{\frac{1}{2}}\mathbf{B}\right) \tag{2.17}$$

where

$$\mathbf{A}_1(x_j - y_k) = \begin{bmatrix} f_1(x_j - y_k) & g_1(x_j - y_k) \\ g_2(x_j - y_k) & f_2(x_j - y_k) \end{bmatrix} \tag{2.18}$$

$$\mathbf{A}_2(y_j - x_k) = \begin{bmatrix} f_2(y_j - x_k) & g_2(y_j - x_k) \\ g_1(y_j - x_k) & f_1(y_j - x_k) \end{bmatrix} \tag{2.19}$$

and

$$\mathbf{B} = \begin{bmatrix} \mathbf{0}_{2M} & \mathbf{A}_1(s_j - r_k) \\ \mathbf{A}_2(r_j - s_k) & \mathbf{0}_{2M} \end{bmatrix} \tag{2.20}$$

The notation in the second summation is the same as that used in (2.7).

Substituting (2.15) in (2.17) shows that the grand partition function at $\Gamma = 4$, $\Xi_4$, satisfies the identity

$$(\Xi_4)^2 = \det\left[\mathbf{1}_{4M} + (\xi^{\frac{1}{2}}(\sin\pi\phi)/M)\mathbf{L}\right] \tag{2.21}$$

where

$$\mathbf{L} = \begin{bmatrix} \mathbf{0}_{2M} & \mathbf{L}_1 \\ \mathbf{L}_2 & \mathbf{0}_{2M} \end{bmatrix} \tag{2.22}$$

$$\mathbf{L}_1 = \begin{bmatrix} \sin^{-1}\pi(j-k+\phi)/M & -(M/\pi)\frac{d}{d\phi}\sin^{-1}\pi(j-k+\phi)/M \\ -(M/\pi)\frac{d}{d\phi}\sin^{-1}\pi(j-k+\phi)/M & (M/\pi)^2\frac{d^2}{d\phi^2}\sin^{-1}\pi(j-k+\phi)/M \end{bmatrix} \tag{2.23}$$

$$\mathbf{L}_2 = \begin{bmatrix} (M/\pi)^2\frac{d^2}{d\phi^2}\sin^{-1}\pi(j-\phi-k)/M & (M/\pi)\frac{d}{d\phi}\sin^{-1}\pi(j-\phi-k)/M \\ (M/\pi)\frac{d}{d\phi}\sin^{-1}\pi(j-\phi-k)/M & \sin^{-1}\pi(j-\phi-k)/M \end{bmatrix} \tag{2.24}$$

## 3. Exact Evaluation of the Grand Partition Function and Correlations

### 3.1 Grand partition function at $\Gamma = 2$

Consider the identity (2.9) with $a = b = 0$. The elements of the upper right hand block are

$$c_{jk} := \sin^{-1} \pi(j - k + \phi)/M \tag{3.1}$$

and consequently have the property

$$c_{jk} = c_{j-k} = -c_{j+M-k}. \tag{3.2}$$

The same property holds for the lower left hand block. Each block is therefore anti-cyclic, and can thus be diagonalized by the transformation

$$\begin{bmatrix} \mathbf{U}^{-1} & \mathbf{0}_M \\ \mathbf{0}_M & \mathbf{U}^{-1} \end{bmatrix} (\mathbf{1}_{2M} + \{\xi^{\frac{1}{2}}(\sin \pi\phi)/M\}\mathbf{J}) \begin{bmatrix} \mathbf{U} & \mathbf{0}_M \\ \mathbf{0}_M & \mathbf{U} \end{bmatrix} \tag{3.3}$$

where

$$\mathbf{U} = \frac{1}{\sqrt{M}} \left[ e^{2\pi i j(k+\frac{1}{2})/M} \right]_{j,k=0,\ldots,M-1} \tag{3.4}$$

is unitary. Use of (3.3) gives

$$\Xi_2 = \det(\mathbf{1}_{2M} + \Lambda) \tag{3.5}$$

where

$$\Lambda = \begin{bmatrix} \mathbf{0}_M & \delta_{jk} s_j(\phi) \\ \delta_{jk} s_j(-\phi) & \mathbf{0}_M \end{bmatrix}_{j,k=1,\ldots,M} \tag{3.6}$$

and

$$s_j(\phi) = \frac{\sin \pi\phi}{M} \xi^{\frac{1}{2}} \sum_{\ell=0}^{M-1} \frac{e^{2\pi i \ell(j-1/2)/M}}{\sin \pi(\phi - \ell)/M} \tag{3.7}$$

The symbol $\delta_{jk}$ denotes the Kronecker delta.

By rearranging the rows and columns in (3.5) we see that $\Xi_2$ factorizes into a product of $2 \times 2$ determinants. Thus

$$\Xi_2 = \prod_{j=1}^{M} \det \begin{pmatrix} 1 & s_j(\phi) \\ s_j(-\phi) & 1 \end{pmatrix} \tag{3.8}$$

Furthermore, we have the summation formula

$$\sum_{\ell=0}^{M-1} \frac{e^{2\pi i \ell(j-\frac{1}{2})/M}}{\sin \pi(\phi - \ell)/M} = \frac{M}{\sin \pi\phi} e^{2\pi i \phi(j-(M+1)/2)/M} \tag{3.9}$$

To obtain (3.9), we begin with the identity

$$\sin^{-1} \pi(\phi - \ell)/M = 2i \lim_{\mu \to 1^-} \sum_{k=0}^{\infty} \mu^k e^{-2\pi i(\phi-\ell)(k+\frac{1}{2})/M} \tag{3.10}$$

and replace the summation over $k$ by

$$k = nM + s, \ n = 0, 1, \ldots, \quad s = 0, 1, \ldots, M-1 \tag{3.11}$$

This gives

$$\sin^{-1} \pi(\phi - \ell)/M = \frac{2i}{1 - e^{-2\pi i \phi}} \sum_{s=0}^{M-1} e^{-2\pi i(\phi-\ell)(s+\frac{1}{2})/M} \tag{3.12}$$

which when substituted in the summation in the left hand side of (3.9) gives the identity.

Substituting (3.9) in (3.7) and (3.8) yields the exact evaluation

$$\Xi_2 = (1+\xi)^M \tag{3.13}$$

Remarkably, this is the grand partition function for a perfect gas on $M$ lattice sites, with the constraint that no two particles may occupy the same site.

### *3.2 n-particle distribution function for like charges*

The dimensionless $n$-particle distribution for $n$ charges of like sign (positive, say) at $\ell_1, \ell_2, \ldots, \ell_n$ is given by

$$\rho_n(\ell_1, \ell_2, \ldots, \ell_n) = (\Xi_2)^{-1} \frac{\delta^n}{\delta a(\ell_1) \ldots \delta a(\ell_n)} \Xi_2(a; 0) \tag{3.14}$$

The identity (2.5) for $\Xi_2(a; b)$ allows the functional derivative to be performed row-by-row in the determinant. We obtain

$$\rho_n(\ell_1, \ell_2, \ldots, \ell_n) = (\Xi_2)^{-1} \det(\mathbf{1}'_{2M} + \{\xi^{\frac{1}{2}}(\sin \pi\phi)/M\}\mathbf{J}) \prod_{1 \le j < k \le n} (1 - \delta_{\ell_k, \ell_j}) \tag{3.15}$$

where $\mathbf{1}'_{2M}$ denotes the identity matrix of dimension $2M \times 2M$ with diagonal entries 0 instead of 1 at positions $\ell_1, \ell_2, \ldots, \ell_n$ and $\mathbf{J}$ is given by (2.10).

The method of evaluating the determinant in (3.15) is the same as that for evaluating (2.9). We find

$$\det(\mathbf{1}'_{2M} + \{\xi^{\frac{1}{2}}(\sin \pi\phi)/M\}\mathbf{J}) = \xi^n (1+\xi)^{M-n} \tag{3.16}$$

and thus

$$\rho_n(\ell_1, \ldots, \ell_n) = \rho^n \prod_{1 \le j < k \le n} (1 - \delta_{\ell_k, \ell_k}) \tag{3.17}$$

where $\rho$ denotes the dimensionless particle density of either species (positive or negative). This is the distribution function of a perfect gas on a lattice of $M$ sites with the constraint that no two particles can occupy the same lattice site.

### *3.3 Grand partition function at* $\Gamma = 4$

The $2 \times 2$ matrices (2.18) and (2.19) have the properties

$$\mathbf{A}_1(j,k) = \mathbf{A}_1(j-k) = -\mathbf{A}_1(j+M-k) \tag{3.18}$$

and

$$\mathbf{A}_2(j,k) = \mathbf{A}_2(j-k) = -\mathbf{A}_2(j+M-k) \tag{3.19}$$

Hence the $2M \times 2M$ matrices

$$\Big[\mathbf{A}_1(j,k)\Big]_{j,k=1,\dots,M} \quad \text{and} \quad \Big[\mathbf{A}_2(j,k)\Big]_{j,k=1,\dots,M} \tag{3.20}$$

are block anticyclic. Analogous to (3.3), the matrix in (2.21) can therefore be diagonalised by the transformation

$$\begin{bmatrix} \mathbf{V}^{-1} & \mathbf{0}_{2M} \\ \mathbf{0}_{2M} & \mathbf{V}^{-1} \end{bmatrix} (\mathbf{1}_{4M} + \{\xi^{\frac{1}{2}}(\sin \pi\phi)/M\}\mathbf{L}) \begin{bmatrix} \mathbf{V} & \mathbf{0}_{2M} \\ \mathbf{0}_{2M} & \mathbf{V} \end{bmatrix} \tag{3.21}$$

where $\mathbf{V}$ is the $2M \times 2M$ unitary block matrix

$$\mathbf{V} = \frac{1}{\sqrt{M}} \left[e^{2\pi i j(k+1/2)/M}\mathbf{1}_2\right]_{j,k=1,\dots,M} \tag{3.22}$$

Application of (3.21) gives

$$(\Xi_4)^2 = \det(\mathbf{1}_{4M} + \mathbf{T}) \tag{3.23}$$

where

$$\mathbf{T} = \begin{bmatrix} \mathbf{0}_{2M} & \delta_{jk}\alpha_j(\phi) \\ \delta_{jk}\beta_j(\phi) & \mathbf{0}_{2M} \end{bmatrix} \tag{3.24}$$

and

$$\alpha_j(\phi) = \begin{bmatrix} s_j(\phi) & -(M/\pi)\frac{d}{d\phi}s_j(\phi) \\ -(M/\pi)\frac{d}{d\phi}s_j(\phi) & (M/\pi)^2\frac{d^2}{d\phi^2}s_j(\phi) \end{bmatrix} \tag{3.25}$$

$$\beta_j(\phi) = \begin{bmatrix} (M/\pi)^2\frac{d^2}{d\phi^2}s_j(-\phi) & (M/\pi)\frac{d}{d\phi}s_j(-\phi) \\ (M/\pi)\frac{d}{d\phi}s_j(-\phi) & s_j(-\phi) \end{bmatrix} \tag{3.26}$$

The quantity $s_j(\phi)$ which occurs in (3.25) and (3.26) is defined by (3.7) and given in closed form by (3.9).

Rearrangement of the rows and columns in (3.23) shows that $(\Xi_4)^2$ factorizes into a product of $4 \times 4$ determinants. Explicitly

$$(\Xi_4)^2 = \prod_{j=1}^{M} \det \begin{bmatrix} \mathbf{1}_2 & \alpha_j(\phi) \\ \beta_j(\phi) & \mathbf{1}_2 \end{bmatrix} = \prod_{j=1}^{M} \det \Big[\mathbf{1}_2 + \alpha_j(\phi)\beta_j(\phi)\Big] \tag{3.27}$$

where $\alpha_j(\phi)$ and $\beta_j(\phi)$ are the $2 \times 2$ matrices given by (3.25) and (3.26) respectively. Straightforward manipulations of (3.27) give the evaluation

$$\Xi_4 = \prod_{k=0}^{[M/2]-1} \{1 - \xi\frac{4\sin^2 \pi\phi}{M^2}[4k^2 - 4k(M-1) + (M-1)^2 - \frac{M^2}{2\sin^2 \pi\phi}] + \xi^2\}$$

$$\times \begin{cases} (1+\xi), & \text{M odd} \\ 1, & \text{M even} \end{cases} \tag{3.28}$$

where [ ] in the product terminal denotes the integer part. The choice $\phi = \frac{1}{2}$ gives agreement with a previous exact result [4, eqn. (3.59)].

## 4. Properties of the Exact Solution

### *4.1 The zeros of the grand partition function*

The grand partition function is a polynomial in the scaled fugacity $\xi$. The order of the polynomial is $M$ (the number of lattice sites on each sublattice) and for a given value of $\Gamma$, the zeros will in general depend on the cutoff $\phi$.

At $\Gamma = 2$, from (3.13) we see that the $M$ zeros all occur at the point $\xi = -1$ independent of the value of $\phi$. At $\Gamma = 4$, from (3.28), the zeros lie at the zeros of the quadratic

$$1 - \xi \frac{4\sin^2 \pi\phi}{M^2}\left[4k^2 - 4k(M-1) + (M-1)^2 - \frac{M^2}{2\sin^2 \pi\phi}\right] + \xi^2 \tag{4.1}$$

for $k = 0, 1, \ldots, [M/2] - 1$, as well as at $\xi = -1$ if $M$ is odd. It is a straightforward exercise to show that all the zeros in this case lie on the unit circle in the complex $\xi$-plane.

For $\phi = 1/2$, it has previously been conjectured [4] that all the zeros of $\Xi_\Gamma(\xi)$ as a function of $\xi$ lie on the negative real axis for $\Gamma < 2$ and on the unit circle for $\Gamma > 2$. The above results provide evidence that the conjecture may also be true for general $0 < \phi < 1$.

### *4.2 The thermodynamic limit*

Further properties of the exact solution at $\Gamma = 4$ are revealed by taking the thermodynamic limit. In the large-M limit the logarithm of (3.28) behaves like a Riemann sum and thus

$$\beta P = \frac{1}{\tau}\int_0^{\frac{1}{2}} \log\left[1 - 4\xi \sin^2 \pi\phi(4t^2 - 4t + 1 - (2\sin^2 \pi\phi)^{-1}) + \xi^2\right] dt \tag{4.2}$$

where $\tau$ denotes the lattice spacing within each sublattice and $\beta P$ has its usual meaning. The change of variables

$$t = \frac{1}{2}\left(1 - \frac{\cos\theta/2}{\sin \pi\phi}\right) \tag{4.3}$$

gives

$$\beta P = \frac{1}{4\tau \sin \pi\phi}\int_{2\pi|\phi - 1/2|}^{\pi} \log(1 - 2\xi\cos\theta + \xi^2)\sin\theta/2 d\theta \tag{4.4}$$

From (4.4) we can read off that the density of zeros of the grand partition function on the unit circle at $\Gamma = 4$ is

$$\begin{cases} \dfrac{\sin\theta/2}{4\sin\pi\phi}, & 2\pi|\phi-\frac{1}{2}| \le \theta \le 2\pi(1-|\phi-\frac{1}{2}|) \\ 0, & \text{otherwise} \end{cases} \tag{4.5}$$

For $\phi = \frac{1}{2}$, we see from (4.5) that the zeros pinch the real axis, and there is thus a phase transition as $\xi$ is varied through one. This phase transition has previously been reported and categorized [6]. However, if $\phi \neq 1/2$, the exact solution reveals that there is no phase transition at $\Gamma = 4$ as the scaled fugacity $\xi$ is varied. The conjectured phase diagram for $\phi = \frac{1}{2}$ has been given in [6]. For $\phi \neq \frac{1}{2}$ it seems that the region $\Gamma < 2$ is a conducting phase for all $\xi$ and the region $\Gamma > 2$ is an insulating phase for all $\xi$. The isotherm $\Gamma = 2$ is the conductor-insulator phase boundary.

## Acknowledgments

Peter J. Forrester was supported financially in this project by the Australian Research Council.

## References

1. J.L. Cardy, in *Phase Transitions and Critical Phenomena, Vol.11, p55*, edited by C. Domb and J.L. Lebowitz, (Academic Press, New York, 1987).

2. J. Bhattacharjee, S. Chakravarty, J.L. Richardson and D.J. Scalapino, *Phys. Rev. B.* **24** (1982) 3862.

3. M. Gaudin, *J. Phys. (Paris)* **46** (1985) 1027.

4. P.J. Forrester, *J. Stat. Phys.* **51** (1988) 457.

5. M.L. Rosinberg, private communication.

6. P.J. Forrester, *J. Stat. Phys.* **59** (1989) 57.

International Journal of Modern Physics B, Vol. 4, No. 5 (1990) 953–968
© World Scientific Publishing Company

# QUANTUM SPIN CHAINS: CONFORMAL INVARIANCE AND INTEGRABILITY

Michael N. BARBER
*Department of Mathematics & Centre for Mathematical Analysis*
*School of Mathematical Sciences, Australian National University*
*Canberra, ACT, Australia.*

and

Murray T. BATCHELOR
*Department of Applied Mathematics*
*Research School of Physical Sciences, Australian National University*
*Canberra, ACT, Australia.*

Recent exact results concerning the behaviour of antiferromagnetic spin chains are reviewed with emphasis on conformal invariance and integrability.

## 1. Introduction

Interesting and challenging problems have been posed by the quantum mechanics of interacting spins since the early days of the quantum theory of the solid state. A key idea – the Bethe Ansatz – that surfaces in a number of papers in these Proceedings, dates, of course, from the 1930's and Bethe's work [1] on the spin-$\frac{1}{2}$ Heisenberg antiferromagnetic chain:

$$H = J \sum \mathbf{s}_i \cdot \mathbf{s}_{i+1}, \tag{1}$$

where $J > 0$. Interest in the behaviour of low dimensional quantum antiferromagnets has been renewed recently by two developments – conformal invariance in statistical mechanics and field theory and high-$T_c$ superconductivity. While it appears that the behaviour of *two*-dimensional quantum antiferromagnets is likely to be of most relevance to understanding high-$T_c$ superconductivity, one-dimensional spin chains are important test beds for many ideas and computational methods. On the other hand, one-dimensional quantum spin chains are of direct relevance to our understanding of conformal invariance and its mathematical and physical implications. In particular, one-dimensional quantum chains yield specific physical realizations of conformally invariant field theories that allow both tests of the basic ideas of conformal invariance and the practical exploitation of these ideas.

In this article we review some of these developments and discuss some of the key ideas involved. Since much of the material is also covered in other papers in these Proceedings, we will be brief in places and try to paint a general overview of a pattern of interrelating ideas that is beginning to emerge.[1]

## 2. Spin-$\frac{1}{2}$ XXZ Model

We begin with the spectrum of the spin-$\frac{1}{2}$ XXZ Hamiltonian:

$$H_{XXZ}(M) = -\frac{1}{2}\sum_{m=1}^{M}\left(\sigma_m^x\sigma_{m+1}^x + \sigma_m^y\sigma_{m+1}^y + \Delta\sigma_m^z\sigma_{m+1}^z\right), \tag{2}$$

where the $\sigma$'s are the Pauli matrices. For the moment, we will assume periodic boundary conditions so that $\sigma_{M+1}^\alpha = \sigma_1^\alpha$, $\alpha = x, y, z$. By now the spectrum of this Hamiltonian is particularly well understood.[2] However, since this Hamiltonian is central to much of what follows, a reiteration of the main features of its spectrum is useful.

The nature of the ground state in the thermodynamic limit $M \to \infty$ is determined by $\Delta$. For $|\Delta| > 1$, the ground state is two-fold degenerate; ferromagnetically ordered for $\Delta > 1$ and antiferromagnetically ordered for $\Delta < 1$.[3] Our interest here will focus on the antiferromagnetic regime and the intermediate regime $|\Delta| \leq 1$, where the model is conformally invariant.

As is well known, the eigenstates of (2) can be found by the Bethe Ansatz [6]. Since the total spin operator $S^z = \sum \sigma_m^z$ commutes with $H_{XXZ}(M)$, we can block diagonalize (2) into sectors labelled by the eigenvalues of $S^z$. Since the sectors with $S^z$ and $-S^z$ are automatically degenerate it is sufficient to consider only sectors with $S^z \geq 0$ and label these sectors by integers $r = \frac{1}{2}(M - S_z) = 0, 1, \ldots, \frac{1}{2}M$, where we shall assume that $M$ is even. States in sector $r$ are specified by $r$ (complex) numbers $\lambda_1, \lambda_2, \ldots \lambda_r$ that satisfy a system of $r$ non-linear equations – the Bethe Ansatz equations. From these equations follow not only bulk properties such as the ground state energy [6] or the gap [7] but also finite lattice corrections of relevance to conformal invariance [8,9,10].

As an example of the typical form of the Bethe Ansatz equations, we quote the case of $\Delta = -1$, for which the Bethe Ansatz equations read

$$\left(\frac{\lambda_j - i/2}{\lambda_j + i/2}\right)^M = -\prod_{k=1}^{r}\left(\frac{\lambda_j - \lambda_k - i}{\lambda_j - \lambda_k + i}\right), \qquad j = 1, 2 \ldots, r. \tag{3}$$

In terms of the $\lambda_j$'s, the associated eigenenergy is then

$$E = \frac{1}{2}M - \sum_{j=1}^{r}\frac{1}{\lambda_j^2 + \frac{1}{4}}. \tag{4}$$

[1] For some other recent reviews, see: Refs [2,3,4].

[2] While this is true, the XXZ Hamiltonian, like the two-dimensional Ising model, continues to surprise and reveal novel features that extend our physical understanding; see [5] for a recent example.

[3] The Heisenberg antiferromagnet (1) corresponds to $\Delta = -1$ as can be seen by making the transformation: $\sigma_m^x \to (-)^m\sigma_m^x$, $\sigma_m^y \to (-)^m\sigma_m^y$, $\sigma_m^z \to \sigma_m^z$.

The spectrum and its finite-size corrections are different depending on whether $-1 \le \Delta \le 1$ or $\Delta < -1$. In both regimes the ground state is determined by the lowest energy solution of the Bethe Ansatz equations with $r = \frac{1}{2}M$, *i.e.*, $S^z = 0$. As mentioned above, the ground state for $\Delta < -1$ is two-fold degenerate in the limit $M \to \infty$, with [6]

$$e_0 = \lim_{M\to\infty} \frac{E}{M} = \frac{1}{2}\cosh\theta - \sinh\theta \left[1 + 4\sum_{n=1}^{\infty} \frac{1}{1+e^{2n\theta}}\right], \tag{5}$$

where $\Delta = -\cosh\theta$. Above this state there is a finite energy gap [7],

$$\Lambda = 2\sinh\theta \prod_{n=1}^{\infty} \left(\frac{1-e^{n\theta}}{1+e^{n\theta}}\right)^2, \tag{6}$$

to the next level, which is the lowest energy level in the sector $r = \frac{1}{2}M - 1$ and is the beginning of a continuous band of $S^z = 2$ spin-wave states. Similar bands of higher $S^z$ states commence at energies $n\Lambda$, $n = 2, 3 \ldots$, above the ground state [11]. On a large but finite lattice, two relatively minor effects occur: the ground state degeneracy is broken to give a gap that varies[4] as $O(e^{-aM})$ as $M \to \infty$ and the continuous bands are discretized by the quantization of the momentum (translation) operator.

The behaviour of the spectrum in the regime $-1 \le \Delta \le 1$ is more subtle and more interesting. In this regime it is convenient to write

$$\Delta = -\cos\gamma. \tag{7}$$

In the thermodynamic limit (2) is massless with a singlet $S^z = 0$ ground state of energy [6]

$$e_0(\gamma) = \frac{1}{2}\cos\gamma - \int_{-\infty}^{\infty} \frac{\sin^2\gamma \; dx}{\cosh(\pi x)[\cosh(2\gamma x) - \cos\gamma]}. \tag{8}$$

The Hamiltonian is also conformally invariant, provided we scale (2) to ensure that the energy-momentum relation is linear with a coefficient (speed of sound) equal to unity [15]. The required scaling parameter is easily found [8,9] from known results [16] to be

$$\zeta = \frac{\pi \sin\gamma}{\gamma}. \tag{9}$$

The asymptotic behaviour of the ground state energy for large $M$ can be determined from the Bethe Ansatz equations [8]. One finds:

$$E_0(M) = M e_0 - \frac{\pi\zeta c}{6M} + o(M^{-1}), \tag{10}$$

where

$$c = 1. \tag{11}$$

---

[4]This exponentially small splitting of asymptotically degenerate levels is to be expected [12] on rather general grounds. However, more complex boundary conditions than periodic can lead to finite lattice gaps that close algebraically; see *e.g.*, Refs [13] and [14].

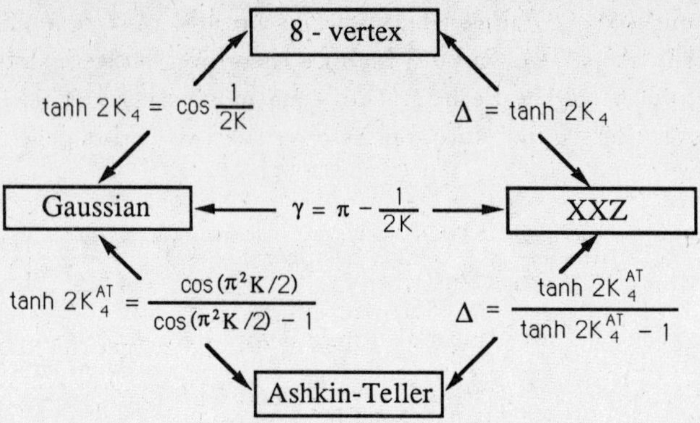


Figure 1. Mappings between the critical lines of the XXZ chain and other models with $c$=1 (after Ref. [9]).

The expansion (10) is precisely that predicted [17,18] by conformal invariance and identifies the line $-1 \leq \Delta \leq 1$ of the XXZ Hamiltonian as a critical line described by a conformal field theory with central charge, $c = 1$.

There are a number of other well-known models, whose critical behaviour is described by a $c = 1$ conformal field theory. Figure 1 indicates the relations between the XXZ model and these other $c = 1$ models as elaborated by Alcaraz *et al.* [9] building on earlier work of Kadanoff and Brown [19], Haldane [20] and den Nijs [21].

Conformal invariance also predicts the structure of the higher energy states, which is determined by the critical operators of the theory [22,23]. For each operator $\mathcal{O}_\alpha$ with anomalous dimension $x_\alpha$ and spin $s_\alpha$, there exists a tower of states in the spectrum with energies

$$E^\alpha_{j,j'}(M) = E_0(M) + \frac{2\pi\zeta(x_\alpha + j + j')}{M} + o(M^{-1}), \tag{12}$$

and momentum

$$P^\alpha_{j,j'}(M) = \frac{2\pi(s_\alpha + j - j')}{M}, \tag{13}$$

where $j, j' = 0, 1, \ldots$.

For the XXZ chain, these predictions have been confirmed by numerical [9] and analytical [10] calculations[5] based on the Bethe Ansatz equations. Explicitly, if we write $n = r - \frac{1}{2}M$ so that the ground state sector corresponds to $n = 0$ then there exist operators $\mathcal{O}_{n,m}$ with dimensions

$$x_{n,m} = n^2 x_p + \frac{m^2}{4x_p}, \tag{14}$$

where $x_p = (\pi - \gamma)/2\pi$. In accord with the mappings depicted in Figure 1, these operators are direct analogs of operators previously identified [19] in the Gaussian

[5]See, also *e.g.*, Refs [24] and [25] and references therein.

model and include the polarization ($\mathcal{O}_{1,0}$) and energy ($\mathcal{O}_{2,0}$) operators of the eight-vertex model and the energy operator ($\mathcal{O}_{0,1}$) of the Ashkin-Teller model. In addition, there exists a marginal operator which corresponds to motion along the critical line and a number of redundant operators [9,21].

However, not all the critical operators of these other $c = 1$ models are represented in the spectrum of the periodic XXZ chain. For example, the magnetic operator of the 8-vertex model and both the polarization and magnetic operators of the Ashkin-Teller model are missing. To recover these operators it is necessary to consider (2) with boundary conditions other than periodic. One then discovers [9] that an *exact* correspondence can be established between levels in the eigenspectrum of the quantum Hamiltonian of the Ashkin-Teller model with periodic boundary conditions and those of the XXZ chain with modified boundary conditions. The most important of these new boundary conditions are *twisted* boundary conditions defined by

$$\sigma^x_{M+1} \pm i\sigma^y_{M+1} = e^{\pm i\phi}(\sigma^x_1 \pm i\sigma^y_1), \quad \sigma^z_{M+1} = \sigma^z_1, \tag{15}$$

where $\phi$ is an arbitrary angle. The Bethe Ansatz is still applicable and the spectrum can be elucidated either numerically [9] or analytically [26]; thereby allowing identification of the "missing" operators. In particular, the conformal tower associated with the Ashkin-Teller polarization operator occurs in the spectrum of a twisted XXZ chain with $\phi = \pi$.

The boundary conditions (15) are important for a second reason: they correspond to the introduction of a "defect" seam. This seam is the precise XXZ equivalence of the seam introduced by Baxter *et al.* [27] in the 6-vertex model in order to establish an exact equivalence between the partition function of the 6-vertex model and those of the $q$-state Potts models. This equivalence can be reformulated in the language of quantum Hamiltonians [28,9]. For $q \leq 4$, for which the Potts models exhibit conformally invariant critical points, this then implies that the ground state energy of the quantum Hamiltonian of the $q$-state Potts model on a chain of $L$ sites with *periodic* boundary conditions is given exactly by the ground state energy of a twisted XXZ Hamiltonian on $2L$ sites with coupling $\Delta = -\frac{1}{2}\sqrt{q} \equiv -\cos\gamma$ and boundary conditions (15) with $\phi = 2\gamma$. This latter quantity can be calculated from the Bethe Ansatz equations and analysed by the now standard methods to yield the asymptotic expansion for the ground state energy of the Potts quantum Hamiltonian:[6]

$$E_0(q; L) = Le^P_0(q) - \frac{\pi\zeta_P c(q)}{6L} + o(L^{-1}). \tag{16}$$

Here

$$\zeta_P = \frac{\pi\sqrt{4-q}}{4\cos^{-1}(\sqrt{q}/2)} \tag{17}$$

is the conformal scale factor for the Potts Hamiltonian and

$$e^P_0(q) = 2e_0(\cos^{-1}\sqrt{q}/2) - \frac{1}{2}\sqrt{q}, \tag{18}$$

[6]Specifically, (16) to (18) refer to the Hamiltonian defined below in (20) but with periodic rather than free end boundary conditions applied.

with $e_0(\gamma)$ given by (8) is the limiting value of the ground state energy per site. Finally, $c(q)$ is the conformal anomaly of the $q$-state Potts model given by

$$c(q) = 1 - \frac{3y^2}{2-y}, \tag{19}$$

where, in accord with common practice, we have set $y = 2\gamma/\pi$. This result is precisely of the form predicted by other approaches [29,17].

Finite lattice calculations of the spectrum can be extended to reveal and/or confirm the complete operator content of the XXZ chain as a $c = 1$ conformal theory [30]. Again the twisted boundary conditions defined in (15) are crucial to the complete resolution of the content. In related work, Alcaraz, Grimm and Rittenberg [31] have been able to project out the spectra of various systems with $c < 1$ from the finite-size scaling spectra of the $XXZ$ model.[7]

## 3. Potts models and Temperley-Lieb Hamiltonians

If we switch from periodic to free boundary conditions, we can make the connection between the XXZ Hamiltonian and Potts models more directly. We begin with the quantum Hamiltonian limit of the $q$-state Potts model [32,28]

$$H_{\rm Potts}(L,\lambda) = -\frac{1}{\sqrt{q}}\left(\sum_{l=1}^{L}\sum_{k=0}^{q-1}\Omega_l^k + \lambda\sum_{l=1}^{L-1}\sum_{k=0}^{q-1} R_l^k R_{l+1}^{q-k}\right) \tag{20}$$

on a chain of $L$ sites with free ends. The operators $\Omega_l$ and $R_l$ at site $l$ obey a $Z(q)$-algebra:

$$\Omega_l R_l = \omega^{-1} R_l \Omega_l, \tag{21}$$
$$\Omega_l R_l^\dagger = \omega R_l^\dagger \Omega_l, \tag{22}$$
$$\Omega_l^q = R_l^q = 1 \tag{23}$$

with $\omega = e^{2\pi i/q}$. The coupling parameter $\lambda$ is the analogue of temperature in the conventional statistical mechanical formulation of the Potts model. The self-dual transition point of the Potts model corresponds to $\lambda = \lambda_c = 1$ with $\lambda > 1$ corresponding to the ordered phase ($T < T_c$).

Now define [33,9] a set of operators $\{U_l\}$ by

$$U_{2l-1} = \frac{1}{\sqrt{q}}\sum_{k=0}^{q-1}\Omega_l^k, \qquad l = 1,2,\ldots,L, \tag{24}$$

$$U_{2l} = \frac{1}{\sqrt{q}}\sum_{k=0}^{q-1} R_l^k R_{l+1}^{q-k}, \qquad l = 1,2,\ldots,L-1, \tag{25}$$

[7] See also the paper by Grimm and Rittenberg in this volume.

in terms of which

$$H_{\text{Potts}}(L,\lambda) = -\sum_{l=1}^{L} U_{2l-1} - \lambda \sum_{l=1}^{L-1} U_{2l}. \tag{26}$$

A straightforward calculation establishes [33,9] that the $U_l$'s satisfy a Temperley-Lieb algebra [34,35]:

$$U_l^2 = \sqrt{q}U_l, \tag{27}$$

$$U_l U_{l\pm1} U_l = U_l, \tag{28}$$

$$[U_l, U_{l'}] = 0, \quad |l-l'|>1. \tag{29}$$

The equivalence of the Potts Hamiltonian to the XXZ chain follows by introducing an alternative representation [34,35] of the Temperley-Lieb operators in terms of Pauli spin operators:

$$U_l = \frac{1}{2}\left(\sigma_l^x\sigma_{l+1}^x + \sigma_l^y\sigma_{l+1}^y\right) + \frac{1}{2}\cosh\theta\left(1-\sigma_l^z\sigma_{l+1}^z\right) + \frac{1}{2}\sinh\theta\left(\sigma_{l+1}^z - \sigma_l^z\right), \tag{30}$$

where $\cosh\theta = \frac{1}{2}\sqrt{q}$. For general $\lambda$, this definition of the $U$'s results in an equivalence between the Potts Hamiltonian and a staggered $XXZ$ chain. For $\lambda = 1$, the staggering vanishes and we obtain the equivalence:

$$H_{\text{Potts}}(L,\lambda=1) \Leftrightarrow H^f_{XXZ}(2L) - \frac{1}{2}(2L-1)\cosh\theta, \tag{31}$$

where

$$H^f_{XXZ}(M) = -\frac{1}{2}\sum_{l=1}^{M-1}\left(\sigma_l^x\sigma_{l+1}^x + \sigma_l^y\sigma_{l+1}^y + \Delta\sigma_l^z\sigma_{l+1}^z\right) + \frac{1}{2}p\left(\sigma_1^z - \sigma_M^z\right) \tag{32}$$

is the Hamiltonian of an $XXZ$ model on a chain of $M$ sites with coupling $\Delta = -\cosh\theta = -\frac{1}{2}\sqrt{q}$ and fields $\frac{1}{2}p$ and $-\frac{1}{2}p$, $p = \sinh\theta = \frac{1}{2}\sqrt{q-4}$, applied to the two free ends. Note that $p$ is real for $q \geq 4$ but is pure imaginary for $q < 4$.

The $XXZ$ Hamiltonian (32) is still solvable by the Bethe Ansatz [36,37]. In particular, the ground state energy can be calculated and, as expected, is the same in the limit $M \to \infty$ as that obtained for the usual periodic boundary conditions [6]. For $q \leq 4$, the spectrum of (32) retains the characteristic structure – gaps that close as $1/M$ – of a conformally invariant theory but now involving various surface exponents [38,36].

An obvious question is now whether there exist other representations of the Temperley-Lieb algebra that via (26) or even more simply $H = -\sum U$ generate Hamiltonians of physical interest. Recently, we have shown [39,40] that there exist isotropic spin-$s$ representations.[8] Explicitly, define the polynomial

$$U(X) = (-1)^{2s}\left[\frac{2^s}{(2s)!}\right]^2 \prod_{S=1}^{2s}\left(X - \frac{1}{2}S(S+1) + s(s+1)\right). \tag{33}$$

[8]These representations are special cases of representations discussed by Owczarek and Baxter [41] in a different context; see also the paper by Pearce in this volume.

Then the set of operators $\{U_m \equiv U(\mathbf{s}_m \cdot \mathbf{s}_{m+1})\}$ satisfy the Temperley-Lieb algebra defined by (27) to (29) with $q = (2s+1)^2$.

Hence, as was done for the Potts Hamiltonian, we can infer a correspondence between the spin-$s$ Hamiltonian

$$H_s = -\sum_{m=1}^{M-1} U(\mathbf{s}_m \cdot \mathbf{s}_{m+1}) \tag{34}$$

on a chain of $M$ sites with free ends and $H^f_{XXZ}$. Explicitly

$$H_s(M) \Leftrightarrow H^f_{XXZ}(M) - \frac{1}{2}(M-1)\cosh\theta, \tag{35}$$

where, in (32), $\Delta$ is now given by $\Delta = -s - \frac{1}{2}$ and $p$ by $\frac{1}{2}\sqrt{(2s+3)(2s-1)}$. As immediate consequences of this correspondence, we can conclude [40] that the thermodynamic limit of the ground state energy per site of (34) can be deduced from (5) and that for all $s > \frac{1}{2}$, $H_s$ is massive. A particular example of the Hamiltonians defined by (34) is the biquadratic spin-1 Hamiltonian

$$H_{\mathrm{bQ}} = -\sum_{m=1}^{M-1} (\mathbf{s}_m \cdot \mathbf{s}_{m+1})^2, \tag{36}$$

which is a special case of a more general spin-1 Hamiltonian that we discuss further in Section 5.

## 4. Higher-spin chains

We turn now to other families of higher-spin chains. The simplest generalization of (1) is to simply replace the $\mathbf{s}_i$'s by spin-$s$ operators. Such a generalization might be expected to have little qualitative effect on the spectrum. However, as first argued by Haldane [42] (see also, Refs [43,4]), this is not the case: the Hamiltonian remains massless only for half-integral spin and develops a gap for integer $s$. Haldane's argument was based on a field-theoretic mapping (valid for large $s$) to the non-linear sigma model. Consequently, his conclusion that the spin-1 antiferromagnetic Heisenberg model was massive was received originally with some skepticism. However, despite this early uncertainty, more direct numerical methods have confirmed a non-zero gap, $\Lambda/J \approx 0.41$ [44]. Neutron scattering experiments on $CsNiCl_3$ (a spin-1 Heisenberg antiferromagnet in which the Ni atoms order in linear chains) have also found evidence of a gap [45].

However not all higher-spin antiferromagnetic Hamiltonians are massive.[9] In particular, *integrable* Hamiltonians arise from the powerful framework of the quantum inverse scattering method developed by Faddeev and coworkers (see *e.g.*, the reviews

[9]For an interesting discussion, see [46], where connections are made between phases in quantum spin chains and recent work on roughening transitions.

[47] and [48]). Within this context, a particular integrable anisotropic spin-1 Hamiltonian was written down by Zamolodchikov and Fateev [49].

Here we confine attention to the isotropic spin-$s$ Takhtajan-Babujian models [50,51] which are solvable by the Bethe Ansatz. These models are defined by the Hamiltonian:

$$H_s^{TB} = -\sum_{n=1}^{N} Q_{2s}(\mathbf{s}_n \cdot \mathbf{s}_{n+1}), \tag{37}$$

where

$$Q_{2s}(x) = \sum_{j=1}^{2s} \left( \sum_{k=1}^{j} \frac{1}{k} \right) \prod_{\substack{l=0 \\ l \neq j}}^{2s} \frac{x - x_l}{x_j - x_l} \tag{38}$$

is a polynomial of degree $2s$ in $\mathbf{s}_n \cdot \mathbf{s}_{n+1}$ with

$$x_l = \frac{1}{2} l(l+1) - s(s+1). \tag{39}$$

For example, for spin-1 we have

$$H_1^{TB} = \frac{1}{4} \sum_{n=1}^{N} \left[ (\mathbf{s}_n \cdot \mathbf{s}_{n+1}) - (\mathbf{s}_n \cdot \mathbf{s}_{n+1})^2 \right]. \tag{40}$$

In general, the Bethe Ansatz equations can be written [50,51]

$$\left( \frac{\lambda_j - is}{\lambda_j + is} \right)^N = -\prod_{k=1}^{r} \frac{\lambda_j - \lambda_k - i}{\lambda_j - \lambda_k + i}, \quad j = 1, 2, \ldots, r \tag{41}$$

where $r = 1, 2, \ldots, Ns$ labels the sectors. The associated eigenenergies are given by

$$E_s^{TB} = -\sum_{j=1}^{r} \frac{s}{\lambda_j^2 + s^2}, \tag{42}$$

where the ground state energy corresponds to the lowest energy solution in the sector $r = Ns$.

From the low-temperature behaviour of the specific heat calculated by Babujian [51], Affleck [18] argued that the Takhtajan-Babujian models are massless with central charge

$$c = \frac{3s}{s+1} \tag{43}$$

and are described at low energies by the SU(2) $k = 2s$ Wess-Zumino-Witten nonlinear sigma models. Equations (41) and (42) look fairly harmless generalizations of the corresponding spin-$\frac{1}{2}$ equations (3) and (4). However, the result (43) has not yet been derived via the methods used to derive the central charge via eqn (10) as in the spin-$\frac{1}{2}$ case. This is because the Bethe Ansatz roots for the ground state are no longer real once $s \geq 1$ but occur as $2s$-strings in the complex $\lambda$-plane. In contrast with the spin-$\frac{1}{2}$ case, where the roots lie along the real axis for finite systems, the string solutions for $s \geq 1$ lie along straight lines only in the infinite-size limit; see *e.g.*, [52]. Recently de Vega and Woynarovich [53] have calculated the finite-size deviation of these roots.[10] It is known that neglecting these deviations gives a value of unity for the central charge *for all* $s$ [54]; effectively corresponding to the finite-size correction associated with the *centres* of the strings. Despite these difficulties with the behaviour of the roots, the result (43) has nevertheless been confirmed for small values of $s$ by solving the Bethe Ansatz equations numerically for relatively large system sizes [55,54,56,57].[11]

Precise numerical estimates, from the finite-size variation of mass gaps, of the anomalous dimensions of associated critical operators is complicated by the presence of logarithmic factors which can cause misconvergence [59]. However, careful analyses involving data from large lattices and specially tuned extrapolation procedures have yielded reasonable agreement with the field theoretic predictions [59,54,56].[12]

Other known families of exactly solvable spin-$s$ chains can also be written in the form[13]

$$H_s^K = \sum_{n=1}^{N} P_{2s}^K(\mathbf{s}_n \cdot \mathbf{s}_{n+1}), \tag{44}$$

where the polynomial

$$P_{2s}^K(x) = \prod_{\substack{l=0 \\ l\neq K}}^{2s} \frac{x - x_l}{x_K - x_l} \tag{45}$$

is the projector onto the eigenspace of $S_T$ with eigenvalue $S = K$.

For $K = 1$, (44) generates the Hamiltonians associated with the permutation models discussed by Lai [61] and Sutherland [62]. For $s = 1$ this family includes the model solved by Uimin [63],

$$H_1^1 = \sum_{n=1}^{N} \left[(\mathbf{s}_n \cdot \mathbf{s}_{n+1}) + (\mathbf{s}_n \cdot \mathbf{s}_{n+1})^2\right], \tag{46}$$

which is known [25] to be critical with central charge $c = 2$. In general these models are solvable via a nested Bethe Ansatz.

In contrast, the $K = 2$ family contains the spin-1 model of Affleck, Kennedy, Lieb and Tasaki [64]

$$H_1^2 = \sum_{n=1}^{N} \left[(\mathbf{s}_n \cdot \mathbf{s}_{n+1}) + \frac{1}{3}(\mathbf{s}_n \cdot \mathbf{s}_{n+1})^2\right]. \tag{47}$$

which has a ground state explicitly constructed from valence bond states and is known to be massive (see, also [65]).

---

[10] See also the paper by de Vega in this volume.

[11] An alternative procedure for calculating finite-size corrections in integrable models, based on setting up integral equations for the corrections rather than the standard procedure of manipulating root densities, has more recently [58] been applied to the spin-1 case and shown to yield the value $c = 1.5$ in agreement with (43).

[12] See, also: Ref. [60], where the similar problem that arises in the 4-state Potts model is discussed in detail.

[13] Apart from some trivial additive and multiplicative factors in the Hamiltonians.

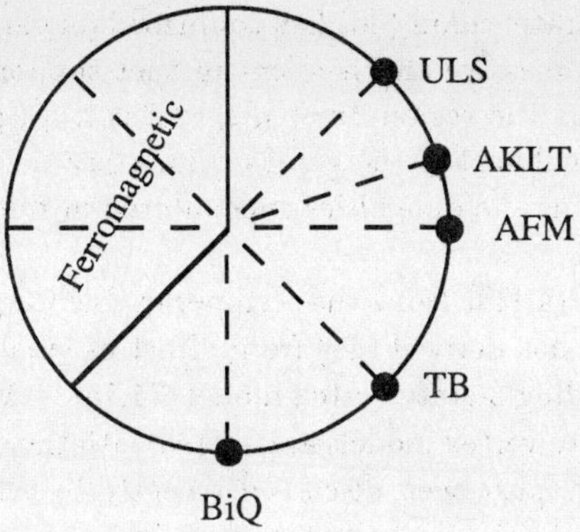


Figure 2. Phase digram of the spin-1 bilinear-biquadratic Hamiltonian (48). The labelling of the points indicated in the antiferromagnetic region correspond to the Uimin-Lai-Sutherland (ULS); Affleck, Kennedy, Lieb and Tasaki (AKLT); Heisenberg antiferromagnet (AFM); Takhtajan-Babujian (TB) and biquadratic (BiQ) models.

## 5. The spin-1 bilinear-biquadratic Hamiltonian

We conclude by focussing on a particular spin-1 Hamiltonian:

$$H = \sum_{n=1}^{N} \left[\cos\omega\ (\mathbf{s}_n \cdot \mathbf{s}_{n+1}) + \sin\omega\ (\mathbf{s}_n \cdot \mathbf{s}_{n+1})^2\right], \tag{48}$$

where $\omega \in [0, 2\pi]$. This model has received considerable attention recently since it includes as special cases all of the known spin-1 results. The nature of the phase diagram in $\omega$, as depicted in Figure 2, has recently been discussed by Papanicolou [66][14]. For $\pi/2 < \omega < 5\pi/4$ the ground state is trivial and ferromagnetic. In the antiferromagnetic regime, $\omega = 7\pi/4$ represents the Takhtajan-Babujian model (40) while $\omega = \pi/4$ corresponds to (46). The Heisenberg chain is represented by $\omega = 0$ with the valence bond model (47) located at $\tan\omega_0 = \frac{1}{3}$. The biquadratic model (36) is located at $\omega = 3\pi/2$.

The existence of a massive phase at $\omega_0$ in the vicinity of the Heisenberg chain is in general agreement with the Haldane conjecture. Whether or not a gap opens up in the spectrum on either side of the critical Takhtajan-Babujian point has been the subject of some recent controversy. In particular, in the region $3\pi/2 < \omega < 7\pi/4$, series [67] and (less confidently) finite-lattice studies [68,69] suggest the presence of a massless phase. However, more extensive finite-lattice calculations tend to suggest a gap [70], in agreement with the prediction of a massive phase by Affleck *et al.* [71,43,64].[15]

More recently, the exact result [39,74,75]

$$\Lambda_{bQ} = 0.173178\ldots \tag{49}$$

[14]See also the paper by Itoyama, McCoy and Perk in this volume.

[15]This controversy has been the subject of further recent discussion [72]. For other studies, see *e.g.*, [73]

for the gap of the biquadratic chain (36) has confirmed an earlier conjecture [71] that the biquadratic model is massive and has lent further support for the existence of a massive phase throughout the region $3\pi/2 < \omega < 7\pi/4$. In particular, if the region is indeed massive, the smallness of the gap (49), or equivalently, the large but finite correlation length, explains the difficulties encountered in the series and finite-lattice studies.

The result (49) follows [39] from the Temperley-Lieb equivalence of Section 4. Independently, Klümper has derived (49) from direct calculations based on inversion relations for a corresponding 3-state vertex model [74,75].[16] More generally, Klümper identified the $(2s+1)$-state vertex models associated with the spin-$s$ Hamiltonian (34). These Hamiltonians have also been discussed recently by Affleck [76] in connection with the dimerization transition in SU($n$) spin chains with $n = 2s+1$.

The general Temperley-Lieb equivalence between the Potts Hamiltonian (26) on the one hand, and the spin-$s$ Hamiltonian (34) on the other, allows exact results to be obtained for *staggered* spin chains [9,39,77]. For the biquadratic model, (36) generalizes to

$$H_{\rm bQ}(M,\lambda) = -\sum_{m=1}^{M-1} \epsilon_m (\mathbf{s}_m \cdot \mathbf{s}_{m+1})^2, \tag{50}$$

where

$$\epsilon_m = \begin{cases} 1, & \text{for } m \text{ odd} \\ \lambda, & \text{for } m \text{ even} \end{cases} \tag{51}$$

controls bond alternation and hence the tendency to dimerization via a possible spin-Peierls transition [2]. In particular for the biquadratic model, two spontaneously dimerized states exchange stability at $\lambda = 1$.

From (50), we observe that the infinite lattice limit of the ground state energy per site, $e_g(\lambda)$, obeys

$$e_g(\lambda) = \lambda e_g(1/\lambda) \tag{52}$$

which corresponds to the duality relation in the Potts model [32]. The equivalence between the non-critical Potts Hamiltonian (26) and (50) follows via the spin-1 representation

$$U_m = (\mathbf{s}_m \cdot \mathbf{s}_{m+1})^2 - 1 \tag{53}$$

which satisfies the Temperley-Lieb algebra, (27) to (29), with $q = 9$.[17] In this way we obtain the equivalence

$$H_{\rm bQ}(2M,\lambda) \Leftrightarrow H_{\rm Potts}(M,\lambda) - \lambda(M-1) - M \tag{54}$$

from which we infer that

$$e_g(\lambda) = \frac{1}{2} e_{Potts}(\lambda) - \frac{1}{2}(1+\lambda), \tag{55}$$

[16] See also the paper by Klümper in this volume.

[17] This representation has recently been generalized to all $q$ [78].

where $e_{Potts}(\lambda)$ is the ground state energy per site of the nine-state Potts model in the thermodynamic limit. This energy is known to have a discontinuous first derivative at $\lambda = 1$ corresponding to the latent heat [79,28]. Hence we find that [39]

$$e_g(\lambda) = e_g(1) + A_\pm|1-\lambda| + O[(1-\lambda)^2], \tag{56}$$

where $e_g(1) = -2.796863\ldots$, $A_+ = -1.9606\ldots$ and $A_- = 0.8362\ldots$. The value for $e_g(1)$ had earlier been conjectured by Parkinson [80]. Higher spin values can be discussed in a similar way; the jump discontinuity in the derivative of the ground state energy increasing with increasing $s$.

One of us (MTB) acknowledges support from the Australian Research Council.

## REFERENCES


1. Bethe H *Z. Phys.* **71** (1931), 205
2. Bonner J C *J. Appl. Phys.* **61** (1987), 3941
3. Affleck I in *Fields, Strings and Critical Phenomena*, Les Houches Lectures 1988 (to be published)
4. Affleck I *J. Phys.: Condens. Matter* **1** (1989), 3047
5. Singh R R P, Fisher M E and Shankar R *Phys. Rev. B* **39** (1989), 2562
6. Yang C N and Yang C P *Phys. Rev.* **150** (1966), 327
7. des Cloizeaux J and Gaudin M *J. Math. Phys.* **7** (1966), 1384
8. Hamer C J *J. Phys. A* **19** (1986), 3335
9. Alcaraz F C, Barber M N and Batchelor M T *Phys. Rev. Lett.* **58** (1987), 771 ; *Ann. Phys. (N.Y.)* **182** (1988), 280
10. Woynarovich F *Phys. Rev. Lett.* **59** (1987), 259
11. Batchelor M T and Hamer C J *J. Phys. A* **23** (1990) (in press)
12. Fisher M E 1969 *J. Phys. Soc. Japan Suppl.* **26** 87
13. Barber M N and Cates M E *Phys. Rev. B* **36** (1987), 2024
14. Abraham D B, Ko L F and Švrakić N M *Phys. Rev. Lett.* **21** (1988), 2393
15. von Gehlen G, Rittenberg V and Ruegg H *J. Phys. A* **19** (1986), 107
16. Johnson J D, Krinsky S and McCoy B M *Phys. Rev. A* **8** (1973), 2526
17. Blöte H W J, Cardy J L and Nightingale M P *Phys. Rev. Lett.* **56** (1986), 742
18. Affleck I *Phys. Rev. Lett.* **56** (1986), 746

19. Kadanoff L and Brown A C *Ann. Phys. (N.Y.)* **121** (1979), 318

20. Haldane F D M *Phys. Rev. Lett.* **45** (1980), 1358

21. den Nijs M P M *Phys. Rev. B* **23** (1981), 6111

22. Cardy J L *Nucl. Phys. B* **270** (1986), 186

23. For a recent review see, Cardy J L in *Phase Transitions and Critical Phenomena* Vol 11, eds. C. Domb and J.L. Lebowitz, (Academic Press, New York, 1987)

24. Karowski M *Nucl. Phys. B* **300** (1988), 473

25. de Vega H J *Int. J. Mod. Phys.* **4** (1989), 2371

26. Hamer C J, Quispel G R W and Batchelor M T *J. Phys. A* **20** (1987), 5677

27. Baxter R J, Kelland S B and Wu F Y *J. Phys. A* **9** (1976), 397

28. Hamer C J *J. Phys. A* **14** (1981), 2981

29. Dotsenko Vl S *Nucl. Phys. B* **235** (1984), 54

30. Alcaraz F C, Baake M, Grimm U and Rittenberg V *J. Phys. A* **21** (1988), L117

31. Alcaraz F C, Grimm U and Rittenberg V *Nucl. Phys. B* **316** (1989), 735

32. Sólyom J and Pfeuty P *Phys. Rev. B* **24** (1981), 218

33. Berkcan E *Nucl. Phys. B* **215** (1983), 68

34. Temperley H N V and Lieb E H *Proc. Roy. Soc. (London) A* **322** (1971), 251

35. Baxter R J 1982 *Exactly Solved Models in Statistical Mechanics* (London: Academic)

36. Alcaraz F C, Barber M N, Batchelor M T, Baxter R J and Quispel G R W *J. Phys. A* **20** (1987), 6397

37. Sklyanin E K *J. Phys. A* **21** (1988), 2375

38. Cardy J L *Nucl. Phys. B* **240** (1984), 514

39. Barber M N and Batchelor M T *Phys. Rev. B* **40** (1989), 4621

40. Batchelor M T and Barber M N *J. Phys. A* **23** (1990), L15

41. Owczarek A L and Baxter R J *J. Stat. Phys.* **49** (1987), 1093

42. Haldane F D M *Phys. Lett. A* **93** (1983), 464 ; *Phys. Rev. Lett.* **50** (1983), 1153

43. Affleck I and Haldane F D M *Phys. Rev. B* **36** (1987), 5291

44. Nightingale M P and Blöte H W J *Phys. Rev. B* **33** (1986), 659

45. Morra R M, Buyers W J L, Armstrong R L and Hirakawa K *Phys. Rev. B* **38** (1988), 543

46. den Nijs M and Rommelse K *Phys. Rev. B* **40** (1989), 4709

47. Faddeev L D *Sov. Sci. Rev. C* **1** (1980), 107

48. Kulish P P and Sklyanin E K *Lecture Notes in Physics* **151** (1982)

49. Zamolodchikov A B and Fateev V A *Sov. J. Nucl. Phys.* **32** (1980), 298 [*Yad. Phys.* **32** (1980), 581]

50. Takhtajan L A *Phys. Lett. A* **87** (1982), 479

51. Babujian H M *Nucl. Phys. B* **215** (1983), 317

52. Dörfel B-D *Fortschr. Phys.* **36** (1988), 281

53. de Vega H J and Woynarovich F, preprint PAR-LPTHE-89-32

54. Alcaraz F C and Martins M J *J. Phys. A* **21** (1988), 4397

55. Avdeev L V and Dörfel B-D *Theor. Math. Phys.* **71** (1987), 528

56. Affleck I, Gepner D, Schulz H J and Ziman T *J. Phys. A* **22** (1989), 511

57. Dörfel B-D *J. Phys. A* **22** (1989), L657

58. Klümper A and Batchelor M T *J. Phys. A* **23** (1990) (in press)

59. Affleck I and Bonner J C , preprint (1989)

60. Hamer C J, Batchelor M T and Barber M N *J. Stat. Phys.* **52** (1988), 679

61. Lai C K *J. Math. Phys.* **15** (1974), 1675

62. Sutherland B *Phys. Rev. B* **12** (1975), 3795

63. Uimin G V *JETP Lett.* **12** (1970), 225

64. Affleck I, Kennedy T, Lieb E H and Tasaki H *Comm. Math. Phys.* **115** (1988), 477

65. Knabe S *J. Stat. Phys.* **52** (1988), 627

66. Papanicolaou N *Nucl. Phys. B* **305** (1988), 367

67. Singh R R P and Gelfand M P *Phys. Rev. Lett.* **61** (1988), 2133

68. Sólyom J *Phys. Rev. B* **36** (1987), 8642

69. Oitmaa J, Parkinson J B and Bonner J C *J. Phys. C* **19** (1986), L595 ; Bonner J C, Parkinson J B, Oitmaa J and Blöte H W J *J. Appl. Phys.* **61** (1987), 4432

70. Chang K, Affleck I, Hayden G W and Soos Z G *J. Phys.: Condens. Matter* **1** (1989), 153

71. Affleck I *Nucl. Phys. B* **265** (1986), 409

72. Affleck I *Phys. Rev. Lett.* **62** (1989), 839 ; Singh R R P and Gelfand M *Phys. Rev. Lett.* **62** (1989), 840 ; see also the corresponding Errata: Singh R R P and Gelfand M *Phys. Rev. Lett.* **63** (1989), 1442 ; Affleck I *Phys. Rev. Lett.* **63** (1989), 1442

73. Schulz H J and Ziman T *Phys. Rev. B* **33** (1986), 6545 ; Blöte H W J and Capel H W *Physica A* **139** (1986), 387 ; Blöte H W J and Bonner J C *Phys. Rev. B* **36** (1987), 2337

74. Klümper A *Europhys. Letters* **9** (1989), 815

75. Klümper A *J. Phys. A* **23** (1990) (in press)

76. Affleck I *J. Phys.: Condens. Matter* **2** (1990), 405

77. Batchelor M T *Phys. Rev. B* **40** (1989), 9303

78. Batchelor M T, Mezincescu L, Nepomechie R I and Rittenberg V, *J. Phys. A* **23** (1990) (in press)

79. Baxter R J *J. Phys. C* **9** (1973), L445

80. Parkinson J B *J. Phys. C* **20** (1987), L1029 ; *J. Phys. C* **21** (1988), 3793

International Journal of Modern Physics B, Vol. 4, No. 5 (1990) 969–978
© World Scientific Publishing Company

# THE MODIFIED XXZ HEISENBERG CHAIN: CONFORMAL INVARIANCE, SURFACE EXPONENTS OF $c < 1$ SYSTEMS AND HIDDEN SYMMETRIES OF FINITE CHAINS

U. GRIMM and V. RITTENBERG
*Physikalisches Institut der Universität Bonn*
*Nußallee 12, 5300 Bonn 1, West-Germany*

The spin-1/2 XXZ Heisenberg chain with two types of boundary terms is considered. For the first type, the Hamiltonian is hermitian but not for the second type which includes the $U_q[SU(2)]$ symmetric case. It is shown that for a certain "tuning" between the anisotropy angle and the boundary terms the spectra present unexpected degeneracies. These degeneracies are related to the structure of the irreducible representations of the Virasoro algebras for $c < 1$.

Some time ago [1] we addressed the problem of universality classes in the case of two-dimensional systems at a second-order phase transition. Assuming that conformal invariance can be used (which is not always the case), one is tempted to define the universality class through a certain modular invariant partition function corresponding to a given central charge of the Virasoro algebra. In the case of the minimal series for example, the modular invariant partition functions were given by Cappelli et al. [2] . A modular invariant partition function does not however specify "who is who" in the list of primary fields. A given primary operator however can be associated with a symmetry breaking operator (corresponding to an order parameter) or to a thermal operator (its anomalous dimension determines the specific heat). Phrased in a different way, the question is: Can two or more systems defined on a lattice and having different symmetries correspond to the same modular invariant partition function. We think that the answer is "yes" and that the complete description of the system needs the knowledge of all the partition functions corresponding to the different boundary conditions (BC) compatible with translational invariance (all *toroidal BC*). In order to get an insight into the problem we have shown that one can project out from the Coulomb gas partition functions, the partition functions of several systems, all of them having the same modular invariant partition function. In this way, for the minimal series we have shown that for each given modular invariant partition function we have four different systems and we called the corresponding series the $p$-state Potts, tricritical $p$-state Potts, the $O(p)$, and the low temperature $O(p)$ models. A similar picture was proposed earlier on quite different grounds by Nienhuis et al. [3] . Behind each of the series are different short-distance algebras. This work was extended later [4] to the case of the $N=1$ superconformal series where for each of the modular invariant partition functions of Cappelli [5] we have found two different models. One remarkable fact is that when applied to the XXZ spin-1/2 Heisenberg chain, for one class of models, the projection mechanism also works for finite chains. A similar phenomenon occurs if the projection mechanism is applied to the Fateev-Zamolodchikov spin-1 quantum chain [4].

In the present lecture we will describe the partition functions of the minimal models described above in the case of *free BC.* As a by-product we will show some "miraculous" properties of the XXZ chain with modified BC. Some of the results presented here were already published [6], some others are new.

The modified XXZ model with $N$ sites is defined by the Hamiltonian

$$H = -\frac{\gamma}{2\pi \sin\gamma} \left\{ \sum_{j=1}^{N-1} \left[ \sigma_j^x \sigma_{j+1}^x + \sigma_j^y \sigma_{j+1}^y - \cos\gamma\, \sigma_j^z \sigma_{j+1}^z \right] + p\sigma_1^z + p'\sigma_N^z \right\} \quad , \tag{1}$$

where $\sigma^x$, $\sigma^y$, and $\sigma^z$ are Pauli matrices and $p$, $p'$ describe the coupling to external fields. The special case $p=p'=0$ corresponds to free BC.

Instead of describing the anisotropy of the Hamiltonian through the parameter $\gamma$ it is useful to define

$$h = \frac{1}{4}\left(1 - \frac{\gamma}{\pi}\right)^{-1} \quad (h \geq \frac{1}{4}) \ . \tag{2}$$

As discussed in Ref. [7], $h$ is related to the compactification radius $R$ of the bosonic string ($h=R^2/2$). It also is convenient to denote

$$p = i \sin\gamma \coth\alpha \quad , \quad p' = i\sin\gamma\coth\alpha' \quad , \tag{3}$$

where $\alpha$ and $\alpha'$ are complex parameters. It turns out that two parametrisations of $\alpha$ and $\alpha'$ leave the spectrum real:

$$\alpha = -\tfrac{i\pi}{2}\psi + \tfrac{i\pi}{2}a \quad , \quad \alpha' = -\tfrac{i\pi}{2}\psi - \tfrac{i\pi}{2}a \tag{4.a}$$

where $\psi$ and $a$ are real (in this case the Hamiltonian is hermitian), and

$$\alpha = -\tfrac{i\pi}{2}\psi + \tfrac{\pi}{2}b \quad , \quad \alpha' = -\tfrac{i\pi}{2}\psi - \tfrac{\pi}{2}b \tag{4.b}$$

where $\psi$ and $b$ are real (in this case the Hamiltonian is *not* hermitian). Note that for $a=b=0$, $\psi=1$ one recovers the case of free BC and that the two BC coincide for $a=b=0$. The case $b\to\infty$ for the BC (4.b) is very special since in this case the Hamiltonian has the quantum algebra $U_q[SU(2)]$ with $q=e^{i\gamma}$ as symmetry.

We notice that the charge operator

$$\hat{Q} = \frac{1}{2}\sum_{j=1}^{N} \sigma_j^z \tag{5}$$

commutes with $H$ and that its eigenvalues $Q$ are integer (half-integer) when $N$ is even (odd). Let $E_{Q;i}(N)$ be the energy levels, $i=0,1,\ldots$, in the charge sector $Q$ of the Hamiltonian with $N$ sites and $E_0^F(N)$ the ground-state energy of the Hamiltonian with free BC. We consider the following quantities:

$$\begin{aligned} \overline{E}_{Q;i} &= \frac{N}{\pi}\left(E_{Q;i}(N) - E_0^F(N)\right) \\ \mathcal{E}_Q(N,z) &= \sum_i z^{\overline{E}_{Q;i}(N)} \\ \mathcal{E}_Q(z) &= \lim_{z\to\infty} \mathcal{E}_Q(N,z) \ . \end{aligned} \tag{6}$$

Using numerical estimates from chains up to 18 sites as well as analytical methods [8] , we have obtained the following ansatz:

$$\mathcal{E}_Q(z) = z^{\frac{(Q+\varphi)^2}{4h}}\Pi_V(z) \quad , \quad \Pi_V(z) = \prod_{m=1}^{\infty}(1-z^m)^{-1} \quad , \tag{7}$$

where $\varphi = 2h(1-\psi)$ independent of $a$ or $b$ (see Eqs. (4.a,4.b)). Notice that $\mathcal{E}_Q \neq \mathcal{E}_{-Q}$. In the $U_q[SU(2)]$ symmetric case the partition functions are very different, one obtains [9] :

$$\mathcal{E}_Q = \mathcal{E}_{-Q} = \sum_{j=|Q|}^{\infty}\left[z^{\frac{(2j+1-h)^2}{4h}} - z^{\frac{(2j+1+h)^2}{4h}}\right]\Pi_V(z) \ . \tag{8}$$

In the expressions (7) and (8), $Q$ is an integer or half-integer. This result is very interesting because from Eq. (7) we learn that the operator content in the charge sector $Q$ is given by a single irrep of a shifted $U(1)$ Kac-Moody algebra [7], the shift $\varphi$ being related to $\psi$. The expression (7) is the starting point of the Feigin-Fuchs construction [10,11] of irreps of Virasoro algebras with $c<1$ starting from irreps with $c=1$ (this is the central charge of the XXZ chain). This construction will allow us to identify the operator content of systems with $c<1$ with free, fixed, and mixed BC. First, we recall that for $c<1$ the central charge is quantized:

$$c = 1 - \frac{6}{m(m+1)} \quad (m = 3,4,\ldots) \ , \tag{9}$$

and for a given $m$, the highest weights $\Delta_{r,s}$ of unitary irreps are

$$\Delta_{r,s} = \frac{[(m+1)r - ms]^2 - 1}{4m(m+1)} \quad , \quad 1 \le r \le m-1 \ , \ 1 \le s \le m \ . \tag{10}$$

The corresponding character functions are:

$$
\begin{aligned}
\chi_{r,s} &= \Omega_{r,s} - \Omega_{r,-s} \\
\Omega_{r,s} &= \sum_{\alpha=-\infty}^{\infty} z^{\frac{[2m(m+1)\alpha+(m+1)r-ms]^2-1}{4m(m+1)}} \Pi_V(z) \ .
\end{aligned} \tag{11}
$$

In order to apply the Feigin-Fuchs procedure, let us assume that we have fixed $h$ and $\varphi$. Instead of considering $E_0^F(N)$ as ground-state energy, we take as ground-state energy $E_{0;0}(N)$, i.e. the ground-state in the charge-zero sector of the Hamiltonian (1) with $\gamma$ , $p$ and $p'$ fixed. Next, instead of considering charges taking values in $\mathbf{Z}$ or $\mathbf{Z}+1/2$, it is convenient to work with charges in $\mathbf{Z}_n$ or $\mathbf{Z}_{n+1/2}$, respectively (the values of $n$ will be fixed later). Thus, instead of Eqs. (6) and (7) we have

$$
\begin{aligned}
\overline{F}_{\overline{q};i} &= \frac{N}{\pi}(E_{\overline{q};i} - E_{0;0}(N)) \\
\mathcal{F}_{\overline{q}}(N,z) &= \sum_i z^{\overline{F}_{\overline{q};i}} \\
\mathcal{F}_{\overline{q}}(z) &= \lim_{N\to\infty} \mathcal{F}_{\overline{q}}(N,z) = \sum_{\alpha=-\infty}^{\infty} z^{\frac{(n\alpha+\overline{q}+\varphi)^2-\varphi^2}{4h}} \Pi_V(z) \ ,
\end{aligned} \tag{12}
$$

where $\overline{q}=0,1,\ldots,n-1$ for $N$ even and $\overline{q}=\frac{1}{2},\frac{3}{2},\ldots,n-\frac{1}{2}$ for $N$ odd. We now choose

$$
h = \frac{n^2}{4m(m+1)} \quad , \quad \varphi = \frac{n}{2m(m+1)} \ , \tag{13}
$$

that is $\psi=1-1/n$ , and get

$$
\mathcal{F}_{\overline{q}}(z) = \sum_{\alpha=-\infty}^{\infty} z^{\frac{[2m(m+1)\alpha+2m(m+1)\overline{q}/n+1]^2-1}{4m(m+1)}} \Pi_V(z) \ . \tag{14}
$$

Since the finite-size scaling spectrum given by Eq. (7) stays positive for any positive $h$, we will assume that we can use the equations above for $0<h<\frac{1}{4}$ also (see Eq. (2)). The four choices of $n$ given in Table 1 divide the domain $h\geq\frac{1}{6}$ into four regions which correspond to the $p$-state Potts models, tricritical $p$-state Potts models, $O(p)$ models, and low temperature $O(p)$ models [12]. Let us first consider the case of the $p$-state Potts model ($n=m+1$). Comparing Eqs. (11) and (14), we have

$$
\mathcal{F}_{-\overline{q}} = \Omega_{1,2\overline{q}+1} \tag{15}
$$

and

$$
\begin{aligned}
\chi_{1,2v+1} &= \mathcal{F}_{-v} - \mathcal{F}_{v+1} && \overline{q} \text{ integer} \qquad (16) \\
\chi_{1,2v} &= \mathcal{F}_{\frac{1}{2}-v} - \mathcal{F}_{\frac{1}{2}+v} && \overline{q} \text{ half-integer} \ . \qquad (17)
\end{aligned}
$$

Table 1. Definition of the models. The arguments of the cosine functions are taken positive. The values of $n$ occur in Eq. (13).

| | $p$-state Potts $0 < p \leq 4$ | Tricritical $p$-state Potts $0 < p \leq 4$ |
|---|---|---|
| Definition of the model | $\frac{1}{2}\sqrt{p} = \cos\pi(1 - \frac{1}{4h})$ | $\frac{1}{2}\sqrt{p} = \cos\pi(\frac{1}{4h} - 1)$ |
| Domain of $h$ | $\frac{1}{2} \geq h \geq \frac{1}{4}$ | $\frac{1}{4} \geq h \geq \frac{1}{6}$ |
| Values of $n$ | $m+1$ | $m$ |

| | $O(p)$ $-2 < p \leq 2$ | Low temperature $O(p)$ $-2 < p \leq 2$ |
|---|---|---|
| Definition of the model | $\frac{1}{2}p = \cos\pi(\frac{1}{h} - 1)$ | $\frac{1}{2}p = \cos\pi(1 - \frac{1}{h})$ |
| Domain of $h$ | $1 \geq h \geq \frac{1}{2}$ | $h \geq 1$ |
| Values of $n$ | $2m$ | $2(m+1)$ |

In order to clarify the physical significance of our results, let us take $m=3$, $n=4$. We find

$$\begin{aligned} \chi_{1,1} &= \mathcal{F}_0 - \mathcal{F}_1 \quad , \quad \Delta = 0 \ , \\ \chi_{1,3} &= \mathcal{F}_{-1} - \mathcal{F}_2 \quad , \quad \Delta = \frac{1}{2} \ . \end{aligned} \tag{18}$$

These are precisely the surface exponents of the Ising model (or, alternatively, the exponents for fixed BC which are the same) [13,14] . If we take $\overline{q}$ half-integer, we have

$$\chi_{1,2} = \mathcal{F}_{-\frac{1}{2}} - \mathcal{F}_{\frac{3}{2}} \quad , \quad \Delta = \frac{1}{16} \ , \tag{19}$$

which corresponds to the case of mixed BC [13] . We now consider $m=5$, $n=6$ (the three-state Potts model) and $\overline{q}$ integer. We obtain

$$\begin{aligned}
\chi_{1,1} &= \mathcal{F}_0 - \mathcal{F}_1 \quad , \quad \Delta = 0 \ , \\
\chi_{1,3} &= \mathcal{F}_{-1} - \mathcal{F}_2 \quad , \quad \Delta = \frac{2}{3} \ , \\
\chi_{1,5} &= \mathcal{F}_{-2} - \mathcal{F}_3 \quad , \quad \Delta = 3 \ .
\end{aligned} \tag{20}$$

These are again the known surface exponents (free BC) or the exponents corresponding to fixed BC. If we take $\overline{q}$ half-integer we find

$$\begin{aligned}
\chi_{1,2} &= \mathcal{F}_{-\frac{1}{2}} - \mathcal{F}_{\frac{3}{2}} \quad , \quad \Delta = \frac{1}{8} \ , \\
\chi_{1,4} &= \mathcal{F}_{-\frac{3}{2}} - \mathcal{F}_{\frac{5}{2}} \quad , \quad \Delta = \frac{13}{8} \ .
\end{aligned} \tag{21}$$

The values $\Delta = \frac{1}{8}$ and $\Delta = \frac{13}{8}$ are indeed the exponents of the three-state Potts model with mixed BC [13]. The case $m=4$ does not correspond to the tricritical Ising model. It has the same modular invariant partition function as the tricritical Ising model but with another distribution of the primary fields between the charge sectors [1].

¿From the above considerations, we will assume that for all values of $m$, Eq. (16) gives us the surface (or fixed BC) exponents and Eq. (17) the exponents for mixed BC. The operator contents for free (fixed) and mixed BC for various models are given in Table 2. Let us derive some known results. For the tricritical Ising model ($m=n=4$), we get

$$\text{free BC} \qquad (0) \oplus \left(\tfrac{3}{2}\right) \tag{22}$$

$$\text{mixed BC} \qquad \left(\tfrac{7}{16}\right) \ . \tag{23}$$

Equation (22) is in agreement with the result of Cardy [13], Eq. (23) is new. In the case of the $O(p)$ models, the leading surface exponent is $\Delta_{2,1}$, again in agreement with Cardy [15] .

Let us proceed with the discussion of our results by means of an example. ¿From the first line of Eq. (20) we see that combining the finite-size scaling spectra of the $\overline{q}=0$ and $\overline{q}=1$ sectors (see Eq. (12) ), disregarding all the doublets (levels which occur in the sectors $\overline{q}=0$ and $\overline{q}=1$), and keeping only the singlets (levels which occur in the $\overline{q}=0$ sector only), we obtain the vacuum sector of the three-state Potts model. This is the general structure of the projection mechanism of the finite-size scaling spectra for $c<1$ from the spectra for $c=1$. We can go one step further and check whether the same projection mechanism also works for a *finite number of sites.* For the example of Eq. (20) this would imply that all the levels of the sector $\overline{q}=1$ are exactly degenerate with levels of the sector $\overline{q}=0$ even for a finite number of sites, and the remaining levels in the sector $\overline{q}=0$ define the spectrum of the vacuum sector of the three-state Potts model. The same picture should be valid for the $\overline{q}=-1$ and $\overline{q}=2$ sectors as well as for the $\overline{q}=-2$ and $\overline{q}=3$ sectors. In Table 3 we give the values

Table 2. The operator content for free (fixed) and mixed boundary conditions and their relation to the XXZ chain for various models. (For the $p$-state Potts case see Eqs. (16) and (17) ).

| | $p$-state Potts | Tricritical $p$-state Potts |
|---|---|---|
| | $\mathcal{F}_{-\overline{q}} = \Omega_{1,2\overline{q}+1}$ | $\mathcal{F}_{\overline{q}} = \Omega_{2\overline{q}+1,1}$ |
| Free (fixed) BC | $\chi_{1,2v+1} = \mathcal{F}_{-v} - \mathcal{F}_{v+1}$ | $\chi_{2v+1,1} = \mathcal{F}_{v} - \mathcal{F}_{-v-1}$ |
| Mixed BC | $\chi_{1,2v} = \mathcal{F}_{-v+\frac{1}{2}} - \mathcal{F}_{v+\frac{1}{2}}$ | $\chi_{2v,1} = \mathcal{F}_{v-\frac{1}{2}} - \mathcal{F}_{-v-\frac{1}{2}}$ |

| | $O(p)$ | Low temperature $O(p)$ |
|---|---|---|
| | $\mathcal{F}_{\overline{q}} = \Omega_{\overline{q}+1,1}$ | $\mathcal{F}_{-\overline{q}} = \Omega_{1,\overline{q}+1}$ |
| Free (fixed) BC | $\chi_{r,1} = \mathcal{F}_{r-1} - \mathcal{F}_{-r-1}$ | $\chi_{1,s} = \mathcal{F}_{1-s} - \mathcal{F}_{1+s}$ |
| Mixed BC | m odd only<br>$\chi_{r,\frac{m-1}{2}} = \mathcal{F}_{\frac{m}{2}-1-r} - \mathcal{F}_{\frac{m}{2}-1+r}$ | m even only<br>$\chi_{\frac{m}{2}-1,s} = \mathcal{F}_{s-\frac{m-1}{2}} - \mathcal{F}_{-s-\frac{m-1}{2}}$ |

of $\overline{F}_{\overline{q};i}$ for four sites ($h=\frac{3}{10}$ , $\psi=\frac{5}{6}$), and $a=\frac{2}{3}$, $a=\frac{1}{6}$, $a=b=0$, $b=1,2$, and $\infty$ . The case $b=\infty$ corresponds to the $U_q[SU(2)]$ symmetric case with $q=\exp(\frac{i\pi}{6})$ . Leaving aside for the moment the case $b=\infty$, we notice the following features:

1. The spectra have doublets and singlets. This suggests a supplementary symmetry in the Hamiltonian. This symmetry goes away if $h$ and $\varphi$ are not "tuned" like in Eq. (13) ($\psi=1-\frac{\gamma}{\pi}$).

2. The doublets occur (like in Table 3 in the sectors $\mathcal{F}_0$ and $\mathcal{F}_1$ for example) in order to allow the projection mechanism to occur, but also between other sectors (this can be seen for a larger number of sites).

3. The doublets "move" if we modify the values of the parameters $a$ and $b$ (like in Table 3) or do not (as one can notice for a larger number of sites).

4. For larger number of sites (eight for the example presented in Table 3), the

degeneracy of a level can be larger than two, the levels are however distributed among the various sectors such that the projection mechanism works.

5. *The projection mechanism works even for a finite number of sites!* The levels left (the singlets) after projecting out the doublets occurring in the corresponding charge sectors give precisely the spectrum of the three-state Potts model with free (fixed) BC (see Table 2 of Ref. [15]).

We now consider the $U_q[SU(2)]$ symmetric case $b=\infty$. In this case the multiplets are much larger. Since $q=\exp(\frac{i\pi}{6})$, the multiplets are $(2j+1)$-dimensional with $j=0,1,2$ for $\overline{q}$ even and $j=\frac{1}{2},\frac{3}{2}$ for $\overline{q}$ odd. The projection mechanism is very different in this case (see Ref. [9], where a superb explanation of the projection mechanism is given). One again obtains the exponents (20) but with different multiplicities: (0) and (3) occur once, but $\left(\frac{2}{3}\right)$ occurs twice.

The same pattern occurs for an odd number of sites and for any value of $m$ as long as $n=m+1$. The projection mechanism does *not* work however for the $O(p)$ $(n=2m)$ or low temperature $O(p)$ $(n=2(m+1))$ models where one does not observe the required doublets in the spectra. The existence of multiplets for $\psi=1-\frac{\gamma}{pi}$ suggesting a discrete non-local symmetry is one of the big mysteries of our work.

The projection mechanism described above can certainly be extended to other systems since the structure of the character expressions of extended algebras is always similar to Eq. (11). The first step in this direction in order to derive the surface exponents of the $N=1$ superconformal model was already made [16].

Table 3. Table of $\overline{F}_{\overline{q};i}(N)$ values (see Eq. (12) ) for $N=4$, $h=\frac{3}{10}$ , $\psi=\frac{5}{6}$ , and several values of $a$ and $b$ (see Eqs. (4.a) and (4.b) . The underlined levels occur in the three-state Potts model with free BC. (The XXZ chain with $2N$ sites gives the $p$-state Potts model with $N$ sites and free BC.)

| $a=\frac{2}{3}$ | | $a=\frac{1}{6}$ | | $a=b=0$ | |
|---|---|---|---|---|---|
| $\overline{q}=0$ | $\overline{q}=1$ | $\overline{q}=0$ | $\overline{q}=1$ | $\overline{q}=0$ | $\overline{q}=1$ |
| 0.000 000 | | 0.000 000 | | 0.000 000 | |
| 0.869 482 | 0.869 482 | 0.573 943 | 0.573 943 | 0.566 777 | 0.566 777 |
| 1.347 175 | 1.347 175 | 1.128 282 | 1.128 282 | 1.124 431 | 1.124 431 |
| 1.407 619 | | 1.407 619 | | 1.407 619 | |
| 2.022 885 | 2.022 885 | 1.693 423 | 1.693 423 | 1.689 669 | 1.689 669 |
| 2.520 290 | 2.520 290 | 1.992 459 | 1.992 459 | 1.980 841 | 1.980 841 |
| $\overline{q}=-1$ | $\overline{q}=2$ | $\overline{q}=-1$ | $\overline{q}=2$ | $\overline{q}=-1$ | $\overline{q}=2$ |
| 0.471 151 | | 0.471 151 | | 0.471 151 | |
| 1.071 362 | | 1.071 362 | | 1.071 362 | |
| 1.671 573 | | 1.671 573 | | 1.671 573 | |
| 2.386 227 | 2.386 227 | 1.928 985 | 1.928 985 | 1.920 189 | 1.920 189 |
| $\overline{q}=-2$ | $\overline{q}=3$ | $\overline{q}=-2$ | $\overline{q}=3$ | $\overline{q}=-2$ | $\overline{q}=3$ |
| 1.806 467 | | 1.806 467 | | 1.806 467 | |

| $b=1$ | | $b=2$ | | $b=\infty$ | |
|---|---|---|---|---|---|
| $\overline{q}=0$ | $\overline{q}=1$ | $\overline{q}=0$ | $\overline{q}=1$ | $\overline{q}=0$ | $\overline{q}=1$ |
| 0.000 000 | | 0.000 000 | | 0.000 000 | |
| 0.485 791 | 0.485 791 | 0.471 833 | 0.471 833 | 0.471 151 | 0.471 151 |
| 1.079 782 | 1.079 782 | 1.071 757 | 1.071 757 | 1.071 362 | 1.071 362 |
| 1.407 619 | | 1.407 619 | | 1.407 619 | |
| 1.668 001 | 1.668 001 | 1.671 294 | 1.671 294 | 1.671 573 | 1.671 573 |
| 1.838 081 | 1.838 081 | 1.808 041 | 1.808 041 | 1.806 467 | 1.806 467 |
| $\overline{q}=-1$ | $\overline{q}=2$ | $\overline{q}=-1$ | $\overline{q}=2$ | $\overline{q}=-1$ | $\overline{q}=2$ |
| 0.471 151 | | 0.471 151 | | 0.471 151 | |
| 1.071 362 | | 1.071 362 | | 1.071 362 | |
| 1.671 573 | | 1.671 573 | | 1.671 573 | |
| 1.823 501 | 1.807 257 | 1.807 257 | 1.806 467 | 1.806 467 | |
| $\overline{q}=-2$ | $\overline{q}=3$ | $\overline{q}=-2$ | $\overline{q}=3$ | $\overline{q}=-2$ | $\overline{q}=3$ |
| 1.806 467 | | 1.806 467 | | 1.806 467 | |

## References


1. F. Alcaraz, U. Grimm, and V. Rittenberg. *Nucl. Phys. B316* (1989) 735.
2. A. Cappelli, C. Itzykson, and J.-B. Zuber. *Nucl. Phys. B280* (1987) 445.
3. B. Nienhuis, E.K. Riedel, and M. Schick. *Phys. Rev. B27* (1983) 5625.
4. D. Baranowski, U. Grimm, V. Rittenberg, and G. Schütz. *to be published.*
5. A. Cappelli. *Phys. Lett. B185* (1987) 82.
6. F. Alcaraz, M. Baake, U. Grimm, and V. Rittenberg. *J. Phys. A: Math. Gen. 22* (1989) L5.
7. M. Baake, Ph. Christe, and V. Rittenberg. *Nucl. Phys. B300* (1988) 637.
8. C.J. Hamer and M.T. Batchelor. *J. Phys. A21* (1988) L173.
9. V. Pasquier and H. Saleur. *Nucl. Phys. B* (1989) *(to be published).*
10. B.L. Feigin and K.B. Fuchs. *Funct. Anal. Appl. 16* (1982) 114.
11. V.S. Dotsenko and V.A. Fateev. *Nucl. Phys. B240* (1984) 312.
12. B. Nienhuis. *J. Stat. Phys. 34* (1984) 731.
13. J.L. Cardy. *Nucl. Phys. B275* (1986) 200.
14. G. v. Gehlen and V. Rittenberg. *J. Phys. A: Math. Gen. 19* (1986) L631.
15. G. v. Gehlen, V. Rittenberg, and T. Vescan. *J. Phys. A: Math. Gen. 20* (1987) 2577.
16. L. Mezincescu, R.I. Nepomechie, and V. Rittenberg. (1990) *to be published.*

International Journal of Modern Physics B, Vol. 4, No. 5 (1990) 979–993
© World Scientific Publishing Company

# C = 1 COSET MODELS AND THEIR FUSION ALGEBRAS+

P.D. JARVIS*[1] and R.B. ZHANG[2]

[1]*Department of Physics, University of Tasmania, Hobart, Tasmania*

[2]*Department of Mathematics, University of Queensland, St. Lucia, Brisbane, Queensland*

Higher symmetry algebras, conjectured empirically in certain c = 1 models in statistical mechanics, are identified with the extended conformal algebras in coset constructions of the maximal hermitian symmetric spaces. The branching functions and their modular properties are derived and associated algebras, compatible with the fusion rules are discussed.

+ Based on Ref [17].

* Alexander van Humboldt Fellow

## 1. Introduction

The study of conformal field theories [1] with $c \geq 1$ turns out to be a very difficult task, mainly due to the fact that such theories generically admit an infinite number of primary fields [2]. However, in some special cases, eg., rational conformal theories [3], there exist extra symmetries apart from the basic conformal invariance, which, restricting the theories to finite spectra, render them tactable. In this paper, we study a particular class of such theories with c = 1, obtained from hermitian symmetric spaces through Goddard–Kent–Olive coset construction [4].

The GKO construction proves to be a powerful tool for constructing unitary conformal theories, explicitly characterizing their primary fields and elucidating the underlying symmetries. In the original work of GKO [4], the minimal models [1][5] were constructed from the coset $SU^{(k)}(2) \times SU^{(1)}(2)/SU^{(k+1)}(2)$, their operator contents were revealed through the branching of unitary representations of $\hat{s}u^{(k)}(2) \oplus \hat{s}u^{(1)}(2)$ under $\hat{s}u^{(k)}(2) \oplus \hat{s}u^{(1)}(2) \supset \hat{s}u^{(k+1)}(2)$; and a similar treatment was also given to the supersymmetric minimal models [5]. Recently, this construction was also extended to the parafermionic theories [6], where the parafermionic fields are identified with certain primary fields which are invariants of the subalgebras [7].

Due to the fact that the resultant theory factorizes if the subalgebra used in GKO construction is nonmaximal, we will consider the so–called maximal hermitian symmetric spaces only, a complete list of which is [8]

$$\begin{aligned}
&SU(p+q)/SU(p) \times SU(q) \times U(1)\\
&SO(2n)/SU(n) \times U(1)\\
&SO(2n)/SO(2n-2) \times U(1)\\
&E_6/SO(10) \times U(1)\\
&E_7/E_6 \times U(1)\\
\text{and}\quad &SO(2n+1)/SO(2n-1) \times U(1)\\
&Sp(2n)/SU(n) \times U(1)\,. \qquad (1)
\end{aligned}$$

The motivation for studying these cosets is two fold. This class of symmetric spaces, interesting on their own merit, deserves a thorough investigation; they also allow us to understand the "$\frac{p}{r}$ – characters" [9] [10] in purely algebraic terms.

The "$\frac{p}{r}$ – character" arises from various c = 1 statistical mechanics models [9] [10]; it is an infinite sum of $\hat{u}(1)$ characters, defined by

$$\mathrm{Ch}^{\rho}_{\xi}(\tau,y) = \sum_{m\in\mathbb{Z}} z^{\rho(m+\xi)^2}\, y^{(m+\xi)}\, \eta^{-1}(\tau), \quad y \in \mathbb{C}, \quad t = e^{i\,2\pi\tau}, \qquad (2)$$

with $\rho = \frac{p}{r}$ (thus the name). In general, $\xi$ can be an arbitrary real number. However, modular properties of (2) indicate that when $\xi$ is irrational, an infinite number of such characters is required in order to form a complete theory, while rational $\xi$'s lead to a finite spectrum. Thus we will consider the latter case only.

Below the GKO construction is briefly reviewed, then applied to the cosets G/H related to the maximal hermitian symmetric spaces G/H × U(1). It is shown explicitly

that the "p/r – characters" (with rational $\xi$'s) arise in the decomposition of level 1 characters of $\hat{g}$ under
$\hat{g} \supset \hat{h} \oplus \hat{u}(1)$ ($\hat{g}$ and $\hat{h}$ are the Kac–Moody algebras corresponding to G and H respectively). The modular properties of these characters and their associated algebraic structures compatible with the fusion rules [11] are discussed, and finally, their correspondence with the conjectured "p/r – algebras" [9] is illustrated by comparison with some known examples.

## 2. Goddard-Kent-Olive Coset Construction

Given an affine Kac–Moody algebra with a basis

$$\hat{g} = \left[T^a_m,\ m \in \mathbb{Z},\ a = 1,2,\cdots,d\right]$$

defined by the commutation relation

$$\left[T^a_m, T^b_n\right] = f^{ab}{}_c\, T^c{}_{m+n} + k(T^a_o, T^b_o)m\, \delta_{m+n,o} \tag{3}$$

with k the central extension, the zero modes of this algebra

$$g = \left[T^a_o,\ a = 1,2,\cdots,d\right]$$

form an ordinary Lie algebra with structure constants $f^{ab}{}_c$. This fact has already been used in the defining relation (3), where $(T^a_o, T^b_o)$ is taken to be the killing form on g for consistency. By properly choosing a basis for $\hat{g}$ we can always transform the killing form into the simple form

$$\left[T^a_o, T^b_o\right] = \delta^{ab}\ ,$$

and in the following it is assumed that we are using this basis. Introduce the collective notation

$$T^a(z) = \sum_{m\in\mathbb{Z}} T^a_m/z^{m+1}, \quad z \in \mathbb{C}. \tag{4}$$

Then the operator product expansion technique from conformal field theory allows us to reformulate the defining relation for $\hat{g}$, (2), into

$$T^a(z)T^b(w) \sim \frac{k\delta^{ab}}{(z-w)^2} + \frac{f^{ab}{}_c T^c(w)}{z - w} + \cdots, \tag{5}$$

where ~ means equality in the sense of operator product expansion and the dots represent terms which are regular when $w \to z$. Let us construct the following operator

$$L^g(z) = \frac{1}{2k + Q_g} : \sum_a T^a(z)\, T^a(z) : \tag{6}$$

$$= \sum_{m\in\mathbb{Z}} L^g_m/z^{m+z},$$

with $Q_g$ the quadratic Casimir of g in the adjoint representation under the normalization that the longest root has length 2. In (6), the normal ordering is to position the positive modes of $T^a(z)$ on the right in each product $T^a(z)T^a(z)$, thus removing the possible singularity arising from the multiplication of two operators at the same point. It can be checked out that

$$L^g(z)L^g(w) \sim \frac{C^k_g/2}{(z-w)^4} + \frac{2L^g(w)}{(z-w)^2} + \frac{\partial_w L^g(w)}{z - w} + \cdots, \tag{7}$$

and

$$L^g(z)T^a(w) \sim \frac{T^a(w)}{(z-w)^2} + \frac{\partial_w T^a(w)}{z - w} + \cdots, \tag{8}$$

i.e., $L^g = \{L^g_m, m \in \mathbb{Z}\}$, while spanning a Virasoro algebra with central charge $C^k_g$ by itself, forms a semiproduct with $\hat{g}$.

For any given untwisted affine Kac–Moody algebra, (6) provides a concrete construction, known as the Sugawara construction, of a related Virasoro algebra. Now one may ask the following question: if $\hat{g} \supset \hat{h}$, with $\hat{h}$ a sub. affine Kac–Moody algebra, does there exist a Virasoro algebra related to the coset G/H? This question is answered by the Goddard–Kent–Olive construction [4] which we are discussing now.

Assume $\hat{h} = \{t^\alpha(z),\ \alpha = 1,2,\cdots,\ \dim h\} \subset \hat{g}$, such that

$$t^\alpha(z)\, t^\beta(w) \sim \frac{k'\delta^{\alpha\beta}}{(z-w)^2} + \frac{\sigma^{\alpha\beta}{}_\gamma\ t^\gamma(z)}{z-w} + \cdots, \tag{9}$$

then the Sugawara construction provides us with a new Virasoro algebra
$L^h = \{L^h{}_m,\ m \in \mathbb{Z}\}$

given by $L^h(z) = \frac{1}{2k' + Q_h} : \sum_\alpha t^\alpha(z)\, t^\alpha(z) : = \sum_{m\in\mathbb{Z}} L^h_m / z^{m+z}$.

Since $\hat{h} \subset \hat{g}$, for all $t^\alpha(z) \in \hat{h}$, we have

$$\begin{aligned} L^g(z)\, t^\alpha(w) &\sim \frac{t^\alpha(w)}{(z-w)^2} + \frac{\partial_w t^\alpha(w)}{z-w} + \cdots \\ &\sim L^h(z) t^\alpha(w)\,. \end{aligned} \tag{10}$$

Let
$$K(z) = L^g(z) - L^h(z)\,; \tag{11}$$

then (10) is equivalent to

$$K(z)\, t^\alpha(w) \sim 0,\ \ \forall\, t^\alpha(w) \in \hat{h}\,, \tag{12}$$

and this immediately leads to

$$K(z)\, L^h(w) \sim 0\,. \tag{13}$$

Therefore

$$L^g(z)\, L^g(w) \sim (K(z) + L^h(z))\,(K(w) + L^h(w))$$
$$\sim K(z)\, K(w) + L^h(z)\, L^h(w)\,.$$

When we move the second term on the right hand side to the left, we arrive at

$$K(z)\, K(w) \sim \frac{(C_g^k - C_h^{k'})/2}{(z - w)^4} + \frac{2K(w)}{(z-w)^2} + \frac{\partial_w K(w)}{z - w} + \cdots, \tag{14}$$

i.e. $K = \{K_m, m \in \mathbb{Z}\}$ spans a Virasoro algebra related to the coset G/H with a central charge

$$C_{g/h} = C_g^k - C_h^{k'} = \frac{2k \dim g}{2k + Q_g} - \frac{2k' \dim h}{2k' + Q_h}\,. \tag{15}$$

It is worth noting that if $\hat{g} \supset \hat{f} \supset \hat{h}$, with $\hat{f}$ and $\hat{h}$ being subalgebras then

$$L = L^g - L^h$$
$$= (L^g - L^f) + (L^f - L^h)\,.$$

$L^g - L^f$ and $L^f - L^h$ form two mutually commuting Virasoro algebras indicating that the resultant conformal theory factorizes according to

$$G/H = G/F \otimes F/H\,.$$

We also want to point out that the central extension k' is uniquely determined by the embedding $\hat{g} \supset \hat{h}$. In general we have

$$k' = \ell_n\, k$$

where $\ell_h$ is the embedding index of $g \supset h$.

### 3. Coset Models with C = 1

In this section we apply the GKO construction to the cosets listed in (1) to obtain c = 1 conformal field theoretical models. To show the correspondence of the models with the "p/r – characters", we analyse the branching of the level 1 characters of $\hat{g}$ under $\hat{g} \supset \hat{h} \oplus \hat{u}(1)$. It turns out that the $\hat{u}(1)$ branching functions derived therefrom are precisely the "p/r – characters" with rational shifts. For simplicity, we will concentrate on the first five cosets only, ignoring SO(2n+1)/SO(2n–1) and Sp(2n)/SU(n), both of which involve non–simply laced algebras (and Sp(2n)/SU(n) does not give rise to c = 1 models). Consider the coset $E_6$/SO(10) as an example. Since $\hat{E}_6 \supset \hat{so}(10) \oplus \hat{u}(1)$ is regular. $\hat{E}_6$ and $\hat{so}(10)$ must be of the same level. Recall that $C^k_{E_6} = \frac{78k}{k+12}$ and $C^k_{D_5} = \frac{45k}{k+8}$, thus we require k equal 1 in order to obtain a c = 1 model. Similar analyses apply to the other cosets and we have that all the Kac–Moody algebras involved are of level 1.

For a simply laced level 1 Kac–Moody algebra $\hat{g}$, the Sugawara construction is extremely simple. Let $g^\mu(z)$, $\mu = 1,2,\cdots$, rank g, be the generators of $\hat{g}$ corresponding to the Cartan subalgebra of g(c.f. (4)) defined in an orthonormal basis such that

$$g^\mu(z)\, g^\nu(w) \sim \frac{\delta^{\mu\nu}}{(z-w)^2} + \cdots ,$$

then

$$L^g(z) = \frac{1}{2} \sum_{\mu=1}^{\text{rank}\,g} : g^\mu(z)\, g^\mu(z): . \tag{16}$$

Similarly for a subalgebra $\hat{h}$ regularly embedded in $\hat{g}$,

$$L^h(z) = \frac{1}{2} \sum_{\mu=1}^{\text{rank}\,h} : h^s(z)\, h^s(z): , \tag{17}$$

where $h^s(z)$, $s = 1,2,\cdots$, rank h, satisfying

$$h^s(z)\, h^t(w) \sim \frac{\delta^{st}}{(z-w)^2} + \cdots ,$$

are orthogonal linear combinations of $g^{\mu}(z)$. Note that for all the cosets we are interested in. rank g − rank h = 1, therefore

$$K(z) = L^{g}(z) - L^{h}(z)$$
$$= \frac{1}{2}{:}T(z)\,T(z){:}\ , \tag{18}$$

where

$$T(z) = \sum_{\mu} \sigma^{\mu}\, g^{\mu}(z)/\sqrt{\sigma^2} \tag{19}$$

is a linear combination of $g^{\mu}$ orthogonal to all the $h^{s}(z)$, which by itself forms a $\hat{u}(1)$ algebra

$$T(z)\,T(w) \sim \frac{1}{(z-w)^2} + \cdots . \tag{20}$$

$\sigma^{\mu}$ can be determined explicitly for the cosets, results are listed in the table.

In order to establish the correspondence with the "p/r − characters", we consider the characters of the level one representations of $\hat{g}$ and their branching under $\hat{g} \supset \hat{h} \oplus \hat{u}(1)$. Given such a character $\chi^{g}$. In general, it branches into

$$\chi^{g} = \sum_{\xi} Ch_{\xi}^{\zeta} \cdot \chi_{\xi}^{h}\ ,$$

where $\chi_{\xi}^{h}$ are level one characters of $\hat{h}$, and $Ch_{\xi}^{\zeta}$ are the branching coefficients. As the notation suggests, they are actually the "p/r − characters".

To illustrate how the procedure works, we again consider $E_6/SO(10)$ as an example. But before getting into details we explain our notations first. We use $\alpha$ to label the simple roots of g, $\beta$ the simple roots of h (a subset of g roots, since h is a regular subalgebra of g), and $\Lambda$ and $\lambda$ the corresponding fundamental weights. In the case of $E_6/SO(10)$,

$E_6$ : 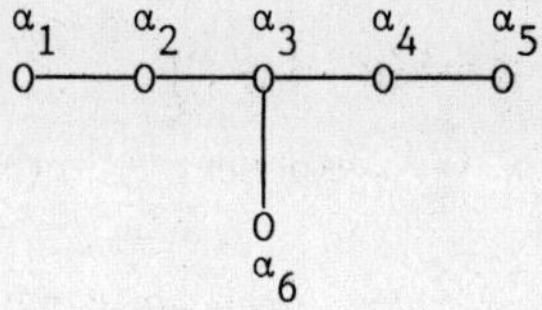

, $D_5$ : 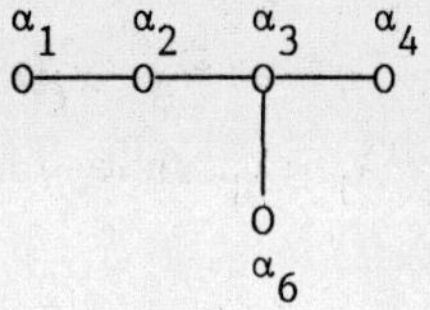

;

clearly $\alpha_1$, $\alpha_2$, $\alpha_3$, $\alpha_4$ and $\alpha_6$ generate $D_5$, while $\sigma = \alpha_5 + \lambda_4$.

There are three level one characters $\chi^{E_6}_{0}$, $\chi^{E_6}_{1}$ and $\chi^{E_6}_{5}$ corresponding to the coset decomposition

$$\Sigma_W = (\Lambda_o + \Sigma_R)\cup(\Lambda_1 + \Sigma_R)\cup(\Lambda_5 + \Sigma_R)$$

of the $E_6$ weight lattice. Consider for example

$$\chi^{E_6}_{1}(z,\theta) = z^{-6/24}\mathrm{tr}[z^{L_0}\exp(i\theta.H_0)] = \sum_{\alpha} z^{(\Lambda_1+\alpha)^2/2} e^{i\theta.(\Lambda_1+\alpha)}\eta^{-6}. \tag{21}$$

In this case any root vector may be written $\alpha = \beta + n\alpha_5$ for integer n, and we have (see for example [16])

$$\Lambda_1 = \lambda_1 + 2(\alpha_5 + \lambda_4)/3$$

for $\beta$ in the root lattice of $D_5$. Thus

$$\theta.(\Lambda_1+\alpha) = \theta.(\beta+\lambda_1-n\lambda_4) + (n+2/3)\theta.(\alpha_5+\lambda_4),$$

$$(\Lambda_1+\alpha)^2/2 = (\Lambda_1+\beta+n\alpha_5)^2/2 = n^2/2 + (\beta^2/2+\beta.(\lambda_1-n\lambda_4)) + \Lambda_1{}^2/2$$

(since $\beta.(\alpha_5+\lambda_4) = 0$). The $\beta$ summation can be recognised as a shifted vacuum character of $D_5$

$$\chi_\xi(z,\theta) = z^{\xi/2}\mathrm{tr}[z^{L_0+\xi.H_0}\exp(i\theta.(H_0+\xi)]$$

where $\xi = \lambda_1 \cdot n\lambda_4$. In the remaining n summation, the z exponent is, in this case

$$n^2(1-\lambda_4{}^2/2) + n\lambda_1.\lambda_4 + (\Lambda_1{}^2-\lambda_1{}^2)/2 = 3(n+2/3)^2/8$$

and defining $y = \exp i\theta.(\alpha_5+\lambda_4)$ we have

$$\chi^{E_6}{}_1 = \sum_{n\in\mathbb{Z}} z^{3(n+2/3)^2/8}\, y^{(n+2/3)}\, \chi_{\lambda_1-n\lambda_4}(z,\theta)\eta^{-1}(\tau) \qquad (22)$$

In $D_5$ the weights $\lambda_1 - (4k+a)\lambda_4$ lie in the cosets $\lambda_1$, $\lambda_4$, $\lambda_0$, and $\lambda_5$ for a = 0,1,2,3 respectively. Finally (c.f.(2))

$$\chi^{E_6}{}_1 = \sum_{a=0}^{3} Ch^6{}_{(a/4+1/6)}(z,y)\, \chi_{(\lambda_1-a\lambda_4)}(z,\theta) \qquad (23)$$

The results for the remaining hermitian symmetric spaces are given in the table below. Notation for roots and weights is as explained above, and standard Dynkin labelling conventions are used(see for example [16]).

| Coset | $\underline{\sigma}$ | $\underline{\rho}$ | $\underline{\xi}$ |
|---|---|---|---|
| $E_6/SO(10)$ | $\alpha_5 + \lambda_4$ | 6 | a/12, a = 0,1,2,···,11 |
| $E_7/E_6$ | $\alpha_6 + \lambda_5$ | 3 | a/6, a = 0,1,2,···,5 |
| SO(2n)/SO(2n–2) | $\alpha_1 + \lambda_2$ | 4 | a/8, a = 0,1,2,···,7 |
| SO(2n)/SU(n) | $\alpha_n + \lambda_{n-2}$ | 2n | a/4n, a=0,1,2,···,4n–1 |
| SU(p+q)/SU(p)×SU(q) (p,q) = 1 | $\alpha_p + \lambda_{p-1} + \mu_1$ | pq(p+q)/2 | a/pq(p+q), a=0,1,2,···,pq(p+q)–1 |
| SU(2p)/SU(p) × SU(p) 2p–1 | $\alpha_p + \lambda_{p-1} + \mu_1$ | p | a/2p, a = 0,1,2,···, |

It should be pointed out that the characters derived from the $\hat{g}$ character decompositons, having the properties that

$$\rho = p,\ 0 < p \in \mathbb{Z}_+,\ \rho = \frac{a}{2p},\ a = 0,1,2,\cdots,2p-1\ , \tag{24}$$

provide the building blocks of the $Ch_\xi^\rho$ characters with $\xi \in \mathbb{R}$ in the sense that every such character can be expressed as a finite sum of $Ch^p_{a/2p}$. In the next section we will study these characters in some detail by examining their Modular properties and the associated fusion algebras [11].

## 4. Fusion Algebras and Extended Symmetries

In order to gain a deeper understanding of the coset models constructed above, we study the fusion algebras associated with the "p/r – characters" and point out the conjectured "p/r – algebras" for some known examples.

In a given conformal field theory, the fields, both primary and secondary close on themselves under the operator product expansion. As the fields can be decomposed into irreducible representations of the symmetry algebra of the theory, i.e.

$$\{A\} = \{[\phi_i],\ i \in I\}\ .$$

the closure of $\{A\}$ requires that the operator product of the fields of two irreducible representations reduces completely into irreducible representations of the symmetry algebra provided that the theory is unitary. This fact is embodied in the fusion algebras, which we will discuss for the coset models.

Set $y = 1$, $Ch^p_{a/2p}$ reduces to

$$\mathrm{Ch}^p_a(\tau) = \sum_{m \in \mathbb{Z}} z^{p(m + a/2p)^2} \eta^{-1}(\tau) \ , \ z = e^{i2\pi\tau}, \ 0 < p \in \mathbb{Z}^+, \ a = 0,1,2,\cdots,2p-1$$

It is easy to show that, under T: $\tau \to \tau+1$,

$$\mathrm{Ch}^p_a(\tau) \longrightarrow \mathrm{Ch}^p_a(\tau+1) = \exp[i2\pi(a^2/4p - 1/24)] \ \mathrm{Ch}^p_a(\tau),$$

thus, the primary field giving rise to this character has conformal dimension

$$\Delta = \frac{a^2}{4p} \, . \tag{25}$$

The modular property under S: $\tau \longrightarrow s - \frac{1}{\tau}$,

$$\mathrm{Ch}^p_a(-\tfrac{1}{\tau}) = \sum_{b=0}^{2p-1} (w)^{ab} \ \mathrm{Ch}^p_b(\tau)/\sqrt{2p} \tag{26}$$

where $w = e^{i2\pi/2p}$, can also be easily established by using the Poisson summation formula. Define

$$S\mathrm{Ch}^p_a(\tau) = \mathrm{Ch}^p_a(-\tfrac{1}{\tau}) = \sum_{b=0}^{2p-1} S^a{}_b \ \mathrm{Ch}^p_b(\tau) \ ,$$

then equation (26) yields

$$S^a_b = \exp(\frac{iab\pi}{p})/\sqrt{2p} \, . \tag{27}$$

Using the matrix S, one can easily obtain the fusion coefficient [11]

$$A_{ab}{}^c = \sum_{f=0}^{2p-1} S_a{}^f S_b{}^f (S^+)_f{}^c / S_o{}^f = \delta^{\ c}_{a+b(\mathrm{mod}\ 2p)} \, ,$$

and this immediately leads to the fusion rule

$$\phi_a \times \phi_b = \phi_{a+b(\mathrm{mod}\ 2p)} \, . \tag{28}$$

(28) indicates that the coset modules we have constructed are the so–called Gaussian models; a quite thorough study of which is given in [13].

Finally we reinforce the foregoing by giving some known, extended symmetries as they appear in this framework. Here the notations of [12] are adopted.

(i) $p = 1,\ a = 0,1$

This case gives directly the $\hat{su}(2)$ Kac–Moody algebra and its level one vacuum ($Ch^1{}_0$) and spinor ($Ch^1{}_1$) characters.

(ii) $p = 2,\ a = 0,1,2,3$

Taking the {0,2} subalgebra produces cosets {0,2} and {0,3} for which

$$Ch^2{}_{0+2} \equiv Ch^{1/2}{}_0,\ Ch^2{}_{1+3} \equiv Ch^{1/2}{}_{1/2}$$

from which we recognise the U(1) content of two free fermions.

(iii) $p = 3,\ a = 0,1,2,3,4,5$

The $Ch^3$ characters with shifts 0, 1/2, ±1/3, ±1/6 corresponds to primary fields generating irreducible representations of the $c = 1$ $Z_4$ parafermion (Fateev–Zamolodchikov [6]) algebra. The ±–primary fields have the same conformal dimension and opposite U(1) charges.

(iv) $p = 6,\ a = 0,1,\ldots,11$

An abundance of subalgebras is possible: consider in particular {0,6} and the cosets {0,6}, {1,7}, {2,8}, {3,9}, {4,10} and {5,11}. Then

$$\mathrm{Ch}^6{}_{1+7} = \sum_m \left(z^{6(m+1/12)^2} + z^{6(m+7/12)^2}\right) = \sum_m \left(z^{3/2(2m+1/6)^2} + z^{3/2(2m+7/6)^2}\right) = \sum_m z^{3/2(m+1/6)^2} .$$

etc., and so we have

$$\mathrm{Ch}^6{}_{0+6} \equiv \mathrm{Ch}^{3/2}{}_0,\ \mathrm{Ch}^6{}_{1+7} \equiv \mathrm{Ch}^{3/2}{}_{1/6},\ \mathrm{Ch}^6{}_{2+8} \equiv \mathrm{Ch}^{3/2}{}_{1/3},$$

$$\mathrm{Ch}^6{}_{3+9} \equiv \mathrm{Ch}^{3/2}{}_{1/2},\ \mathrm{Ch}^6{}_{4+10} \equiv \mathrm{Ch}^{3/2}{}_{-1/3},\ \mathrm{Ch}^6{}_{5+11} \equiv \mathrm{Ch}^{3/2}{}_{-1/6},$$

corresponding precisely to the irreducible characters of the $N = 2$ superconformal algebra at $c = 1$ [14].

## 5. Conclusion

It is well known that in certain cases ((i), (ii) and (iv) above) $c = 1$ extended algebras can be represented by a single boson [15]. In the present coset method this field appears as a one–dimensional projection amongst a set of bosons representing the Cartan subalgebra of $\hat{g}$. The same algebraic construction operates for all the extended algebras with characters (2) and exponents $\rho \equiv p =$ integer (see table at the end of Section 3).

## References


[1] A.A. Belavin, A.M. Polyakov and A.B. Zamolodchikov, Nucl. Phys. B241 (1984) 333.

[2] J.L. Cardy, Nucl. Phys. B275 (1986) 200.

[3] See for example G. Moore and N. Seiberg, Comm. Math. Phys. 123 (1989) 177 and references there in.

[4] P. Goddard, A. Kent and D. Olive, Comm. Math. Phys. 103 (1986) 105.

[5] D. Friedan, Z. Qin and S. Shenker, Phys. Rev. Lett. 52 (1984) 1575, Vertex Operators in Mathematics and Physics, ed. by J. Lepowsky, S. Mandelstam and I.M. Singer, (Springer–Verlag, New York, 1985).

[6] A.B. Zamolodchikov and V.A. Fateev, Sov. Phys. JETP 62 (1985) 215.

[7] P. Goddard and A. Schwimmer, Phys. Lett B206 (1988) 62.

[8] S. Helgason, Differential Geometry, Lie Groups and Symmetric Spaces. (Academic Press, 1978).

[9] M. Baake, P. Christie and V. Rittenberg, Nucl. Phys. B300 (1988) 637.

[10] P.A. Pearce and K.A. Seaton, Exact Solution of Cyclic Solid–on–Solid Lattice Models, Melbourne Uni Preprint (1989).

[11] E. Verlinde, Nucl. Phys. B300 (1988) 360.

[12] B. Blok and S. Yankielowicz, Nucl. Phys. B315 (1989) 25.

[13] R. Dijkgraaf, E. Verlinde and H. Verlinde, Comm. Math. Phys. 115 (1988) 649; P. Ginsparg, Nucl. Phys. B295 (1988) 153.

[14] F. Ravanini and S.K. Yang, Phys. Lett. B193 (1987) 202.

[15] G. Waterson, Phys. Lett. B171 (1986) 77.

[16] M.R. Bremner, R.V. Moody and J. Patera, Tables of Dominant Weight Multiplicities for Representations of Simple Lie Algebras, (Dekker, 1985).

[17] P.D. Jarvis and R.B. Zhang, Phys. Lett. B225 (1989) 112.

International Journal of Modern Physics B, Vol. 4, No. 5 (1990) 995 – 1001
© World Scientific Publishing Company

# LEVEL CROSSING, SPONTANEOUS PARITY VIOLATION, HIGH $T_c$ SUPERCONDUCTIVITY MECHANISMS AND THE CHIRAL POTTS CHAIN

Hiroshi ITOYAMA

Barry M. McCOY

*Institute for Theoretical Physics, State University of New York at Stony Brook, Stony Brook, NY 11794-3840, U.S.A.*

and

Jacques H.H. PERK

*Physics Department, Oklahoma State University, Stillwater, Oklahoma 74078-0444, U.S.A.*

We use recent results on the integrable Chiral Potts chains, the $t$-$J$ model, and the bilinear-biquadratic spin-1 model to discuss the level crossing-spontaneous parity violating mechanism for high $T_c$ superconductivity.

Recently there has been a remarkable convergence of ideas from the integrable Chiral Potts models [1,2], the $t$-$J$ model used in high $T_c$ superconductivity [3,4], the bilinear-biquadratic spin 1 model [5,6,7] and quantum field theory [8,9]. The fundamental physical principle which connects these problems together is the fact that if the off diagonal elements of a quantum Hamiltonian are not all non-positive then the phenomena of ground state level crossing and spontaneous violation of parity can occur. These phenomena are forbidden when the off diagonal elements do satisfy a non-positivity condition. In this note we make explicit some of these connections.

Conditions of non-positivity enter quantum physics in the form of Sturm-Liouville theory [10] and, more generally, as the theorem of Perron-Frobenius [11]. This theorem states that if the off diagonal elements of a matrix are all non-positive and if the matrix is not in a block diagonal form then the ground state eigenvalue is non-degenerate. This is a somewhat awkward theorem to use because its very statement requires the introduction of a basis into the vector space. Nevertheless it is of great use because for many commonly considered hamiltonians a basis can be found for which there is a non-positivity property. Indeed, the most common hamiltonian

$$\mathcal{H} = -\frac{1}{2m}\sum_{\ell=1}^{N}\frac{\partial^2}{\partial x_\ell^2} - \sum_{j<\ell} V(x_j - x_\ell) \tag{1}$$

manifestly satisfies the non-positivity constraint.

A most common use of Perron-Frobenius for a translationally invariant hamiltonian like (1) where the interaction energy is diagonal is to conclude that if the non-interacting system ($V = 0$) has zero total momentum in the ground state then the ground state of the interacting system must have zero momentum since conserved quantum numbers (such as total momentum) of the ground state cannot change because ground states cannot cross.

However, there has recently been much attention given to systems which do not obey the condition of Perron-Frobenius. Three notable examples are as follows.

1) The $N$ state integrable Chiral Potts chain

$$\mathcal{H} = -\sum_{j=1}^{\mathcal{N}}\sum_{n=1}^{N-1}[\bar{\alpha}_n(X_j)^n + \alpha_n(Z_j Z_{j+1}^\dagger)^n] \tag{2}$$

where $X_j$ and $Z_j$ are $N^{\mathcal{N}} \times N^{\mathcal{N}}$ matrices defined in a direct product notation by

$$X_j = I_N \otimes \ldots \otimes X^{jth} \otimes \ldots \otimes I_N \tag{3a}$$

$$Z_j = I_N \otimes \ldots \otimes Z^{jth} \otimes \ldots \otimes I_N \, , \tag{3b}$$

where $I_N$ is the $N \times N$ identity matrix, $Z$ and $X$ are $N \times N$ matrices whose elements are

$$Z_{jk} = \delta_{j,k}\exp[2\pi i(j-1)/N], \quad X_{j,k} = \delta_{j,k+1} \pmod N \, , \tag{4}$$

and

$$\begin{aligned} \alpha_n &= \exp[i(2n-N)\phi/N]/\sin(\pi n/N)\, , \\ \bar{\alpha}_n &= \lambda\exp[i(2n-N)\bar{\phi}/N]/\sin(\pi n/N)\, , \\ \text{with} \quad \cos\phi &= \lambda\cos\bar{\phi}\, . \end{aligned} \tag{5}$$

This hamiltonian is Hermitian for $\lambda, \phi$ and $\bar{\phi}$ real, but only possesses the additional property of having non-positive off diagonal elements at $\phi = \bar{\phi} = 0$ and $\lambda = 1$.

2) The $t$-$J$ model of mobile impurities in the spin $\frac{1}{2}$ antiferromagnet [3,4]. In the basis of +,− for spin up or down and 0 for vacancies this Hamiltonian is

$$\mathcal{H} = -t\sum_{\ell,n.n.}\left\{P_{0+}^{(\ell)} + P_{0-}^{(\ell)}\right\} + J\sum_{\text{spin pairs}} \vec{\sigma}\cdot\vec{\sigma}' \tag{6}$$

where $P_{0+}^{(\ell)}(P_{0-}^{(\ell)})$ permutes 0 on the site $\ell$ with the $+(-1)$ on the nearest neighbor sites and $\vec{\sigma}$ and $\vec{\sigma}'$ are Pauli spin matrices on nearest neighbor sites. If we use the standard basis for the Pauli spin matrices $\vec{\sigma}$, the second term clearly has positive off diagonal elements for $J > 0$ and the first term has negative off diagonal elements for $t > 0$. Thus Perron-Frobenius will not exclude level crossing in the ground state.

3) The bilinear-biquadratic spin 1 model [5,6,7]

$$\mathcal{H} = \sum \{\cos\theta(\vec{S}\cdot\vec{S}') + \sin\theta(\vec{S}\cdot\vec{S}')^2\} \tag{7}$$

where $\vec{S}$ and $\vec{S}'$ are spin one matrices on nearest neighbor sites.

If we choose the basis for $\vec{S}$ such that at the $\ell^{th}$ site $\vec{S}$ is

$$S^z = \begin{pmatrix} 0 & 0 & 0 \\ 0 & 1 & 0 \\ 0 & 0 & -1 \end{pmatrix}, \qquad S^x = 2^{-1/2}\begin{pmatrix} 0 & e^{-\pi i/4} & e^{\pi i/4} \\ e^{\pi i/4} & 0 & 0 \\ e^{-\pi i/4} & 0 & 0 \end{pmatrix},$$

$$S^y = 2^{-1/2}\begin{pmatrix} 0 & e^{\pi i/4} & e^{-\pi i/4} \\ e^{-\pi i/4} & 0 & 0 \\ e^{\pi i/4} & 0 & 0 \end{pmatrix} \tag{8}$$

we see that the off diagonal elements are non-positive for $\pi \le \theta \le 5\pi/4$. If we send $S^x \to -S^x$ and $S^y \to -S^y$ on alternate sites we have non-positivity for $\frac{3\pi}{2} \le \theta \le 2\pi$. If we use

$$S^x = i\begin{pmatrix} 0 & 0 & 0 \\ 0 & 0 & -1 \\ 0 & 1 & 0 \end{pmatrix}, S^y = i\begin{pmatrix} 0 & 0 & 1 \\ 0 & 0 & 0 \\ -1 & 0 & 0 \end{pmatrix}, S^z = i\begin{pmatrix} 0 & -1 & 0 \\ 1 & 0 & 0 \\ 0 & 0 & 0 \end{pmatrix} \tag{9}$$

we find this non-positivity holds for $5\pi/4 \le \theta \le 3\pi/2$.

However for $0 < \theta < \pi$ neither representation makes all off diagonal elements non-positive and again Perron-Frobenius will not apply.

Once Perron-Frobenius cannot be used the possibility of ground state level crossing is not readily ruled out. We will examine the consequences of the absence of Perron-Frobenius for the above 3 models.

In the integrable Chiral Potts chain exact computation is possible because the transfer matrix for the associated 2 dimensional Chiral Potts statistical mechanical model obeys a star-triangle equation with a spectral variable lying on a Riemann surface of ge nus $> 1$ [1]. Moreover, the transfer matrix eigenvalues satisfy functional equations first found for $N = 3$ [2] and recently extended to general values of $N$ [13]. The solution of these functional equations has so far been done in detail only for $\phi = \bar{\phi} = \pi/2$ [2]. We here discuss the consequences of these computations and refer the reader to ref. 1, 2, and 13 for the details of computation.

The hamiltonian (2) is not invariant under parity so the question of spontaneous breaking of parity cannot be studied here. However, the phenomenon of level crossing can be studied in great detail.

As a function of $\lambda$ the system has four phases. There is a duality between $\lambda$ and $\lambda^{-1}$ and the ground state energy per site satisfies $e(1/\lambda) = e(\lambda)/\lambda$. Consistent with this duality the 3 values of $\lambda$ which separate the four phases are $\lambda^I, 1$, and $1/\lambda^I$ where numerically $\lambda^I = .90292\ldots$ .

In the small $\lambda$ phase $0 < \lambda < \lambda^I$ the spectrum of excitations has a gap, the correlations are monotonic, and the momentum of the ground state $P^0$ is zero. By duality this is also true for $\lambda > 1/\lambda^I$.

For $\lambda^I < \lambda < 1$, however, the ground state energy is given by the solution of an integral equation, the excitations are gapless, the momentum of the ground state is not zero, the correlations are oscillatory, and level crossing occur s. More precisely, for a finite chain of $\mathcal{N}$ sites the ground state momentum is [2]

$$e^{-iP^0} = \prod_{k=1}^{m_p} \left( \frac{1+\omega^2 v_k}{1+\omega v_k} \right) \tag{10}$$

where the $m_p$ real numbers $v_k$ satisfy

$$\left( \frac{\omega^2 + v_j}{\omega + v_j} \right)^{\mathcal{N}} = -\omega^{\mathcal{N}} \prod_{k=1}^{m_p} \left( \frac{v_j - \omega^2 v_k}{v_j - \omega v_k} \right) \tag{11}$$

$\lim m_p/\mathcal{N} > 0$ is a function of $\lambda$ determined to minimize the energy, and $\omega = e^{2\pi i/3}$. It is most important to remark that even though the ground state energy per site has a perfectly well defined limit as $\mathcal{N} \to \infty$ that this momentum of the ground state does not have such a limit but that the eigenvalues of the shift operator execute a cycle on the unit circle.

To discuss excited states for $\lambda^I < \lambda < 1$ we must consider the differences

$$\Delta E_{\mathcal{N}} = \text{excited energy } - E^0_{\mathcal{N}} \tag{12}$$

and

$$\Delta P_{\mathcal{N}} = \text{momentum of excited state } - P^0_{\mathcal{N}} \,. \tag{13}$$

Even though the individual momenta in (13) have no limit as $\mathcal{N} \to \infty$ both limits

$$\Delta E = \lim_{\mathcal{N}\to\infty} \Delta E_{\mathcal{N}} \quad \text{and} \quad \Delta P = \lim_{\mathcal{N}\to\infty} \Delta P_{\mathcal{N}} \tag{14}$$

do exist and the dependence of $\Delta E$ on $\Delta P$ is determined by the solution of an integral equation [2]. When $\Delta P \to 0$ we find that $\Delta E$ vanishes linearly in $\Delta P$. This means there is a massless excitation in the theory and allows us to define a speed of sound (or speed of light, if you will). However, because the original hamilitonian is not parity invariant

$$\Delta E(\Delta P) \neq \Delta E(-\Delta P) \,. \tag{15}$$

In other words the speeds of right and left moving sound are different. In the language of organic chemistry the system has optical activity.

It is of particular interest to note how the two speeds of sound behave near $\lambda^I$. Using the definitions

$$V_{R,L} = \lim_{\Delta P \to 0^+} \Delta E(\pm \Delta P)/\Delta P \tag{16}$$

it is found [2] that as $\lambda \to \lambda^I$

$$V_{R,L} = C(\lambda - \lambda^I)^{1/2} + C_{R,L}(\lambda - \lambda^I) + O(\lambda - \lambda^I)^{3/2} \tag{17}$$

Thus the difference $V_R - V_L$ vanishes more rapidly than does $V_R$ or $V_L$ separately. This has important consequences if we attempt to study the system by the methods of continuum field theory.

To discuss the field theory limit of the Chiral Potts model in the phase $\lambda^I < \lambda < 1$ we must recognize as argued by Howes, Kadanoff, and den Nijs [11] that correlation functions such as $\left\langle Z_0 Z_n^\dagger \right\rangle$ will oscillate as $e^{iQn}$. The phase $Q$ vanishes as $\lambda \to \lambda^I$. This oscillation is on the scale of the lattice spacing as long as $Q > 0$. Thus to get a field theory limit we must consider the limit $\lambda \to \lambda^I$ and $n \to \infty$ such that $Qn$ is fixed. In such a continuum limit there will be massless excitations with a rescaled non-zero speed of sound, but, because of (17) there will be no optical activity. We then conclude that the reason optical activity is never encountered in the field theory with parity violation or chiral bosons is because it is a physical effect which vanishes in the continuum limit.

Armed with these insights gained from the study of the one-dimensional Chiral Potts chain we now proceed to examine the $t$-$J$ model and the spin 1 bilinear-biquadratic model. Our philosophy is that the physics of level crossing which is explicitly computed in the solvable one-dimensional model [1,2] is generic in the sense that the two-dimensional Ising model is generic for second order phase transitions. Of course the $t$-$J$ model and the bilinear-biquadratic models have hamiltonians which are invariant under parity. However, the work of [3] argues that in the $t$-$J$ models in two dimensions a spiral order arises which spontaneously breaks parity. This would be forbidden if the "potential energy term" $J$ had non-positive off diagonal elements because for $J = 0$ the ground state obviously has a parity invariant ground state with zero momentum and no spiral order. All these effects come from level crossings of the ground state.

Work on the bilinear-biquadratic spin 1 model in 2 or 3 dimensions has not been as extensive as that on the $t$-$J$ model, but recently Papanicolaou [7] has given strong arguments from a semi-classical analysis that very peculiar phases occur for $\pi/4 < \theta < \pi/2$. At $\theta = \pi/4$ and $\theta = \pi/2$ the model possesses an additional SU(3) symmetry and for $\pi/4 < \theta < \pi/2$ it is argued that the system is in a "semi-ordered" phase. It is easy to see that at $\theta = \pi/2$ the ground state is extremely degenerate. If we now piece together 1) the perturbation theory arguments of Howes, Kadanoff, and den Nijs [12], 2) the spiral order of the related two dimensional $t$-$J$ model, and 3) the fact that Perron-Frobenius does not apply for $0 < \theta < \pi$ we conclude that the most likely interpretation of this semi-ordered phase is that for some region of $\theta$ inside $\pi/4 < \theta < \pi/2$ there is an incommensurate phase with a spiral order. Considering that the phase with $5\pi/4 < \theta < 3\pi/2$ was identified as having a nematic order [7] a spiral order would most naturally be regarded as a cholesteric phase of a liquid crystal.

Once having argued that spiral phases should be regarded as a cholesteric phase the question of the behavior of the system for $T > 0$ may lead to most interesting phenomena. This is because the most dramatic property of cholesteric liquid crystals is the existence of a narrow ($1^0 K$) "blue phase" which separates the cholesteric phase from the disordered high temperature phase [14]. In the "blue phase" there seem experimentally to be many first order phase transitions which seem to have a great potentiality of being related to level crossings.

It should also be remarked that the RVB models of Anderson [15] also violate Perron-Frobenius. In particular both the Heisenberg antiferromagnet on the triangular lattice and the Heisenberg antiferromagnet on the square lattice with next nearest neighbor interactions have the property of not having all non-positive off diagonal elements. The bilinear-biquadratic model at $\tan\theta = 1/3$ was solved in ref. [6] and shown to be a valence bound state. This state has a mass gap and does not violate $P$ (but does have a degenerate ground state on the finite lattice). This situation corresponds to the phase $0 < \lambda < \lambda^I$ in the Chiral Potts chain. From this point of view the difference between the RVB model [14] and the parity violating models of ref. [8] and [9] correspond to the several different phases the bilinear-biquadratic model can attain in the range $0 < \theta < \pi/2$.

We conclude by comparing our discussion here with the recent field theory treatment of anyon superconductivity by Chen, Wilczek, Witten, and Halperin [8]. The property of spontaneous violation of parity is a key property of the anyon model. As such level crossing transitions as a function of $1/n$ should be involved. Howe ver, in ref. [8] the statement is made above eq. (7.30) that even in the parity broken state the expectation value of the momentum operator vanishes. This seems to us to be based on the continuum notion that parity can be violated without having optical activity. In our above discussion the spiral order exists on a microscopic scale even though the excitations are massless. Hence, as mentioned above, any attempt to get a continuum one length scale field theory description will force a limit where the pitch of the spiral vanishes. This is the limit where the optical activity vanishes for the Chiral Potts model. Clearly, it would be desirable to treat the anyon model on a lattice to see these effects. Moreover, then we could define a crystal momentum for a finite anyon lattice just as we do for the Chiral Potts chain and, just as for the Chiral Potts chain, this momentum of the ground state will not possess a thermodynamic limit. It is this elusive ground state momentum which is the most fundamental aspect of the parity violating order of the system.

## Acknowledgements

We are pleased to acknowledge fruitful discussions with Prof. I. Affleck, M. Hatsuda, Prof. M. den Nijs, Dr. B. Shraiman, and Dr. A. Sevrin. Two of us (B.M.M. and J.H.H.P.) are pleased to acknowledge the hospitality of R. J. Baxter, M. N. Barber, and the Center for Mathematical Analysis in Canberra where this work was presented. This research was supported in part by NSF Grant #DMR-8803678.

## References


1. H. Au-Yang, B.M. McCoy, J.H.H. Perk, and M.L. Yan, *Phys. Letts.* **A123** (1987) 219; R.J. Baxter, J.H.H. Perk. and H. Au-Yang, *Phys. Letts.* **A128** (1988) 138.

2. G. Albertini, B.M. McCoy, and J.H.H. Perk, *Phys. Letts.* **A135** (1989) 159; R.J. Baxter, *Phys. Letts.* **A133** (1988) 185; G. Albertini, B.M. McCoy, and J.H.H. Perk in "Advanced Studies in Pure Mathematics" vol. 19, Kinokuniya-Academic 1989; G. Albertini, B.M. McCoy, and J.H.H. Perk, *Phys. Letts. A* **139** (1989) 204.

3. B.I. Shraiman and E.D. Siggia, *Phys. Rev. Lett.* **61** (1988) 467; **62** (1989) 1564.

4. S.A. Trugman, *Phys. Rev.* **B37** (1988) 1597; S. Schmitt-Rink, C.M. Varma, and A.E. Ruckenstein, *Phys. Rev. Lett.* **60** (1988) 2793; C.L. Kane, P.A. Lee, and N. Read, *Phys. Rev.* **B39** (1989) 6880.

5. B. Sutherland, *Phys. Rev.* **B12** (1975) 3795; A.B. Zamolodchikov, and V.A. Fateev, *Yadernaya Fiz.* **32** (1980) 581 [*Sov. J. Nucl. Phys.* **32** (1980) 298]; K. Chang, I. Affleck, G.W. Hayden, and A.G. Soos, *J. Phys. Condens. Matter 1* (1989) 153; I. Affleck, ibid, 3047; M.N. Barber and M.T. Batchelor, preprint CMA-R07-89.

6. I. Affleck, T. Kennedy, E.H. Lieb, and H. Tasaki, *Phys. Rev. Lett.* **59** (1987) 799; *Comm. Math. Phys.* **115** (1988) 477.

7. N. Papanicolaou, *Nucl. Phys.* **305** [FS23] (1988) 367.

8. Y-H. Chen, F. Wilczek, E. Witten, and B.I. Halperin, *Int. J. Mod. Phys. B* **3** (1989) 1001.

9. X.G. Wen and A. Zee, *Phys. Rev. Letts.* **62** (1989) 2873.

10. C. Sturm, *Jour. de Math. Pure et Appl.* (1)1(1836) 106; J. Liouville, ibid. (1) 1 (1836) 253; (1)2(1837)16; see also "Methods of Theoretical Physics", P.M. Morse and H. Feshbach, McGraw-Hill, New York, 1953, vol. 1, pp. 719-726.

11. O. Perron, *Math Ann.* **64**(1907) 248; G. Frobenius, *S.B. Deutsch. Akad. Wiss. Berlin, Math-Nat. Kl.* (1908) 471, 511; ibid. (1912) 456; see also F.R. Gantmacher, "Matrix Theory", Chelsea Publishing Co. 1959, vol. 2.

12. S. Howes, L.P. Kadanoff, and M. den Nijs, *Nucl. Phys.* **B215** [FS7] (1983) 169.

13. V.V. Bazhanov and Yu.G. Stroganoff, *J. Stat. Phys.* ( to be published).

14. D.C. Wright and N.D. Mermin, *Rev. Mod. Phys.* **61** (1989) 385.

15. P.W. Anderson, *Science* **235** (1987) 1196.

International Journal of Modern Physics B, Vol. 4, No. 5 (1990) 1003 – 1037
© World Scientific Publishing Company

# DIFFERENTIAL EQUATIONS FOR QUANTUM CORRELATION FUNCTIONS

A.R. Its, A.G. Izergin, V.E. Korepin, N.A. Slavnov

*Leningrad Branch of the Steklov Mathematical Institute, Academy of Sciences of the U.S.S.R., Fontanka 27, Lomi, Leningrad U.S.S.R. 191011*

The quantum nonlinear Schrödinger equation (one dimensional Bose gas) is considered. Classification of representations of Yangians with highest weight vector permits us to represent correlation function as a determinant of a Fredholm integral operator. This integral operator can be treated as the Gelfand-Levitan operator for some new differential equation. These differential equations are written down in the paper. They generalize the fifth Painlève transcendent, which describe equal time, zero temperature correlation function of an impenetrable Bose gas. These differential equations drive the quantum correlation functions of the Bose gas. The Riemann problem, associated with these differential equations permits us to calculate asymptotics of quantum correlation functions. Quantum correlation function (Fredholm determinant) plays the role of $\tau$ functions of these new differential equations. For the impenetrable Bose gas space and time dependent correlation function is equal to $\tau$ function of the nonlinear Schrödinger equation itself. For a penetrable Bose gas (finite coupling constant $c$) the correlator is $\tau$-function of an integro-differentiation equation.

## 0. Introduction

The nonlinear Schrödinger (NS) equation is very well known :

$$\begin{aligned} i\partial_t\psi &= -\partial_x^2\psi + 2c\psi^+\psi\psi \\ i\partial_t\psi^+ &= \partial_x^2\psi^+ - 2c\psi^+\psi^+\psi \end{aligned} \tag{0.1}$$

The Hamiltonian of the model is equal to

$$H = \int_{-\infty}^{\infty} dx(\partial_x\psi^+\partial_x\psi + c\psi^+\psi^+\psi - h\psi^+\psi) \tag{0.2}$$

where $h$ denotes the chemical potential. The operator for the number of particles $Q$ and momentum $P$ are equal to

$$\begin{aligned} Q &= \int_{-\infty}^{\infty} dx\psi^+(x)\psi(x) \\ P &= -\frac{i}{2}\int_{-\infty}^{\infty} dx(\psi^+(x)\partial_x\psi - (\partial_x\psi^+(x))\psi(x)) , \end{aligned} \tag{0.3}$$

In the quantum case the commutation relations are

$$[\psi(x,t),\psi^+(y,t)]=\delta(x-y),[\psi(x,t),\psi(y,t)]=0\;. \tag{0.4}$$

The bare vacuum is defined by

$$\psi(x)|0\rangle=0\;;\quad \langle 0|\psi^+(x)=0 \tag{0.5}$$

Another name of the model is the one-dimensional Bose gas. The model is exactly solvable [1]. We shall consider a ground state $|\Omega\rangle$ of the model with finite density $D$. In this case momenta of the particles fill a one-dimensional Fermi sphere $[-q,q]$ ($q$ is the Fermi momenta). Thermodynamics of the model at temperature $T>0$ was constructed in [2]. We shall be interested in thermodynamics only for an impenetrable Bose gas ($c=\infty$). In this case the spectrum of the Hamiltonian is equivalent to a free fermion one. In the state of thermodynamic equilibrium the distribution of particles in the momentum space are given by the Fermi weight

$$\rho(\lambda)=\frac{1}{2\pi}\left(1+e^{(\lambda^2-h)/T}\right)^{-1} \tag{0.6}$$

In this case the density $D$ is equal to

$$D=\frac{1}{2\pi}\int_{-\infty}^{\infty}\frac{d\lambda}{1+e^{(\lambda^2-h)/T}}\;;\qquad c=\infty \tag{0.7}$$

We shall consider different correlation functions in the model. First we shall consider the impenetrable Bose gas ($c=\infty$). In this case we shall calculate the field correlator

$$\langle\psi(x_2,t_2)\psi^+(x_1,t_1)\rangle\equiv\frac{\langle\Omega|\psi(x_2,t_2)\psi^+(x_1,t_1)|\Omega\rangle}{\langle\Omega|\Omega\rangle} \tag{0.8}$$

This is the zero temperature case. Recall that $|\Omega\rangle$ is the ground state with finite density $D$ and

$$\psi(x,t)=e^{iHt}\psi(x,0)e^{-iHt}\;. \tag{0.9}$$

We shall calculate also finite-temperature correlation functions

$$\langle\psi(x_2,t_2)\psi^+(x_1,t_1)\rangle_T=\frac{tr(e^{-H/T}\psi(x_2,t_2)\psi^+(x_1,t_1))}{tr(e^{-H/T})}\;. \tag{0.10}$$

We shall consider also many point correlators. It should be noted that the zero-temperature equal time correlator was calculated in [3-5]. It was proved that it is associated with the fifth Painlève transcendent. There exist models similar to the impenetrable Bose gas (Ising model, XY model) where different correlators were calculated [6-9]. For a penetrable Bose gas ($0<c<\infty$) the problem of correlators is especially interesting, because the model is not equivalent to free fermions, nevertheless we succeed in calculating correlators in this case (see Sections 4, 8).

Let us define the simplest correlation function which we calculate here for finite $c$. In the ground state of the model particles move in the Brownian way. There exist some probability

$$P(x) \tag{0.11}$$

that on the interval $[-x, x]$ there will be no particles (empty place). This probability is the simplest correlation function and we shall calculate it here.

Calculation of correlation functions consists of two steps. First we represent correlation function as a Fredholm determinant (minor) of some integral operator. (These integral operators are of very special form, they form infinite dimensional group, see Part II.) Then we treat this integral operator as a Gelfand-Levitan [10] operator of some new differential equation and derive this differential equation. It follows that correlation functions (Fredholm determinant) play the role of $\tau$-functions [11] for this new differential equation. A Riemann problem [11] can be associated with this differential equation, which enables us to calculate the asymptotics of correlation functions.

The plan of the paper is as follows. In Part I we represent correlation functions as Fredholm determinants. In Sect. 1 we present the Lenard formula. In Sect. 2 we derive the Fredholm determinant for time dependent correlation functions. In Sect. 3 we present generalization to the case of many fields. We use the Quantum Inverse Scattering Method (QISM) [12-16] in Sect. 4 where we derive a Fredholm determinant for finite coupling constant. The algebraic part of the derivation is associated with enumeration of representations of Yangians with highest weight vector [17-21].

In Part II, starting from Fredholm determinants we derive differential equations which drive correlation functions. Correlation functions (Fredholm determinants of integral operators) are equal to $\tau$-functions of these differential equations, and integral operators play the role of Gelfand-Levitan operators. This approach is close to [22-25][42,43]. First we consider the impenetrable Bose gas $c = \infty$. In Sect. 5 we construct a differential equation for the equal time temperature correlator. This is the case most close to the fifth Painlève transcendent [5,26]. We get a kind of Painlève equation with partial derivatives. This is a new integrable equation. The place of this equation in the general scheme is discussed in [27]. All ideas are explained in this case. All other cases are treated similarly. In Sect. 6 we consider the time and space dependent correlator (0.8) (zero temperature case). In this case the fifth Painlève equation is deformed into a nonlinear Schrödinger equation (not selfajoint). In Sect. 7 time-temperature correlators (0.10) are considered. We derive a three-dimensional completely integrable system of differential equations (based on NS equation) for this correlation function. Maybe it is interesting from the point of view of integrable differential equations. The most interesting results are in Sect. 8, where we derive an integro-differential equation for correlation function at finite coupling constant $c$.

In Part III we consider asymptotics of correlation functions. All differential equations derived in Part II are associated with some Riemann problem. This permits us to calculate the long distance asymptotics of correlation functions. In Sect. 9 we do this for the example of the equal time finite temperature correlator. In Sect. 10 we present some formulas of current $\psi^+_{(x)}\psi_{(x)}$ correlators, obtained together with N.M. Bogoliubov. Many point space and time dependent finite temperature correlators of currents of the impenetrable Bose gas can be represented as a determinant of matrix.

# I. Correlation Functions as Fredholm Determinants

In this part we shall represent different correlation functions as Fredholm determinants (minor) of some integral operators. We shall begin with the impenetrable Bose gas $c = \infty$. In the Sect. 4 of this part we shall derive a Fredholm determinant for finite $c(0 < c < \infty)$. The derivation is similar to the derivation of the determinant formula for norms of Bethe wave function [16].

## 1. Lenard Formula

First let us consider equal time finite temperature field correlators

$$\langle\psi^+(z)\psi(-z)\rangle_T = \frac{tr(e^{-H/T}\psi^+(z)\psi(-z))}{tr(e^{-H/T})} \tag{1.1}$$

by $z$ here we denote distance. Let us make a rescaling and introduce new variables

$$\beta = \frac{h}{T}\ ; \quad x = z\sqrt{T}\ . \tag{1.2}$$

The reason is that correlator depends only on two variables $x$ and $\beta$

$$\langle\psi^+(z)\psi(-z)\rangle_T = \sqrt{T}g(x,\beta)\ . \tag{1.3}$$

To define $g$ let us introduce an integral operator $K$, it acts on the real axis. The kernel $K(\lambda,\mu)$ is equal to :

$$K(\lambda,\mu) = \sqrt{\vartheta(\lambda)}\frac{\sin x(\lambda-\mu)}{(\lambda-\mu)}\sqrt{\vartheta(\mu)}\ . \tag{1.4}$$

Here $\vartheta(\lambda)$ is Fermi weight

$$\vartheta(\lambda) = (1+\exp\{\lambda^2-\beta\})^{-1}\ . \tag{1.5}$$

It is convenient to describe properties of the operator $K$ in terms of functions

$$f_i(\lambda,x,\beta) \qquad (i=\pm)$$

(the dependence on $x$ and $\beta$ will be suppressed). The functions $f_\pm(\lambda)$ are defined as solutions of linear integral equations :

$$f_i(\lambda) - \gamma\int_{-\infty}^{\infty}K(\lambda,\mu)f_i(\mu)d\mu = e_i(\lambda)\ ; \quad i=\pm\ . \tag{1.6}$$

Here $\gamma$ is a real parameter and the functions $e_i(\lambda)$ are given as

$$e_{\pm}(\lambda) = \sqrt{\vartheta(\lambda)} \exp\{\pm i\lambda x\} \,. \tag{1.7}$$

Of primary importance will be the matrix of "potentials" $B_{ik}(x,\beta)$ :

$$B_{ik}(x,\beta) = \gamma \int_{-\infty}^{\infty} e_i(\lambda) f_k(\lambda) d\lambda \,. \tag{1.8}$$

It is a real symmetric matrix with two independent matrix elements $B_{++}$ and $B_{--}$

$$\overline{B_{++}} = B_{++} = B_{--} \,; \quad \overline{B_{+-}} = B_{+-} = B_{-+} \,. \tag{1.9}$$

Now all the notation necessary to express correlators in terms of the operator $K$ has been introduced. The field correlator $\langle \psi^+(z)\psi(-z)\rangle_T = \sqrt{T} g(x,\beta)$ can be represented as

$$g(x,\beta) = \frac{1}{4} B_{++}(x,\beta) \det(1-\gamma K)/_{\gamma=2/\pi} \,. \tag{1.10}$$

The last factor here is a Fredholm determinant. In the paper [27] it is proved that formula (1.10) can be transformed to Lenard formula by means of Fourier transformation. Correlators of many fields can be represented in the similar way [27]. It is interesting to mention that $\det(1-\gamma K)$ has meaning as correlator at any complex value of parameter $\gamma$. Let us consider

$$\langle \exp(\alpha Q(z))\rangle_T = \frac{tr(e^{-H/T} e^{\alpha Q(z)})}{tr(e^{-H/T}} \,. \tag{1.11}$$

Here $\alpha$ is an arbitrary complex parameter and $Q(z)$ is the operator of the number of particles on the interval $[-z,z]$

$$Q(z) = \int_{-z}^{z} \psi^+(y)\psi(y)dy \,. \tag{1.12}$$

In [28], [29] it is shown that

$$\begin{aligned} \langle e^{\alpha Q(z)}\rangle_T &= \det(1-\gamma K) \\ \gamma &= \frac{1-e^{\alpha}}{\pi} \,. \end{aligned} \tag{1.13}$$

Especially interesting is the value $e^{\alpha} = 0$, $\gamma = \frac{1}{\pi}$ it describes the probability (0.11) that on the interval $[-z,z]$ there will be no particles of the Bose gas. This case $\gamma = \frac{1}{\pi}$ also describes the level spacing probability distribution function which appears in the theory of random matrices [30, 5]. There is an interesting connection between formulae (1.10) and (1.13). From [31] we know the fermionization formula :

$$\psi_B(x) = \psi_\varphi(x) \exp\left\{ i\pi \int_{-\infty}^{x} \psi_\varphi^+(y)\psi_\varphi(y)dy \right\} \,. \tag{1.14}$$

In the l.h.s. we have the canonical Bose field (of impenetrable Bose gas), $\psi_\varphi$ is canonical Fermi field. So (1.14) shows that for the field correlator (1.1) $\alpha = i\pi$.

Formula (1.13) $\gamma = \frac{1-e^{\alpha}}{\pi}$ leads to the correct value of $\gamma = \frac{2}{\pi}$ for field correlator (1.10). One should mention that for zero temperature case $T = 0$ the operator in (1.4) turns in

$$K_0(\lambda, \mu) = \frac{\sin x(\lambda - \mu)}{\lambda - \mu} .$$

It acts on the interval $[-q, q]$. The probability of empty place (0.11) is equal to

$$P(\lambda) = \det\left(1 - \frac{1}{\pi} K_0\right) . \tag{1.15}$$

In Sect. 4 we shall generalize this representation to finite coupling constant. In Sect. 5 we shall derive differential equation which drive equal time finite temperature correlation function, based on representation (1.3) and (1.10).

## 2. Time Dependent Correlator

First let us consider the zero temperature $T = 0$ case. In this case the ground state $|\Omega\rangle$ is a Fermi sphere $[-q, q]$, with Fermi momentum $q$

$$\langle \psi^+_{(2)} \psi_{(1)} \rangle = \frac{\langle \Omega | \psi^+_{(2)} \psi_{(1)} | \Omega \rangle}{\langle \Omega | \Omega \rangle} .$$

In the paper [32] it was proved that :

$$\langle \psi(x_2, t_2) \psi^+(x_1, t_1) \rangle = e^{iht_{21}} \left( \frac{1}{2\pi} G(t_{12}, x_{12}) + \frac{\partial}{\partial \alpha} \right) \det(1 + V_0)/_{\alpha=0} \cdot \tag{2.1}$$

Here

$$t_{12} = t_1 - t_2 \ , \quad x_{12} = x_1 - x_2 \ ,$$

$$G(t_{12}, x_{12}) = \int_{-\infty}^{\infty} d\mu e^{it_{12}\mu^2 - ix_{12}\mu} . \tag{2.2}$$

The last factor in (2.1) is a Fredholm determinant with the kernel

$$V_0(\lambda, \mu) = \left[ \frac{E(\lambda) - E(\mu)}{(\lambda - \mu)\pi^2} - \frac{\alpha}{2\pi^3} E(\lambda).E(\mu) \right] \times e^{\frac{i}{2} t_{21}(\lambda^2 + \mu^2)} \times e^{-\frac{i}{2} x_{21}(\lambda + \mu)} . \tag{2.3}$$

This integral operator acts on the interval $[-q, q](\lambda, \mu \in [-q, q])$. The function $E(\lambda)$ is defined as follows :

$$E(\lambda) = \fint_{-\infty}^{\infty} d\mu \frac{\exp\{it_{12}\mu^2 - ix_{12}\mu\}}{\mu - \lambda} . \tag{2.4}$$

Now let us discuss temperature - time correlators

$$\langle \psi(x_2, t_2) \psi^+(x_1, t_1) \rangle_T = \frac{tr(e^{-H/T} \psi(x_2, t_2) \psi^+(x_1, t_1))}{tr(e^{-H/T})} . \tag{2.5}$$

They are given by a similar formula

$$\langle\psi(x_2,t_2)\psi^+(x_1,t_1)\rangle_T = e^{iht_{21}}\left(\frac{1}{2\pi}G(t_{12},x_{12}) + \frac{\partial}{\partial\alpha}\right)\det(1+V_T)/_{\alpha=0}\,. \tag{2.6}$$

Now the integral operator $V_T$ acts on the whole real axis and the kernel $V_T(\lambda,\mu)$ can be expressed in terms of $V_0$

$$V_T(\lambda,\mu) = \sqrt{\vartheta(\lambda)}V_0(\lambda,\mu)\sqrt{\vartheta(\mu)}\,. \tag{2.7}$$

Here $\vartheta(\lambda) = (1+\exp\{\lambda^2-\beta\})^{-1}$ . Another interesting object is the Euclidean time correlator. Let us denote $\tau = it$

$$\psi(x,\tau) = e^{H\tau}\psi(x,0)e^{-H\tau} \tag{2.8}$$

$$\langle\psi(x_2,\tau_2)\psi^+(x_1,\tau_1)\rangle = \frac{tr(e^{-H/T}\psi(x_2,\tau_2)\psi^+(x_1,\tau_1))}{tr(e^{-H/T})}\,. \tag{2.9}$$

Let us write down the Fredholm determinant formula for it

$$\langle\psi(x_2,\tau_2)\psi^+(x_1,\tau_1)\rangle_T = e^{h\tau_{21}}\left(\frac{1}{2\pi}\tilde{G}(\tau_{12},x_{12}) + \frac{\partial}{\partial\alpha}\right)\det(1+\tilde{V}_T)/_{\alpha=0}\,. \tag{2.10}$$

Here $\tilde{V}_T$ is an integral operator on the whole real axis, with the kernel

$$\tilde{V}_T(\lambda,\mu) = \frac{e^{(\tau_{21}(\lambda^2+\mu^2)-ix_{21}(\lambda-\mu))/2}}{\sqrt{1+e^{(\lambda^2-h)/T}}\sqrt{1+e^{(\mu^2-h)/T}}} \times \left\{\frac{\tilde{E}(\lambda)-\tilde{E}(\mu)}{\pi^2(\lambda-\mu)} - \frac{\alpha}{2\pi^3}\tilde{E}(\lambda).\tilde{E}(\mu)\right\}\,. \tag{2.11}$$

Here $x_{12} = x_1 - x_2$ ; $0 \le \tau_{21} = \tau_2 - \tau_1 \le \frac{1}{T}$ ; $\tau_{12} = -\tau_{21}$

$$\begin{aligned}\tilde{G}(\tau_{12},x_{12}) &= \int_{-\infty}^{\infty} d\mu e^{-\tau_{21}\mu^2 - ix_{12}\mu}\\ \tilde{E}(\mu) &= \fint_{-\infty}^{\infty}\frac{d\lambda}{\lambda-\mu}e^{-\tau_{21}\lambda^2 - ix_{12}\lambda}\,.\end{aligned} \tag{2.12}$$

In Sections 6, 7 we shall derive differential equations, based on these Fredolm determinants.

## 3. Manypoint Field Correlators

In the paper [33] many field correlator is represented as the Fredholm minor :

$$\begin{aligned}&\langle\prod_{k=1}^{N}\psi(x_{2k},t_{2k})\psi^+(x_{2k-1},t_{2k-1})\rangle = e^{ih\sum_{k=1}^{N}(t_{2k}-t_{2k-1})}\\ &\cdot\prod_{k=1}^{N}\left(\frac{1}{2\pi}G(t_{2k-1}-t_{2k},x_{2k-1}-x_{2k}) + \frac{\partial}{\partial\alpha_k}\right)\det\mathcal{K}/_{\alpha_j=0}\,.\end{aligned} \tag{3.1}$$

Here $\mathcal{K}$ is a Fredholm integral operator acting on the interval $[-q,q]$. The kernel $\mathcal{K}(\lambda_0,\lambda_N)$ of this operator is equal to

$$\begin{aligned}\mathcal{K}(\lambda_0,\lambda_N) &= \int_{-\infty}^{\infty} d\lambda_1 \dots d\lambda_{N-1} \prod_{m=1}^{N} \Big[\delta(\lambda_m-\lambda_{m-1}) + K_m(\lambda_m,\lambda_{m-1})\Big] \\ &\times \exp\left(i\{t_{2m}\lambda_m^2 - t_{2m-1}\lambda_{m-1}^2 + x_{2m-1}\lambda_{m-1} - x_{2m}\lambda_m\}\right). \end{aligned} \quad (3.2)$$

Here

$$K_m(\lambda,\mu) = \frac{E_m(\lambda)-E_m(\mu)}{\pi^2(\lambda-\mu)} - \frac{\alpha_m}{2\pi^2}E_m(\lambda)\cdot E_m(\mu) \quad (3.3)$$

and

$$E_m(\lambda) = \fint_{-\infty}^{\infty} \frac{d\mu}{\mu-\lambda} \exp\left\{i\mu^2(t_{2m-1}-t_{2m}) - i\mu(x_{2m-1}-x_{2m})\right\} . \quad (3.4)$$

These formulae represent the correlation function at zero temperature. Integral operator $\mathcal{K}$ can be represented in the form

$$\mathcal{K}(\lambda,\mu) = I + V(\lambda,\mu) . \quad (3.5)$$

The kernel of $V$ has canonical form

$$V(\lambda,\mu) = \left(\frac{1}{\lambda-\mu}\right)\sum_{j=1}^{2N} e_j(\lambda)\tilde{e}_j(\mu) . \quad (3.6)$$

This will permit us to derive differential equations which drive correlation function (3.1). To represent the finite temperature correlation function as a Fredholm minor one should replace in (3.1) the operator $\mathcal{K}$ by an operator acting on the whole real axis, its kernel being equal to :

$$\mathcal{K}_T(\lambda,\mu) = I + V_T(\lambda,\mu) \quad (3.7)$$

and

$$V_T(\lambda,\mu) = \sqrt{\vartheta(\lambda)}V(\lambda,\mu)\sqrt{\vartheta(\mu)}.$$

The kernel $V(\lambda,\mu)$ can be taken from (3.5).

## 4. Finite Coupling Constant Correlator

We shall discuss the penetrable Bose gas (finite coupling constant $0 < c < \infty$) in terms of the Quantum Inverse Scattering Method. In terms of this method, the model can be associated with the monodromy matrix

$$T(\lambda) = \begin{pmatrix} A(\lambda) & B(\lambda) \\ C(\lambda) & D(\lambda) \end{pmatrix} \quad (4.1)$$

It is matrix $2\times 2$ with matrix elements being quantum operators (functions of $\psi,\psi^+$). Their commutation relations are given by the $R$-matrix

$$R(\lambda,\mu)T(\lambda)\otimes T(\mu) = T(\mu)\otimes T(\lambda)R(\lambda,\mu) . \quad (4.2)$$

The $R$-matrix is a $4 \times 4$ matrix

$$R(\lambda,\mu) = \begin{pmatrix} f(\mu,\lambda) & & & \\ & g(\mu,\lambda) & 1 & \\ & 1 & g(\mu,\lambda) & \\ & & & f(\mu,\lambda) \end{pmatrix} . \tag{4.3}$$

Here

$$g(\mu,\lambda) = \frac{ic}{\mu-\lambda} \; ; \quad f(\mu,\lambda) = \frac{\mu-\lambda+ic}{\mu-\lambda} \tag{4.4}$$

The bare vacuum $|0\rangle$ plays the role of the highest weight vector :

$$C(\lambda)|0\rangle = 0 \; ; \quad A(\lambda)|0\rangle = a(\lambda)|0\rangle \; ; \quad D(\lambda)|0\rangle = d(\lambda)|0\rangle \; . \tag{4.5}$$

The complex valued functions $a(\lambda)$ and $d(\lambda)$ for NS model are equal to $a(\lambda) = \exp\{\frac{-i\lambda L}{2}\}$, $d(\lambda) = \exp\{\frac{i\lambda L}{2}\}$ ($L$ is the length of periodical box). For the finite coupling constant case $0 < c < \infty$, the ground state of the Bose gas with finite density $D$ (for zero temperature) is still a one-dimensional Fermi sphere $[-q, q]$ ($q$ is the Fermi momentum). Detailed explanations are given for example in [34] (see also references there)). For calculation of correlation functions it is necessary to enumerate all $T(\lambda)$ (4.1) with commutation relations given by (4.2), (4.3) and with highest weight vector (4.5). This is connected with the problem of enumeration of representation of Yangians (associated with $s\ell_2$) with highest weight vector [18]. Solution of this problem can be taken from [21]: there are no restrictions on the functions $a(\lambda)$ and $d(\lambda)$ - they are arbitrary functions. To solve the recursion relations for correlation functions in terms of determinants one should use variational derivatives $\frac{\delta}{\delta a(\lambda)}$, $\frac{\delta}{\delta d(\lambda)}$ with respect to arbitrary functions $a(\lambda)$ and $d(\lambda)$. They act from one irreducible representation of $Y(s\ell_2)$ with highest weight vector into another. This leads to the appearance of dual quanum fields

$$\hat{\varphi}(\lambda) = \hat{a}(\lambda) + \int_{-q}^{q} d\nu \left( \ell n \frac{c^2}{c^2 + (\lambda-\nu)^2} \right) \hat{a}^+(\nu) \tag{4.6}$$

acting in auxiliary Fock space. Here $\hat{a}$ and $\hat{a}^+$ are canonical Bose fields

$$[a(\lambda), a^+(\mu)] = \delta(\lambda-\mu) \; ; \quad [a(\lambda), a(\mu)] = [a^+(\lambda), a^+(\mu)] = 0 \tag{4.7}$$

in auxiliary Fock space, and $|0\rangle$ in Fock vacuum of the new space :

$$\hat{a}(\lambda)|0\rangle = 0 \; ; \quad \langle 0|\hat{a}^+(\lambda) = 0 \; . \tag{4.8}$$

It is remarkable that the dual fields (4.6) commute :

$$[\hat{\varphi}(\lambda), \hat{\varphi}(\mu)] = 0 \; , \tag{4.9}$$

so they can be treated as classical functions. This construction permits us to represent correlation functions as Fredholm determinants [28],[29]. Let us recall that the simplest correlation function is the probability $P(x)$ (0.11) that on the interval

$[-x, x]$ there will be no particles. Let us write it down as a Fredholm determinant using the dual quantum fields $\varphi(\lambda)$ (4.6)

$$P(x) = \frac{\langle 0|\det(I+V)|0\rangle}{\det(I - \frac{1}{2\pi}K)} . \tag{4.10}$$

Both integral operators $V$ and $K$ act on the interval $[-q, q]$. The kernel of $K(\lambda, \mu)$ is equal to

$$K(\lambda, \mu) = \frac{2c}{c^2 + (\lambda - \mu)^2} . \tag{4.11}$$

The kernel of $V(\lambda_1, \lambda_2)$ is expressed in terms of the dual fields:

$$V(\lambda_1, \lambda_2) = \frac{-c}{2\pi}\left\{\frac{e^{ix\lambda_1 + \frac{1}{2}\varphi(\lambda_1)} \cdot e^{-ix\lambda_2 - \frac{1}{2}\varphi(\lambda_2)}}{(\lambda_1 - \lambda_2)(\lambda_1 - \lambda_2 + ic)} + \frac{e^{ix\lambda_2 + \frac{1}{2}\varphi(\lambda_2)} \cdot e^{-ix\lambda_1 - \frac{1}{2}\varphi(\lambda_1)}}{(\lambda_2 - \lambda_1)(\lambda_2 - \lambda_1 + ic)}\right\} \tag{4.12}$$

This is the answer. Remember that in the limit $c \to \infty$ ($q$-fixed)

$$\begin{gathered} \hat{\varphi}(\lambda) \to 0\ ; \quad K \to 0\ ; \quad c \to \infty \\ V(\lambda_1, \lambda_2) \to -\frac{1}{\pi}\frac{\sin x(\lambda_1 - \lambda_2)}{\lambda_1 - \lambda_2} . \end{gathered} \tag{4.13}$$

This is in accordance with (1.15). Let us rewrite $V(\lambda_1, \lambda_2)$ in the form

$$V(\lambda_1, \lambda_2) = i\int_0^\infty ds e^{-sc}\left\{\frac{e_+(\lambda_1)e_-(\lambda_2) - e_+(\lambda_2)e_-(\lambda_1)}{\lambda_1 - \lambda_2}\right\} \tag{4.14}$$

$$e_\pm(\lambda) = \sqrt{\frac{c}{2\pi}}\exp\left\{\pm(ix\lambda + is\lambda + \frac{1}{2}\varphi(\lambda))\right\} . \tag{4.15}$$

Now the integral operator $V$ for finite coupling constant is represented in a form similar to $c = \infty$ but we have an additional variable $s$.

# II. Differential Equations for Correlation Functions

Now we shall use Fredholm determinants (of integral operators obtained in Part I) to derive differential equations, which drive correlation functions,. One should note that all Fredholm integral operators which occur have very special kernels $V(\lambda_1, \lambda_2)$. The product $(\lambda_1 - \lambda_2)V(\lambda_1, \lambda_2)$ is equal to the sum of one-dimensional projectors

$$(\lambda_1 - \lambda_2)V(\lambda_1, \lambda_2) = \sum_{j=1}^{N} e_j(\lambda_1)E_j(\lambda_2) \tag{II.0}$$

(for finite coupling constant the sum is replaced by integration). Functions $e_j(\lambda), E_j(\lambda)$ should also obey the constraint $\sum_{j=1}^N e_j(\lambda)E_j(\lambda) = 0$.Such an integral operator one can call "completely integrable". The product of two such operators has the same form. Let

$$\begin{aligned} V^{(1)}(\lambda_1,\lambda_2) &= \sum_{j=1}^{N} \frac{e_j^{(1)}(\lambda_1)E_j^{(1)}(\lambda_2)}{\lambda_1-\lambda_2} \\ V^{(2)}(\lambda_1,\lambda_2) &= \sum_{k=1}^{M} \frac{e_k^{(2)}(\lambda_1)\cdot E_k^{(2)}(\lambda_2)}{\lambda_1-\lambda_2} \,. \end{aligned} \qquad \text{(II.1)}$$

The product of two such operators is equal to

$$\begin{aligned} &\int_{-q}^{q} V^{(1)}(\lambda_1,\nu)V^{(2)}(\nu,\lambda_2)d\nu = V^{(3)}(\lambda_1,\lambda_2) \\ &= \frac{1}{\lambda_1-\lambda_2}\left\{\sum_{j=1}^{N} e_j^{(1)}(\lambda_1)E_j^{(3)}(\lambda_2) + \sum_{K=1}^{M} e_K^{(3)}(\lambda_1)E_K^{(2)}(\lambda_2)\right\} \,. \end{aligned} \qquad \text{(II.2)}$$

Here

$$\begin{aligned} E_j^{(3)}(\lambda) &= \sum_{K=1}^{M} E_K^{(2)}(\lambda)\int_{-q}^{q}\frac{d\nu}{\nu-\lambda}E_j^{(1)}(\nu)e_K^{(2)}(\nu) \\ e_j^{(3)}(\lambda) &= \sum_{j=1}^{N} e_j^{(1)}(\lambda)\int_{-q}^{q}\frac{d\nu}{\lambda-\nu}E_j^{(1)}(\nu)e_K^{(2)}(\nu) \,. \end{aligned} \qquad \text{(II.3)}$$

This is a very important property, which shows that the resolvent of this operator can be constructed universally, as follows. Let us take kernel operator (II.0). We introduce the resolvent $R$ in the following way

$$(I+V)(I-R)=I\,;\quad (I+V)R=V\,. \qquad \text{(II.4)}$$

Let us write down the integral equation for the kernel of the resolvent $R(\lambda_1,\lambda_2)$:

$$R(\lambda_1,\lambda_2)+\int_{-q}^{q}V(\lambda_1,\nu)R(\nu,\lambda_2)d\nu = V(\lambda_1,\lambda_2)\,. \qquad \text{(II.5)}$$

To write down $R(\lambda_1,\lambda_2)$ in explicit form let us introduce functions $f_j^L(\lambda)$ and $f_j^R(\lambda)$ by the following integral equations :

$$f_j^R(\lambda)+\int_{-q}^{q}V(\lambda,\mu)f_j^R(\mu)d\mu = e_j(\lambda) \qquad \text{(II.6)}$$

$$f_j^L(\lambda)+\int_{-q}^{q}V(\mu,\lambda)f_j^L(\mu)d\mu = E_j(\lambda)\,. \qquad \text{(II.7)}$$

Then one can prove the theorem that

$$R(\lambda_1,\lambda_2)=\frac{\sum_{j=1}^{N} f_j^L(\lambda_1)f_j^R(\lambda_2)}{\lambda_1-\lambda_2}\,. \qquad \text{(II.8)}$$

Formulae (II.2)-(II.8) show that operators $(1+V)$ (here $V$ is given by (II.0) or (II.1) form infinite dimensional group. This permits us to derive differential equations (from integral one), its Lax representation and the associated Riemann problem. We shall do this in more detail for equal time finite-temperature correlators in Sect. 5. All other cases are treated in a similar way. We shall construct differential equations for which these "completely integrable" integral operators play the role of Gelfand-Levitan operators. One should mention that direct quantization of the Gelfand-Levitan equation was proposed in [35]. It will be interesting to understand the connections between these two approaches.

## 5. Finite-Temperature Equal Time Correlator

Let us come back to Sect. 1. The finite-temperature equal time correlators of an impenetrable Bose gas were described as Fredholm determinants $\det(I - \gamma K)$ of an integral operator $K$ (on the real axis)

$$K(\lambda, \mu) = \sqrt{\vartheta(\lambda)} \frac{\sin x(\lambda - \mu)}{\lambda - \mu} \sqrt{\vartheta(\mu)} = \frac{e_+(\lambda)e_-(\mu) - e_-(\lambda)e_+(\mu)}{2i(\lambda - \mu)} = K(\mu, \lambda) . \quad (5.1)$$

Here

$$e_\pm(\lambda) = \sqrt{\vartheta(\lambda)} e^{\pm i\lambda x} \quad (5.2)$$

and

$$\vartheta(\lambda) = (1 + \exp(\lambda^2 - \beta))^{-1} .$$

Functions $f_\pm(\lambda)$ are constructed by (1.6)

$$f_\pm(\lambda) - \gamma \int_{-\infty}^{\infty} K(\lambda, \mu) f_\pm(\mu) d\mu = e_\pm(\lambda) . \quad (5.3)$$

Let us define the inverse operator as follows

$$(I - \gamma K)(I + \gamma R) = I ; \quad (I - \gamma K)R = K . \quad (5.4)$$

The kernel of the resolvent $R$ is equal to

$$R(\lambda, \mu) = \frac{f_+(\lambda)f_-(\mu) - f_-(\lambda)f_+(\mu)}{2i(\lambda - \mu)} = R(\mu, \lambda) . \quad (5.5)$$

To construct a nonlinear differential equation for $B_{++}$ (see (1.8)), first let us construct a Lax representation of this equation. Let us differentiate (5.3) with respect to $x$, the result can be written in the vector form :

$$\frac{\partial}{\partial x} \vec{f} = (i\lambda\sigma_3 + Q)\vec{f} . \quad (5.6)$$

Here $\vec{f}$ is a two-component vector-function

$$\vec{f}(\lambda) = \begin{pmatrix} f+(\lambda) \\ f_-(\lambda) \end{pmatrix} \quad (5.7)$$

and $Q$ is $2\times 2$ matrix

$$Q=\begin{pmatrix} 0 & B_{++} \\ B_{++} & 0 \end{pmatrix} . \tag{5.8}$$

Here we used that $\partial_x K(\lambda,\mu)$ is sum of one-dimensional projectors

$$2\partial_x K(\lambda,\mu)=e_+(\lambda)e_-(\mu)+e_-(\lambda)e_+(\mu) .$$

We shall consider (5.6) as the $L$ operator; to construct the $M$ operator we apply $(2\lambda\partial_\beta+\partial_\lambda)$ to the equation (5.3). This gives the $M$ operator

$$(2\lambda\partial_\beta+\partial_\lambda)\vec{f}=(ix\sigma_3-i\partial_\beta V)\vec{f} \tag{5.9}$$

Here

$$V=\begin{pmatrix} B_{+-} & -B_{++} \\ B_{++} & -B_{+-} \end{pmatrix} \tag{5.10}$$

and we used $(2\lambda\partial_\beta+\partial_\lambda)\vartheta(\lambda)=0$. Write now the compatibility conditions for the system (5.6) and (5.9)

$$[\partial_x-i\lambda\sigma_3-Q,2\lambda\partial_\beta+\partial_\lambda-i\lambda\sigma_3+i\partial_\beta V]=0 \tag{5.11}$$

at any $\lambda$. We arrive at the equations

$$\begin{aligned} \partial_x B_{+-} &= B_{++}^2 \\ \partial_\beta B_{++}^2 &= 1+\partial_x\left(\frac{\partial_x\partial_\beta B_{++}}{2B_{++}}\right) . \end{aligned} \tag{5.12}$$

Here $\partial_\beta$ is a derivative with respect to our independent variable $\beta=\frac{h}{T}$. Let us denote by $\sigma(x,\beta)$

$$\sigma(x,\beta)=\ell n\det(I-\gamma K) . \tag{5.13}$$

One can show [27] that $\sigma|_{x=0}=0$

$$\partial_x^2\sigma=-B_{++}^2 , \quad \partial_x\sigma=-B_{+-} \tag{5.14}$$

and $\sigma$ itself satisfies the differential equation

$$(\partial_\beta\partial_x^2\sigma)^2=-4(\partial_x^2\sigma)[2x\partial_\beta\partial_x\sigma+(\partial_\beta\partial_x\sigma)^2-2\partial_\beta\sigma] . \tag{5.15}$$

The initial data for this equation can be extracted from the Fredholm determinant itself

$$\begin{aligned} \lim_{\beta\to-\infty} B_{++} &= \lim_{\beta\to-\infty}\sigma=0 \\ B_{++} &= \gamma\int\vartheta(\lambda)d\lambda+x\gamma^2\left(\int\vartheta(\lambda)d\lambda\right)^2+O(x^2) \\ x &\to 0 \\ \sigma &= -x\gamma\int\vartheta(\lambda)-\frac{x^2}{2}\left(\int\vartheta(\lambda)d\lambda\right)^2+O(x^3) . \end{aligned} \tag{5.16}$$

This fixes the solution uniquely [27] and describes completely the correlation function (1.10), (1.3). It is interesting to mention that at $T = 0$ the value $\sigma = \ell n \det(I - \gamma K)$ depends only on the product of variables $x$ and $\sqrt{\beta}$; $\tau = x\sqrt{\beta} = x\sqrt{h}$. Equation (5.15) is conveniently rewritten for a function

$$\sigma_0(\tau) = \tau \frac{\partial}{\partial \tau} \ell n \det(I - \gamma K)$$

One has

$$(\tau\sigma_0'')^2 = -4(\tau\sigma_0' - \sigma_0)(4\tau\sigma_0' + (\sigma_0')^2 - 4\sigma_0) . \tag{5.17}$$

This is the famous fifth Painlève transcendent of [5]. To conclude this section, let us present a matrix Riemann problem, associated with these equations [27]. Consider the $2 \times 2$ matrix $\chi(\lambda)$ function of $\lambda$.

1) it is holomorphic at $Im(\lambda) > 0$ and $Im(\lambda) < 0$

2) $\chi(\infty) = I = \begin{pmatrix} 1 & 0 \\ 0 & 1 \end{pmatrix}$.

3) $\det \chi(\lambda) \neq 0 \ \forall \lambda$

4) the boundary values $\chi^{\pm}(\lambda)$ at the real axis $\lambda \in R$ are related

$$\chi^-(\lambda) = \chi^+(\lambda) G(\lambda) \; ; \quad Im\lambda = 0 \tag{5.18}$$

$$G(\lambda) = I + \pi\gamma \begin{pmatrix} e_+(\lambda)e_-(\lambda) & -e_+^2(\lambda) \\ e_-^2(\lambda) & -e_+(\lambda)e_-(\lambda) \end{pmatrix} \tag{5.19}$$

This Riemann problem permits one to calculate the long distance asymptotics of correlators. A Riemann problem can be associated with each of the differential equations constructed below. Let us mention also that differential equations for many point correlators are constructed in [27].

## 6. Time Dependent Correlator

Here we shall take the time correlator of Sect. 2 and derive a differential equation which drives this correlator. This will be the classical nonlinear Schrödinger equation. So we shall study $\langle \psi(x_2, t_2)\psi^+(x_1, t_1)\rangle$, $(c = \infty)$ for Euclidean time

$$t_1 - t_2 = 2i\tau \; ; \quad \tau > 0 \; ; \quad x_1 - x_2 = 2x \; ; \quad x > 0 . \tag{6.1}$$

Let us evaluate the derivative in (2.1)

$$\langle \psi(2)\psi^+(1)\rangle \equiv \langle \psi(x_2, t_2)\psi^+(x_1, t_1)\rangle = \frac{-1}{2\pi} e^{2\tau q^2} b_{++} \det(I + V_0) . \tag{6.2}$$

Here $V_0$ is the integral operator on the interval $[-q, q]$ (for $b_{++}$ see (6.10)):

$$f_{\pm}(\lambda) + \int_{-1}^{q} V_0(\lambda, \mu) f_{\pm}(\mu) d\mu = e_{\pm}(\lambda) . \tag{6.3}$$

The kernel $V_0(\lambda,\mu)$ can be represented in the form :

$$V_0(\lambda,\mu) = \frac{e_+(\lambda)e_-(\mu) - e_+(\mu)e_-(\lambda)}{\lambda-\mu} . \tag{6.4}$$

Here

$$\begin{aligned} e_-(\lambda) &= \frac{1}{\pi}e^{\tau\lambda^2+ix\lambda} ; \quad e_+(\lambda) = e_-(\lambda)\cdot E(\lambda) \\ E(\lambda) &= \fint \frac{d\nu}{\nu-\lambda}e^{-2\tau\nu^2-2ix\nu} . \end{aligned} \tag{6.5}$$

The resolvent $R$ is defined as usual

$$(I+V)(I-R) = I ; \quad (I+V)R = V . \tag{6.6}$$

Its kernel can be represented in terms of functions $f_\pm$ (6.3)

$$R(\lambda,\mu) = \frac{f_+(\lambda)f_-(\mu) - f_+(\mu)f_-(\lambda)}{\lambda-\mu} . \tag{6.7}$$

It is important to introduce potentials

$$B_{jk} = \int_{-q}^{q} d\mu e_j(\mu)f_k(\mu) ; \quad j,k = \pm \tag{6.8}$$

$$C_{jk} = \int_{-q}^{q} d\mu \ \mu e_j(\mu)f_k(\mu) . \tag{6.9}$$

Let us introduce also $b_{++}$

$$\begin{aligned} b_{++} &= B_{++} - G \\ G &= \int_{-\infty}^{\infty} d\mu e^{-2\tau\mu^2-2ix\mu} . \end{aligned} \tag{6.10}$$

Let us mention now some symmetry properties : First of all $V$ is symmetric :

$$V(\lambda,\mu) = V(\mu,\lambda) ; \quad R(\lambda,\mu) = R(\mu,\lambda) \tag{6.11}$$

$$B_{jk} = B_{kj} . \tag{6.12}$$

Second, there exists a complex involution

$$\bar{e}_-(\lambda) = e_-(-\lambda) ; \quad \bar{e}_+(\lambda) = -e_+(-\lambda) ; \quad \bar{f}_\pm(-\lambda) = \mp f_\pm(\lambda) \tag{6.13}$$

$$\overline{V(\lambda,\mu)} = V(-\lambda,-\mu) ; \quad \overline{R(\lambda,\mu)} = R(-\lambda,-\mu) . \tag{6.14}$$

So $B_{++}, b_{++}$ and $B_{--}$ are real

$$\overline{B}_{++} = B_{++} ; \overline{b}_{++} = b_{++} ; \overline{B}_{--} = B_{--} \tag{6.15}$$

but $B_{+-}$ is pure imaginary

$$\overline{B}_{+-} = -B_{+-} . \tag{6.16}$$

Third, we mention also that

$$\begin{aligned} \overline{C}_{++} &= -C_{++}; \quad \overline{C}_{--} = -C_{--}; \quad \overline{C}_{+-} = C_{+-}; \\ \overline{C}_{-+} &= C_{-+}; \qquad C_{-+} = C_{+-} + B_{++}B_{--} - B_{+-}^2 . \end{aligned} \tag{6.17}$$

### *6.1 Lax Representation*

In this subsection we shall derive a Lax representation, for a nonlinear differential equation, which will drive the correlator (6.2). First we take the solutions of (6.3) $f_\pm(\lambda)$ and define vector-function

$$\vec{f}(\lambda) \equiv \begin{pmatrix} f_+(\lambda) \\ f_-(\lambda) \end{pmatrix}$$

It is easy to differentiate (6.3) with respect to $x$, because $\frac{\partial}{\partial x} V_0(\lambda, \mu)$ is equal to the sum of one-dimensional projectors (similar to Sect. 5). The result is the $L$ operator :

$$\left( \frac{\partial}{\partial x} + i\lambda\sigma_3 - 2iQ \right) \vec{f} = 0 \tag{6.18}$$

Here

$$Q = \begin{pmatrix} 0 & b_{++} \\ B_{--} & 0 \end{pmatrix} \tag{6.19}$$

see (6.10), (6.8). In a similar way one can differentiate (6.3) with respect to time $\tau$. This gives the $M$ operator :

$$\left( \frac{\partial}{\partial \tau} + \lambda^2\sigma_3 - 2\lambda Q - V \right) \vec{f} = 0\ . \tag{6.20}$$

Here

$$V = \begin{pmatrix} 2b_{++}B_{--} & i\partial_x b_{++} \\ -i\partial_x B_{--} & -2B_{--}b_{++} \end{pmatrix} . \tag{6.21}$$

Compatibility condition for (6.18) and (6.20) leads to :

$$\left[ \frac{\partial}{\partial x} + i\lambda\sigma_3 - 2iQ ; \frac{\partial}{\partial \tau} + \lambda^2\sigma_3 - 2\lambda Q - V \right] = 0\ . \tag{6.22}$$

This gives the nonlinear Schrödinger equation

$$\begin{cases} \frac{\partial}{\partial \tau} B_{--} = -\frac{1}{2}\frac{\partial^2}{\partial x^2} B_{--} - 4B_{--}^2 b_{++} \\ \frac{\partial}{\partial \tau} b_{++} = \frac{1}{2}\frac{\partial^2}{\partial x^2} b_{++} + 4b_{++}^2 B_{--} \end{cases} \tag{6.23}$$

Initial data $\tau = 0$ for these equations can be extracted from the description [5] of equal time correlator $\langle \psi^+(x)\psi(-x) \rangle$. This also gies the expression for $C^{++}$ and $C_{--}$

$$\begin{cases} C_{++} = \frac{i}{2}\partial_x B_{++} + B_{++} \cdot B_{+-} - 2GB_{+-} \\ C_{--} = -\frac{i}{2}\partial_x B_{--} - B_{+-}B_{--} \end{cases} \tag{6.24}$$

Direct differentiation of (6.8) (using (6.3)) gives also

$$\frac{\partial}{\partial x} B_{+-} = 2iB_{--}b_{++}\ . \tag{6.25}$$

This permits us to express matrix $V$ (6.21) in terms of the matrix $U$

$$U = \begin{pmatrix} -B_{+-} & b_{++} \\ -B_{--} & B_{+-} \end{pmatrix} \tag{6.26}$$

$$V = i\partial x U \,. \tag{6.27}$$

*6.2 Expression for the correlation function in terms of NS equation*

The field correlator (6.2) is equal to the product of two factors $b_{++}$ and $\det(I+V_0)$. One can get a differential equation for $b_{++}$ by substituting one of eq. (6.23) into another

$$2\frac{\partial}{\partial\tau}\left(\frac{2\dot{b}_{++}-b''_{++}}{b^2_{++}}\right)+\frac{\partial^2}{\partial x^2}\left(\frac{2\dot{b}_{++}-b''_{++}}{b^2_{++}}\right)+b_{++}\left(\frac{2\dot{b}_{++}-b''_{++}}{b^2_{++}}\right)^2=0\,. \tag{6.28}$$

To characterize $\det(I+V_a)$ let us introduce the notation

$$\sigma = \ell n \det(I+V_0)\,. \tag{6.29}$$

At $\tau=0$, function $\sigma(x)$ was described in [5]. Derivatives of $\sigma(x,\tau)$ can be expressed in terms of solutions of the NS equation (6.23)

$$\frac{\partial\sigma}{\partial x} = -2iB_{+-}\,; \qquad \frac{\partial^2\sigma}{\partial x^2} = 4b_{++}B_{--} \tag{6.30}$$

$$\frac{\partial\sigma}{\partial x} = -2GB_{--} - 2(C_{+-}+C_{-+})\,. \tag{6.31}$$

Straightforward calculations involving (6.3), (6.8), (6.9) show also that

$$\begin{aligned} \frac{\partial}{\partial x}(C_{+-}+C_{-+}) &= (B_{++}-2G)\frac{\partial B_{--}}{\partial x} - B_{--}B'_{++} \\ i\frac{\partial B_{+-}}{\partial\tau} &= b_{++}\frac{\partial B_{--}}{\partial x} - B_{--}\frac{\partial b_{++}}{\partial x}\,. \end{aligned} \tag{6.32}$$

This set of equations completely describes the correlator $\langle\psi(x_2,t_2)\psi^+(x_1,t_1)\rangle$. The associated Riemann problem can be constructed similar to Sect. 5. This permits us to evaluate asymptotics.

## 7. Finite Temperature Time Correlator

Here we shall consider the temperature time correlator of an impenetrable Bose gas $c=\infty$, $\langle\psi(x_2,t_2)\psi^+(x_1,t_1)\rangle_T$. The Fredholm determinant representation can be extracted from Sect. 2. Let us consider Euclidean time

$$t_1-t_2=2i\tau\,; \quad \tau>0\,; \quad x_1-x_2=2x\,; \quad x>0\,. \tag{7.1}$$

Here we shall use the notation of Sect. 6. We shall use also the variable $\beta = \frac{h}{T}$, the same as in (1.2) and Sect. 5, and the Fermi weight

$$\vartheta(\lambda) = (1 + \exp\{\lambda^2 - \beta\})^{-1} . \tag{7.2}$$

So

$$\langle\psi(x_2, t_2)\psi^+(x_1, t_1)\rangle_T = \frac{-1}{2\pi} e^{2\tau\beta} b_{++} \det(I + V_T) . \tag{7.3}$$

Here $V_T$ is an integral operator on the whole real axis

$$f_\pm(\lambda) + \int_{-\infty}^{\infty} V_T(\lambda, \mu) f_\pm(\mu) d\mu = e_\pm(\lambda) \tag{7.4}$$

with the kernel:

$$V_T(\lambda, \mu) = \frac{e_+(\lambda)e_-(\mu) - e_+(\mu)e_-(\lambda)}{(\lambda - \mu)} = V_T(\mu\lambda) \tag{7.5}$$

and

$$e_-(\lambda) = \frac{1}{\pi}\sqrt{\vartheta(\lambda)} e^{\tau\lambda^2 + ix\lambda} \ ; \ e_+(\lambda) = e_-(\lambda)E(\lambda) \tag{7.6}$$

$$E(\lambda) = \fint_{-\infty}^{\infty} \frac{d\nu}{\nu - \lambda} e^{-2\tau\nu^2 - 2ix\nu} \ ; \ G = \int_{-\infty}^{\infty} d\mu e^{-2\tau\mu^2 - 2ix\mu} . \tag{7.7}$$

The functions $f_\pm(\lambda)$ are defined by (7.4) and the kernel of the resolvent is given by (6.7). The potentials

$$B_{ik} = \int_{-\infty}^{\infty} e_i(\mu) f_k(\mu) d\mu \ ; \ C_{ik} = \int_{-\infty}^{\infty} d\mu e_i(\mu) f_k(\mu)\mu \tag{7.8}$$

now depend on three variables $x, \tau\beta$. The symmetry properties (6.11)-(6.17) are still valid.

### *7.1 Lax representation*

Let us define a vector-function, as usual

$$\vec{f}(\lambda) = \begin{pmatrix} f_+(\lambda) \\ f_-(\lambda) \end{pmatrix} . \tag{7.9}$$

Differentiate (7.4) with respect to $x, \tau$; also apply the operator $(2\lambda\partial_\beta + \partial_\lambda)$ (as in (5.9)). In this way we get three operators

$$\hat{L}(\lambda)\vec{f} = 0 \ ; \quad \hat{M}(\lambda)\vec{f} = 0 \ ; \quad \hat{N}(\lambda)\vec{f} = 0 . \tag{7.10}$$

The operators $L$ and $M$ are the same as in (6.18) and (6.20)

$$\hat{L}(\lambda) = \frac{\partial}{\partial x} + i\lambda\sigma_3 - 2iQ \tag{7.11}$$

$$\hat{M}(\lambda) = \frac{\partial}{\partial \tau} + \lambda^2 \sigma_3 - 2\lambda Q - V \; . \tag{7.12}$$

Formulae (6.19), (6.21), (6.27) and (6.26) are still valid. The operator $\hat{N}$ is similar to (5.9), but now we have

$$\hat{N}(\lambda) = 2\lambda \frac{\partial}{\partial \beta} + \frac{\partial}{\partial \lambda} + 2\lambda \tau \sigma_3 + ix\sigma_3 - 4\tau Q - 2\partial_\beta U \; . \tag{7.13}$$

Here

$$Q = \begin{pmatrix} 0 & b_{++} \\ B_{--} & 0 \end{pmatrix} ; \quad U = \begin{pmatrix} -B_{+-} & b_{++} \\ -B_{--} & B_{+-} \end{pmatrix} \tag{7.14}$$

*7.2 Three dimensional completely integrable differential equation*

All three operators $L, M, N$ should commute at arbitrary value of spectral parameter $\lambda, [L(\lambda), M(\lambda)] = 0, [L(\lambda, N(\lambda)] = 0$ and $[M(\lambda), N(\lambda)] = 0$. Now let us write down nonlinear differential equations for the potentials $B_{ik}$ and $C_{ik}$. First of all (6.23)-(6.25) are still valid. To write down a complete set of equations we introduce the notation :

$$g_- = e^{-2\tau\beta} B_{--} \; ; \quad g_+ = e^{2\tau\beta} b_{++} \tag{7.15}$$

$$n(x, \tau, \beta) \equiv g_+ g_- \; ; \quad p(x, \tau, \beta) \equiv g_- \partial_x g_+ - g_+ \partial_x g_- \; . \tag{7.16}$$

First of all the NS equations are valid

$$\begin{aligned} &\text{(i)} \quad \partial_\tau g_+ = 2_\beta g_+ + \tfrac{1}{2}\partial_x^2 g_+ + 4g_+^2 g_- \\ &\text{(ii)} \quad \partial_\tau g_- = -2_\beta g_- - \tfrac{1}{2}\partial_x^2 g_- - 4g_-^2 g_+ \\ &\text{(iii)} \quad 2\dot{n} \; = p' \end{aligned} \tag{7.17}$$

The first equation containing the $\beta$ derivative looks like

$$\frac{\partial_\beta \partial_x g_+}{g_+} = \frac{\partial_\beta \partial_x g_-}{g_-} \equiv \varphi(x, \tau, \beta) \; . \tag{7.18}$$

Here we define a new function $\varphi(x, \tau, \beta)$. The two remaining equations look like

$$\begin{aligned} \partial_\tau \varphi + 4\partial_\beta p &= 0 \\ \partial_x \varphi + 2 + 8\partial_\beta n &= 0 \end{aligned} \tag{7.19}$$

Initial data $\tau = 0$ for this equations can be extracted from equal time temperature correlator $\langle \psi^+(x)\psi(-x)\rangle_T$, see Sections 1, 5. So (7.17)-(7.19) are a complete set of equations for $b_{++}$ and $B_{--}$. Other potentials are defined in terms of these two. The derivatives of $B_{+-}$ are equal to :

$$\begin{aligned} \partial_x B_{+-} &= 2in \; , \partial_\tau B_{+-} = ip \\ \partial_\beta B_{+-} &= -\frac{i}{2}x - \frac{i}{4}\varphi \end{aligned} \tag{7.20}$$

The potentials $C_{ik}$ are still given by the expression (6.24), (6.32).

### 7.3 *Expression for correlation function*

The correlator (7.3) is equal to the product of $g_{++}$ and $e^{\sigma}$

$$\sigma = \ell n \det(I + V_T)\,. \tag{7.21}$$

It is easy to get an equation involving only $g_+$ because $g_-$ can be expressed, from (7.17)(i)

$$8g_- = \frac{2\dot{g}_+ - 4_\beta g_+ + \partial_x^2 g_+}{g_+^2}\,. \tag{7.22}$$

One can substitute this expression in all other equations of (7.17)-(7.19) to get a complete set of equations for $g_+$. At $\tau = 0, \sigma$ was described at Sect. 5, its derivatives can be expressed in terms of the potentials $B_{ik}, C_{ik}$

$$\frac{\partial \sigma}{\partial x} = -2iB_{+-}; \frac{\partial \sigma}{\partial \tau} = -2GB_{--} - 2(C_{+-} + C_{-+}) \tag{7.23}$$

$$\begin{aligned} \frac{\partial \sigma}{\partial \beta} &= -2\tau\partial_\beta(C_{+-} + C_{-+}) - 2ix\partial_\beta B_{+-} - 2\tau B_{--}\partial_\beta B_{--} + \\ &\quad +2\tau(B_{++} - 2G)\partial_\beta B_{--} + 2(\partial_\beta B_{++})(\partial_\beta B_{--}) - 2(\partial_\beta B_{+-})^2 \end{aligned} \tag{7.24}$$

This completely defines the correlation function. The corresponding Riemann problem (similar to Sect. 5) also can be constructed.

## 8. Integro-Differential Equation for the Finite Coupling Correlator

Here we shall derive trhe integro-differential equation which drives the simplest correlator at finite coupling $0 < c < \infty$. This is probability of absence of particles on the interval $[-x, x]$ [see (4.10, (0.11)] ;

$$P(x) = \frac{\langle 0|\det(I+V)|0\rangle}{\det(I - \frac{1}{2\pi}K)}\,. \tag{8.1}$$

Only the numerator depend on the distance $x$. So we shall investigate here

$$\det(I + V)\,. \tag{8.2}$$

The integral operator $V$ acts on the interval $[-q, q]$, its kernel $V(\lambda_1, \lambda_2)$ is given by (4.12), it depends on the dual quantum field $\varphi(\lambda)$ (4.6), but it commutes at different values of spectral parameters $[\varphi(\lambda), \varphi(\mu)] = 0$ so we shall treat $\varphi(\lambda)$ and $V(\lambda_1, \lambda_2)$ (4.12) as classical unknown functions. Let us use (4.14) as representations for

$$\begin{aligned} V(\lambda_1, \lambda_2) &= i\int_0^\infty ds(e^{-sc})\frac{e_+(\lambda_1|s)e_-(\lambda_2|s) - e_+(\lambda_2|s)e_-(\lambda_1|s)}{\lambda_1 - \lambda_2} \\ e_\pm(\lambda|s) &= \sqrt{\frac{c}{2\pi}}\exp\left\{\pm(ix\lambda + is\lambda + \frac{1}{2}\varphi(x))\right\}\,. \end{aligned} \tag{8.3}$$

One should emphasize that this $V_\infty$ has the canonical form (II.0) with $\sum_{j=1}^N$ replaced by the integral $\int_0^\infty ds$, so the formulae of previous section can be generalized. Let us define the resolvent in the standard way

$$(I+V)(I-R)=I \qquad (I+V)R=V$$
$$R(\lambda_1,\lambda_2)+\int_{-q}^{q} V(\lambda_1,\lambda_3)R(\lambda_3,\lambda_2)d\lambda_3 = V(\lambda_1,\lambda_2)\,. \tag{8.4}$$

It is symmetric

$$V(\lambda_1,\lambda_2)=V(\lambda_2,\lambda_1)\,;\quad R(\lambda_1,\lambda_2)=R(\lambda_2,\lambda_1)\,. \tag{8.5}$$

To write down an explicit formulae for $R$, we introduce functions $f_\pm(\lambda|s)$ :

$$f_\pm(\lambda|s)+\int_{-q}^{q} V(\lambda,\mu)f_\pm(\mu|s)d\mu = e_\pm(\lambda|s)\,. \tag{8.6}$$

The kernel of the resolvent can be represented in the form :

$$R(\lambda,\mu)=i(\lambda-\mu)^{-1}\int_0^\infty dse^{-sc}\{f_+(\lambda|s)f_-(\mu|s)-f_-(\lambda|s)f_+(\mu|s)\}\,. \tag{8.7}$$

Now define potentials $B_{ik}(s,t)$

$$B_{ik}(s,t)=\int_{-q}^{q} d\mu e_i(\mu|s)f_k(\mu|t)\,,\quad i,k=\pm \tag{8.8}$$

not it is the kernel of matrix integral operator. It is symmetric

$$B_{ik}(s,t)=B_{ki}(t,s)\,. \tag{8.9}$$

Differentiate (8.6) with respect to $x$, to construct the $L$ operator (it is an integral operator):

$$\frac{\partial}{\partial x}\vec f(\lambda|t)=i\lambda\sigma_3\vec f(\lambda|t)+2\int_0^\infty dse^{-sc}Q(t,s)\vec f(\lambda|s)\,. \tag{8.10}$$

Here we introduce the 2 component vector $\vec f$ and matrix $Q$

$$\vec f(\lambda|s)=\begin{pmatrix} f_+(\lambda|s)\\ f_-(\lambda|s)\end{pmatrix};\quad Q(t|s)=\begin{pmatrix}0 & B_{++}(s,t)\\ B_{--}(t,s) & 0\end{pmatrix}; \tag{8.11}$$
$$Q(t,s)=Q(s,t)\,.$$

Now let us differentiate (8.6) with respect to $q$. We find

$$\frac{\partial}{\partial q}\vec f(\lambda|t)+i\int_0^\infty dse^{-sc}\mathcal{U}(t,s)\vec f(\lambda|s)=0, \tag{8.12}$$

$$\mathcal{U}(t,s)=\frac{A_+(s,t)}{\lambda-q}+\frac{A_-(s,t)}{\lambda+q} \tag{8.13}$$

Here

$$A_+(s,t) = \begin{pmatrix} f_-(q|s)f_+(q|t) & -f_+(q|s)f_+(q|t) \\ f_-(q|s)f_-(q|t) & -f_+(q|s)f_-(q|t), \end{pmatrix}, \tag{8.14}$$

$$i\int_0^\infty dse^{-sc}A_+(t,s)A_+(s,u) = 0,$$

$$A_-(s,t) = \begin{pmatrix} f_-(-q|s)f_+(-q|t) & -f_+(-q|s)f_+(-q|t) \\ f_-(-q|s)f_-(-q|t) & -f_+(-q|s)f_-(-q|t)\ . \end{pmatrix} \tag{8.15}$$

Finally we write the compatibility condition for equations (8.10) and (8.12). In this way we get

$$\begin{aligned} \frac{\partial}{\partial q}B_{++}(s,t) &= f_+(q|s)f_+(q|t) + f_+(-q|s)f_+(-q|t), \\ \frac{\partial}{\partial q}B_{--}(s,t) &= f_-(q|s)f_-(q|t) + f_-(-q|s)f_-(-q|t), \end{aligned} \tag{8.16}$$

$$\begin{aligned} \frac{\partial}{\partial x}f_+(q|t) &= iqf_+(q|t) + 2\int_0^\infty dse^{-sc}B_{++}(t,s)f_-(q|s), \\ \frac{\partial}{\partial x}f_-(q|t) &= -iqf_-(q|t) + 2\int_0^\infty dse^{-sc}B_{--}(t,s)f_+(q|s), \\ \frac{\partial}{\partial x}f_+(-q|t) &= -iqf_+(-q|t) + 2\int_0^\infty dse^{-sc}B_{++}(t,s)f_-(-q/s), \\ \frac{\partial}{\partial x}f_-(-q|t) &= iqf_-(-q|t) + 2\int_0^\infty dse^{-sc}B_{--}(t,s)f_+(-q|s)\ . \end{aligned} \tag{8.17}$$

This is a complete set of equations for 6 unknown functions ;

$$B_{++}(s,t)\ ;\ B_{--}(s,t)\ ;\ f_+(q|s)\ ;\ f_-(q|s)\ ;\ f_+(-q|s)\ ;\ f_-(-q|s)\ .$$

The associated Riemann problem can be constructed in a way similar to Sect. 5. So (8.16) and (8.17) is our integro-differential equation, which drives the correlation function at finite coupling constant.

### *8.1 Correlation function*

To study the space dependence of $P(x)$ (8.1), we shall express $\det(I+V)$ (8.2) in terms of solutions of (8.16) and (8.17). First of all we have as usual

$$\frac{\partial}{\partial x}\ell n\det(I+V) = -2\int_0^\infty dse^{-sc}B_{+-}(s,s) \tag{8.18}$$

Differentiating (8.6) and (8.8) with respect to $x$ and $q$ we obtain

$$\frac{\partial}{\partial x}B_{+-}(s,s) = 2\int_0^\infty dte^{-ct}B_{++}(s,t)B_{--}(t,s) \tag{8.19}$$

$$\frac{\partial}{\partial q}B_{+-}(s,s) = f_+(q|s)f_-(q|s) + f_+(-q|s)f_-(-q|s)\,. \tag{8.20}$$

Even more important is the derivative with respect to $q$. (We know that $\ell n \det(I+V)=0$ at $q=0$)

$$\frac{\partial}{\partial q}\ell n \det(I+V) = R(q,q) + R(-q,-q)\,. \tag{8.21}$$

These are diagonal values of the resolvent on the edges. From (8.7) we have

$$R(q,q) = i\int_0^\infty ds e^{-sc}\left[f_-(q|s)\partial_\lambda f_+(\lambda|s)|_{\lambda=q} - f_+(q|s)\partial_\lambda f_-(\lambda|s)|_{\lambda=q}\right]\,. \tag{8.22}$$

Here r.h.s. can be calculated from (8.12) (8.13) at $\lambda = q$.

$$\begin{aligned} R(q,q) &= i\int_0^\infty ds e^{-sc}[f_-(q|s)\partial_q f_+(q|s) - f_+(q|s)\partial_q f_-(q|s)] \\ &\quad + \frac{1}{2q}\left\{\int_0^\infty ds e^{-sc}[f_+(q|s)f_-(-q|s) - f_-(q|s)f_+(-q|s)]\right\}^2 \\ R(-q,-q) &= i\int_0^\infty ds e^{-sc}[f_+(-q|s)\partial_q f_-(-q|s) - f_-(-q|s)\partial_q f_+(-q|s)] \\ &\quad + \frac{1}{2q}\left\{\int_0^\infty ds e^{-sc}[f_+(q|s)f_-(-q|s) - f_-(q|s)f_+(-q|s)]\right\}^2 \end{aligned} \tag{8.23}$$

In this way we have expressed $\frac{\partial}{\partial q}\ell n \det(I+V)$ in terms of the solution of the system (8.17). To finish the program one should evaluate the long distance asymptotics of $\det(I+V)$ at $x \to \infty$ and make normal ordering with respect to quantum operators $\hat{a}(\lambda), \hat{a}^+(\mu)$ entering $\varphi(\lambda)$ (4.6) (that is, we must calculate the expectation value with respect to the Fock vacuum $|0\rangle$ in dual Fock space).

### *8.2 The operator Riemann-Hilbert problem*

In this section we present the Riemann-Hilbert problem which corresponds to the integrable non-linear system (8.16), (8.17). The latter is a system of integral-differential equations, which is why the corresponding Riemann-Hilbert problem will be an integral-operator-value problem. One can say that the constructions of this section are the natural generalizations of the techniques of [5,27] to the infinitely-dimensional case.

Consider the integral-operator-valued function at real $\lambda$ ($\mathbb{R}$ is real axis):

$$G(\lambda) = I + \theta(q^2 - \lambda^2)g(\lambda)\,, \qquad \lambda \in \mathbb{R}\,,$$

where the matrix-kernel $g(\lambda|s,s')$ of the integral operator $g(\lambda)$ is given by:

$$g(\lambda|s,s') = +2\pi\begin{pmatrix} e_+(\lambda|s)e_-(\lambda|s'), & -e_+(\lambda|s)e_+(\lambda|s') \\ e_-(\lambda|s)e_-(\lambda|s'), & -e_-(\lambda|s)e_+(\lambda|s') \end{pmatrix}$$

$$e_\pm(\lambda|s) = \sqrt{\frac{c}{2\pi}}\exp\{\pm i\lambda(x+s) \pm \frac{1}{2}\varphi(\lambda)\}$$

All our operators are integral operators in the $L_2 \subset \mathbb{R}^+, e^{-sc}$. We propose that the operator $G(\lambda)$ as the "conjugation matrix"for an infinite dimension Riemann-Hilbert problem which is the problem to construct the integral operator-valued function $\chi(\lambda)$ with the following properties:

1. $\chi(\lambda)$ is analytic for $Im\lambda > 0$ or $Im\lambda < 0$.
2. $\chi^-(\lambda) = \chi^+(\lambda)G(\lambda), Im(\lambda) = 0$, and $\chi^\pm$ are the boundary values of function $\chi(\lambda)$ as $\lambda \in \mathbb{R} \pm i0$.
3. $\chi(\lambda) \to I$ as $\lambda \to \infty$.

In terms of the corresponding kernels the properties 1-3 can be rewritten in the following way:

P1. $\chi(\lambda|s,s')$ is an analytic function of $\lambda$ in the half planes $Im\lambda > 0$ or $Im\lambda < 0$ for all $s, s'$.

P2. $\chi^-(\lambda|s,s') = \int_0^\infty ds'' e^{-s''c}\chi^+(\lambda|s,s'')G(\lambda|s'',s') = \chi^+(\lambda|s,s') + 2\pi\theta(q^2 - \lambda^2) \times$ $\times \int_0^\infty e^{-s''c} ds'' \chi^+(\lambda|s,s'')g(\lambda|s''s')$.

P3. $\chi(\lambda|s,s') = \begin{pmatrix} 1 & 0 \\ 0 & 1 \end{pmatrix} \delta(s-s')e^{s'c} + \lambda^{-1}\psi_1(s,s') + \ldots$, as $\lambda \to \infty$.

Suppose that the solution of the Riemann-Hilbert problem 1-3 exists and unique. Then we shall show that the function

$$\psi(\lambda) = \chi(\lambda)E(\lambda),$$

where

$$E(\lambda|s,s') = \delta(s-s')e^{i\lambda(x+s')\sigma_3}e^{s'c},$$

satisfies the integral-operator-valuc linear system in the form of (8.10),(8.12). We begin with the $x$-equation. Applying the standard reasoning (see,m for example [27]) based on Liouville's theorem and on the $x$ independence of the conjugation integral operator,

$$G_0(\lambda) = E^{-1}(\lambda)G(\lambda)E(\lambda) ,$$

we have the equality,

$$\partial_x\psi(\lambda)\psi^{-1}(\lambda) = i\lambda I_0 + U_0 , \tag{8.24}$$

where $I_0(s,s') = \delta(s-s')e^{s'c}\sigma_3$ and $U_0$ is independent of $\lambda$. Rewriting (8.24) in the form,

$$\partial_x\psi(\lambda|s,s') = i\lambda\sigma_3\psi(\lambda|s,s') + \int_0^\infty e^{-s''c}U_0(s,s'')\psi(\lambda|s'',s')ds'' . \tag{8.25}$$

We see that

$$B_{ik}(s,s') = \int_{-q}^{q} e_i(\lambda|s)f_k(\lambda|s')d\lambda, \quad U_0(s,s') = i[\psi_1(s,s'),\sigma_3] , \tag{8.26}$$

where $\psi_1(s,s')$ is the first coefficient in the series P3. So we obtained the first equation of the system (8.10). Note that we also have the equalities

$$\begin{aligned} B_{++}(s,s') &= -i[\psi_1(s,s')]_{12} \,, \\ B_{--}(s,s') &= +i[\psi_1(s,s')]_{21} \end{aligned} \tag{8.27}$$

$$U_0(s,s') = 2\begin{pmatrix} 0 & +B_{++} \\ +B_{--} & 0 \end{pmatrix}(s,s') \,.$$

Let us now study the $q$-derivative of the function $\psi$. To do this it is convenient to consider the integral operator valued function $\psi_0(\lambda)$ with the kernel,

$$\psi_0(\lambda|s,s') = \left[\delta(s-s')e^{s'c} - \frac{ic}{2\pi}\ell n\frac{\lambda+q}{\lambda-q}\begin{pmatrix} 0 & 1 \\ 0 & 0 \end{pmatrix}\right]\begin{pmatrix} e^{-\varphi(\lambda)/2} & 0 \\ -e^{-\varphi(\lambda)/2} & e^{\varphi(\lambda)/2} \end{pmatrix} .$$

It is easy to check that for all real $\lambda$ the following conjugation condition is satisfies.

$$\psi_0^-(\lambda) = \psi_0^+(\lambda)G_0(\lambda) \,.$$

Then we can represent the integral operator $\psi(\lambda)$ in the neighbourhood of the interval $[-q,q]$ in the form,

$$\psi(\lambda) = \hat{\psi}(\lambda)\psi_0(\lambda) \,, \tag{8.28}$$

where $\hat{\psi}(\lambda)$ is singlevalued and analytic in that neighbourhood. Note that the representation (8.28) is just the infinite dimension analogue of the basic formulae of the article [27]. Just as in this article we conclude from (8.28) that,

$$i\partial_q\psi(\lambda)\psi^{-1}(\lambda) = \frac{A_+}{\lambda-q} + \frac{A_-}{\lambda+q} \,, \tag{8.29}$$

where

$$A_\pm = i\lim_{\lambda\to\pm q}(\lambda\mp q)\hat{\psi}(\lambda)\partial_q\psi_0(\lambda)\psi_0^{-1}(\lambda)\hat{\psi}^{-1}(\lambda) \,. \tag{8.30}$$

Note that the kernels of the integral operators $\psi_0^{-1}(\lambda)$ and $\psi^{-1}(\lambda)$ are described by

$$\begin{aligned} \psi_0^{-1}(\lambda|s,s') &= \begin{pmatrix} e^{\varphi(\lambda)/2} & 0 \\ e^{-\varphi(\lambda)/2} & e^{-\varphi(\lambda)/2} \end{pmatrix} \\ &\quad\cdot\left[\delta(s-s')e^{s'c} + \frac{ic}{2\pi}\ell n\left(\frac{\lambda+q}{\lambda-q}\right)\cdot\begin{pmatrix} 0 & 1 \\ 0 & 0 \end{pmatrix}\right] \end{aligned} \tag{8.31}$$

$$\psi^{-1}(\lambda|s,s') = \begin{pmatrix} \psi_{22}(\lambda|s',s) & -\psi_{12}(\lambda|s',s) \\ -\psi_{21}(\lambda|s',s) & \psi_{11}(\lambda|s',s) \end{pmatrix} \tag{8.32}$$

The first of these equalities is obvious, the second one follows from the identity,

$$G^{-1}(\lambda) = I - \theta(q^2-\lambda^2)g(\lambda) \,.$$

This identity is a direct corollary of the important relation

$$g^2(\lambda) = 0 \ .$$

Note also that from (8.31) we have

$$[\partial_q \psi_0(\lambda)\psi_0^{-1}(\lambda)](s,s') = -\frac{ic}{2\pi}\left(\frac{1}{\lambda-q} + \frac{1}{\lambda+q}\right)\begin{pmatrix} 0 & 1 \\ 0 & 0 \end{pmatrix} \tag{8.33}$$

Returning to the integral operators $A_+$ we obtain from (8.30-8.33) that

$$\begin{aligned}
A_\pm(s,s') &= \frac{c}{2\pi}\int_0^\infty\int_0^\infty ds''ds'''e^{-(s''+s''')c}\hat{\psi}(\pm q|s,s'')\begin{pmatrix} 0 & 1 \\ 0 & 0 \end{pmatrix}\hat{\psi}^{-1}(\pm q|s'''s') \\
&= \frac{c}{2\pi}\int_0^\infty ds''e^{-s''c}\hat{\psi}(\pm q|s,s'')\begin{pmatrix} 0 & 1 \\ 0 & 0 \end{pmatrix}\int_0^\infty ds''e^{-s''c}\hat{\psi}^{-1}(\pm q|s'',s') \\
&= \frac{c}{2\pi}\int_0^\infty ds''e^{-s''c}\psi(\pm q|s,s'')\begin{pmatrix} e^{\varphi(\pm q)/2} & 0 \\ e^{-\varphi(\pm q)/2} & 0 \end{pmatrix} \\
&\quad \times \begin{pmatrix} -e^{-\varphi(\pm q)/2} & e^{\varphi(\pm q)/2} \\ 0 & 0 \end{pmatrix}\cdot\int_0^2 ds''e^{-s''c}\psi^{-1}(\pm q|s'',s') \\
&= \begin{pmatrix} f_+(\pm q|s)f_-(\pm q|s'), & -f_+(\pm q|s)f_+(\pm q|s') \\ f_-(\pm q|s)f_-(\pm q|s'), & -f_-(\pm q|s)f_+(\pm q|s') \end{pmatrix},
\end{aligned}$$

where we set:

$$\begin{aligned}
f_+(\lambda|s) &= \int_0^\infty ds'e^{-s'c}[\chi_{11}(\lambda|s,s')e_+(\lambda|s') + \chi_{12}(\lambda|s,s')e_-(\lambda|s')] \\
f_-(\lambda|s) &= \int_0^\infty ds'e^{-s'c}[\chi_{21}(\lambda|s,s')e_+(\lambda|s') + \chi_{22}(\lambda|s,s')e_-(\lambda|s')]
\end{aligned} \tag{8.34}$$

By this formulae the reconstruction of the Lax representation (8.12) from the Riemann-Hilbert problem 1-3 is finished.

Our last task is to show that the formulae (8.27) and (8.34) give us just the same functions $B_{++}(s,s'), B_{--}(s,s')$ and $f_+(\lambda|s)$ as we had in the previous section. To do this let us consider the singular integral equation equivalent to the problem 1-3. It has the form

$$\chi^+(\lambda) = I - 1/2\pi i\int_{-q}^q \frac{d\mu}{\mu-\lambda-i0}\chi^+(\mu)g(\mu) \tag{8.35}$$

Set

$$\Phi^+(\lambda) = \chi^+(\lambda)T(\lambda) \ , \tag{8.36}$$

where

$$\begin{aligned}
T(\lambda|s,s') &= \tau(\lambda|s)\delta(s-s')e^{+s'c} \ , \\
\tau(\lambda|s) &= \begin{pmatrix} 1 & e_+(\lambda|s) \\ 0 & e_-(\lambda|s) \end{pmatrix} .
\end{aligned}$$

Using the function $\Phi^+(\lambda)$ equation (8.35) can be rewritten in the following way :

$$\Phi^+(\lambda) = T(\lambda) - 1/2\pi i \int_{-q}^{q} \frac{d\mu}{\mu - \lambda - i0} \Phi^+(\mu) T^{-1}(\mu) g(\mu) T(\lambda) .$$

Note that

$$\begin{aligned}(T^{-1}(\mu)g(\mu)T(\lambda))(\mu,\lambda|s,s') &= \tau^{-1}(\mu|s)g(\mu|s,s')\tau(\lambda|s') \\ &= +2\pi \begin{pmatrix} 0\,, & 0 \\ e_-(\mu|s'), & e_-(\mu|s')e_+(\lambda|s') - e_+(\mu|s')e_-(\lambda|s') \end{pmatrix} .\end{aligned}$$

So, for the matrix elements $\Phi^+_{ij}(\lambda|s,s')$ we have:

$$\begin{aligned}
\Phi^+_{11}(\lambda|s,s') &= \delta(s-s')e^{s'c} + i\int_{-q}^{q}\frac{d\mu}{\mu-\lambda-i0}\int_0^\infty ds'' e^{-s''c}\Phi^+_{12}(\mu|s,s'')e_-(\mu|s'), \\
\Phi^+_{12}(\lambda|s,s') &= e_+(\lambda|s)\delta(s-s')e^{s'c} \\
&\quad + i\int_{-q}^{q}\frac{d\mu}{\mu-\lambda}\int_0^\infty ds'' e^{-s''c}\Phi^+_{12}(\mu|s,s'')[e_-(\mu|s')e_+(\lambda|s') - e_+(\mu|s')e_-(\lambda|s')], \\
\Phi^+_{21}(\lambda|s,s') &= +i\int_{-q}^{q}\frac{d\mu}{\mu-\lambda-i0}\int_0^\infty ds'' e^{-s''c}\Phi^+_{22}(\mu|s,s'')e_-(\mu|s'), \qquad (8.37) \\
\Phi^+_{22}(\lambda|s,s') &= e_-(\lambda|s)\delta(s-s')\exp(s'c) \\
&\quad + i\int_{-q}^{q}\frac{d\mu}{\mu-\lambda}\int_0^\infty ds'' e^{-s''c}\Phi^+_{22}(\mu|s,s'')[e_-(\mu|s')e_+(\lambda|s') - e_+(\mu|s')e_-(\lambda|s')]
\end{aligned}$$

Introduce the functions

$$f_{ij}(\lambda|s) = \int_0^\infty ds' e^{-s'c}\Phi^+_{ij}(\lambda|s,s') .$$

Then we obtain from (8.37) that

$$\begin{aligned}
f_{12}(\lambda|s) &= e_+(\lambda|s) - \int_{-q}^{q} d\mu V(\lambda,\mu) f_{12}(\mu|s) , \\
f_{22}(\lambda|s) &= e_-(\lambda|s) - \int_{-q}^{q} d\mu V(\lambda,\mu) f_{22}(\mu|s) .
\end{aligned}$$

were

$$V(\lambda,\mu) = +i\int_0^\infty ds e^{-sc}\frac{e_+(\lambda|s)e_-(\mu|s) - e_+(\mu|s)e_-(\lambda|s)}{\lambda-\mu} .$$

So we come to our basic integral operator (8.3). In particular we have the following equalities

$$\begin{aligned}
f_+(\lambda|s) &= f_{12}(\lambda|s) , \\
f_-(\lambda|s) &= f_{22}(\lambda|s) . \qquad (8.38)
\end{aligned}$$

Taking into account the definition (8.36) of the function $\Phi^+(\lambda)$ we see that (8.38) coincides with (8.34).

To prove (8.27) note that due to (8.35) the matrix coefficient $\psi_1(s,s')$ can be expressed in the following way :

$$\psi_1(s,s') = 1/2\pi i \int_{-q}^{q} d\mu \int_0^\infty ds'' e^{-s''c} \chi^+(\mu|s,s'') g(\mu|s'',s') ds'' .$$

Therefore for the corresponding matrix elements we have

$$\begin{aligned}
[\psi_1(s,s')]_{11} &= -i\int_{-q}^{q} d\mu f_+(\mu|s)e_-(\mu|s') = -iB_{-+}(s',s) \\
[\psi_1(s,s')]_{12} &= +i\int_{-q}^{q} d\mu f_+(\mu|s)e_+(\mu|s') = +iB_{++}(s',s) \\
[\psi_1(s,s')]_{21} &= -i\int_{-q}^{q} d\mu f_-(\mu|s)e_-(\mu|s') = +iB_{--}(s',s) \\
[\psi_1(s,s')]_{22} &= +i\int_{-q}^{q} d\mu f_-(\mu|s)e_+(\mu|s') = +iB_{+-}(s',s) .
\end{aligned} \tag{8.39}$$

Note also that the symmetric property

$$B_{ik}(s,s') = B_{ki}(s',s)$$

is the direct corollary of the formula (8.32)

Now we finish the discussion of the Riemann-Hilbert problem which corresponds to the case $0 < c < \infty$. In another paper we will use this Riemann-Hilbert problem for the calculation of the asymptotics of $\det(I+V)$ as $x \to \infty$ (as it was done in [27] for the case $c = \infty$).

# III. Asymptotics of Correlation Functions

This part consists of two sections. In Sect. 9 we study long distance asymptotics of equal time temperature correlator of the impenetrable Bose gas. In Sect. 10 we study density $\psi^+(x,t)\psi(x,t)$ (current) correlations for the impenetrable Bose gas (it is equal to a matrix determinant), we study also $\frac{1}{c}$ corrections.

## 9. Temperature Equal Time Correlator

Following Sect. 1 and 5 we shall study asymptotics of equal time temperature correlators (1.1), (1.10)

$$\langle \psi^+(z)\psi(-z)\rangle_T = \frac{\sqrt{T}}{4} B_{++} \det(I - \gamma K) . \tag{9.1}$$

Let us use our variables (1.2)

$$\beta = \frac{h}{T} ; \qquad x = z\sqrt{T} . \tag{9.2}$$

The associated Riemann problem approach leads to the following asymptotic formula $z \to \infty, x \to \infty$.

$$\begin{aligned}\langle \psi^+(z)\psi(-z)\rangle_T &= \sqrt{\frac{T}{\pi}}\rho_\infty e^{-xc(\beta)} \exp\left\{-\frac{1}{2}\int_\beta^\infty dt(\partial_t c(t))^2\right\} \\ &\times \left[1 - 2\,\mathrm{Re}\,\frac{i\alpha^2(\lambda_1)}{\lambda_0\lambda_1(\lambda_0+\lambda_1)^2} e^{2ix\lambda_1}\left\{(\lambda_0^2+\lambda_1^2)\sin\theta + 2i\lambda_0\lambda_1\cos\theta\right\}\right].\end{aligned} \tag{9.3}$$

Here we took $\rho_\infty$ from [4],[5] :

$$\rho_\infty = \pi e^{1/2} 2^{-1/3} A^{-6} .$$

$A$ is the Glaisher constant. The function $C(t)$ is given by

$$C(t) = \frac{1}{\pi}\int_{-\infty}^{\infty} d\mu\; \ell n\left|\frac{e^{\mu^2-t}+1}{e^{\mu^2-t}-1}\right| . \tag{9.4}$$

In (9.3) $\lambda_0 = \sqrt{\beta}$

$$\lambda_1 = \sqrt{\beta + 2\pi i}\;; \quad \mathrm{Im}\,\lambda_1 > 0\;; \quad \mathrm{Re}\,\lambda_1 > 0\,. \tag{9.5}$$

The function $\alpha(\lambda)$ is given by

$$\alpha(\lambda) = \sqrt{\frac{\lambda+\lambda_0}{\lambda-\lambda_0}} \exp\left\{\frac{i}{2\pi}\int_{-\infty}^{\infty} \frac{d\mu}{\mu-\lambda} \ell n\left|\frac{e^{\mu^2-\beta}+1}{e^{\mu^2-\beta}-1}\right|\right\} . \tag{9.6}$$

The function $\theta$ is given by

$$\theta(x,\beta) = 2\lambda_0 x - \frac{\pi}{2} + \frac{1}{\pi}\fint_{-\infty}^{\infty} \frac{d\mu}{\mu-\lambda}\ell n\left|\frac{e^{\mu^2-\beta}+1}{e^{\mu^2-\beta}-1}\right| . \tag{9.7}$$

These formulae give us the long distance asymptotics of the equal time temperature correlator. The zero temperature case was considered in [5].

Another important correlator (at $c = \infty$) of the impenetrable Bose gas is $P(x)$ (0.11), (1.15), (4.10), (4.13), the probability of absence of particles in the interval $[-x, x]$. It is given by (1.15)

$$P(z) = \det(I - \gamma K)/_{\gamma=1/\pi} \tag{9.8}$$

here $z$ and $x$ are related by (1.2). The associated Riemann problem gives us

$$\begin{aligned}\ell n P(x) &= -xC_0(\beta) + \frac{1}{2}\int_{-\infty}^{\beta} (\partial_t C_0(t))^2\, dt \\ &+ e^{-2x^2+2\beta}\cdot\left[\frac{-1}{16\pi x^2} + \frac{C_0(\beta)}{8\pi x^3} + \ldots\right] + e^{-4x^2+4\beta}\left[\frac{1}{2^{10}\pi^2 x^4} + \ldots\right] + \ldots\end{aligned} \tag{9.9}$$

Here

$$C_0(t) = \frac{1}{\pi}\int_{-\infty}^{\infty} d\lambda \ell n\left(1 + e^{\beta-\lambda^2}\right) . \tag{9.10}$$

At zero temperature, the asymptotics are different

$$\ell n\; P(x) = -\frac{(xq)^2}{2} - \frac{1}{8}\;\ell n\; 2qx\,. \tag{9.11}$$

## 10. Current Correlators

Firt we study time correlators of currents $j(x,t) = \psi^+(x,t)\psi(x,t)$ for the impenetrable Bose gas at finite temperature as well as at zero temperature. Special attention is paid to the asymptotics, and the $\frac{1}{c}$ corrections are derived. We would like to emphasize the contribution of N.M. Bogoliubov in this section. First let us consider the zero temperature case. The density of the gas is given by

$$D = \frac{q}{\pi} . \tag{10.1}$$

Here $q$ is the Fermi momentum. Let us introduce also two functions

$$\mathcal{P}(t,x) = \frac{1}{2\pi} \int_{-1}^{q} d\lambda e^{it(\lambda^2-h)-i\lambda x} \tag{10.2}$$

$$\begin{aligned} \mathcal{H}(t,x) &= \frac{1}{2\pi}\int_{-\infty}^{\infty} e^{-it(\lambda^2-h)+ix\lambda} d\lambda - \frac{1}{2\pi}\int_{-q}^{q} d\lambda e^{-it(\lambda^2-h)+ix\lambda} \\ h &= q^2 . \end{aligned} \tag{10.3}$$

The $N$-current correlator is equal [36],[37] to the determinant of an $N \times N$ matrix :

$$\langle j(x_N,t_N) j(x_{N-1},t_{N-1}) \cdot \ldots \cdot j(x_2,t_2) j(x_1,t_1)\rangle = \det_N M_{ik} \ ; \quad c = \infty \tag{10.4}$$

where the matrix $M_{iK}$ is constructed as follows :

$$\begin{aligned} M_{ii} &= D \ ; \quad M_{ki} = \mathcal{P}(t_i - t_k; x_i - x_k) \quad k < i \\ M_{ki} &= -\mathcal{H}(t_k - t_i, x_k - x_i) \quad k > i . \end{aligned} \tag{10.5}$$

A similar formula is valid for finite temperature $T > 0$, where the distribution function of particles in momentum space is given by the Fermi weight (0.6)

$$\rho(\lambda) = \left(\frac{1}{1 + e^{(\lambda^2-h)/T}}\right)\frac{1}{2\pi} . \tag{10.6}$$

The density is equal to $D = \int d\lambda \rho(\lambda)$ [see (0.7)]. The functions $\mathcal{P}_T$ and $\mathcal{H}_T$ are defined now as

$$\mathcal{P}_T(t,s) = \int_{-\infty}^{\infty} d\lambda \rho(\lambda) e^{it(\lambda^2-h)-i\lambda x} \tag{10.7}$$

$$\mathcal{H}_T(t,x) = \frac{1}{2\pi}\int_{-\infty}^{\infty} e^{-it(\lambda^2-h)+ix\lambda} d\lambda - \int_{-\infty}^{\infty} \rho(\lambda) e^{-it(\lambda^2-h)+ix\lambda} d\lambda . \tag{10.8}$$

In this notation formula (10.4) is still valid. Euclidean time correlators are also given by similar formulae. Let us replace $t \to -i\tau$ in (10.4) and demand that

$$\frac{1}{T} > \tau_{k+1} - \tau_k > 0 . \tag{10.9}$$

It is possible to change the operator ordering in the l.h.s. of (10.4). Using fermionization (1.14) and denoting by : : normal ordering with respect to fermions one obtains

$$\langle : j(x_N,t_N)\cdot\ldots\cdot j(x_1,t_1):\rangle = \det_N m_{ik}\,, \quad c=\infty\,. \tag{10.10}$$

The matrix elements of $m_{ik}$ are

$$m_{ik} = \mathcal{P}(t_i-t_k, x_i-x_k)\,. \tag{10.11}$$

It is interesting to look how $\frac{1}{c}$ corrections change these formulae. Let us write down the $\frac{1}{c}$ correction for the equal time zero temperature $n$-point current correlator

$$\begin{aligned}\langle : j(x_1)\cdot\ldots\cdot j(x_n):\rangle = \Bigg\{1 + \\ +\frac{1}{c}\sum_{m=1}^{n}\left[\frac{2q}{\pi}+\frac{1}{2}\sum_{\ell=1}^{n}\varepsilon(\tilde{x}_\ell-\tilde{x}_m)\left(\frac{\partial}{\partial\tilde{x}_\ell}-\frac{\partial}{\partial\tilde{x}_m}\right)+\frac{1}{\pi}\frac{\partial}{\partial\alpha}\frac{\partial}{\partial\tilde{x}_m}\right]\Bigg\} \\ \times\det_n\left(\frac{\sin q\tilde{x}_{jk}}{\pi\tilde{x}_{jk}}+\alpha\tau(\tilde{x}_{jm},\tilde{x}_{km})\right)\Big/_{\alpha=0}\,.\end{aligned} \tag{10.12}$$

Here $\varepsilon(x)$ is the sign function, and

$$\begin{aligned}\tilde{x} &= x\left(1+\frac{2q}{\pi c}\right)\,; \quad \tilde{x}_{jk}=\tilde{x}_j-\tilde{x}_k \\ \tau(\tilde{x},\tilde{y}) &= \frac{1}{2\pi i}\int_{-q}^{q}d\lambda_1\,d\lambda_2\frac{e^{i\lambda_1\tilde{x}}-e^{i\lambda_2\tilde{y}}}{\lambda_1-\lambda_2}\,.\end{aligned} \tag{10.13}$$

Let us consider also the $1/c$ correction to the two point time dependent field correlator (zero temperature):

$$\begin{aligned}\langle : j(x,t)j(0,0):\rangle &= D^2+\left(\frac{1+\frac{4q}{\pi c}}{2\pi}\right)\int_\Lambda d^2\lambda\, e^{it(\lambda_1^2-\lambda_2^2)} \\ &\times\ (\cos\lambda_{12}\tilde{x})\left[1+\frac{\lambda_{12}}{\pi c}\fint_{-q}^{q}d\lambda_3\left(\frac{1}{\lambda_{31}}-\frac{1}{\lambda_{32}}\right)\right]+0\left(\frac{1}{c^2}\right) \\ \tilde{x} &= x\left(1+\frac{2q}{\pi c}\right)\,; \qquad \lambda_{jk}=\lambda_j-\lambda_k\,.\end{aligned} \tag{10.14}$$

The intgration domain $\Lambda$ is defined by $\Lambda=\{|\Lambda_1|>q, |\lambda_2|<q\}$.

Let us study the asymptotics in different directions in $x,t$ plane. Simplest example is the two-point correlator

$$\langle : j(x_2,t_2)j(x_1,t_1):\rangle = D^2-\mathcal{P}(t_{21},x_{21})\mathcal{P}(t_{12},x_{12})\,. \tag{10.15}$$

An important role is played by the Fermi velocity $v=2q$. In the space like directions $x_{21}\to\infty, t_{21}\to\infty, |x_{21}/t_{21}|>v$ integration by parts gives the complete asymptotic expansion

$$\mathcal{P}(t,x)=\frac{e^{iqx}}{2\pi i}\sum_{n=0}^{\infty}\frac{(-2it)^n(2n-1)!!}{(x+vt)^{2n+1}}-\frac{e^{-iqx}}{2\pi i}\sum_{n=0}^{\infty}\frac{(-2it)^n(2n-1)!!}{(x-vt)^{2n+1}}\,. \tag{10.16}$$

This is in agreement with conformal results [38-40]. In time-like directions the asymptotics are different. One should represent $\mathcal{P}$ as follows :

$$\mathcal{P}(t,x) = \frac{1}{2\pi}\int_{-\infty}^{\infty} e^{it(\lambda^2-h)-i\lambda x}d\lambda - \frac{1}{2\pi}\left(\int_{-\infty}^{-q} + \int_{q}^{\infty}\right) e^{it(\lambda^2-h)-i\lambda x}d\lambda \ . \tag{10.17}$$

The first term here is a decaying wave packet ($t = t_{21}\,, x = x_{21}$):

$$\frac{1}{2\pi}\sqrt{\frac{i\pi}{t_{21}}}\exp\left\{-\frac{ix_{21}^2}{4t_{21}} - iht_{21}\right\} \ . \tag{10.18}$$

Integration by parts of two other terms in (10.17) gives an expansion similar to (10.16). So the difference between space and time asymptotics is in the wave packet. A similar difference in space and time asymptotics in the $XY$ model was found in [8]. Combination of (10.4), (10.10) and (10.16), (10.18) gives the complete asymptotic expansion for the $N$ current correlator.

Now let us discuss the asymptotics of temperature correlators. In (10.7) we move the integration contour down in the complex plane and obtain the asymptotic expansion

$$\mathcal{P}(t_{21}\,, x_{21}) = T \ \mathrm{Re} \sum_{K=0}^{\infty} \frac{1}{i\lambda_K} e^{-\pi T t_{21}(1+2K) + ix_{21}\lambda_K} \tag{10.19}$$

$$\lambda_K = \sqrt{h + i\pi T(1+2_K)} \ ; \quad \mathrm{Re}\ \lambda_K \geq 0 \ ; \quad Im\lambda_K \geq 0 \ ; \quad K = 0,1,\ldots \ . \tag{10.20}$$

This gives us the complete asymptotic expansion of $N$-current correlator (10.4), (10.10). For small temperature the series (10.19) is a geometric one. We can summarise

$$\mathcal{P}(t_{21}\,, t_{21}) = \left(\frac{T}{v}\right) \frac{\sin qx_{21}}{sh\frac{\pi T}{v}(x_{21} + vt_{21})} \tag{10.21}$$

where $v = 2q$. This is also in agreement with conformal results [38-40].

To conclude, let us emphasize once more that the problem of calculating correlation functions for the quantum Nonlinear Schrödinger equation is a rich interesting and important problem [41].

## Acknowledgments

We are grateful to L.D. Faddeev, A.S. Fokas, M.J. Ablowitz, T. Miwa, J.H.H. Perk, B.M. McCoy and N.Yu. Reshetigkhin for discussions.

## References

1. Mathematical Physics in One Dimension. Lieb E.JH. Mattis D.C. (eds.) New York, London : Academic Press 1966.

2. Yang C.N., Yang C.P. Thermodynamics of a one-dimensional system of bosons with repulsive delta-function interaction. *J. Math. Phys.* **10** 1115-1122 (1969).

3. Lenard A. One-dimensional impenetrable Bosons in Thermal equilibrium. *J. Math. Phys.* **7** 1268-1272 (1965).

4. Vaidya H.G., Tracy C.A. *Phys. Rev. Lett.* **42** (1979) 3.

5. Jimbo M., Miwa T., Mori Y., Sato M.: Density matrix of an impenetrable Bose gas and the fifth Painlève transcendent. *Physica* **1D**, 80-158 (1980).

6. McCoy B.M., Wu T.T. The two-dimensional Ising model. Cambridge, Mass. Harvard University Press 1973.

7. McCoy B.M., Perk J.H.H., Shrock R.E. Time-dependent correlation functions of the transverse Ising chain at the critical magnetic field. *Nucl. Phys.* **B220** FS8 (1983) 35-47.

8. Perk J.H.H., Capel H.W., Quispel G.R.W., Nijhoff F.W. Finite-temperature correlations for the Ising Chain in a transverse field. *Physica* **123 A** (1984) 1-49.

9. Au-Yang H., Perk J.H.H. Critical Correlations in a $z$-invariant Inhomogenous Ising Model. *Physica* **144 A** (1987), 44-104.

10. Gelfand I.M., Levitan R.M. *Dokl. Akad. Nauk SSSR* **88**, 593-596 (1953).

11. Faddeev L.D., Takhtajan L.A. "Hamiltonian Approach to the Soliton Theory","Nauka", Moscow, 1986.

12. Faddeev L.D. *Sov. Sci. Rev. Math. Phys.* **C1**, 107-160 (1981).

13. Faddeev L.D. Sklyanin E.K., *Dokl. Akad. Nauk SSSR* **243**, 1430-1433 (1978).

14. Izergin A.G. Korepin V.E. Lattice Versions of Quantum field theory Models in Two Dimensions. *Nucl. Phys.* **B205** FS5 (1982), 401-413.

15. Izergin A.G., Korepin V.E. The most general $L$ operator for the $R$-matrix of the XXX model. *Letts. Math. Phys.* **8** (1984), 259-265.

16. Korepin V.E. Calculation of Norms of Bethe wave function. *Commun. Math. Phys.* **86**, 391-418 (1982).

17. Drinfel'd V.G. Hopf algebras and the quantum Yang-Baxter equations. *Dokl. Akad. Nauk SSSR* **283** (1985) (Russian) = *Soviet Math. Dokl.* **32** (1985) 254-256.

18. Drinfel'd V.G. A new realization of Yangians and quantized affine algebras. *Dokl. Akad. Nauk. SSSR* **296** (1987) Russian = *Soviet Math. Dokl.* **36** (1988) 212-216.

19. Tarasov V.O. On the structure of quantum $L$-operator for the $R$-matrix of the XXZ model. *Teor.Mat.Fiz.* **61** (1984), 163-173 (Russian) = *Theor. Math. Phys.* **61** (1984).

20. Tarasov V.O. Irreducible monodromy matrices for the $R$-matrix of the XXZ-model and lattice local quantum Hamiltonians. *Teor. Mat. Fiz.* **63** (1985), 175-196 (Russian) = *Theor.Math.Phys.* **363** (1985).

21. Korepin V.E. The analysis of the bilinear relation of the six-vertex model. *Dokl. Acad. Nauk SSSR* **265** (1982), 1361-1564 (Russian) = *Soviet Phys. Dokl.* **27** (1982), 612-613.

22. Krein M.G. *DAN SSSR* p. 97 (1954), 21-24.

23. Krein M.G. *DAN SSSR* p. 105 (1955), 433-436.

24. Fokas A.S., Ablowitz M.J. Linearization of the Karteweg-de Vries and Painlève II Equations. *Phys. Rev. Lett.* **47** (1981), 1096-1100.

25. Santini P.M., Ablowitz M.J., Fokas A.S. The direct linearization of a class of nonlinear evaluation equation. *J. Math. Phys.* **25** (1984) 2614-1619.

26. Ablowitz M.J., Segur H. Solitons and the Inverse Scattering Transform SIAM Philadelphia 1981.

27. Its A.R., Izergin A.G., Korepin V.E. Completely integrable differential equations for temperature correlation functions of an impenetrable Bose gas. Submitted to *Phys. Lett. A.*; Preprints Intern. Centre. Theor. Phys. Trieste, Italy. IC/89/107, IC/89/120.

28. Korepin V.E. Dual Field Formulation of Quantum Integrable Modes. *Commun. Math. Phys.* **113**, 177-190 (1987).

29. Korepin V.E. Generating Functional for correlation functions of Nonlinear Schrödinger Model. *Functional Analis i Prilogh.* (Russian) **23**, 15-23 (1989).

30. Mehta M.L. Random Matrices (Academic Press, New York, 1967).

31. Creamer, D.B., Thacker H.B., Wilkinson D. *Phys. Rev.* **D21**, 1523-1526 (1980).

32. Korepin V.E., Slavnov N.A. Time Correlation function of Impenetrable Bose gas as Fredholm minor. Submitted to *Commun. Math. Phys.*

33. N.A. Slavnov "Manypoint time field correlation functions of impenetrable Bose gas". Submitted to *Commun. Math. Phys.*

34. Bogoliubov N.M., Izergin A.G., Korepin V.E. "Quantum Inverse Scattering Method and Correlation Functions". *Lecture Notes in Physics* **242**, 220-316 (1985).

35. Smirnov F.A. *Dokl. Akad. Nauk SSSR* **262**, 78-83 (1982).

36. Bogoliubov N.M., Korepin V.E. Proceedings of II Soviet-Italian Symposium of the Math. Problem of Stat. Phys. Lvov, 1985.

37. Bogoliubov N.M., Korepin V.E. *Zap. Nauch. Semin. LOMI* **156** (1986) 26-36.

38. Bogoliubov N.M., Izergin A.G., Reshetikhin N.Yu. "Finite-size effects and infrared asymptotics of the correlation functions in two dimensions" *J.Phys. A. Math. Gen.* **20** (1987) 5361-5369.

39. Bogoliubov N.M., Izergin A.G., Korepin V.E. "Critical exponents for Integrable Models" *Nuclear Physics* **B275** FS17 (1986) 687-705.

40. Izergin A.G., Korepin V.E., Reshetikhin N.Yu. "Conformal Dimensions in Bethe Ansatz" Stony Brook ITP-SB-88-47.

41. Hatfield B., Yau S.-T. "An exchange symmetry expansion for the 2-point correlation function of the nonlinear Schrödinger model" *Nuclear Physics* **B305** FS23 (1988), 16-32.

42. Nijhoff, I.F.W., Capel, H.W., Wiersma, G.L. and Quispel, G.R.W. "Linearising integral transformation and partial differential equations" *Phys. Lett.* **103A** (1984) p293.

43. Nijhoff, F.W., "Linear integral transformations and hierarkies of integrable nonlinear evolution equations" *Physica* **D31** (1986) p339.

International Journal of Modern Physics B, Vol. 4, No. 5 (1990) 1039 – 1047
© World Scientific Publishing Company

# OFF THE CRITICAL POINT BY PERTURBATION (II): THE ISING MODEL SPIN-SPIN CORRELATOR IN A MAGNETIC FIELD

Vl. S. Dotsenko

*Landau Institute for Theoretical Physics*, 117334 *Moscow, USSR*

An extension of the analytic regularization technique based on the conform 1 theory is suggested for the case of the spin–spin correlation function of the Ising model in a magnetic field, $<\sigma_0\sigma_R>_h = F(t)/(R)^{1/4}$, $t = hR^{15/8}$. Several first terms of the expansion of the scaling function $F(t)$ are given.

The derivation of an analytic theory for the off-critical scaling limit of critical phenomena in two dimensions is gaining importance at present. Two basic approaches are integrable deformations of conformal field theories, initiated by Zamolodchikov, and, in the context of generically off-critical integrable lattice models, the identification of the Virasoro algebra and the coset structures, found basically by the Kyoto group. From another point of view: the monodromy, or the duality transformation matrices of the conformal theory correlation functions have been related to the solutions of the Yang-Baxter equation (see [1] and related references therein).

A somewhat straightforward approach was attempted in [2] by defining a regularization technique, based on conformal theory. This technique allows an unambiguous definition of the perturbation series expansion for the off-critical correlation functions. The perturbation is with respect to the part of the Hamiltonian which takes the theory off-criticality. This approach provides at least a way to calculate the short distance expansion of the correlation functions. (In this context short distance means $R/R_c << 1$ where $R_c$ is the correlation length).

In [2] the regularization technique was defined and shown to give the correct results in the case of the Ising model taken slightly off the critical temperature $T_c$, i.e. perturbed by the Hamiltonian

$$H_\tau = \tau\int d^2x\epsilon(x) \tag{1}$$

where $\tau = (T-T_c)/T_c$.

We recall that the spin-spin correlation function, which we shall denote here as $<\sigma_0\sigma_R>_\tau$, is then represented in the form:

$$<\sigma_0\sigma_R>_\tau = \frac{<\sigma_0\,\sigma_R\exp\{\tau\int d^2x\epsilon(x)\}>_0}{<\exp\{\tau\int d^2x\epsilon(x)\}>_0}, \tag{2}$$

where $<\ldots>_0$ is the correlation function (expectation value) defined by the critical point conformal theory. Expansion of (2) in powers of $\tau$ gives:

$$<\sigma_0\sigma_R>_\tau = <\sigma_0\sigma_R>_0 + \tau\int d^2x<\sigma_0\sigma_R\epsilon_x>_0 + \\ + \frac{\tau^2}{2}\int d^2x_1\int d^2x_2<\sigma_0\sigma_R\epsilon_1\epsilon_2>_0 + \ldots, \tag{3}$$

where the correlation functions inside the integrals are the connected ones.

The off-critical scaling limit of the two-point function is usually represented in the form:

$$<\sigma_0\sigma_R>_\tau = \frac{F_\tau(t)}{(R)^{2\Delta_\sigma}}, \tag{4}$$

where $F_\tau(t)$ is the scaling function, and

$$t = \tau R^{2-\Delta_\epsilon} \tag{5}$$

is the scaling variable with $\Delta_\epsilon$ the dimension of the perturbing operator, $\epsilon(x)$ in (2), conjugate to the "perturbing field" $\tau$. Since in this case $\Delta_\epsilon = 1$, we have

$$t = \tau R\,. \tag{6}$$

The R dependence can be scaled out of the integrals in (3) and then the expansion (3) defines the expansion of the scaling function $F_\tau(t)$ for small t. The first two terms were found in [2] to have the form:

$$F_\tau(t) = 1 + \pi t \log\left[\frac{b}{|t|}\right] + \frac{\pi^2}{4}t^2 + \ldots \tag{7}$$

and were checked with the expansion given in [3] where the function $F_\tau(t)$ has been calculated exactly for this particular theory by other means. (Note that the definition of the scaling variable in [3] is different by a factor of $1/2\pi$, i.e. $t = t_0/2\pi$). The constant b in (7) is an adjustable parameter presumed to be unique for the whole series.

In general, the expansion of $F_\tau(t)$ involves integer powers of t multiplied by logarithmic factors. In principle further terms could be calculated in (7) but it gets progressively more difficult to compute them and it has not been encouraging at this time to do the calculation since the result is known anyway [3]. The second term in (7) was already a nontrivial check of the technique and we relied on that in [2].

More encouraging is to try to extend the technique to other theories. In case of the Ising model, perturbed off-criticality by a magnetic field, the spin-spin correlation function takes the form:

$$\langle \sigma_0 \sigma_R \rangle_h = \frac{\langle \sigma_0\, \sigma_R \exp\{h\int d^2x\,\sigma(x)\}\rangle_0}{\langle \exp\{h\int d^2x\,\sigma(x)\}\rangle_0} \tag{8}$$

The expansion analogous to (3) is

(9) $$< \sigma_0 \sigma_R >_h = < \sigma_0 \sigma_R >_0 + \frac{h^2}{2} \int d^2x_1 \int d^2x_2 < \sigma_0 \sigma_R \sigma_1 \sigma_2 > + \\ + \ldots$$

Again, scaling the R dependence out of the integrals provides an expansion of the scaling function $F_h(t)$. This is defined in the same way as in (4) but with the scaling variable now having the form:

(10) $$t = hR^{2-\Delta_\sigma} = hR^{15/8}$$

since the dimension of the perturbing operator in (8) is $\Delta_\sigma = 1/8$.

We recall at this point some details of the technique. The trouble with series like (3) or (9) is that the integrals involved are in general powerlike divergent. The problem is to find a uniquely defined technique which renders the integrals finite and so defines the exact expansion of the scaling function as in (7). This means that the regularization of the integrals amounts to an exact resummation technique with possibly a few constants left as adjustable parameters but unique for the whole series.

The way the integrals in (3) were defined in [2] was by using the Coulomb gas representation for the correlation functions of the conformal theory of the Ising model. Then a shift was made of the whole conformal theory by shifting the central charge from its Ising model value:

$$c = \frac{1}{2} \to c = \frac{1}{2} + \epsilon$$

(The shift of the conformal theory by a small amount can be implemented in the Coulomb gas representation because it is defined for a continuous range of c).

As result all the dimensions of the operators are shifted and it becomes possible to define the integrals by analytic continuation. Finally the limit $\epsilon \to 0$ is taken. As usual, the original logaarithmic divergences manifest themselves as poles in $\epsilon$:

(11) $$\frac{1}{(\epsilon)^n}$$

It was suggested in [2] that a unique replacement for all orders was:

$$\frac{1}{\epsilon} = A \log \frac{b}{|t|}\,. \tag{12}$$

The coefficient A can be fixed by the direct calculation of the leading logarithmic divergence in the first term in (3), leading to the first term in (7), while b remains an adjustable parameter. What is specific and unique about this analytic regularization (by analytic continuation in c) is that it preserves the conformal structure of the theory at all steps.

To check the regularization, the second order integral in (3) was calculated in [2]. Superficially, as it appears in (3), this might lead to an expression of the form:

$$t^2\left(C_2^2\left(\log \frac{b}{|t|}\right)^2 + C_2^1 \log \frac{b}{|t|} + C_2^0\right) \tag{13}$$

However, by direct analysis of the integral it was possible to check that $C_2^2 = 0$, see [4]. While the coefficients $C_2^1$ and $C_2^0$ had never been recovered from the perturbation expansion, use of the c continuation technique defined them both uniquely and correctly. Although this calculation does not constitute a complete check of the technique, we think that it is a rather nontrivial check. The main aspect that warrants further checking is the handling of logarithms in higher orders, by the prescription in (12).

As remarked above, it seems more useful to find some other application of the technique to make the effort worthwhile. For the Ising model perturbed slightly off $T_c$, the scaling function $F_\tau(t)$ of the correlator $<\sigma_0\sigma_R>_\tau$ in (4) has an expansion involving integer powers of t multiplied by logarithms of t. One explanation of this analyticity of $F_\tau(t)$ up to logarithms, is as a consequence of the integer dimension of the perturbing operator $\epsilon$, namely $\Delta_\epsilon = 1$. In fact $\epsilon$ is a free field in case of the Ising model. A related fact is that the $\epsilon$ operators, arising from the perturbing exponent in (2), do not produce any new primary operators in an operator product expansion because of the identity:

(14) $$\epsilon\epsilon \to I$$

This is different in the case of the Ising model in the magnetic field and the correlator $<\sigma_0\sigma_R>_h$ in (8). First, it is known [5] that the associated scaling function now involves terms nonanalytic in the field h, and so in t by the definition (10). By a simple scaling analysis, the expansion of the scaling function $F_h(t)$ is found to start as ([5], see also [2]):

(15) $$F_h(t) = 1 + \text{const}\ (t) + \ldots$$

On the other hand, if regularized as described in [2] (and sketched above) the expansion (9) for $<\sigma_0\sigma_R>_h$ would lead to the expansion of $F_h(t)$ which is analytic in t (again up to logarithms). This problem was remarked on in [2].

Second, we notice that unlike $\epsilon$ in (14) the perturbing operator $\sigma$ in (8) does produce a new primary field in operator product expansions since:

(16) $$\sigma\sigma \to I + \epsilon$$

To account for this new operator, which presumably enters the Hamiltonian dynamically anyway, we suggest extending the perturbing Hamiltonian in (8) with a new piece:

(17) $$\sim \int d^2x\,\epsilon(x)$$

Since there is only one perturbing field involved - the magnetic field h, the coefficient of (17) should be a scaling expression of h, like $h^\nu$, times a dimensionless coefficient. The exponent $\nu$ is easily fixed by dimensional analysis to be

(18) $$\nu = \frac{2-\Delta_\epsilon}{2-\Delta_\sigma} = \frac{8}{15}$$

Equation (8) is now replaced by the expression

$$\text{(19)} \qquad <\sigma_0\sigma_R>_h = \frac{<\sigma_0\,\sigma_R \exp\{h\int d^2x\sigma(x)+a(h)^{8/15}\int d^2x\epsilon(x)\}>}{<\exp\{h\int d^2x\sigma(x)+a(h)^{8/15}\int d^2x\epsilon(x)\}>}$$

where the effective Hamiltonian is supposed to account for the non perturbative (nonanalytic) effect. Here a is a dimensionless constant, which is to be a new adjustable parameter. The expansion for small h now reaches in increasing powers of h;

$$\begin{aligned} <\sigma_0\sigma_R>_h &= <\sigma_0\sigma_R>_0 + ah^{8/15}\int d^2x <\sigma_0\sigma_R\epsilon_x>_0 + \\ &+ \frac{a^2}{2}h^{16/15}\int d^2x_1\int d^2x_2 <\sigma_0\sigma_R\epsilon_1\epsilon_2>_0 + \\ &+ \frac{a^3}{3!}h^{24/15}\int d^2x_1\int d^2x_2 <\sigma_0\sigma_R\epsilon_1\epsilon_2\epsilon_3>_0 + \\ &+ \frac{1}{2}h^2\int d^2x_1\int d^2x_2 <\sigma_0\sigma_R\sigma_1\sigma_2>_0 \\ &+ \frac{a^4}{4!}h^{32/15}\int d^2x_1\int d^2x_2\int d^2x_3\int d^2x_4 <\sigma_0\sigma_R\epsilon_1\epsilon_2\epsilon_3\epsilon_4>_0 \\ &+ \ldots \end{aligned} \qquad (20)$$

The integrals appearing in this expansion can be calculated by using the analytic regularization. However, we can go one step further. Since the whole scheme is going to be either consistent or not, we may assume that the higher order terms in the expansions (3), (7) for $<\sigma_0\sigma_R>_\tau$ are given correctly by the analytic regularization as for the second order term. Consequently, they should agree with those in [3]. On the other hand, the corresponding integrals (those involving three and four $\epsilon$ operators) also appear in (20) and we can thus evaluate them from the results of [3]. This finally leads to the following expansion of the scaling function $F_h(t)$:

$$\begin{aligned} F_h(t) &= 1 + \pi a t^{8/15}\Omega + \frac{\pi^2}{4}a^2t^{16/15} + \frac{\pi^3}{4}a^3t^{24/15}\Omega \\ &+ Bt^2 + \frac{\pi^4}{16}a^4t^{32/15}(-\Omega^2+\Omega+\frac{1}{8}) \\ &+ O(t^{38/15}) \end{aligned} \qquad (21)$$

Here $\Omega = \log(b/|t|)$ and t is given by (10). The coefficient B of the $t^2$ term can be obtained explicitly by evaluating the $h^2$ integral in (20) with the technique of [2]. (So far we have only obtained it as a convergent numerical series, which at least allows it to be calculated numerically).

The expansion (21) involves one more adjustable parameter, a. However, the number of terms derived would be sufficient for fixing its value, if the shape of $F_h(t)$ was found either numerically by computer simulations or experimentally.

We stress again that the key aspects of the generalization of the analytic regularization (the resummation) technique discussed in this paper and in [2] are suggested above is that it appears to be defined in a unique way and, from the start, it respects the conformal structure of the theory. For these reasons we expect it to have a good chance of being correct.

We conclude, as in [2], by pointing out problems of extending the technique further to other theories. In most other cases, for example in $Z_3$ theory, there are also irrelevant (in terms of the renormalization group) primary operators. We do not see any reasons why they should not be included in the effective Hamiltonian of the analytic resummation technique, if they are generated by the operator products of the perturbing operator. However, if they are included, the fact that they have dimensions $\Delta > 2$ to terms, by the dimensional analysis,to terms in the effective Hamiltonian with negative powers of the perturbing field in front, cf. (18). In order to obtain the expansion of the scaling function in the expected form must occur in the perturbation expansion. This implies that, in principle, an infinite number of terms of the perturbation series will contribute. It means that to any particular t-power term of the expansion of $F(t)$. We do not know at present how to overcome this difficulty.

I am grateful for the hospitality extended to me at the Centre for Mathematical Analysis at ANU, Canberra, where this work was finished.

**REFERENCES**

1. G. Felder, Braid Matrices and Structure Constants for Minimal Conformal Models, ETH preprint, February 1989.

2. Vl. S. Dotsenko, Nucl. Phys. *B314* (1989) 687.

3. T. T. Wu, B. M. McCoy, C. A. Tracy and E. Barouch, Phys. Rev. *B13* (1976) 316.

4. J. B. Zuber and C. Itzykson, Phys. Rev. *D15* (1977) 2875.

5. A. M. Polyakov, Zh. Eksp. Teor. Fiz. *57* (1969) 271.
[JETP *30* (1970) 151]

International Journal of Modern Physics B, Vol. 4, No. 5 (1990) 1049 – 1057
© World Scientific Publishing Company

# HYPERGEOMETRIC–TYPE INTEGRALS AND THE $\mathfrak{sl}(2,\mathbf{C})$ KNIZHNIK–ZAMOLODCHIKOV EQUATION

Etsuro DATE

*Department of Mathematics, College of General Education, Kyoto University, Kyoto 606, Japan*

Michio JIMBO

*Department of Mathematics, Faculty of Science, Kyoto University, Kyoto 606, Japan*

Atsushi MATSUO and Tetsuji MIWA

*Research Institute for Mathematical Sciences, Kyoto University, Kyoto 606, Japan*

## 1. Introduction

Since the early stage of development in conformal field theory, it has been noticed that the correlation functions of primary fields satisfy differential equations of hypergeometric-type [3]. Apart from their physical implications, these equations exhibit intriguing mathematical properties. Most notable among such is the connection between the monodromy of these differential equations and braid representations germane to Hecke algebras and quantum groups [13,12,9]. In view of the recent development in the theory of hypergeometric functions [1,2,10], the mathematical structures of these equations warrant further investigation.

The aim of this paper is to construct integral representation of solutions to the $\mathfrak{sl}(2,\mathbf{C})$ Knizhnik-Zamolodchikov equation [11]. Our result is a generalization of Christe-Flüme's result [4] on the 4 point functions to the $n$ point functions. It is a consequence of simple calculation based on Aomoto's result as we shall explain below.

In [6] Dotsenko-Fateev initiated the study of integral representations of correlation functions. Using the vertex operator construction due originally to Feigin-Fuchs [8], they derived integrals representing 4-point functions in the minimal models. In the Wess-Zumino-Witten models, integral representation of 4-point functions has been obtained by Christe-Flüme [4] for the gauge symmetry algebra $\mathfrak{g} = \mathfrak{sl}(2,\mathbf{C})$. Since this is the case we are concerned with in this paper, let us recall here the relevant differential equations derived by Knizhnik-Zamolodchikov, the KZ equation [11]. For $j \in \frac{1}{2}\mathbf{Z}_{\geq 0}$, let $V_j$ denote the $(2j+1)$-dimensional irreducible representation of $\mathfrak{g}$. Given a sequence $j_1, \cdots, j_n \in \frac{1}{2}\mathbf{Z}_{\geq 0}$, the $\mathfrak{sl}(2, \mathbf{C})$ KZ equation is the following system of differential equations for a function $\psi(z_1, \cdots, z_{n-1})$ taking values in $\mathrm{Hom}_{\mathfrak{g}}(V_{j_1} \otimes \cdots \otimes V_{j_{n-1}}, V_{j_n})$ :

$$d\psi = \nu \sum_{1 \leq a < b \leq n-1} \Omega_{ab}^{*} \psi d \log(z_a - z_b). \tag{1.1}$$

Here $\Omega_{ab}^{*}\psi = \psi \circ \Omega_{ab}$ is the pullback of $\psi$ by

$$\Omega_{ab} = \frac{1}{2}\rho_a(h)\rho_b(h) + \rho_a(e)\rho_b(f) + \rho_a(f)\rho_b(e),$$
$$h = \begin{pmatrix} 1 & 0 \\ 0 & -1 \end{pmatrix}, e = \begin{pmatrix} 0 & 1 \\ 0 & 0 \end{pmatrix}, f = \begin{pmatrix} 0 & 0 \\ 1 & 0 \end{pmatrix},$$

and $\rho_a(X) = 1 \otimes \cdots \otimes \overset{\overset{a}{\vee}}{X} \otimes \cdots \otimes 1$ signifies the action of $\mathfrak{g}$ on the $a$-th component. The parameter $\nu$ is related to the level $l \in \mathbf{Z}_{\geq 0}$ via $\nu = \frac{1}{l+2}$. Equation (1.1) however makes sense for arbitrary complex parameters $\nu, j_1, \cdots, j_n$ (taking $V_j$ to be the irreducible module with the highest weight $2j$). We shall henceforth adopt this point of view.

In [2] Aomoto studied integrals of the form

$$\int \cdots \int \prod_{i=1}^{m} dt_i \prod_{i<j} (t_i - t_j)^{\nu} \prod_{i=1}^{m} \prod_{a=1}^{n-1} (t_i - z_a)^{\lambda_a}$$
$$\times \frac{1}{t_1 - z_{a_1}} \cdots \frac{1}{t_m - z_{a_m}}. \tag{1.2}$$

Christe-Flüme's integral has precisely the form (1.2) with $n = 4$ and

$$\lambda_a = -\nu j_a.$$

The choice of cycles for the integration is such that the integral is symmetric with respect to permutations of the integration variables $t_1, \cdots, t_m$. This implies that the order of $a_1, \cdots, a_m$ is irrelevant and the integral depends only on $\alpha = (\alpha_1, \cdots, \alpha_{n-1})$ where

$$\alpha_a = \#\{i \mid a_i = a\}.$$

We shall accordingly denote the integral by $I_\alpha(z)$.

Aomoto derived the following linear constraints for $I_\alpha(z)$;

$$\sum_a (\lambda_a + \frac{\alpha_a \nu}{2}) I_{\alpha_1, \cdots, \alpha_a + 1, \cdots, \alpha_{n-1}} = 0 \tag{1.3}$$

for any $\alpha \in (\mathbf{Z}_{\geq 0})^{n-1}$ such that $|\alpha| = m - 1$, and also the following linear differential equations;

$$dI_\alpha = \sum_{a \neq b} \alpha_a (\lambda_a + \frac{\alpha_b \nu}{2})(I_\alpha - I_{\alpha^{ab}}) d \log(z_a - z_b) \tag{1.4}$$

where $\alpha^{ab}$ is obtained from $\alpha$ by changing $\alpha_a$ to $\alpha_a - 1$ and $\alpha_b$ to $\alpha_b + 1$.

Suppose that $m = j_1 + \cdots + j_{n-1} - j_n$. Consider the irreducible $\mathfrak{g}$-module $V_{j_a}^*$ with the lowest weight $-2j_a$. The notation $V_j^*$ comes from the fact that there exists a dual pairing between $V_j$ and $V_j^*$ such that

$$(Xv, v^*) = -(v, Xv^*),$$

for $v \in V_j$, $v^* \in V_j^*$ and $X \in \mathfrak{g}$. Choose a lowest weight vector $v_{j_a}^*$ in $V_{j_a}^*$, and consider the following vector in $V_{j_1}^* \otimes \cdots \otimes V_{j_{n-1}}^*$:

$$\begin{aligned} w^*(z) &= \prod_{1\le a<b\le n-1} (z_a - z_b)^{2j_a j_b \nu} \\ &\times \sum_{|\alpha|=m} I_\alpha(z) \frac{e^{\alpha_1}}{\alpha_1!} v_{j_1}^* \otimes \cdots \otimes \frac{e^{\alpha_{n-1}}}{\alpha_{n-1}!} v_{j_{n-1}}^*. \end{aligned}$$

Because of (1.3) this is a lowest weight vector with weight $-2j_n$, i.e.,

$$fw^*(z) = 0, \quad hw^*(z) = -2j_n w^*(z).$$

Therefore we can define $\psi(z) \in \mathrm{Hom}_{\mathfrak{g}}(V_{j_1} \otimes \cdots \otimes V_{j_{n-1}}, V_{j_n})$ by

$$(\psi(z)v,\ f^k v_{j_n}^*) = (v, f^k w^*(z))$$

for any $k \geq 0$ and $v \in V_{j_1} \otimes \cdots \otimes V_{j_{n-1}}$. Then Aomoto's equation (1.4) for $I_\alpha(z)$ with $\lambda_a = -\nu j_a$ turns into the KZ equation (1.1).

In this paper we present the whole derivations for the above statements. In Section 2 we review Aomoto's work and reproduce the proof of (1.3) and (1.4). In Section 3 we construct $w^*(z)$ and interpret (1.3) and (1.4) as the KZ equation.

We thank A.B. Zamolodchikov for explaining us the boson representation of the affine Lie algebra $\widehat{\mathfrak{sl}}(2, \mathbf{C})$. After completing the work we learned about the preprint of Feigin- Frenkel [7], in which a complete study of the boson representation of the affine Lie algebras is given. Two of us (M.J. and T.M.) also thank V. Dotsenko for explaining how to apply Feigin-Frenkel's result to the integral representation of correlation functions. We stress the point that the method in this paper is simple and direct, and uses no particular technique like the Feigin-Frenkel type bosonization.

Besides the above acknowledgement we would like to thank K. Aomoto, I.M. Gelfand, Y. Kanie, M. Kashiwara, T. Kohno, V. Rittenberg and A. Tsuchiya for discussions and useful informations. Two of us (M.J. and T.M.) are grateful to the colleagues in the Centre for Mathematical Analysis, the Australian National University, for hospitality extended to them during their stay in June-July, 1989, which was most enjoyable and exciting.

## 2. Hypergeometric–Type Integrals and Related Linear Differential Equations

We start from the following setting; the projection

$$\begin{array}{cccc} \pi : \mathbf{C}^{m+n-1} & \longrightarrow & \mathbf{C}^{n-1} & \\ \Cup & & \Cup & , \\ (t_1,\ldots,t_m,z_1,\ldots,z_{n-1}) & \longmapsto & (z_1,\ldots,z_{n-1}) & \end{array}$$

and the multi-valued $m$-form

$$\Phi dt_1 \wedge \cdots \wedge dt_m = \prod_{1\le i<j\le m}(t_i-t_j)^{\nu}\prod_{i=1}^{m}\prod_{a=1}^{n-1}(t_i-z_a)^{\lambda_a}dt_1\wedge\cdots\wedge dt_m.$$

The aim of this section is to derive linear differential equations for the integrals of hypergeometric type:

$$\int \varphi\Phi dt_1 \wedge \cdots \wedge dt_m. \tag{2.1}$$

Here $\varphi$ is a rational function of $t_1,\cdots,t_m,\ z_1,\cdots,z_{n-1}$ such that its poles are contained in the union of hyperplanes:

$$D = \bigcup_{i<j}\{t_i=t_j\}\bigcup_{i,a}\{t_i=z_a\}.$$

This problem was formulated and solved by Aomoto [2]. The case $m=1$ and $n=4$ reduces to the Gauss hypergeometric equation. The case of $m\ge 1$ and $n=4$ was independently studied by Christe and Flüme [4]. They established the equivalence between the linear differential equations for these integrals and the KZ equation.

We shall not go into details about the precise meaning of the integrals (2.1), i.e., how to specify the cycles on which they are integrated (cf. [2]). We manipulate those integrals assuming the following properties:

(i) $$\int(\sigma^*\varphi)\Phi dt_1\wedge\cdots\wedge dt_m = \int\varphi\Phi dt_1\wedge\cdots\wedge dt_m, \tag{2.2}$$

where $\sigma$ is an arbitrary permutation of $\{1,\cdots,m\}$ and

$$(\sigma^*\varphi)(t_1,\cdots,t_m,z_1,\cdots,z_{n-1}) = \varphi(t_{\sigma(1)},\cdots,t_{\sigma(m)},z_1,\cdots,z_{n-1}).$$

(ii) $$\int(\nabla_{t_i}\varphi)\Phi dt_1\wedge\cdots\wedge dt_m = 0. \tag{2.3}$$

(iii) $$\frac{\partial}{\partial z_a}\left(\int\varphi\Phi dt_1\wedge\cdots\wedge dt_m\right) = \int(\nabla_{z_a}\varphi)\Phi dt_1\wedge\cdots\wedge dt_m,$$

where

$$\nabla_{t_i} = \frac{\partial}{\partial t_i}+\frac{\partial}{\partial t_i}\log\Phi,\quad \nabla_{z_a} = \frac{\partial}{\partial z_a}+\frac{\partial}{\partial z_a}\log\Phi.$$

Aomoto chose $\varphi$ in (2.1) for each $A = (a_1, \ldots, a_m)$ as follows:

$$\varphi_A = \prod_{i=1}^{m} (t_i - z_{a_i})^{-1}.$$

We define $\alpha = (\alpha_1, \ldots, \alpha_{n-1})$ by

$$\alpha_a = \sharp\{i \mid a_i = a\}.$$

Then the integral $\int \varphi_A \Phi dt_1 \wedge \cdots \wedge dt_m$ depends only on $\alpha$ and not on the particular choice of $A$. Therefore we denote it by $I_\alpha(z)$. Even though the integrals $I_\alpha(z)$ are not linearly independent, it is transparent to use all $I_\alpha(z)$ and to write both the linear constraints (1.3) and the differential equations (1.4) among them.

Let us derive these equations. For rational functions $\psi$, $\varphi$ we write $\psi \equiv \varphi$ if and only if the equality

$$\int \psi \Phi dt_1 \wedge \cdots \wedge dt_m = \int \varphi \Phi dt_1 \wedge \cdots \wedge dt_m$$

follows from (2.2) and (2.3). For example,

$$\left( \frac{\lambda_{a_i} - 1}{t_i - z_{a_i}} + \sum_{j(\neq i)} \frac{\nu}{t_i - t_j} + \sum_{b(\neq a_i)} \frac{\lambda_b}{t_i - z_b} \right) \varphi_A \equiv 0,$$

because the left hand side is $\nabla_{t_i} \varphi_A$. Using this we obtain

$$\begin{aligned} \nabla_{z_a} \varphi_A &= \left( \sum_{\substack{i \\ a_i = a}} \frac{1 - \lambda_a}{t_i - z_a} + \sum_{\substack{i \\ a_i \neq a}} \frac{-\lambda_a}{t_i - z_a} \right) \varphi_A, \\ &\equiv \sum_{\substack{i \\ a_i = a}} \left( \sum_{j(\neq i)} \frac{\nu}{t_i - t_j} + \sum_{b(\neq a_i)} \frac{\lambda_b}{t_i - z_b} \right) \varphi_A + \sum_{\substack{i \\ a_i \neq a}} \frac{-\lambda_a}{t_i - z_a} \varphi_A. \end{aligned}$$

We rewrite each term by using the identities

$$\frac{1}{t_i - z_a} \cdot \frac{1}{t_i - z_b} = \frac{1}{z_a - z_b} \left( \frac{1}{t_i - z_a} - \frac{1}{t_i - z_b} \right),$$

$$\begin{aligned} &\frac{1}{t_i - t_j} \cdot \frac{1}{t_i - z_a} \cdot \frac{1}{t_j - z_b} + \frac{1}{t_j - t_i} \cdot \frac{1}{t_j - z_a} \cdot \frac{1}{t_i - z_b} \\ &= \frac{-1}{z_a - z_b} \left( \frac{1}{t_i - z_a} - \frac{1}{t_i - z_b} \right) \left( \frac{1}{t_j - z_a} - \frac{1}{t_j - z_b} \right), \end{aligned}$$

and we obtain (1.4). For (1.3), start from

$$\begin{aligned} 0 &\equiv \left( \frac{\partial}{\partial t_m} + \frac{\partial}{\partial t_m} (\log \Phi) \right) \prod_{1 \leq i \leq m-1} (t_i - z_{a_i})^{-1}, \\ &= \left( \sum_{j(\neq m)} \frac{\nu}{t_m - t_j} + \sum_a \frac{\lambda_a}{t_m - z_a} \right) \prod_{1 \leq i \leq m-1} (t_i - z_{a_i})^{-1}. \end{aligned}$$

Using the identity

$$\frac{1}{t_m - t_j} \cdot \frac{1}{t_j - z_{a_j}} + \frac{1}{t_j - t_m} \cdot \frac{1}{t_m - z_{a_j}} = \frac{1}{t_j - z_{a_j}} \cdot \frac{1}{t_m - z_{a_j}}$$

for the first term, we obtain (1.3).

## 3. Identification with the KZ Equation

Our goal in this section is to identify Aomoto's differential equation with the KZ equation. We shall employ the following notations. Let $\mathfrak{g} = \mathfrak{sl}(2, \mathbf{C})$. We denote its standard basis by $e = \begin{pmatrix} 0 & 1 \\ 0 & 0 \end{pmatrix}, f = \begin{pmatrix} 0 & 0 \\ 1 & 0 \end{pmatrix}, h = \begin{pmatrix} 1 & 0 \\ 0 & -1 \end{pmatrix}$. For $j \in \mathbf{C}$ let $V_j$ be the unique irreducible $\mathfrak{g}$ -module with highest weight $2j$. Thus for $j \in \frac{1}{2}\mathbf{Z}_{\geq 0}$ $V_j$ is $(2j+1)$ dimensional, and for $j \notin \frac{1}{2}\mathbf{Z}_{\geq 0}$ $V_j$ coincides with the Verma module. Let $V_j^*$ denote the restricted dual $\mathfrak{g}$-module of $V_j$ (i.e. the direct sum of the vector spaces dual to the weight spaces of $V_j$). It is an irreducible lowest weight $\mathfrak{g}$ -module via $\langle X v^*, v\rangle = -\langle v^*, Xv\rangle$ for $X \in \mathfrak{g}, v^* \in V_j^*$ and $v \in V_j$.

From now on we fix a sequence of complex numbers $j_1, \cdots, j_n$ such that

$$j_n = j_1 + \cdots + j_{n-1} - m, \ m \in \mathbf{Z}_{\geq 0}. \tag{3.1}$$

For $1 \leq a < b \leq n-1$ we set

$$\Omega_{ab} = \frac{1}{2}\rho_a(h)\rho_b(h) + \rho_a(e)\rho_b(f) + \rho_a(f)\rho_b(e) \tag{3.2}$$

where $\rho_a(X) = 1 \otimes \cdots \otimes \overset{\overset{a}{\vee}}{X} \otimes \cdots \otimes 1$ denotes the $\mathfrak{g}$-action on the $a$-th component of the tensor product $V_{j_1} \otimes \cdots \otimes V_{j_{n-1}}$. Following the line of [13], we formulate the KZ equation for a function $\psi(z_1, \cdots, z_{n-1})$ taking values in $\mathrm{Hom}_{\mathfrak{g}}(V_{j_1} \otimes \cdots \otimes V_{j_{n-1}}, V_{j_n})$. Physically $j_a$ stands for (half) the weight of the primary field at the point $z_a$, and we have taken $z_n$ to be $\infty$. The KZ equation reads as follows:

$$d\psi = \nu \sum_{1 \leq a < b \leq n-1} \Omega_{ab}^* \psi \ d\log(z_a - z_b). \tag{3.3}$$

Here $\Omega_{ab}^*\psi = \psi \circ \Omega_{ab}$ denotes the pull-back of $\psi$ by $\Omega_{ab}$ (3.2). In the original setting, $j_a$ runs over the set $\{0, 1/2, 1, \cdots, l/2\}$ with $l \in \mathbf{Z}_{\geq 0}$ being the level of the representation, and $\nu = \frac{1}{l+2}$. In the sequel we regard $j_1, \cdots, j_n, \nu$ as free complex parameters. For simplicity we consider either of the two cases

(1) $j_1, \cdots, j_n \notin \frac{1}{2}\mathbf{Z}_{\geq 0}$,

(2) $j_1, \cdots, j_n \in \frac{1}{2}\mathbf{Z}_{\geq 0}$.

Let

$$M_{-j_n} = \{v^* \in V_{j_1}^* \otimes \cdots \otimes V_{j_{n-1}}^* \mid fv^* = 0, \ hv^* = -2j_n v^*\}. \tag{3.4}$$

We have then isomorphisms of vector spaces

$$\begin{aligned} \mathrm{Hom}_{\mathfrak{g}}(V_{j_1}\otimes\cdots\otimes V_{j_{n-1}}, V_{j_n}) &\xrightarrow{\sim} \mathrm{Hom}_{\mathfrak{g}}(V^*_{j_n}, V^*_{j_1}\otimes\cdots\otimes V^*_{j_{n-1}}) \\ &\xrightarrow{\sim} M_{-j_n}. \end{aligned} \tag{3.5}$$

The first is canonical. The second is defined by fixing a lowest weight vector $v^*_{j_n}\in V^*_{j_n}$ and sending $\varphi\in\mathrm{Hom}_{\mathfrak{g}}(V^*_{j_n}, V^*_{j_0}\otimes\cdots\otimes V^*_{j_{n-1}})$ to $\varphi(v^*_{j_n})$. It is an isomorphism in the case (1) or (2). Using the identification (3.5), the KZ equation (3.3) can be reformulated for an $M_{-j_n}$-valued function $w^*(z_1,\cdots,z_{n-1})$. Denoting the action of $\Omega_{ab}$ on $V^*_j\otimes\cdots\otimes V^*_{j_{n-1}}$ by the same $\Omega_{ab}$ we have

$$dw^*(z) = \nu \sum_{1\le a<b\le n-1} \Omega_{ab}(w^*(z))d\log(z_a - z_b). \tag{3.6}$$

Choose a lowest weight vector $v^*_{j_a}\in V^*_{j_a}$ for each $a=1,\cdots,n-1$. For a multi-index $\alpha=(\alpha_1,\cdots,\alpha_{n-1})$ we set

$$v^{*(\alpha)} = \frac{e^{\alpha_1}}{\alpha_1!}v^*_{j_1}\otimes\cdots\otimes\frac{e^{\alpha_{n-1}}}{\alpha_{n-1}!}v^*_{j_{\alpha_{n-1}}}.$$

Using the rules $[f,e^k] = -ke^{k-1}(h+k-1)$ and $h\,v^*_{j_a} = -2j_a\,v^*_{j_a}$, we obtain

$$\begin{aligned} \rho^*_a(h)v^{*(\alpha)} &= (-2j_a + 2\alpha_a)v^{*(\alpha)}, \\ \rho^*_a(e)v^{*(\alpha)} &= (\alpha_a+1)v^{*(\alpha+\varepsilon_a)}, \\ \rho^*_a(f)v^{*(\alpha)} &= (2j_a-\alpha_a+1)v^{*(\alpha-\varepsilon_a)}. \end{aligned} \tag{3.7}$$

Here $\rho^*_a$ denotes the $\mathfrak{g}$-action on the $a$-th component of $V^*_{j_1}\otimes\cdots\otimes V^*_{j_{n-1}}$ and $\varepsilon_a = (0,\cdots,\overset{a}{\check{1}},\cdots 0)$. Thus an element of $V^*_{j_1}\otimes\cdots\otimes V^*_{j_{n-1}}$ of weight $-2j_n = -2(j_1+\cdots+j_{n-1}-m)$ can be written as

$$w^* = \sum_{|\alpha|=m} c_\alpha v^{*(\alpha)}, \quad c_\alpha\in\mathbf{C}. \tag{3.8}$$

In the case of (2), we understand $c_\alpha = 0$ if $\alpha_a > 2j_a$ for some $a$.

**Proposition 3.1.** *In the notation of (3.8), $w^*$ belongs to $M_{-j_n}$ (3.4) if and only if*

$$\sum_{a=1}^{n-1}(-2j_a+\beta_a)c_{\beta+\varepsilon_a} = 0 \tag{3.9}$$

*holds for any $\beta$ with $|\beta| = m-1$.*

*Proof.* Using (3.7), we have

$$\begin{aligned} fw^* &= \sum_{a=1}^{n-1}\rho^*_a(f)w^* \\ &= \sum_{|\alpha|=m}\sum_{a=1}^{n-1} c_\alpha(2j_a-\alpha_a+1)v^{*(\alpha-\varepsilon_a)} \\ &= \sum_{|\beta|=m-1}\sum_{a=1}^{n-1} c_{\beta+\varepsilon_a}(2j_a-\beta_a)v^{*(\beta)}, \end{aligned}$$

from which follows the assertion.

**Proposition 3.2.** *The action of $\Omega_{ab}$ on the vector $w^*$ (3.8) is given as follows:*

$$\begin{aligned}(\Omega_{ab} - 2j_a j_b)w^* &= \sum_{|\alpha|=m} \{\alpha_a(-2j_b + \alpha_b)(c_\alpha - c_{\alpha-\varepsilon_a+\varepsilon_b}) \\ &\quad +\alpha_b(-2j_a + \alpha_a)(c_\alpha - c_{\alpha-\varepsilon_b+\varepsilon_a})\}v^{*(\alpha)}. \end{aligned} \tag{3.10}$$

*Proof.* Direct calculation using (3.2), (3.7) shows

$$\begin{aligned}\Omega_{ab}w^* &= \sum_{|\alpha|=m} c_\alpha\{2(-j_a + \alpha_a)(-j_b + \alpha_b)v^{*(\alpha)} \\ &\quad + (\alpha_a + 1)(2j_b - \alpha_b + 1)v^{*(\alpha+\varepsilon_a-\varepsilon_b)} \\ &\quad + (\alpha_b + 1)(2j_a - \alpha_a + 1)v^{*(\alpha+\varepsilon_b-\varepsilon_a)}\} \\ &= \sum_{|\alpha|=m} \{2(-j_a + \alpha_a)(-j_b + \alpha_b)c_\alpha \\ &\quad + \alpha_a(2j_b - \alpha_b)c_{\alpha-\varepsilon_a+\varepsilon_b} \\ &\quad + \alpha_b(2j_a - \alpha_a)c_{\alpha-\varepsilon_b+\varepsilon_a}\}v^{*(\alpha)},\end{aligned}$$

whence follows the result.

Comparing (3.9), (3.10) with (1.3), (1.4) we conclude as follows.

**Theorem 3.3.** *Let*

$$w^*(z_1, \cdots, z_{n-1}) = \prod_{1\le a<b\le n-1} (z_a - z_b)^{2j_a j_b \nu} \sum_{|\alpha|=m} I_\alpha(z_1, \cdots, z_{n-1})v^{*(\alpha)}.$$

*Then $w^*(z)$ belongs to $M_{-j_n}$ and satisfies the KZ equation (2.6) if and only if the $I_\alpha$ satisfy Aomoto's linear relation (1.3) and differential equation (1.4).*

## References


1. K. Aomoto, On the structure of integrals of power product of linear functions, *Sci. Papers Coll. Gen. Ed. Univ. Tokyo* **27** (1977) 49-61.

2. K. Aomoto, Gauss-Manin connection of integrals of difference products, *J. Math. Soc. Japan* **39** (1987) 191-208.

3. A. A. Belavin, A. N. Polyakov and A. B. Zamolodchikov, Infinite conformal symmetries in two dimensional quantum field theory, *Nuclear Physics*, **B241** (1984) 333-380.

4. P. Christe, R. Flüme, The four point correlations of all primary operators of the $d = 2$ conformally invariant SU(2) $\sigma$-model with Wess-Zumino term, *Nuclear Physics* **B282** (1987) 466-494.

5. P. Christe, Structure and applications of the Wess-Zumino-Witten model in two dimensions, Ph. D. thesis, Bonn. Univ.

6. Vl. S. Dotsenko, V. A. Fateev, Conformal algebra and multipoint correlation functions in 2D statistical models, *Nuclear Physics* **B240** (1984) 312-348.

7. B. Feigin, E. Frenkel, Representations of Affine Kac-Moody algebras and bosonization, preprint (1989); Representations of Affine Kac-Moody algebras, bosonization and resolutions, preprint (1989); Affine Kac-Moody algebras and semi-infinite flag manifolds, preprint (1989).

8. B. L. Feigin, D. B. Fuchs, Representations of the Virasoro algebra, in: Seminar on supermanifolds, Reports of the Dept. Math. Stockholm Univ., (1986), to be published in: Representation of infinite-dimensional Lie groups and Lie algebras. N. Y. Gordon and Breach.

9. G. Felder, J. Frölich, G. Keller, Braid matrices and the structure constants for the minimal conformal models, preprint (1989).

10. V. A.Vasil'ev, I. M. Gel'fand, A. V. Zelevinskii, General hypergeometric functions on complex Grassmannians, *Funct. Anal. Prilozhen.* **21** (1987) 19-31.

11. V.G. Knizhnik, A. B. Zamolodchikov, Current algebra and Wess-Zumino models in two dimensions, *Nuclear Physics* **B247** (1984) 83-103.

12. T. Kohno, Monodromy representations of braids groups and Yang-Baxter equations, *Ann. Inst. Fourier* **37**(4) (1987) 139-160.

13. A. Tsuchiya, Y. Kanie, Vertex operators in conformal field theory on $\mathbf{P}^1$ and monodromy representstions of Braid groups, *Adv. Stud. Pure. Math* **16** (1988) 297-372.

International Journal of Modern Physics B, Vol. 4, No. 5 (1990) 1059 – 1068
© World Scientific Publishing Company

# SIMPLE EXAMPLES OF CONFORMAL FIELD THEORIES

A.L. CAREY

*Department of Pure Mathematics, University of Adelaide, G.P.O. Box 498*
*Adelaide 5001, Australia*

and

K.C. HANNABUSS

*Balliol College, Oxford OX1 3BJ, U.K*

## 1. Introduction

The main aim of these notes is to record some examples of conformal field theories in terms which are familiar to 'old-fashioned' quantum field theorists. Basically what we show here is that many of the 'examples' which have appeared in the literature can be easily understood in terms of representations of the algebras of the canonical commutation (CCR) and anticommutation relations (CAR). The representations we describe here are of the sort called quasi-free that is, there is no interaction. The interest in them lies in their connections with other mathematical topics (Riemann surfaces and theta functions in particular). Indeed as we shall see the $n$-point (or correlation functions) of these representations possess holomorphy properties far more interesting than those arising from representations of the CCR or CAR so far considered.

Our motivation lies not in the direction of interacting field theories in two dimensions as such but in the conformal invariance of two dimensional models in statistical mechanics. It has been our view (and one shared by others [12]) that an understanding of these matters may be facilitated by considering field theories on the spectral curves associated with various transfer matrices. In order for this view to make sense it is essential that we provide some meaning for the term 'field theory on a spectral curve'. We will, on the basis of our examples below give a general discussion of what we mean by this in the final section.

As far as the existing literature is concerned (with the exception of the general framework proposed by Segal [15]) it seems that little attention has been paid to the usual questions an old-fashioned quantum field theorist might ask. Principally we are thinking of the questions: what is the Hilbert space of states, how do the fields act and do the $n$-point functions which have been explicitly given satisfy the usual positivity requirements of a field theory?

This work is an extension of previous collaborations with Mike Eastwood and John Palmer whom we thank for many discussions.

## 2. Algebras

In this section we introduce briefly the fermionic (CAR) and bosonic (CCR) algebras (see [3] for more details). The former, denoted $A(H)$, is constructed by starting with a Hilbert space $H$ and introducing the Fock space $\wedge H$ (the completion in the obvious Hilbert space structure of the exterior algebra over $H$). The vector $\Omega = 1 \oplus 0 \oplus 0 \ldots$ we call the vacuum. For each $h \in H$ define the operator $a^*(h)$ on $\wedge H$ by

$$a^*(h) : h_1 \wedge h_2 \wedge h_3 \ldots \wedge h_n \to h \wedge h_1 \wedge h_2 \wedge h_3 \ldots \wedge h_n.$$

The Hilbert space adjoint of this operator is denoted $a(h)$ and together these operators satisfy the familiar relations

$$a^*(g)a(h) + a(h)a^*(g) = < g, h >,$$

$$a(h)a(g) + a(g)a(h) = 0.$$

Irreducible representations of $A(H)$ may be obtained by considering orthogonal projections $P$ on $H$. Each such projection defines a representation $\pi_P$ by

$$\pi_P(a(h)) = a(Ph) + a^*(\overline{(1 - P)h}).$$

(Here the bar denotes a complex conjugation commuting with $P$.) Quasifree representations of the CAR algebra are all obtained by embedding $H$ as a subspace of a bigger Hilbert space $K$, taking an irreducible representation of $A(K)$ given by some projection $P$ and considering the restriction of this representation to $A(H)$. The cyclic subrepresentation generated from $\Omega$ is quasifree and is completely determined by the operator $QPQ$ where $Q$ is the orthogonal projection onto $H$ (cf [13],[1]).

To construct the CCR algebra one starts with a real inner product space $S$ equipped with a non-degenerate symplectic form $\sigma$. Then the CCR algebra is the C*-algebra generated by elements $\delta_f, f \in S$ satisfying

$$\delta_f \delta_g = e^{-i\sigma(f,g)/2} \delta_{f+g},$$

$$\delta_f^* = \delta_{-f}.$$

Representations of this algebra are obtained by considering complex structures $J$ on $S$ which preserve $\sigma$ and are such that the sesquilinear form

$$(f, g) = \sigma(f, Jg) + i\sigma(f, g)$$

together with $J$ equips $S$ with a complex pre-Hilbert space structure. (This situation always arises by taking a complex pre-Hilbert space and equipping it with the symplectic form given by the imaginary part of the inner product.) Then $J$ determines a state on the C*-algebra which is given on the generators by

$$\omega(\delta_f) = e^{-\|f\|^2/4}.$$

The quasifree states of interest to us will be those given by taking a positive operator $T$ on $S$ and introducing the state

$$\omega_T(\delta_f) = e^{-\|f\|^2/4} e^{-(f,Tf)/2}.$$

## 3. The Examples

### 3.1 *The Krichever-Segal-Wilson example*

This example is well known and we include it to avoid confusion with our second example. Here one takes a Riemann surface $\Sigma$ and fixes a point $p$ on the surface. Consider a complex line bundle $L$ over $\Sigma$ together with a local co-ordinate $z$ around $p$. With the convention that $p$ is the point at infinity in this co-ordinate we can now associate with this data a subspace $W$ of $L^2(S^1)$. By choosing a local trivialisation of $L$ restricted to a neighbourhood of the unit disc (in the z co-ordinate) at $p$ this subspace may be identified with the the closure of the space of meromorphic sections of $L$ which are holomorphic except at $p$.

Then we define a representation of the CAR algebra over $L^2(S^1)$ as in the previous section using the orthogonal projection $P$ onto $W$.

### 3.2 *Bordered Riemann surfaces*

This example seems to have been confused with the first one in the literature. They are definitely related and we will discuss the connection below however in order to illustrate clearly the distinction and to make contact with our earlier work [6], we will separate out the genus one case for special mention.

#### 3.2.1 *The torus*

This is an elaboration of the discussion on pages 132 and 133 of Fay [10]. We regard the torus as constructed from two annuli by joining along the boundary. Equivalently (at least topologically) this is the same as joining two cylinders to form the torus.

We let $R$ be the annulus (in the complex plane) $\{w : e^{-\beta/2} < |w| < 1\}$ and $\bar{R}$ be the annulus $\{w : 1 < |w| < e^{\beta/2}\}$ where $\beta > 0$. We may now form the torus $\Sigma = R \sqcup \partial R \sqcup \bar{R}$ by identifying the boundaries of $R$ and $\bar{R}$. Note that $\Sigma$ possesses an anticonformal involution: $w^\flat = 1/\bar{w}$.

With respect to the canonical homology basis

$$A = \{e^{\beta/2+i\theta}, 0 \le \theta \le 2\pi\}, B = \{x \in \mathbb{R} : e^{-\beta/2} < x < e^{\beta/2}\}$$

we have the period matrix as just $\beta$ so we may think of the torus as the complex plane modulo the lattice $2\pi\mathbb{Z} + i\beta\mathbb{Z}$

Introduce the four spin structures (square roots of the canonical bundle) $L^\alpha$ where $\alpha = (0,0), (1/2,0), (0,1/2), (1/2,1/2)$ denotes the so-called theta characteristics corresponding to each spin structure. These line bundles have sections which may be regarded as functions on the complex plane with specific equivariance properties under the action of the generators of the lattice. Specifically these are: periodic under both generators, periodic under one and anti-periodic under the other and antiperiodic under both respectively. The Hilbert spaces for the fermionic algebras are the completion of the space of smooth sections of $L^\alpha|_{\partial R}$ in the $L^2$ norm. (To understand this definition it is perhaps necessary to point out that if $K$ denotes the canonical bundle with transition functions $z_{\gamma\delta}$ then $|K|$ is the bundle with transition functions $|z_{\gamma\delta}|$ so that if $f$ a section of a square root of $K$ then $|f|^2$ is a section of $|K|$. So it makes sense to integrate $|f|^2$ on the two boundary circles and thus the $L^2$ norm is meaningful.) We denote these spaces by $H^\alpha$. The appropriate representation of the CAR algebra $A(H^\alpha)$ is in the first three cases given by the projection $P^\alpha$ onto the subspace of $H^\alpha$ which is the closure of the subspace obtained by restriction to the boundary of holomorphic sections. (For the last spin structure case we indicate the problem with this construction in the next subsection.)

To see that this is an interesting thing to do we need to give these projections explicitly. In each case they are given by the Szego kernel which in Fay's notation [10] is written $\sigma_\alpha(\bar{x}, y)$. We will write this down explicitly in just one case: that covered by our earlier work on KMS states on the loop group [6]. The other cases are similar and lead to similar conclusions.

Explicitly, on restriction to the $A$ cycle in the boundary of $R$ we have

$$\sigma_{(0,0)}(e^{-i\xi}, e^{i\phi}) = \theta_3(\phi - \xi)\theta_1'(0)[\theta_3(0)\theta_1(\phi - \xi)]^{-1}\sqrt{dx}\sqrt{d\bar{y}}$$

where $x = e^{i\phi}, y = e^{i\xi}$ and the factor $\sqrt{dx}\sqrt{d\bar{y}}$ is a 1/2-form in each variable, the result of the fact that the definition of the Szego kernel involves the prime form which is constructed from a quotient of the theta function in the denominator by sections of $L^{(1/2,1/2)}$. Note that our notation for the theta functions is classical [15] and differs from that of Fay.

On the other hand in [6] we gave a novel proof using quantum field theory of the (well known) identity:

$$\sum_n e^{-\beta(n+1/2)}(1 + e^{-\beta(n+1/2)})^{-1} e^{in(\phi-\xi)} = \theta_3(\phi - \xi)\theta_1'(0)[\theta_3(0)\theta_1(\phi - \xi)]^{-1} e^{i(\xi-\phi)/2}.$$

Using this identity let us compute the effect of applying the projection defined by the Szego kernel to a pair of sections $(f\sqrt{dx}, g\sqrt{dx})$ in our Hilbert space and restricting the result to the $A$ cycle. (Note that $f$ and $g$ are $L^2$ functions, with $f$ defined on the $A$ cycle and $g$ on the other cycle.) After some calculation the end result is

$$\sum_n [e^{-\beta(n+1/2)}(1 + e^{-\beta(n+1/2)})^{-1} e^{in\phi} \hat{f}_n \sqrt{dx}$$

$$+e^{-\beta(n+1/2)/2}(1+e^{-\beta(n+1/2)})^{-1}e^{in\phi}\hat{g}_n\sqrt{dx}]$$

where the Fourier transform is

$$f(\phi)=(2\pi)^{-1/2}\sum_n \hat{f}_n e^{in\phi}.$$

The significance of this formula is that it represents the action on $L^2(S^1)\oplus L^2(S^1)$ of the projection operator $P_\beta$ given by

$$P_\beta=\begin{pmatrix} A_\beta & A_\beta^{1/2}(1-A_\beta)^{1/2} \\ A_\beta^{1/2}(1-A_\beta)^{1/2} & 1-A_\beta \end{pmatrix}$$

where

$$A_\beta=\frac{e^{-\beta h}}{1+e^{-\beta h}}$$

and $h$ is the operator on the Fourier transform given by multiplication of the $n^{th}$ Fourier coefficient by $(n+1/2)$. Now this projection defines the quasifree KMS state on the C*-algebra $A(L^2(S^1))$ (with two point function given by the operator $A_\beta$) corresponding to the one parameter group of automorphisms $\tau_t$ with $\tau_t(a(f))=a(e^{ith}f)$ for each $f\in L^2(S^1)$. This situation can be described succinctly by saying that the effect of considering quantum field theory on a Riemann surface of genus one is to consider thermal states on the appropriate algebra (we deal with bosons later).

One question we would like to answer is what is the analogue of this KMS condition for field theories on higher genus surfaces?

### *3.2.2 The higher genus case*

Following the same procedure as in genus one we consider a Riemann surface $\Sigma$ which is formed from two copies of a planar region, say $R$ and $\bar{R}$ which we take to be discs with $g$ (=genus of $\Sigma$) punctures. The punctures we take to be non-overlapping small discs whose boundary circles $\Gamma_j$ $(j=1,2,\dots g)$ become the $A$-cycles on forming $\Sigma=R\sqcup\partial R\sqcup\bar{R}$ by identifying the boundaries of $R$ and $\bar{R}$ [10, p108]. There is an anticonformal map $\flat$ on $\Sigma$ which takes $R$ to $\bar{R}$ and fixes $\partial R$.

Once again we consider even spin structures (i.e. even square roots of the canonical bundle) on $\Sigma$ and hence the CAR algebra $A(H)$ where $H$ is the Hilbert space of $L^2$ sections of the bundle $L|_{\partial R}$. Once again the Szego kernel defines a projection from $H$ to the subspace consisting of boundary values of holomorphic sections of $L|_R$. The two point function can be given explicitly as the Szego kernel (cf [9,2] for example and references therein). Note that in some of the literature this example is treated as though it were the same as the Krichever-Segal-Wilson example. This confusion arises because on restriction to one of the boundary components $\Gamma_i$ one can treat the problem as that of a state on the CAR algebra over $L^2(S^1)$ where it is not clear that the state is quasifree just from the formula for the Szego kernel.

Finally let us note that this representation may be understood from the more canonical viewpoint of [5]. Let $U$ and $\bar{U}$ be neighbourhoods in $\Sigma$ of $R$ and $\bar{R}$ respectively. The Mayer-Vietoris sequence for $\Sigma = U \cup \bar{U}$ gives:

$$0 \to \Gamma(\Sigma, \mathcal{O}(L)) \to \Gamma(U, \mathcal{O}(L)) \oplus \Gamma(\bar{U}, \mathcal{O}(L)) \to \Gamma(R \cap \bar{U}, \mathcal{O}(L)) \to H^1(\Sigma, \mathcal{O}(L)) \to 0$$

(the last arrow being a conseqence of the fact that $U$ and $\bar{U}$ are Stein). The notation here is standard (see [11]), $\Gamma(V, \mathcal{O}(L))$ denoting the sections of the sheaf of germs of holomorphic sections of $L|_V$ for each open set $V$ in $\Sigma$, and $H^1(\Sigma, \mathcal{O}(L))$ the first cohomology group with co-efficients in $\mathcal{O}(L)$. Generically, for any even spin structure (for $g > 1$), the first and last terms in this exact sequence are zero so, after choosing canonical sections and identifying our spaces with spaces of functions, we see that we have a splitting of the space of meromorphic functions on $\Sigma$ which are holomorphic on $U \cap \bar{U}$ into a direct sum of the spaces consisting of those meromorphic functions on $\Sigma$ which are holomorphic on $U$ and $\bar{U}$ respectively. By passing to the completion we obtain the splitting above given by the Szego kernel. Note however that this argument fails when the spin structure is odd as one can then find holomorphic sections and hence the splitting is not canonical.

On the other hand we may use this argument when $R$ is not planar (see Subsection 3.5). Thus one may construct quantum fields on a Riemann surface $\Sigma = R \sqcup \bar{R} \sqcup \partial R$ where now $R$ is a non-compact surface with non-zero genus. It is likely however that explicit computations will only be possible in the planar case.

### *3.3 The bosons or CCR algebra*

Continuing as in Section 3.2.2 we take our symplectic space $S$ to be the space of real parts of the space of (non-constant) holomorphic functions on $R$. Thus $S$ consists of smooth harmonic functions on $R$. We define a complex inner product on the space of holomorphic functions by

$$< f, g >= \int_R d\bar{f} \wedge dg = \int_{\partial R} \bar{f} dg.$$

The imaginary part of this inner product gives the symplectic form on S. One may identify the two point function for the Fock representation of the CCR algebra over $S$ (specifed by the state $\omega(\phi) = e^{-\|f\|^2/4}$ where $\phi \in S$ with $\phi = \bar{f} + f$) by introducing the Bergman kernel of $\Sigma$ [10, p 126]. This is the bilinear differential form $B(x, y)$ ( a 1-form in each variable) with the property that

$$df(x) = \int_{R \times R} B(x, y) \wedge d\phi(y).$$

That is, the integral operator with this kernel is the projection from the space of (complex) smooth exact one forms onto the space of exact holomorphic one forms, and hence $B(x, y)$ can be thought of as the two point 'function'. Actually on exact one forms $B$ takes a particulary simple form:

$$B(x,y) = \frac{i}{2\pi}\frac{\partial^2}{\partial x \partial \bar{y}} \ln E(x,\bar{y}) dx d\bar{y}$$

where the bar denotes the action of $\flat$ and $E$ is the prime form [10, p. 20 and p. 126].

For the purposes of comparison with our earlier work [6] it is worth computing this kernel in the genus one case. Firstly we have

$$\int_{R\times R} d\bar{f} \wedge B(x,y) \wedge df(y) = \int_{\partial R} \bar{f}(x) \int_{\partial R} B(x,y) f(y).$$

By identifying $\partial R$ as in Section 3.2.1 with two copies of the circle we obtain, in terms of Fourier transforms:

$$\frac{1}{2\pi}\sum_k [\frac{k e^{-\beta k}}{1-e^{-\beta k}} |\hat{f}_k^{(1)}|^2 + \frac{k e^{-\beta k/2}}{1-e^{-\beta k}} |\hat{f}_k^{(2)}|^2]$$

where $f^{(j)}, j = 1, 2$ denotes respectivley the restriction of $f$ to the $A$ cycle and to the other boundary circle.

Once again we obtain the two point function of a quasi-free KMS state, this time on the CCR algebra. Thus the effect of the Riemann surface is simply to introduce a thermal state into the quantum field theory.

### *3.4 Hyperelliptic curves*

There is another example which is described in [5] which requires that the Riemann surface have extra structure. The main constraints are the existence of a fixed point free involution and a special form for the divisor of the canonical bundle. These constraints are satisfied for some hyperelliptic curves of odd genus. However the construction of the resulting quantum field theories is more complicated than that detailed above (although the basic requirement of a decomposition of the curve into the union of two open sets is the same). We refer the reader to [5] for details.

### *3.5 Remarks*

The preceding doubling construction can be applied to any Riemann surface $R$ with boundary. The end result is a Riemann surface $\Sigma$ of genus $2g+n-1$ where $g$ is the genus of $R$ and $n$ is the number of boundary components. In particular applied to the Krichever-Segal-Wilson example it produces a Riemann surface of genus $2g$. It is the Szego kernel for this doubled surface $\Sigma$ which gives the projection onto the subspace of $L^2(S^1)$ described in Section 3.1. Hence there is no way to get the classical theta functions associated to a surface of genus one using the method of Section 3.1.

## 4. Analogue of the Loop Group and Some Speculation

There are other harmonic functions of interest besides those in $S$. Let $\{\alpha_j | j = 1, \ldots g\}$ denote the normalised abelian differentials on our Riemann surface $\Sigma$. (Thus

$\int_{A_i} \alpha_j = 2\pi\delta_{ij}$, $\int_{B_i} \alpha_j = \tau_{ij}$ where $(\tau_{ij})$ is the period matrix.) Introduce the functions $u(x) = \sum_j n_j \int^x \alpha_j$ for $n_j$ integers, which are multivalued on $\Sigma$. However the functions $e^{iu}$ are single valued on $R$ and from the construction of $\Sigma$ we have $\alpha_j^b = \bar{\alpha}_j$ and so

$$e^{iu(x)}\overline{e^{iu(x^b)}} = 1$$

for $x \in \partial R$. Hence functions of the form $e^{iu}$ have modulus one on $\partial R$. Thus on the circles $\Gamma_l$,$e^{iu}$ is the analogue of the function $e^{i\theta}$ on the unit circle. It follows that the appropriate analogue of the loop group of U(1) in the higher genus case is the group $\mathcal{G}$ of functions on $R$ of the form $e^{iu+const+\phi}$ where $\phi \in S$.

Now a favourite pastime in this game is to look for forms of the boson-fermion correspondence. So far for Riemann surfaces this has been established at the level of $n$-point functions to varying degrees of precision by several arguments. From a mathematical viewpoint it would be more useful to establish this in a precise analytic form. Analogues of the results proved in [7],[14] or [6] are what we have in mind. Namely take the representers of products of 'special' loops ('blips' in Segal's terminology) which depend on certain parameters. Taking these parameters to zero leads to singular 'blips' however by multiplying the representers by parameter dependent constants one can arrange for the products of the blips to approximate products of the CAR generators at least in the sense of weak convergence on a dense domain. This vertex operator construction has proved very useful in elucidating the representation theory of infinite dimensional groups (cf [14],[7]) and it would be interesting to explore its implications in the present context.

We anticipate that one may produce analogues of these blips on a Riemann surface. In addition we expect to obtain some interesting formulae for theta functions by considering the representations of this generalised loop group.

It is clear that the group $\mathcal{G}$ acts as automorphisms of the CAR algebra via the map which sends $a(f)$ to $a(\gamma f)$ for $f \in H$ and $\gamma \in \mathcal{G}$. We believe that this automorphism is implementable in the representation of $A(H)$ described above. On restriction to the elements $e^{i\sum_j n_j \int^x \alpha_j + const}$ we expect to see some interesting generalisations of KMS functionals. Finally these representations of $A(H)$ should be generically quasi-equivalent to those of $A(\oplus_1^g L^2(S^1))$ obtained using the familiar Hardy space decomposition of $L^2(S^1)$ to define the projection which determines the representation.

The general structure of these examples should now be clear. One is given a Riemann surface $\Sigma$ and a covering by two open sets $U_1$ and $U_2$ and a line bundle $L$ over $\Sigma$. Mayer-Vietoris and a knowledge of the cohomology with co-efficients in the sheaf of germs of holomorphic sections of $L$ tells us when the space of holomorphic sections of $L|_{U_1 \cap U_2}$ splits into a direct sum of sections holomorphic over $U_1$ and $U_2$ respectively. If $L$ is a spin structure there is a natural pre-Hilbert space structure on the sections of $L|_{U_1 \cap U_2}$ given by integrating over a suitable path in $U_1 \cap U_2$.This together with the splitting given above yields a Fock representation of the associated fermion algebra. Boson test functions may be defined as real parts of functions holomorphic on $U_1$. Finally the loop group analogue (necessary for a global construction of vertex operators) will be essentially the same as the space of holomorphic 1-forms on $R$. (Strictly speaking the latter define the 'currents' and the non-exact forms have to be regarded as introducing 'winding numbers' or 'soliton sectors'). The example studied in [5] strongly suggests that the relationship between the sets $U_1$ and $U_2$ and the divisor of $L$ plays a role in the equivalence of different quantum field models constructed in this way.

Besides those questions mentioned above the first named author and John Palmer [12] have conjectured that solvable statistical mechanical models also give rise to field theories on a Riemann surface. In this case the surface arises as the spectral curve of Baxter's commuting family of transfer matrices [4]. Joint work with Brian Davies based on [8] is under way to confirm this in simple cases.

## Acknowledgments

Both authors thank Dai Evans and Aubrey Truman for an invitation to Swansea in 1988 where we were able to exchange ideas on the examples described here. These notes and the work in progress described above are a result of the first author's stay at the Centre for Mathematical Analysis at the Australian National University during 1989 and we thank the organisers Michael Barber and Rodney Baxter for the opportunity to work on these matters in such a fruitful environment.

## References

1. H. Araki, On quasi-free states of the CAR and Bogoliubov automorphisms, *Publ. RIMS* **6** (1971) 385–442.

2. L. Alvarez-Gaume, C. Gomez, G. Moore and C. Vafa, *Nucl. Phys.* **B**303 (1988) 455.

3. O. Bratteli and D. W. Robinson, Operator algebras and quantum statistical mechanics II, Springer, New York/Berlin, 1979.

4. R. J. Baxter, Exactly solved models in statistical mechanics (1982) Academic Press London.

5. A. L. Carey, M. G. Eastwood, K. C. Hannabuss, Riemann surfaces, Clifford algebras and infinite dimensional groups. *Commun. Math. Phys.* (to appear).

6. A. L. Carey and K. C. Hannabuss, Temperature states on loop groups, theta functions and the Luttinger model. *J. Func. Analysis*, **75** (1987) 128–160.

7. A. L. Carey and S. N. M. Ruijsenaars, On fermion gauge groups, current algebras and Kac-Moody algebras, *Acta Applicandae Math.* **10** (1987) 1–86.

8. B. Davies, Corner transfer matrices and the Ising model. A. N. U. preprint 1988.

9. T. Eguchi, Chiral Bosonisation on a Riemann surface, in Conformal field theory and superstrings, World Scientific, Singapore 1987, and references therein. See also C. Vafa, *Phys. Lett.* **B**190 (1987) 47, and N. Ishibashi, Y. Matsuo, H. Ooguri, Soliton equations and free fermions on Riemann surfaces, *Modern Phys. Letts.* **A2** (1987) 119–132.

10. J. D. Fay, Theta functions on Riemann surfaces, Lecture Notes in Mathematics, **352** Springer, Berlin, 1973.

11. R. C. Gunning, Lectures on Riemann Surfaces, Princeton Mathematical Notes, Princeton University Press, Princeton 1966.

12. J. Palmer, private communication.

13. R. Powers and E. Stormer, Free states of the canonical commutation relations, *Comm. Math. Phys.* **16** (1970) 1–33.

14. A. N. Pressley and G. B. Segal, Loop Groups, Oxford University Press, Oxford (1986).

15. G. B. Segal, Oxford lectures and preprint in preparation.

16. H. J. S. Smith, Collected Mathematical papers, Chelsea, New York, 1984.

International Journal of Modern Physics B, Vol. 4, No. 5 (1990) 1069–1118
© World Scientific Publishing Company

# C*-ALGEBRAIC METHODS IN STATISTICAL MECHANICS AND FIELD THEORY

David E. EVANS

*Department of Mathematics and Computer Science*
*University College of Swansea*
*Swansea SA2 8PP Wales, United Kingdom*

We survey the recent work in non-commutative operator algebras (especially AF-algebras, those which are inductive limits of finite dimensional C*-algebras) and which arise in studying critical phenomena in classical statistical mechanics and conformal field theory, from a C*- or topological viewpoint, rather than a von Neumann algebra/measure theoretic one.

Non-commutative C*-algebras, in particular AF algebras, arise naturally when using the transfer matrix formalism with solvable lattice models in two dimensions. An AF (approximately finite dimensional) algebra is an inductive limit of finite dimensional C*-algebras [Bra]. For example, a commutative C*-algebra C(X), the continuous functions on a compact Hausdorff space X, is AF if and only if X is second countable and totally disconnected. We let $M_p$ denote the C*-algebra of $p \times p$ complex matrices. Any finite dimensional C*-algebra is (isomorphic to) a multi matricial algebra:

$$M_{p_1} \oplus \cdots\cdots \oplus M_{p_r} \tag{1}$$

represented symbolically by the n vertices:

$$\begin{matrix} p_1 & p_2 & \dots & \dots & p_r \\ \cdot & \cdot & & & \cdot \end{matrix} \tag{2}$$

recording only the sizes $p_1, \dots, p_n$ of the matrix algebras $M_{p_1}, \dots, M_{p_n}$ respectively. An embedding of one multi matricial algebra $A_1$ say as in (2) in to another say $A_2$ as in (3):

$$\begin{matrix} q_1 & q_2 & \dots & \dots & q_3 \\ \cdot & \cdot & & & \cdot \end{matrix} \tag{3}$$

is (up to an inner automorphism of $A_2$), given by the multiplicities $\Delta(i,j)$ of the embedding of each simple factor $M_{p_i}$ of $A_1$ in $M_{q_j}$ of $A_2$. This can all be represented graphically as:

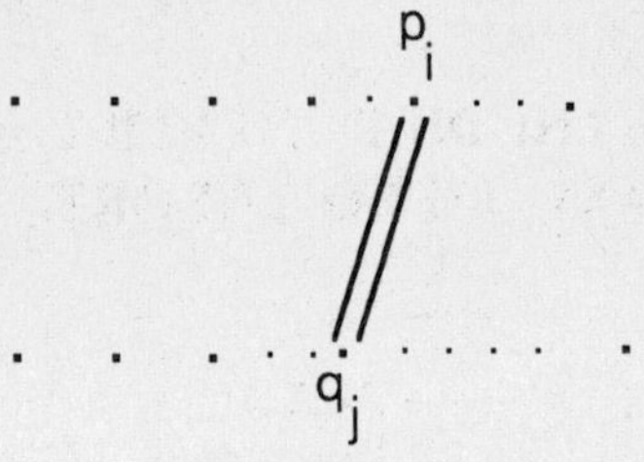


Figure 1

where we draw $\Delta(i,j)$ lines from the upper vertex i (in the set $\Sigma_1$ of vertices corresponding to the algebra $A_1$ to the lower vertex j (in the set of vertices $\Sigma_2$ corresponding to the algebra $A_2$). The graph obtained from an inductive system lim $A_n$ by adjoining the graphs describing the embedding of $A_n$ in $A_{n+1}$ is called a Bratteli diagram. Moreover starting from a single graph $\Gamma$ with distinguished vertex *, we can construct an AF algebra $A(\Gamma)$, by iteration, but starting with the complex numbers at *. The corresponding Bratteli diagram $\hat{\Gamma}$ consists of one-sided infinite paths in $\Gamma$ starting at *, where the size of the matrix algebras for vertex i at level m is the number of paths in $\hat{\Gamma}$ from * to i of length m. Thus for the Dynkin diagram $A_l$ :

Figure 2

with distinguished vertex * = 1, we obtain the following Bratteli diagrams $\hat{A}_l$

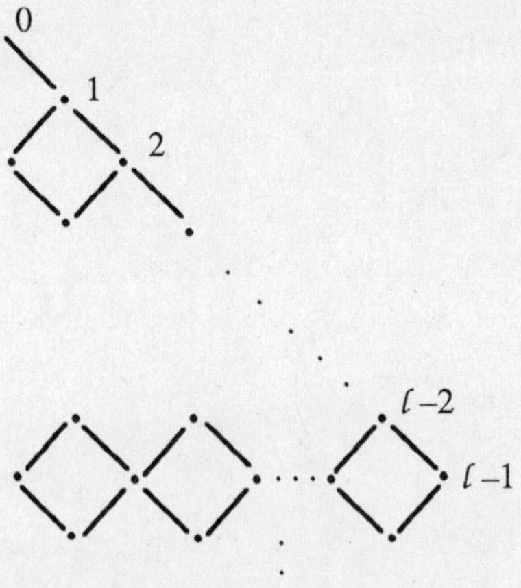


Figure 3 : $\hat{A}_l$

One can associate to a graph $\Gamma$ a lattice model on a two dimensional lattice L in the following way [ABF,Pas,JMO]. A configuration of the two dimensional classical statistical mechanical model is a distribution $\sigma = (\sigma_\alpha, \alpha \in L^{(0)})$ of the vertices $\Gamma^{(0)}$ of $\Gamma$ over the vertices $L^{(0)}$ of L, in such a way that $(\sigma_\alpha, \sigma_\beta)$ , are joined on the graph $\Gamma$ (i.e. $(\sigma_\alpha, \sigma_\beta) \in \Gamma^{(1)}$, the edges of $\Gamma$) if $\alpha,\beta$ are nearest neighbours in L. Let $P$ denote the compact Hausdorff space of all configurations:

$$P = \{\sigma = (\sigma_\alpha, \alpha \in L^{(0)}) : (\sigma_\alpha, \sigma_\beta) \in \Gamma^{(1)} \text{ if } (\alpha,\beta) \in L^{(1)}\} \tag{4}$$

Of course the Hamiltonian has also to be specified but usually we will not do so explicitly here. For the Dynkin diagram $A_3$ with vertices labelled as $\{\cdot, \pm\}$ and square lattice L, one obtains two copies of the Ising model:

```
±       ±       ±
·   ±   ·   ±   ·
±   ·   ±   ·   ±
·   ±   ·   ±   ·
±   ·   ±   ·   ±
```

Figure 4

by placing the frozen spin $\cdot$ on the even or odd sublattices of L. The Q-state Potts model [Pot,Bax2] is a generalisation of the Ising model where the spin at any lattice site can point in any one of Q directions. The relevant graph is $\Gamma_Q$ as in Figure 5.

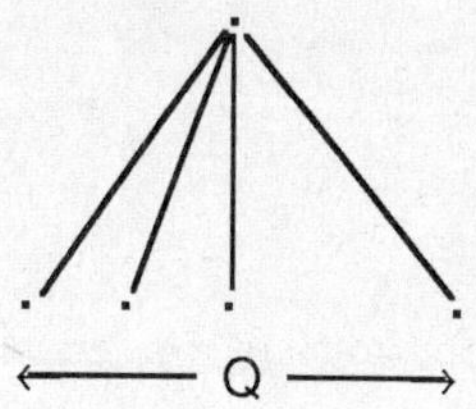


Figure 5

In this case the algebra $A(\Gamma_Q)$ is the UHF algebra $F_Q = \otimes M_Q$, which is an infinite tensor product of the matrix algebra $M_Q$. Thus $A(A_3)$ is the Pauli algebra. This one sided path algebra $A(\Gamma)$ is relevant for models on half lattice $L = \mathbb{N} \times \mathbb{Z}$ , whilst for the lattice $L = \mathbb{Z}^2$ , infinitely extended in all directions one needs an appropriate bilateral algebra.

The classical observables are C(*P* ), the commutative C*-algebra of all continuous functions on the compact Hausdorff configuration space, and at each inverse temperature space $\beta$ we may be interested in the simplex $K_\beta$ of equilibrium states, given say by solutions to the DLR equations or the variational principle:

$$\text{minimise ( energy - T entropy)} \tag{5}$$

where T is the temperature [Dob,LaR,BR]. In the algebraic approach, one uses the transfer matrix formalism to transform the setting to that of a one dimensional quantum model, represented by a non-commutative C*-algebra, an AF algebra (possibly $A(\Gamma)$ or its two sided version or the C*-algebra generated by the local transfer matrices). For each inverse temperature $\beta$, one looks for a map from equilibrium states in $K_\beta$ to linear functionals on *A* and a map $F \to F_\beta$ from (local) classical observables in C(*P* ) to the quantum algebra *A* such that one can recover the classical expectation values or correlation functions from a knowledge of the quantum ones alone:

$$\mu(F) = \phi_\mu(F_\beta) \tag{6}$$

We summarise here the main features of this procedure for the nearest neighbour two dimensional square Ising model on $L = \mathbb{Z}^2$ [Pir, LS, EL1, AE, EL2,CaE, Kui]. In this case the quantum algebra *A* is the Fermion algebra, whilst the local transfer matrices generate the even part. Let us fix some boundary conditions and then for each inverse temperature $\beta$, let $\phi_\beta$ denote the corresponding *state* on A . In general the map $\phi_\mu$ in (6) is not a state; positivity of $\phi_\mu$ is related to reflection positivity of $\mu$ [Kui]. Then if $\beta_c$ denotes the inverse critical temperature of Onsager, there exist [EL2] automorphisms $\upsilon_\beta$ of *A* $(\beta \neq \beta_c)$, which do not depend on the boundary conditions, and real analytic in $\beta \neq \beta_c$ such that

$$\phi_\beta = \begin{cases} \phi_\infty \circ \upsilon_\beta & \beta > \beta_c \\ \phi_o \circ \upsilon_\beta & \beta < \beta_c \end{cases} \tag{7}.$$

The structure at zero and infinite temperatures is accessible. In fact $\phi_o = \otimes_{\mathbb{Z}} \omega_\Omega$, where $\Omega = \begin{pmatrix}1\\1\end{pmatrix}/\sqrt{2}$ is the disordered state, and for + or − boundary conditions $\overset{+}{\phi}_\infty = \otimes_{\mathbb{Z}} w_{\begin{pmatrix}1\\0\end{pmatrix}}$, $\overset{-}{\phi}_\infty = \otimes_{\mathbb{Z}} w_{\begin{pmatrix}0\\1\end{pmatrix}}$ respectively, and $\phi_\infty = \frac{1}{2}(\overset{+}{\phi}_\infty + \overset{-}{\phi}_\infty)$ for free or periodic boundary conditions. Thus with periodic or free boundary conditions, we conclude from (7) that $\phi_\beta$ is pure for $0 \leq \beta < \beta_c$ (in fact also for $\beta = \beta_c$ [AE] by different methods), and is a mixture of two inequivalent pure states $\phi_b^{\pm}$ for $\beta > \beta_c$. Moreover for $\beta < \beta_c < \beta'$, there exists an endomorphism $\alpha : A \to \pi(A_+)''$ where $\pi$ denotes the GNS representation of $\phi_\beta$ such that $\overset{+}{\phi}_\beta = \overset{-}{\phi}_\beta \circ \alpha$, and $\overline{\phi}_\beta$ denotes the natural extension of $\phi_\beta$ to $\pi(A)''$ [CaE]. Since $\{\upsilon_\beta : \beta \neq \beta_c\}$ can be chosen (real) analytic in $\beta$ [Ara,CaE] we can conclude that $\langle F\rangle^{\pm}$ is analytic in $\beta \neq \beta_c$, where $\langle\,\rangle^{\pm}$ denotes the classical states corresponding to + and - boundary conditions respectively. This will follow from the formula $\langle F\rangle^{\pm}_{\beta} = \phi^{\pm}_{\beta}(F_\beta) = \phi^{+}_{\infty} \circ \upsilon_\beta(F_\beta)$. The strategy here is that the family $\{\upsilon_\beta : \beta \neq \beta_c\}$, which is independant of the boundary conditions, is first constructed so that (7) holds for free or periodic boundary conditions. One then needs to show that $\langle F\rangle^{+} = \phi^{+}_{o}\upsilon_\beta(F_\beta)$, for the same family $\{\upsilon_\beta : \beta \neq \beta_c\}$. It is striking to note that this analyticity could of course not be shown in this way for the algebra C(*P*), using transformations of *P* as *P* is totally disconnected, but the non-commutative C*-algebra A has continuous symmetries. As another application of this theory, we note the following result of Araki [Ara] about the classical model, again proved by a detailed analysis of the quantum model. Here translation in the horizontal direction in $\mathbb{Z}^2$ for the classical model is denoted by $\tau_{(1,0)}$, and if $J_1$ and $J_2$ denote interaction constants in the horizontal and vertical directions repectively, let $K_i = \beta J_i$, and define K* from K by sh(2K)sh(2K*) =1.

$$\text{Let } \psi = \begin{cases} \langle\,\rangle, & \beta < \beta_c \quad \delta = 2(K^*_1 - K_2) \\ \langle\,\rangle^{\pm}, & \beta > \beta_c \quad \delta = 4(K_2 - K^*_1) \end{cases} \tag{8}$$

Then

(9)
$$\lim_{l\to\infty} e^{|l|\delta}[\psi(F_1\tau_{(l,0)}(F_2)) - \psi(F_1)\psi(F_2)] = 0$$

for all $F_1, F_2$ local.

(10) for any $\varepsilon > 0$, there exist $F_{\varepsilon,1}, F_{\varepsilon,2}$ such that

$$\lim_{l\to\infty} e^{|l|(\delta+\varepsilon)}[\psi(F_1\tau_{(l,0)}F_2) - \psi(F_1)\psi(F_2)] = \infty.$$

Consider a model on the square lattice $L = \mathbb{Z}^2$, with nearest neighbour interactions. We can define the row-to-row transfer matrix [Bax2] as the partition function $T_{\alpha\beta}$ of the sublattice in Figure 6:

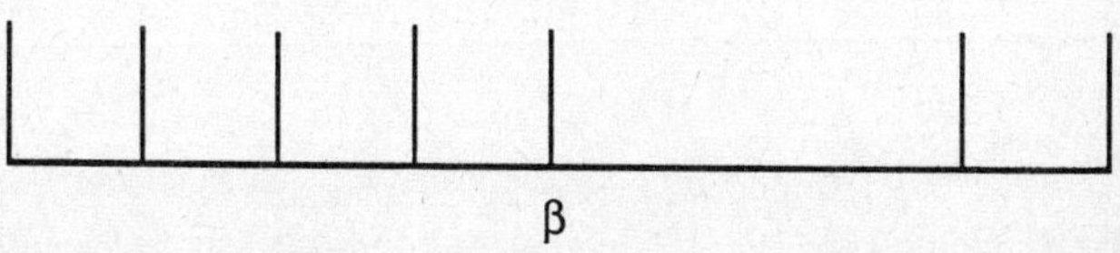


Figure 6

with configurations $\alpha$ and $\beta$ on the upper and lower rows, respectively, and considering interactions along only the edges drawn. A configuration $\sigma$ on the rectangular sublattice of Figure 7:

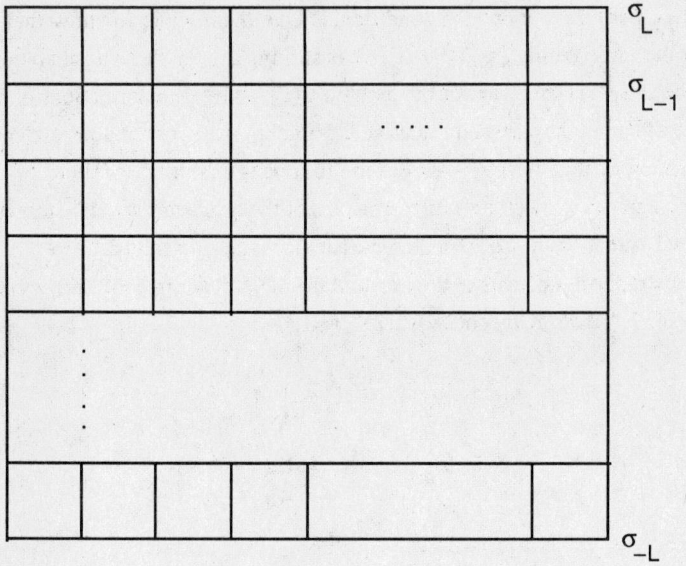


Figure 7

can be broken up into a column $\begin{pmatrix} \sigma_L \\ \cdot \\ \cdot \\ \sigma_{-L} \end{pmatrix}$ of configurations $\sigma_i$ on the sublattice along the $i^{th}$ row. Since we have nearest neighbour interactions, $H(\sigma) = \Sigma H_i(\sigma_i, \sigma_{i+1})$, where $H_i$ is the energy of the $i^{th}$ sublattice as in Figure 6. Thus the Boltzman weight of the configuration $\sigma$ is

$$\exp(-\beta H(\sigma)) = \prod_{-L}^{L} \exp(-\beta H_i(\sigma_i, \sigma_{i+1}))$$

$$= T_{\sigma_{-L}\sigma_{-L+1}} T_{\sigma_{-L+1}\sigma_{-L+2}} \ldots\ldots\ldots T_{\sigma_{L-2}\sigma_{L-1}} T_{\sigma_{L-1}\sigma_L} \tag{11}$$

and the partition function of the sublattice in Figure 8,

Figure 8

with configurations α and β on the upper and lower rows, respectively is $T^N_{\alpha\beta} = <\alpha|T^N|\beta>$, or trace $T^N$ for boundary conditions periodic in the vertical direction. In a similar way the classical expectation value <F> of a local observable F can be expressed as $<\alpha|T^N F_\beta)|\beta>$, or trace($T^N F_\beta$), which leads us to consider states on non-commutative C*-algebras (6) [Pir,LS,EL1,AE,EL2,CaE,Kui]. The Boltzman weight of the configuration in Figure 6 can be written as in Figure 9.

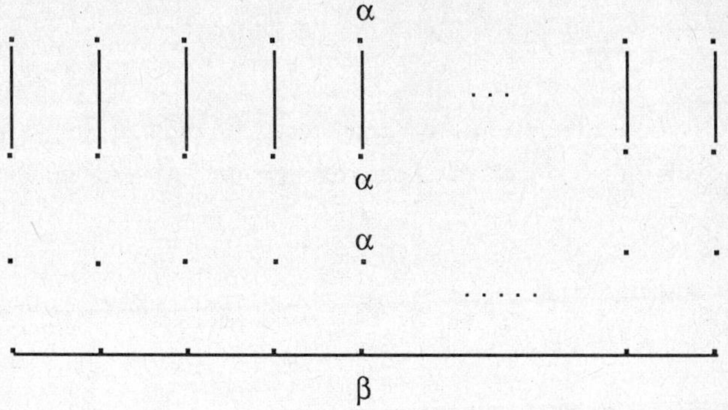


Figure 9

i.e. T = VW, where V,W are the Boltzman weights of the sublattices v,w respectively, with W diagonal. It is instructive to see this in detail in the one-dimensional Ising model, using periodic boundary conditions. Here the energy for the finite lattice is given by

$$H(\sigma) = -\sum_{-L}^{L} J\sigma_i\sigma_{i+1}. \tag{12}$$

We have to compute the classical expectation value by the Gibbs' formula:

$$\langle F\rangle_L = \sum_\sigma F(\sigma)\exp[-\beta H(\sigma)]/\sum_\sigma \exp[-\beta H(\sigma)] \tag{13}$$

where the summations are over all configurations $\sigma \in \{\pm\}_{-L}^{L}$. First we compute the partition function

$$Z = \sum \exp[-\beta H(\sigma)] = \sum_\sigma \prod_{-L}^{L} \exp[\beta J\sigma_i\sigma_{i+1}]$$

$$= \sum V(\sigma_{-L},\sigma_{-L+1})\ldots\ldots V(\sigma_{L-1},\sigma_L)V(\sigma_L,\sigma_{-L}) \tag{14}$$

if $V(\sigma, \sigma') = \exp(\beta J\sigma\sigma')$, and $\sigma,\sigma' \in \{\pm\}$ are configurations on one lattice site. Thus with periodic boundary conditions, we can identify Z as the trace of a certain power of V, regarded as a 2x2 matrix. Namely $Z = \mathrm{tr}(V)^{2L+1}$. In a similar way, we can write $\sum F(\sigma)\exp[-\beta H(\sigma)] = \mathrm{tr}\, F_\beta V^{2L+1}$, for a certain $F_\beta \in M_2$. Thus letting $L \to \infty$,

$$\langle F\rangle = \lim_{L\to\infty} \mathrm{tr}\, F_\beta V^{2L+1}/\mathrm{tr}\, V^{2L+1} = \langle F_\beta\Omega,\Omega\rangle \tag{15}$$

where $\Omega = \begin{pmatrix}1\\1\end{pmatrix}/\sqrt{2}$ is the unit positive eigenvector of V corresponding to the largest eigenvalue. This is only correct for non-zero temperature. At zero temperature, one has to be more careful. Now

$$V = \begin{pmatrix} e^{K} & e^{-K} \\ e^{-K} & e^{-K} \end{pmatrix} = e^{K}(\mathbb{1} + e^{-2K}\sigma_z) = \sinh(2K)e^{K^*\sigma_z}, \tag{16}$$

where $\sigma_z = \begin{pmatrix} 0 & 1 \\ 1 & 0 \end{pmatrix}$, $\sigma_x = \begin{pmatrix} 1 & 0 \\ 0 & -1 \end{pmatrix}$. Thus, with $J > 0$ for ferromagnetic interactions, V

behaves more like a scalar multiple of the identity as $K \to \infty$ (equivalently $K^* \to 0$), and so the largest eigenspace becomes degenerate. Thus

$$<F>_\infty = \mathrm{tr}\, F_\infty = \frac{1}{2}(\omega_{\binom{1}{0}} + \omega_{\binom{0}{1}})(F_\infty) \tag{17}$$

Consequently

$$\phi_\beta = \begin{cases} \omega_\Omega & \beta \neq \infty \\ \frac{1}{2}(\omega_{\binom{1}{0}} + \omega_{\binom{0}{1}}) & \beta = \infty \end{cases} . \tag{18}$$

In 2-dimensions $\mathbb{Z}^2$, with periodic (toroidal) boundary conditions the transfer matrix V has two components: T = VW, where up to scalars:

$$V = \exp K_1^* \Sigma \sigma_z^j \tag{19}$$

arises from the transfer direction, vertical interactions and

$$W = \exp K_2 \Sigma \sigma_x^j \sigma_x^{j+1} \tag{20}$$

is diagonal, arising from horizontal interactions. The spin flip $+ \leftrightarrow -$ in the classical model gives gradings both to the classical algebra $C(P) = C(\{\pm\}^{\mathbb{Z}^2})$ and to the quantum algebra A which in the case of the two dimensional Ising model is the Pauli algebra $\overset{\infty}{\underset{-\infty}{\otimes}} M_2$ through the symmetry $\Theta = \overset{\infty}{\underset{-\infty}{\otimes}} \mathrm{Ad}(\sigma_z^j)$. The even algebra $A_+ = \{x \in A : \Theta(x) = x\}$ is generated by the local transfer matrices, or equivalently by $\sigma_z^j$ and $\sigma_x^j \sigma_x^{j+1}$, $j \in \mathbb{Z}$. High temperature-low temperature duality of Kramers and Wannier gives an automorphism $\kappa$ of the even algebra $A_+$, by

$$\kappa(\sigma_z^j) = \sigma_x^j \sigma_x^{j+1}, \quad \kappa(\sigma_x^j \sigma_x^{j+1}) = \sigma_z^{j+1} \tag{21}$$

where

$$\phi_o \circ \kappa = \phi_\infty \text{ on } A_+. \tag{22}$$

In fact for any fixed but arbitrary boundary condition, and any two inverse temperatures $\beta$, $\beta'$ one can find [EL2] an automorphism $\upsilon$ of the even Pauli algebra such that

$$\phi_\beta \circ \upsilon = \phi_{\beta'} \text{ on } A_+. \tag{23}$$

If $\beta' < \beta_c < \beta$, where $\beta_c$ is the critical temperature of Onsager, there is an obstruction to lifting $\upsilon$ to the whole algebra A. This obstruction is measured via a mod 2 index [EL2] and gives rise to the phase transition in the classical model.

As we have already noted, the structure at zero and infinite temperature is clear. In order to pull out this information into the regions $(0, T_c)$ and $(T_c, \infty)$, it is better not to work with the Pauli algebra alone, but to introduce an auxiliary algebra, a Fermi algebra $A^F$ over $l^2(\mathbb{Z})$, generated by annihilation and creation operators $\{c_i, c_i^* : i \in \mathbb{Z}\}$ satisfying the canonical anti-commutation relations

$$[c_i, c_j]_+ = [c_j^*, c_i^*]_+ = 0 \ , \ [c_j^*, c_i^*] = \delta_{ij} \tag{24}$$

Formally, the Jordan Wigner transformation, involving the Pauli algebra $A^P = C^* (\sigma_x^i, \sigma_z^i : i \in \mathbb{Z})$ and the Fermi algebra $A^F = C^* (c_i : i \in \mathbb{Z})$ is

$$\sigma_z^j = 2c_j^* c_j - 1 \tag{25}$$

$$\sigma_x^j = t\, s_j\, (c_j + c_j^*)) \tag{26}$$

where the tail has been decomposed into

$$t = \prod_{-\infty}^{0} \sigma_z^i \ , \quad s_j = \prod_{1}^{j-1} \sigma_z^i \qquad (j > 1, \text{similarly for } j \le 1) \tag{27}$$

However $t$ does not converge in either the Pauli or Fermi algebra. Araki [Ar] considered the outer autermorphism(s) of period two, both denoted by $\Theta_-$, on $A^P$ and $A^F$ which are implemented formally by t :

$$\Theta_- (\sigma_z^j) = \sigma_z^j \ , \quad \Theta_-(\sigma_x^j) = \begin{cases} \sigma_x^j & j \ge 1 \\ -\sigma_x^j & j < 1 \end{cases} \tag{28}$$

$$\Theta_- (c_j) = \begin{cases} c_j & j \ge 1 \\ -c_j & j < 1 \end{cases} \tag{29}$$

Take any representation of the Fermi algebra $A^F$, where $\Theta_-$ is represented by a self adjoint unitary t, so that $\Theta_-(x) = t\,x\,t$, $x \in A^F$. Then with this t, define (25-26) $\sigma_x^j$, and

$\sigma_z^j$ in terms of $t$ and $\{c_i\}$, and so embed the Pauli algebra into the C*-algebra $\hat{A}$ generated by $A^F$ and $t$, namely the crossed product $A^F \times_{\Theta} \mathbb{Z}_2$. $\hat{A}$ is graded, by a symmetry $\Theta$ so that $c_i$ are odd, $t$ is even. This compatible with the grading of $A^P$ with $\sigma_t^i$ even and $\sigma_x^j$ odd. Their even sub-algebras $A_+^F$, $A_+^P$ are generated by even monomials in $c_i$ and $c_i^*$, and by $\sigma_t^i$, $\sigma_x^i\,\sigma_x^{i+1}$ respectively, and the embedding (25-26) of $A^P$ in $\hat{A}$ identifies $A_+^F$ and $A_+^P$ as

$$\sigma_z^i = 2c_i^*c_i - 1 \tag{30}$$

$$\sigma_x^i\sigma_x^{i+1} = (c_i - c_i^*)\,(c_{i+1} + c_{i+1}^*) \tag{31}$$

In this way we have the following picture:

$$\begin{array}{ccc} & \hat{A} & \\ \subset & & \supset \\ A^P & & A^F \\ \supset & & \subset \\ & A_+^P = A_+^F & \end{array}$$

Figure 10

The state $\varphi_\beta$ on $A^P$ arises from the even transfer matrix, and so is even, $\varphi_\beta = \varphi_\beta \circ \Theta$. Thus $\varphi_\beta$ can be tranformed to an even state $\omega_\beta = \omega_\beta \circ \Theta$ on $A^F$ such that $\omega_\beta | A_+^F = \varphi_\beta | A_+^P$. The state $\omega_\beta$ is a quasi-free state being constructed from the quadratic Hamiltonian log T. It is in fact a Fock state, and so is always pure for temperature. Moreover the automorphisms $\nu$ linking any two temperatures at the even algebra level (23), lift to automorphisms $\bar{\nu}$ of the Fermi algbera $A^F$ such that $\omega_\beta \circ \bar{\nu} = \omega_{\beta'}$. Thus the manifestation of the phase transition in the Fermi algebra picture is not immediately clear (c.f. [Pir]).

The approach in [AE] was to first derive an abstract criterion to decide when an even state $\varphi$ of $A^P$ is pure in terms of properties of the associated even state $\tilde{\varphi}$ of $A^F$. This involves the equivalence and non-equivalence of $\tilde{\varphi}$, $\tilde{\varphi} \circ \Theta_-$ on $A^F$, and $\tilde{\varphi} | A_+$, $\tilde{\varphi} \circ \Theta_-$ $) | A_+$ on $A_+^F$. When applied to the Ising model, we have to decide when the two Fock states $\omega_\beta$ and $\omega_\beta \circ \Theta_-$ are equivalent on $A^F$, and their restrictions on $A_+^F$. It was shown in [AE] that the latter involved a certain mod 2 index. The presence of the mod 2 index arises from the following fact. Suppose $\pi$ is the representation of $A^F$ on a Fock space F, regarded as a space of antisymmetric tensors. Then F splits as $F_+ \oplus F_-$ where $F_+$, $F_-$ are the subspaces of even and odd tensors respectively. The mod 2 index enters because the two representations of $A_+$ on $F_+$ and $F_-$ are not equivalent [AE, Lemma 4.2]. The automorphisms $\{\nu_\beta : \beta \neq \beta_c\}$ were constructed in [EL2] by obtaining a criterion allowing us to decide when the restriction of a Bogoliubov automorphism from $A^F$ to $A_+$ extends to a graded automorphism of $A^P$. This again involved a mod 2 index. The relationship between the mod 2 index computations of [EL2], the Araki-Evans mod 2 index between projections [AE] and the Atiyah-Singer mod 2 index for Fredholm operators [AS] was explained in [CaE].

Note also that $\lim_{M\to\infty} T_M^{it}(\cdot)T_M^{-it} = \alpha_t$ defines a dynamics on $A_+$ which can be lifted by the methods of [EL2] to a strongly continuous one-parameter group on A which has an unique ground state for $\beta < \beta_c$ and two extremal ground states $\phi_\beta^{\pm}$ for $\beta > \beta_c$ [Ara, AM].

We have noted that the local transfer matrix for the Ising model is described algebraically by the operators $\sigma_x^j$ and $\sigma_x^j \sigma_x^{j+1}$ in $\otimes M_2$. These self adjoint unitaries are determined by the projections on their +1 eigenspace, and so are given by a sequence $\{e_j\}$ of projections:

$$e_{2i} = (1 + \sigma_z^i)/2 \qquad e_{2i+1} = (1 + \sigma_x^i \sigma_x^{i+1})/2 \tag{32}$$

which satisfy the (Temperley-Lieb) algebra:

$$e_n e_m = e_m e_n, \ |m-n| \geq 2; \quad e_n e_{n\pm1} e_n = \tau e_n \tag{33}$$

with $\tau^{-1} = 2$. (Indeed, $2(e_m + e_n - e_m e_n - e_n e_m) = 1$ holds). In the Q-state Potts model, the local transfer matrix T = VW is described in terms of a sequence $\{e_i\}$ of projections in the UHF algebra $F_Q = \otimes_* M_Q$, which satisfy the Temperley-Lieb algebra (31) with $\tau^{-1} = Q$, where $V_1 = \exp 2 K_1 \Sigma e_{2i+1}$, $V_2 = \exp 2 K_2 \Sigma e_{2i}$. The transfer matrix formalism for the ABF models (at criticality) produce the same algebra with $\tau^{-1} = 4\cos^2(\pi/(l+1))$, and more generally for the $\Gamma$-models with $\tau^{-1} = ||\Gamma||^2$ where $||\Gamma||$ denotes the norm of the incidence matrix of the graph [Ocn,Sun,AKW,Pas,BO].

In September 1983, I noticed the connection, using this transfer matrix formalism, between subfactors on hand and statistical mechanics on the other. Namely, that the same Temperley-Lieb algebra that appeared in the Potts model was now appearing in Jones' work on subfactors of the hyperfinite $II_1$-factor in the theory of von Neumann algebras, and that the representation of the Temperley-Lieb algebra in the Fermion algebra found by Pimsner and Popa was the same as that obtained by Temperley-Lieb [TL] when they showed that the Potts model at criticality was equivalent to an ice-type or 6-vertex model [Evans, letters to Jones, Pimsner-Popa, December 1983] [Jon1][PP]. Using their representation, Pimsner and

Popa could identify the shift $\kappa: e_n \to e_{n+1}$ as a non-commutative Bernoully shift. In the statistical mechanical picture the shift $\kappa$ is Krammers-Wannier high-temperature/low-temperature duality - see the remarks of Onsager on page 123 of [Ons]. These connections between statistical mechanics and von Neumann algebras was further intrugued by the discovery early in the summer of 1984 by Jones [Jon2] of the one-variable knot polynomial, using the Markov trace on $C^*(e_n)$ and the representation of the braid group. Thus one immediately had the connection between statistical mechanics and knots, in particular invariants of knots/links and partition functions of statistical mechanical models. In the Autumn of 1985, I became intrigued by the connection between the discrete sequence of representations of the Virasoro algebra with central charge c less than one, and the discrete sequence of values of the index of subfactors less than four. We will digress a little before returning to this.

The $F_Q$ representation of the Temperley-Lieb algebra (33) mentioned above arises in the Q-state standard Potts model as follows. Define Q x Q unitary matrices U, V by

$$U = \begin{bmatrix} 1 & 0 & 0 \dots 0 \\ 0 & q & 0 \dots 0 \\ 0 & 0 & q^2 \dots 0 \\ . & . & . & . \\ . & . & . & . \\ . & . & . & . \\ 0 & 0 & 0 \dots q^{Q-1} \end{bmatrix} \quad V = \begin{bmatrix} 0 & 1 & 0 \dots 0 & 0 \\ 0 & 0 & 1 \dots 0 & 0 \\ 0 & 0 & 0 \dots 0 & 0 \\ . & . & . & . & . \\ . & . & . & . & . \\ . & . & . & . & . \\ 0 & 0 & 0 \dots 0 & 1 \\ 1 & 0 & 0 \dots 0 & 0 \end{bmatrix} \tag{34}$$

where $q = e^{\pi i/Q}$, so that $VU = qUV$, and $U, V$ generate $M_Q$. Define a sequence of unitaries $W_i$ in $M_Q \otimes M_Q \otimes M_Q \otimes \cdots$ by

$$W_{2i+1} = 1 \otimes \cdots 1 \otimes U^{-1} \otimes U \otimes 1 \otimes \cdots \tag{35}$$

where the factor U appears as the $(i+1)^{\text{th}}$ factor and

$$W_{2i} = 1 \otimes \cdots 1 \otimes V \otimes 1 \otimes \cdots \tag{36}$$

with V appearing as the $i^{\text{th}}$ factor so that

$$W_j^Q = 1 \tag{37}$$

and

$$W_i W_{i+1} = q W_{i+1} W_i \tag{38}$$

$$W_i W_j = W_j W_i \qquad |i-j| \geq 2 \tag{39}$$

The symmetric group $S_Q$ acts on $\mathbb{C}^Q$ by permuting basis vectors and so induces a product action on $F_Q = \otimes M_Q$. In particular, there is a $\mathbb{Z}_Q$ action on $F_Q$ by $\otimes \mathrm{Ad}(V) = \prod_i \mathrm{Ad}(W_{2i})$. The Temperley-Lieb operators in $F_Q$ referred to above are the spectral projections of $W_i$ corresponding to eigenvalue 1, namely

$$e_i = (1 + W_i + W_i^2 + \cdots W_i^{Q-1})/Q \tag{40}$$

In particular

$$e_{2i-1} = \cdots\, 1 \otimes \underset{\text{i-(i+1) factors}}{g} \otimes 1 \otimes \;\cdots \tag{41}$$

$$e_{2i} = \cdots 1 \otimes \underset{\text{ith factor}}{f} \otimes 1 \otimes \cdots \tag{42}$$

where g, f are projections in $\otimes M_Q$ and $M_Q$ respectively given by

$$g = \sum_{i=1}^{Q} E_{ii} \otimes E_{ii} \;, \qquad f = \sum_{i,j=1}^{Q} E_{ij}/Q \tag{43}$$

if $\{E_{ij} : i,j = 1, 2, \cdots Q\}$ are matrix units for $M_Q$. The family $\{e_i\}$ is $S_Q$ invariant, and the $\{W_i\}$ generate the $\mathbb{Z}_Q$ fixed point algebra of $F_Q$:

$$C^*(e_i) \subset F_Q^{S_Q} \subset F_Q^{\mathbb{Z}_Q} = C^*(U_i) \tag{44}$$

Moreover

$$\text{for } Q = 2: \quad C^*(e_i) = F_2^{S_2} = F_2^{\mathbb{Z}_2} \tag{45}$$

$$\text{for } Q = 3: \quad C^*(e_i) = F_3^{S_3} \subsetneq F_3^{\mathbb{Z}_3} \tag{46}$$

and both inclusions in (44) are strict for $Q \geq 4$. This formulation is related to that of generalised Clifford algebras [Mor] as follows. The formulae

$$\Gamma_i = W_1 W_2 \cdots W_i \;, \quad W_i = \Gamma_{i-1}^{-1} \Gamma_i^1 \tag{47}$$

is the generalised Jordan-Wigner transformation required to transform between (37-39) and

$$\Gamma_i^n = 1\,, \qquad \Gamma_i \Gamma_j = q\,\Gamma_j \Gamma_i \qquad i<j \tag{48}$$

(For bilateral infinite sequences, one treats the infinite tail as in (26-29) ). Temperley and Lieb [TL] showed that at criticality, the local Potts partition function on a planar graph L can be identified with the partition function of an ice type or 6 vertex model on a dual or medial graph $L'$. This identification corresponds algebraically to a representation of the Temperley-Lieb relations (33) in a Pauli algebra $F_2$. This representation exists for all $\tau^{-1} = Q > 0$ (not necessary integral Q) but is only a *-representation if $Q \geq 4$. It was independently rediscovered by Pimsner and Popa [PP] who used it to express the Kramers-Wannier [KW] high temperature - low temperature duality $\kappa_Q : e_i \to e_{i+1}$ as a non commutative Bernoulli shift in $F_2$ if $Q \geq 4$.

Note that the algebra $\{W_i\}$ of (37-39) also features in the Chiral Potts model, including the non hermitian operator

$$H = -\sum_{i=1}^{r-1} \alpha_j W_{2j} - \sum_{i=1}^{r} \gamma_j W_{2j-1} \tag{49}$$

See [ BPA,Bax3,Bax4,Bax5].

We have already noted in (30) how to identify the even Pauli algebra with the even Fermi algebra, and in (32) with the Temperley Lieb algebra for $\tau^{-1} = 2$. In particular, we identify

$$e_{2j-1} = c_j^* c_j$$

$$e_{2j} = [1 + (c_j - c_j^*)(c_{j+1} + c_{j+1}^*)]/2 \tag{50}$$

and

$$c_j^* c_j = e_{2j-1}\,, \quad c_j c_{j+1}^* = e_{2j-1}(1 - e_{2j})\, e_{2j+1}. \tag{51}$$

For completeness, we relate this to the Clifford algebra formalism. Let $\{\gamma_n\}$ satisfy the Clifford relations

$$\{\gamma_n, \gamma_m\} = 2\delta_{nm} \qquad \gamma_n^* = \gamma_n \tag{52}$$

so that $\{\gamma_n\}$ generate the Clifford algebra over $\ell^2_{\mathbb{R}}(\mathbb{Z})$. Taking a complex structure on $\ell^2_{\mathbb{R}}(\mathbb{Z})$ with real basis $\eta_n$ so that $i\,\eta_{2n} = \eta_{2n+1}$, $i\,\eta_{2n+1} = -\eta_{2n}$, then $c_n = \gamma_{2n} + i\,\gamma_{2n+1}$ are creation operators satisfying the canonical anti-commutation relations. Then $U_n = i\,\gamma_n + \gamma_{n+1}$ satisfy the relations (37-39), namely

$$U_m^2 = 1 \tag{53}$$

$$U_m U_n = U_n U_m \,, \quad |m - n| \geq 2 \tag{54}$$

$$U_n U_{n+1} = -U_{n+1} U_n \tag{55}$$

so that $e_n = (1+U_n)/2$ are the Temperley-Lieb operators.

It is convenient when describing AF algebras to use a canonical description of matrix units labelled by paths from the Bratteli diagram [Eva1,Eva2]. If i is a vertex in $\Sigma_m$ at level m , and j in $\Sigma_n$ at level n of a Bratteli diagram ($m \leq n$), let *Path*(i,j) denote the set of paths in the Bratteli diagram of length n-m with initial vertex i and final vertex j. Then let $A^{ij}$ = End $l^2$(*Path*(i,j)), with matrix units $e_{\mu\nu}$ conveniently labelled by the paths $\mu,\nu$ in *Path*(i,j). We then let $A[m,n] = \oplus A^{ij}$, where the summation is over all vertices i in $\Sigma_m$ at level m , and j in $\Sigma_n$ at level n. If $m' \leq m \leq n \leq n'$, we can embedd A[m,n] in A[m',n'] by

$$e_{mn} \text{ Æ } \sum_{a,b} e_{amb,anb} \tag{56}$$

for $\mu,\nu \in$ Path (i,j) where the summation is over all $\alpha \in$ Path (i', i) $\beta \in$ Path (j,j'), i' on level m', j' on level n'. Then we can identify $A(\Gamma)$ with $\lim\limits_{\rightarrow} A[m,n]$, indeed $A[0,n] = A_n$, the multi-matricial algebra at the nth level of the Bratteli diagram $\hat{\Gamma}$. Let $\Delta$ denote the incidence matrix of the graph $\Gamma$. Moreover suppose $(\phi_\rho : \rho \in \Gamma^{(o)}) > 0$ is a solution to

$$x f_v = \mathrm{S}_{w \text{Œ} G^{(0)}} D(v,w) f_w \tag{57}$$

If $\eta \in \Gamma^{(o)}$ we can identify Path(($\eta$,n-1)),($\eta$,n+1)) with $t(\eta) = \{\rho \in \Gamma^{(o)} : (\rho,\eta) \in \Gamma^{(1)}\}$, if there exists a path of length n-1 from * to $\eta$ , and the empty set otherwise. Then we define (Temperley-Lieb) operators $e_n$ in $\oplus$ End $l^2(\eta) \subset A[n-1, n+1]$ to be given by the sum of rank one projections $X(\eta) \otimes X(\eta)^*$ in $l^2(t(\eta)$ , where $X(\eta)$ is given by the weights of $\{\phi_\rho^{1/2} : \rho \in t(\eta)\}$. More precisely, $X(\eta) = (\phi_\rho/\phi_\eta\, x)^{1/2} : \rho \in t(\eta))$ defines a unit vector in $l^2(t(\eta))$ . Then for each $\eta$ in $\mathbb{N}$, let

$$e_n = \sum_\eta \sum_{\rho,\rho'} X(\eta)(\rho)\, X(\eta)(\rho')\, e_{\rho,\rho'} \tag{58}$$

where the summation is overall $\eta \in \Gamma^0$, $\rho, \rho' \in t(\eta)$ such that there exists a path of length n-1 from * to $\eta$ . Then the sequence $\{e_n\}$ satisfy the Temperley-Lieb relations (33) for $\tau^{-1} = x^2$ [AKW,Ocn,Pas,Sun].

We will discuss some aspects of index theory of subalgebras from a C*- or topological view point, and see that it is a far richer theory than the von Neumann algebra / measure theoretic one. (Recall that C*-algebras often provide finer invariants than von Neumann algebras for studying geometric objects or dynamical systems[]). K-(homology) theory provides powerful invariants for C*-algebras, especially AF algebras [Eff1,Bla1]. One can associate to a C*-algebra A a pair $(K_0(A), K_1(A))$ of abelian groups. If A is a *-algebra, define an equivalence relation ~ on P(A), the projections in A namely $\{e \in A : e = e^2 = e^*\}$ by $e \sim f$ if there exists a (partial isometry) v in A taking e to f, i.e. $e = v^* v$, $f = v v^*$. We can define an addition on equivalence classes of projections by

$$[e] \oplus [f] = \left[ \begin{pmatrix} e & 0 \\ 0 & f \end{pmatrix} \right] \tag{59}$$

if we work in $M_\infty(A)$, the *-algebra of infinite matrices over A. If A is a unital C*-algebra, $K_0(A)$ is the enveloping Grothendiek completion of the abelian semigroup $D(A) = P(M_\infty(A))/\sim$. Thus formally at least $K_0(A) = \{[e] - [f]\}$, consists of differences of equivalence classes of projections, with distinguished cone or sub-semigroup $K_0(A)_+ = \{[e]\}$, consisting of the equivalence classes of projections themselves. For example, for a matrix algebra, $A = M_n$, $P(A)/\sim \; \simeq [0, 1, 2, \cdots, n]$, $P(M_\infty(A))/\sim \; \simeq \mathbb{N}$, and so the Grothendiek completion is $K_0(M_n) \simeq \mathbb{Z}$, with $K_0(M_n)_+ \simeq \mathbb{N}$.

For a commutative C*-algebra C(X), the group $K_0(C(X)) = K_0(X)$ is the enveloping group (Grothendieck completion) of the semi-group of vector bundles over the compact vector space X. By Swan's theorem, vector bundles over X can be identified, via their space of sections with finitely generated projective modules. For a C*-algebra A, the finitely generated projective modules can be identifed with projections in (matrix algebras over) A. If e is a projection in $M_n(A)$, the $n \times n$ matrices over A, then $eA^n$ is a finitely generated projective module over A. Elliott showed the power of $K_0$ as a tool for studying *non-commutative* C*-algebras, when he showed that $(K_0(A),[1])$ is a complete invariant for unital AF algebras, where $K_0(A)$ is regarded as an ordered group, and [1] denotes the equivalence class of the identity. (We will not need to mention here the classification of not necessarily unital algebras). For example for the UHF algebra $F_Q = \otimes_{\mathbb{N}} M_Q$, which is an infinite tensor product of the matrix algebra $M_Q$, the invariant is $(\mathbb{Z}[1/Q],1)$, where $\mathbb{Z}[1/Q]$ denotes the Q-adic rationals, with positive cone $\mathbb{Z}[1/Q] \cap [0,\infty)$. Thus $\otimes_{\mathbb{N}} M_Q$ and $\otimes_{\mathbb{N}} M_{Q'}$ are seen to be non isomorphic unless Q and Q' have the same prime factors. However the completion of $\otimes_{\mathbb{N}} M_Q$ in the weak topology in its unique tracial state is the unique hyperfinite $II_1$ factor R (in particular is independent of Q), which has $(K_0(R),[1]) = (R,1)$. The $K_1$- group of a (unital) C*-algebra is defined in a similar way as the abelian group of unitaries modulo the connected component of the identity. For an AF algebra $K_1(A) = 0$, whilst $K_1(C(\mathbb{T})) \simeq \mathbb{Z}$, generated by the unitary $u(z) = z$, $z \in \mathbb{T}$.

Suppose M is the hyperfinite $II_1$ factor, and N is a subfactor of M. We recall the following important facts from the theory of index of subfactors:

N is automatically hyperfinite, and so if infinite dimensional is isomorphic to M [Con].

The index of N in M is finite if and only if M is a finitely generated N-module. In which case the module is automatically projective and one can define the index $\tau^{-1} = [M,N]$ as the value of the module M in $K_0(N) = R$ [PP].

If the index is finite, let e be the projection of M on N in the Hilbert space $L^2(M)$ of the unique trace on M. Then consider the algebra <M,e> generated by M and e on $L^2(M)$ [Chr,Ska,Jon1]. Then <M,e> is hyperfinite and iterating this procedure with the inclusion M <M,e>, Jones obtained a sequence of projections satisfying the Temperley-Lieb relations (33), where $\tau^{-1} = [M,N]$ .

If $[M,N] = 2$, then N is the fixed point algebra $M^{\mathbb{Z}_2}$, for an unique outer $\mathbb{Z}_2$-action on M. That N is the fixed point algebra of M under an outer action is due to Goldman [Gol]; uniqueness was shown by Connes [Con].

This should be contrasted with the situation for C*-algebras:

For a UHF algebra $F_Q = \otimes_{\mathbb{N}} M_Q$, with $Q > 1$, there exists a C*-algebra A, with $K_1(A) \neq 0$, (in particular A is *not* AF), such that $A \otimes F_Q \simeq F_Q$ [EK]. This settles a question mentioned by Effros in [Eff2].

For any compact group $G \neq \{1\}$, $Q > 1$, there exists an action of G on $F_Q$, such that $K_1(F_Q{}^G) \neq 0$. In particular the fixed point algebra $F_Q{}^G$ is *not* AF, and so the action cannot be locally representable in the sense of [HR]. In fact $F_Q$ cannot be expressed as an inductive limit lim $B_n$ of finite dimensional G covariant algebras, $B_n^p$.

The existence of such exotic non locally representable actions was first shown by Blackadar [Bla2] for symmetries on the Fermion algebra $F_2$, Kumjian [Kum] for symmetries on a certain AF algebra associated with Bunce Deddens algebras, and Bratteli, Elliott, Evans and Kishimoto [BEEK1] for a finite group G on the UHF algebra $F_{|G|}$. In these constructions the AF algebra A was expressed in a non-standard way as an inductive limit

$$A_1 \to A_2 \to \cdots \tag{60}$$

$A = \lim A_n = \overline{\cup A_n}$ where each C*-algebra $A_n$ is *not* AF. In Blackadar's case, $A_n = C(\mathbb{T}, M_{4^n})$, and the embedding of $A_n$ in $A_{n+1}$ involves certain winding operationson the torus $\mathbb{T}$. In Kumjian's case, $A_n$ is the crossed product of $C(\mathbb{T}, M_{2n})$ by the flip $z \to \bar{z}$ on $\mathbb{T}$. The crossed product $C(\mathbb{T}, M_{2n}) \times \mathbb{Z}_2$ can be identified with the subalgebra of $C(I, M_{2n+1})$ consisting of continuous functions $f: I \to M_{2n+1}$, on the closed unit interval $I$, such that $f(0)$, respectively $f(1)$, lie in subalgebras of $M_{2n+1}$ isomorphic to $M_{2n} \oplus M_{2n}$. The embedding of $A_n$ in $A_{n+1}$ involved a certain folding operation on the interval. In [BEEK1] inductive limits of algebras $A_n$ of the following kind were considered, with embeddings using folding operations on the interval. Each $A_n$ consisted of a subalgebra of $C(I, M_{m_n})$, such that $f(0)$, respectively $f(1)$ commuted with certain abelian subalgebras $C_n^0$ respectively $C_n^1$ of $M_{m_n}$. A systematic investigation has begun of C*-algebras which are inductive limits of algebras $A_n$, where each $A_n$ is constructed as follows from data $(B, B_0, B_1)$, depending on n: If $B$, $B_0$, $B_1$ are AF C*-algebras, with embeddings $\iota_0$ respectively $\iota_1$ of $B^0$, respectively $B^1$ in B, we let A denote the C*-algebra

$$\{f \in C(I: B) : f(i) \in \iota_i(B_i), i = 0,1\} \tag{61}$$

The algebra A is represented diagramatically as

$$B_0 \underset{}{\overset{B}{\rule{5em}{0.4pt}}} B_1 \tag{62}$$

where the fibre over the open interval $(0,1)$ is B, and over the end points 0 and 1 are isomorphic to $B_0$ and $B_1$ respectively. A folding operation will embed each $A_n$ in $A_{n+1}$. The K-theory of the algebras $A_n$ can be computed using the ideal of functions which vanish at the end points, together with the 6-term short exact sequence of K-theory.

$$K_0(A) \simeq \{(x_0, x_1) \in K_0(B_0) \approx K_0(B_1) : \iota_*^0(x_0) = \iota_*^1(x_1)\} \tag{63}$$

$$K_1(A) \simeq K_0(B)/(\iota_*^0 K_0(B_0) - \iota_*^1 K_0(B_1)) \tag{64}$$

Note that $K_1(A)$ can be non trivial even for simple matrix algebras B, $B_0$, $B_1$ (c.f. [BEEK1]). For example, for $B = M_Q$, $B_0 = B_1 = \mathbb{C}1 \subset B$, it follows that $K_1(A) \simeq \mathbb{Z}_Q$, with generator given by the unitary:

$$t \to e^{2\pi i t} \oplus 1 \oplus 1 \oplus \ldots 1 \oplus 1 \in M_Q \tag{65}$$

Consider the case $B^n \simeq N^{\otimes n}$, where N is a full matrix algebra $M_Q$. Let $E_1$ and $E_2$ be projections of N such that $E_1 + E_2 = 1$, with $E_1$ of rank one. Define embeddings [BEEK1] $\lambda_n^i$ of N in $B^n$ for $i = 0, 1$ by

$$\lambda_1^0 = \lambda_1^0 = \text{id} \tag{66}$$

$$\lambda_n^1(x) = 1^{\otimes(n-1)} \otimes x \tag{67}$$

$$\lambda_n^0(x) = \lambda_{n-1}^0(x) \otimes E_1 + \lambda_n^1(x) \otimes E_2 . \tag{68}$$

[BEEK2] took the algebras $\lambda_n^i(C)'$ at the endpoints, where C is a maximal abelian subalgebra of N, and [EK] considered the situation where $B_n^i = \lambda_n^i(N)$ and

$$A_n = \{f \in C(I, B^n) : f(i) \in \lambda_n^i(N)', i = 0,1\}. \tag{69}$$

We embed $A_n$ in $A_{n+1}$ using a folding operation on the unit interval I = [0, 1] , as follows. Let $\varphi_0$, $\varphi_1$ be the maps on the interval I defined by

$$\varphi_o(t) = t/2, \qquad \varphi_1(t) = 1 - t/2 \tag{70}$$

Then embed $A_n$ in $A_{n+1}$ by:

$$\psi_n(a)(t) = a^\circ\varphi_1(t) \otimes E_1 + a^\circ \varphi_2(t) \otimes E_2 \tag{71}$$

and consider the inductive limit $A = \lim A_n$. In the case $Q = 2$, the embedding

$$f(t) \to \begin{pmatrix} f(t/2) & 0 \\ 0 & f(1-t/2) \end{pmatrix} \tag{72}$$

The embedding of $K_1(A_n)$ in $K_1(A_{n+1})$ is best seen by taking a projection p in $M_{Q^n}$ and considering the unitary $u \in A_n$ :

$$u(t) = \begin{cases} e^{4\pi it}p + 1 - p & 0 \leq t \leq \frac{1}{2} \\ 1 & \frac{1}{2} \leq t \leq 1 \end{cases} \tag{73}$$

Then $\psi(u)(t) = e^{2\pi it}\, p\otimes E_1 + 1\otimes 1 - p\otimes E_1$ in $A_{n+1}$, and since $E_1$ has rank one, we see that the map of $K_1(A_n)$ in $K_1(A_{n+1})$ corresponds to the identity on $\mathbb{Z}_Q$, under the identification $K_1(A_n) \simeq \mathbb{Z}_Q$ (64). In this way we see that $K_1(A) = \mathbb{Z}_Q$. To see that

$A\otimes F_Q$ is AF (hence isomorphic to the UHF $F_Q$ algebra by Elliott's theorem, as it will have the K-theory of the Fermion algebra) one uses Bratteli's criterion. Namely it is enough to show that any finite number of elements in A can be approximated by a finite dimensional C*-subalgebra of A. Thus for $a_1,..,a_r$ in $A_n$ and $\varepsilon > 0$, it is enough to find m and and a finite dimensional C*-subalgebra G of $A_{n+m}\otimes M_Q$ such that the distance from $\{a_1,..,a_r\}$ to G is less than $\varepsilon$. Subdividing the unit interval finely enough, the functions $a_1,..,a_r$ are approximately constant. Thus after iteration of the folding embeddings, their images $\tilde{a}_1,..,\tilde{a}_r$ in $A_{n+m}$, for m sufficiently large, become approximately constant along the whole interval. Then the constant map $g_i : t \to \tilde{a}_i(0)$, satisfies the correct constraint at 0, but not at 1. The sub-algebras at the end points are both isomorphic to $M_{Q^{n+m-1}}$ However in order to use the embedding as we have indicated in (72), we must have the subalgebras at the endpoints of () embedded in different positions of $M_{Q^{n+m}}$. Thus, $g_i(1) = \tilde{a}_i(0) \in M_{Q^{n+m-1}} \otimes M_Q \otimes 1$ which we can flip to $M_{Q^{n+m-1}} \otimes 1 \otimes M_Q$. In this way, by connecting the identity to the flip on $M_Q \otimes M_Q$ along the interval from 0 to 1, we can deform the constant map $g_i$ so that the deformation $\tilde{g}_i$ in $C(I, M_{Q^{n+m+1}})$ satisfies

$$\tilde{g}_i(0) = \tilde{a}_i(0) \otimes 1 \in B^0_{n+m} \otimes 1 \tag{74}$$

$$\tilde{g}_i(1) \in M_{Q^{n+m-1}} \otimes 1 \otimes M_Q = B^1_{n+m} \otimes M_Q \tag{75}$$

Since the elements $\tilde{g}_i \in A_{n+m}\otimes M_Q$ generate a finite dimensional C*-algebra, we have approximated $a_1,..,a_r$ by a finite dimensional subalgebra of $A_{n+m}\otimes M_Q$

Brown and Pedersen [BP] have recently introduced a notion of real rank for C*-algebras. Real rank zero is a generalisation of being total disconnected, thus all AF algebras are of real rank zero. (A C*-algebra is of real rank zero if the invertible self adjoint elements are dense in all invertible elements. This is equivalent to the self adjoint elements of finite spectrum being dense in all self adjoint elements [BP]). It would seem with all the examples constructed so far that the fixed point algebra of an AF algebra under a compact groups is of real rank zero.

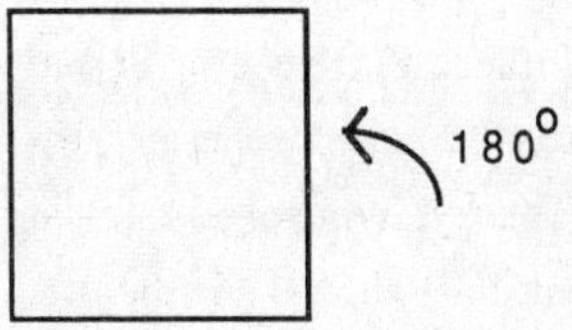


Figure 11

Other exotic symmetries have recently been studied on non-commutative torii [BEEK]. Consider first the classical 2-torus $\mathbb{T}^2 = \{(z_1,z_2) : z_i \in \mathbb{C}, |z_i| = 1\}$, under the identification $(z_1,z_2) \to (\bar{z}_1\ \bar{z}_2)$. The symmetry corresponds to rotation through $\pi$ in $\mathbb{R}^2/\mathbb{Z}^2$ in Figure 11, and so it is enough to consider the triangle indicated in Figure 12, with the identifications on the edges as shown:

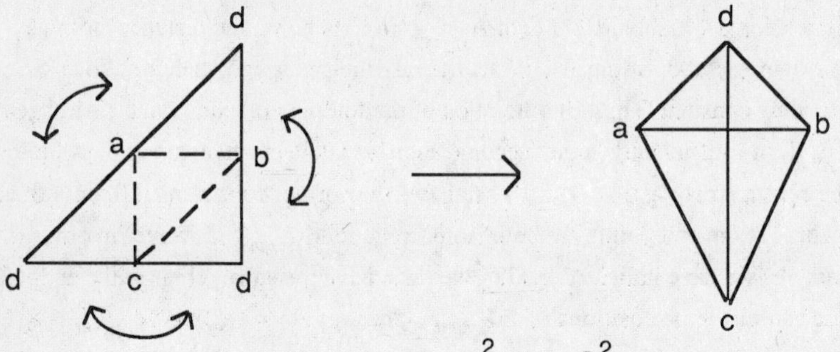


Figure 12 : $\mathbb{T}^2/\mathbb{Z}_2 = S^2$

The orbifold $\mathbb{T}^2/\mathbb{Z}_2$ can be recognised as a 2-sphere but with four singular points - the fixed points of the symmetry. The C*-algebra $C(\mathbb{R}^2)$ is the universal C*-algebra generated by two unitaries U and V which commute: UV = VU, where we can identify U and V with the co-ordinate maps $(z_1,z_2) \to z_i$ on $\mathbb{T}^2$. The fixed point algebra $C(\mathbb{T}^2)^{\mathbb{Z}_2}$. is identified with the continuous functions $C(S^2)$ on a 2-sphere $S^2$. Next consider the non-commutative or quantum torus, and the symmetry corresponding to conjugation in the classical case. If $q \in \mathbb{T}$, then $A_q$, the universal C*-algebra generated by two unitaries U and V satisfying UV = qVU, is called a non-commutative or quantum torus. We can define a symmetry on $A_q$ by $U \to U^{-1}$, $V \to V^{-1}$, since the pair $(U^{-1},V^{-1})$ satisfy the same relations as (U,V), and $A_q$ is the universal C*-algebra for those relations. The fixed point algebra $B_q = A_q^{\mathbb{Z}_2}$ is better thought of as a non-commutative (or quantum) toroidal orbifold rather than a non-commutative sphere, since the four fixed points in $\mathbb{T}^2$ play a key role, especially when computing the K-groups. The original motivation for the study in [BEEK2] was whether or not $B_q$ is AF when q is *not* a root of unity. There is also interest in these algebras since the operators $U+U^{-1} + \lambda(V+V^{-1})$ where $\lambda$ is real q not a root of unity, are the discrete Schroedinger operators with almost periodic potentials, which are known to have Cantor spectra (at least) for generic $\lambda$ and q [BeS,CEY]. Now if q is not a root of unity, $A_q$ is the unique C*-algebra generated by two unitaries U and V satisfying the relation UV = qVU [EH]. The following algebraic characterisation of $B_q$ was obtained in [BEEK2]. If q is not a root of unity then $B_q$ is the unique C*-algebra generated by two self-adjoint elements A and B satisfying the three independant relations:

$$BA^2 + A^2B = (q+q^{-1})ABA - (q-q^{-1})^2 B, \quad AB^2 + B^2A = (q+q^{-1})BAB - (q-q^{-1})^2A,$$

$$BABA = (q^2+1- q^{-2}))\, ABAB - (q+q^{-1})A^2 B^2 - (q+q^{-1}- q^3- q^{-3})(A^2 + B^2 -2). \tag{76}$$

The correspondence between this abstract algebra and $B_q$ is given by $A = U+U^{-1}$, $B = V+V^{-1}$. The first two relations can be explained via the action of $SL(2,\mathbb{Z})$ on $A_q$ which reduces to its usual action on $\mathbb{R}^2$ when $q = 1$. Under this action the matrix $\begin{pmatrix} -1 & 0 \\ 0 & -1 \end{pmatrix}$ corresponds to the basic symmetry $U \to U^{-1}$, $V \to V^{-1}$. Since this matrix generates the centre of $SL(2,\mathbb{Z})$, we have that $PSL(2,\mathbb{Z})$ acts on the fixed points $B_q = A_q^{\mathbb{Z}_2}$. The two basic generators of $PSL(2,\mathbb{Z})$ act as follows on $B_q$. First, $\begin{pmatrix} 0 & -1 \\ 1 & 0 \end{pmatrix}$ merely interchanges A and B. However $\begin{pmatrix} 1 & 1 \\ 0 & 1 \end{pmatrix}$ sends $A \to A$, and $B \to q^{1/2}(1-q^2)^{-1}[A,B]_q$, where $[X,Y]_q$ is the twisted commutator $XY-qYX$. Similarly, the inverse automorphism sends $A \to A$, and $B \to q^{-1/2}(1-q^{-2})^{-1}[A,B]_q{-1}$. Composing, we see $B = -(q-q^{-1})^{-2}[A,[A,B]_q{-1}]_q$, which is the first relation above.

Let $P_q = P(U,V)$ denote the polynomial algebra in U and V. Then when q is a root of unity but $q^2 \neq 1$, the enveloping C*-algebra of $P_q^{\mathbb{Z}_2}$, exists and is canonically isomorphic to $B_q$ [BEEK3], however the enveloping C*-algebra of $P_q^{\mathbb{Z}_2}$ does not exist when $q^2 = 1$.

It is known [Kum2] that:

$$K_0(B_q) = \mathbb{Z}^6 \text{ if } q^2 \neq 1, \ (\mathbb{Z}^5 \text{ when } q = -1, \ \mathbb{Z}^2 \text{ when } q = 1)$$

$$K_1(B_q) = 0 \tag{77}$$

with the latter not giving away any information about whether $B_q$ is AF or not, as $K_1$ always vanishes for an AF algebra. These K-groups can easily be computed [BEEK3] when q is a root of unity using the following simple description of $B_q$ in that case. If $q = e^{2piP/Q}$, where P and Q are relatively prime positive integers, then $B_q$ is a subalgebra of the C*-algebra $C(S^2, M_Q)$ of continuous functions from the 2-sphere $S^2$ into the algebra $M_Q$. The subalgebra is determined [BEEK3] up to isomorphism as follows. There are four distinct points $\omega_0, \omega_1, \omega_2, \omega_3$ in $S^2$ and to each point $\omega_i$ is associated a projection $E_i$ in $M_Q$. The dimension of $E_i$ is as follows:

$$\dim(E_i) = (Q-1)/2, \text{ when Q is odd}$$

$$\dim(E_0) = (Q-2)/2, \text{ and } \dim(E_i) = Q/2 \text{ for } i = 1,2,3, \text{ when Q is even.} \tag{78}$$

Then the subalgebra consists of those functions $f \in C(S^2,M_Q)$ such that $f(\omega_i)$ commutes with $E_i$ for $i = 0,1,2,3$. Thus when $q = e^{2\pi iP/Q}$, the isomorphism class of $B_q$ depends only on Q but *not* on P, although the non-commutative torii $A_q$ and $A_{q'}$ are isomorphic if and

only if $q' = q$, or $q' = q^{-1}$ [Rie,HKS]. The algebra $A_{\exp(2\pi iP/Q)}$ is a homogeneous algebra over the 2-torus $\mathbb{T}^2$ with fibre $M_Q$, but the obstruction to isomorphism for the non-commutative torii is eliminated for the non-commutative orbifolds by the behaviour of the four singular points. Consequently, the bundle for $B_{\exp(2\pi iP/Q)}$ is trivial. The K-theory of $A_{\exp(2\pi iP/Q)}$ can then be computed using the ideal of functions which vanish at all the $\omega_i$, together with the
6-term short exact sequence of K-theory. More informally to see $K_0(B_q) = \mathbb{Z}^6$ if $q^2 \neq 1$, a projection in $B_q$ will give an element in $\mathbb{Z}^8$ say $\{(d_1^i, d_2^i) : 1=0,1,2,3\}$ with three constraints $d_1^i + d_2^i$ = constant, which gives us $\mathbb{Z}^5$; the additional element of $K_0(B_q)$ being provided by the Bott element on the 2-sphere. (If $p \in C(S^2, M_Q)$ is a projection, then $(d_1^i, d_2^i) = (\dim(p(w_i)E_i), \dim(p(w_i)(1-E_i))$ is in $\mathbb{Z}^8$, where the dimension is computed in $M_Q$, with $d_1^i + d_2^i$ = constant due to the homotopy invariance of dim f(w) for all $w \in S^2$). To understand the K-theory and traces of $B_q$ better, it is convenient [BEEK3] to bring the crossed product $C_q = A_q \times \mathbb{Z}_2$ into the game. When $q = e^{2\pi iP/Q}$, and P and Q are relatively prime positive integers $C_q$ is isomorphic a subalgebra of the C*-algebra $C(S^2, M_{2Q})$ of continuous functions from the 2-sphere $S^2$ into the algebra $M_{2Q}$ determined as follows. There are four distinct points $\omega_0, \omega_1, \omega_2, \omega_3$ in $S^2$ and an orthogonal projection E in $M_{2Q}$ of dimension Q, such that the subalgebra consists of those functions $f \in C(S^2, M_{2Q})$ such that $f(\omega_i)$ commutes with E for $i = 0,1,2,3$. For any q, the algebraic crossed product $P_q \times \mathbb{Z}_2$ is the *-algebra generated by three unitaries U,V and W satisfying the relations $VU = qUV$, $WU = U^{-1}W$, $WV = V^{-1}W$, and $W^2 = 1$, its completion is the crossed product $C_q = A_q \times \mathbb{Z}_2$. There is a canonical continuous tracial state $\tau$ on $A_q \times \mathbb{Z}_2$ given by

$$\tau(q^{nm/2}U^nV^m) = 1 \text{ if } n = m = 0, \text{ and } 0 \text{ otherwise} \tag{79}$$

$$\tau(q^{nm/2}U^nV^mW) = 0 \tag{80}$$

There are four trace functionals $\tau_{p_1p_2}$ where $p_1,p_2 \in \{\text{even, odd}\}$ defined on $P_q \times \mathbb{Z}_2$ by :

$$\tau_{p_1p_2}(q^{nm/2}U^nV^m) = 0, \tag{81}$$

$$\tau_{p_1p_2}(q^{nm/2}U^nV^mW) = 1 \tag{82}$$

if parity(n) = $p_1$ and parity(m) = $p_2$, and 0 otherwise.

When q is not a root of unity, the space of trace functionals on $P_q \times \mathbb{Z}_2$ is five dimensional [BEEK2,3], spanned by $\tau$ and the $\tau_{p_1p_2}$, with the only continuous trace functionals on $C_q$ (respectively $A_q$) being scalar multiples of $\tau$ (respectively $\tau|A_q$). It is however remarkable, when q is a root of unity, that the other four trace functionals are continuous and extend to $C_q$. The algebraically defined trace functionals $\tau_{p_1p_2}$ are related to the following geometrically defined trace functionals $\tilde{\tau}_{p_1p_2}$ which are clearly *continuous*. There is a convenient labelling of the four singular points $\omega_i$ by $\{\text{even,odd}\}^2$, i.e. $\omega_{p_1p_2}$. Define four continuous trace functionals on $C_q$ by: $\tilde{\tau}_{p_1p_2}(f) = Tr_{2Q}((2E-1)f(\omega_{p_1p_2}))/Q$, where $Tr_{2Q}$ is the un-normalised trace on $M_{2Q}$, and E is the rank Q projection in $M_{2Q}$ as above. Then, adpopting a proper sign convention, the algebraically defined trace functionals are related to the geometrically defined trace functionals by [BEEK3]:

$$\tau_{p_1p_2} = (-1)^{p_1p_2}\tilde{\tau}_{p_1p_2} \quad \text{if Q is even and}$$

$$\tau_{nm} = (-1)^{pnm}\{\sum_{p_1p_2} (-1)^{np_1+mp_2}\tilde{\tau}_{p_1p_2}\} \text{ when Q is odd} \qquad (83).$$

In particular the trace functionals $\tau_{p_1p_2}$ are continuous when q is not a root of unity. The five trace functionals $\tau$ and $\tau_{p_1p_2}$ separate the following five projections in $C_q$ written down by Nest [Nes]:

$$1,\ (1+W)/2,\ (1+UW)/2,\ (1+VW)/2,\ (1+q^{1/2}UVW)/2. \qquad (84)$$

These projections, together with the Rieffel projection show that $K_0(C_q)$ contains $\mathbb{Z}^6$ even when q is not a root of unity. Nest [Nes] has computed the cyclic cohomology of these non-commutative toroidal orbifolds. In the course of this he has shown that all derivations on the smooth elements of $B_q$ (smooth with respect to the $\mathbb{T}^2$ torus action on the rotation algebra $A_q$) are approximately inner if q is not a root of unity (c.f. [PS] for the conjectured strucure of derivations on AF algebras).

We now return to our AF algebra A(Γ) related to a graph Γ. We have noted how the transfer matrix formalism yields the one-dimensional Pauli algebra setting for the two-dimensional Ising model. More precisely, it yields naturally the even part of the Pauli-algebra, as

$$C^*(e_2,e_3,....) = C^*(\sigma_z^i, \sigma_x^i\sigma_x^{i+1}) = (\otimes M_2)^{\mathbb{Z}_2}$$

We can achieve the Pauli algebra, by adjoining the odd element $\sigma_x^i$, or $e_1 = (1+\sigma_x^i)/2$. Thus

$$C^*(e_1,e_2,....) = C^*(\sigma_z^i, \sigma_x^i, \sigma_x^i\sigma_x^{i+1}) = \otimes M_2$$

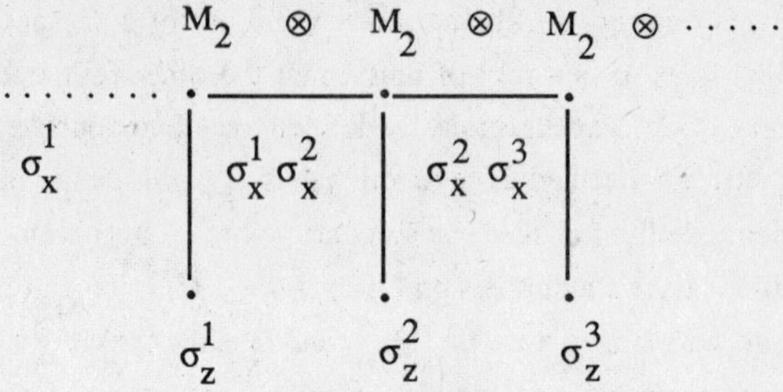


Figure 13

Thus in general we have a hierarchy of AF algebras:

$$C^*(e_2, e_3, \ldots.) \to C^*(e_1, e_2, \ldots.) \to A(\Gamma) \tag{85}$$

with equality in the final inclusion if $\Gamma = A_l$, and the first algebra arises from the local transfer matrices. It is therefore natural to consider two problems. One is to understand embedding of path algebras $A(\Gamma_1) \to A(\Gamma_2)$ more systematically. The other is to consider the embedding $A(\tau) \subset A(\Gamma)$, and the question of how we can adjoin some extra operators to the Temperley-Lieb projections $\{e_n\}$ and relations additional to (33), to obtain the full path algebra $A(\Gamma)$. (Note that $A(\Gamma)$ is generated by $A(\tau)$ and a finite dimensional algebra, at least under some mild conditions on $\Gamma$ [CoE]). These quesion were discussed in detail in [EG2] for the following T-shaped graphs:

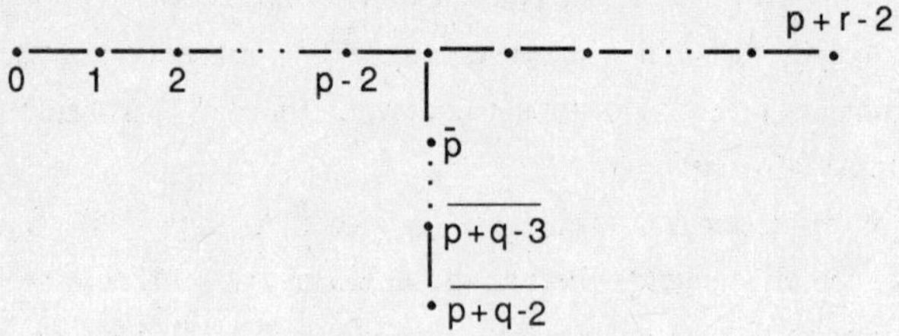


Figure 14 : $T_{p,q,r}$

although the methods should also work for the following graphs, which include the star shaped graphs of Haagerup and Schou [HS]:

$\Gamma$

Figure 15

Shearer [She] has shown that the norms of incidence matrices of such graphs are dense in $[\sqrt{(2+\sqrt{5})}, \infty]$ ; the limit points in the range $\leq \sqrt{(2+\sqrt{5})}$ of norms of incidence matrices are known from Hofman [Hof].

Let $p \geq 2$, $\tau > 0$, and $e_1, e_2, \ldots, e_{\bar{p}}$ a sequence of projections satisfying

$$e_n e_m = e_m e_n, \quad m,n = 1,2, \ldots, |m - n| \geq 2 \tag{86}$$

$$e_n e_{n\pm1} e_n = \tau e_n \tag{87}$$

$$e_n e_{\bar{p}} = e_{\bar{p}} e_n, \ n \neq p \ ; \ (e_1 e_{\bar{p}} = 0, \text{ if } p = 2) \tag{88}$$

$$e_{\bar{p}} e_p e_{\bar{p}} = \tau e_{\bar{p}} \tag{89}$$

$$e_p e_{\bar{p}} e_p = \tau (1 - e_1 \vee \ldots \vee e_{p-2}) e_p. \tag{90}$$

Let $A(\tau,p) = C^*(1, e_1, e_2, \ldots, e_{\bar{p}})$. Then $A(\tau,p)$ is non-trivial [EG2] if and only if

$$\beta = \tau^{-1/2} \in \{ \|T_{p,2,r}\| : r \geq 1\} \cup [\|T_{p,2,\infty}\|, \infty) \tag{91}$$

where $T_{p,2,1} = A_{p+1}$. If $p = 1$, then the sequence of graphs is $T_{1,2,r} = A_{r+1}$, and so this reduces to Jones/Wenzl [Jon1,Wen1] classification of values of $\tau$ for which there exists an infinite sequence of projections satisfying the Temperley-Lieb relations (33). Jones needed a Markov trace on $A(\tau)$; Wenzl removed this assumption. It is striking to note that throwing in extra projections and relations gives a rigidity above index four. Moreover there exists [EG2] a surjective *-homomorphism

$$A(T_{p,2,r}) \approx \mathbb{C}(1 - e_1 \vee \ldots \vee e_{p+r-2} \vee e_{\bar{p}}) \to A(\tau,p) \text{ when } \beta = \|T_{p,2,r}\| \tag{92}$$

$$A(T_{p,2,\infty}) \to A(t,p) \text{ when } \beta \geq \|T_{p,2,r}\|. \tag{93}$$

If $r < \infty$, i.e. $\beta = || T_{p,2,r} || < || T_{p,2,\infty} ||$, then (92) is automatically an isomorphism. Unlike the case p=1 in [Jon1], we do not need a Markov trace on $A(\tau)$ to construct the above homomorphism. Suppose however there exists a (Markov) trace tr on $A(\tau,p)$ such that

$$\mathrm{tr}\,(xe_n) = \tau\,\mathrm{tr}\,x, \quad x \in A(\tau,p)_n \tag{94}$$

where

$$A(\tau,p)_n = \begin{cases} C^*(1,e_1,e_2,\ldots,e_{n-1}) & n < p \\ C^*(1,e_1,e_2,\ldots,e_{n-1}e_{\bar{p}}) & n \geq p \end{cases} \tag{95}$$

We can then deduce that if $\beta = ||T_{p,2,r}||$, $1 \leq r < \infty$ then $1 = e_1 \vee \ldots \vee e_{p+r-2} \vee e_{\bar{p}}$ and $A(\tau,p) \simeq A(T_{p,2,r})$, moreover if $\beta \geq ||T_{p,2,\infty}||$, then $A(\tau,p) \simeq A(T_{p,2,\infty})$.

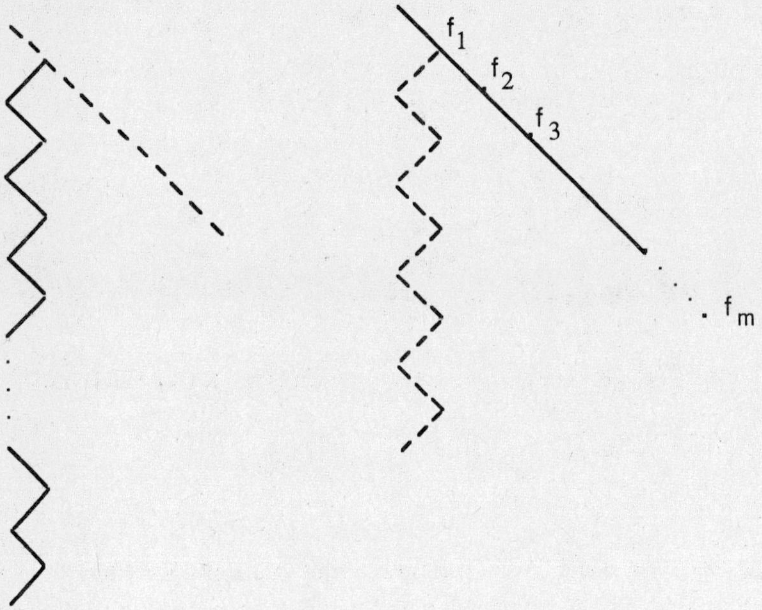


Figure 16

The construction of the homomorphism from the path algebra to the algebra defined in terms of generators and relations requires the construction of matrix units indexed by finite paths in the Bratteli diagram. We will indicate here this construction for the A-series, as this is already interesting. When $\Gamma = A_\infty$, the functions $\{\varphi_v\}$ of (57) are the Chebyshev polynomials $\varphi_n = S_n \in \mathbb{Z}[x]$ of the second kind satisfying

$$S_r = xS_{r-1} - S_{r-2}, \quad S_o = 1, \quad S_{-1} = 0. \tag{96}$$

The two basic paths that need to be considered first are shown in Figure 16. The projection corresponding to the left hand path is clearly $e_1e_3e_5...e_{2n+1}$. Label the projections corresponding to the right hand paths by $f_m$, so that $f_1 = 1$, $f_2 = 1-e_1$. Then working in the path algebra A[m-1,m+1], we have $f_m = f_{m+1} + i$, where i.is the projection on the right hand path in the diamond below:

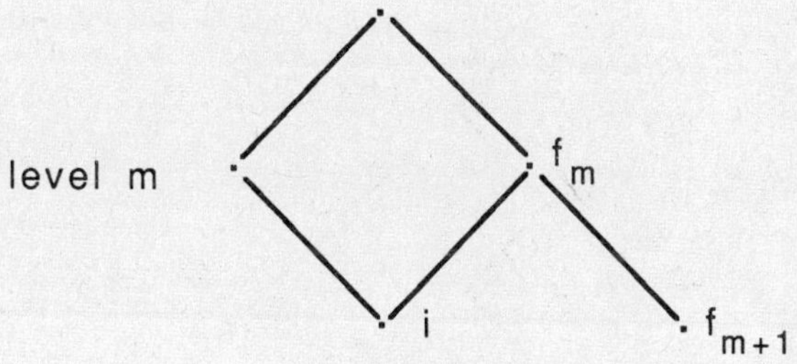


Figure 17

Then $f_{m+1}e_m = 0$, so that $f_me_mf_m = i\,e_m\,i$. Computing in the $2 \times 2$ matrix algebra End $l^2(t(m-1))$, we see that $ie_mi$ must be scalar multiple of the rank one projection i. In fact, from (58) we see that $ie_mi = (S_m/\beta S_{m-1})\,i$. Thus we recover Wenzl's formula [Wen1]:

$$f_{m+1} = f_m - (\beta S_{m-1}/S_m)\, f_me_mf_m = (f_m - i).$$

In this way we inductively define the projections $f_m$ corresponding to the right hand paths. Indeed $f_m = 1 - e_1 \vee ... \vee e_{m-1}$. Since $e_m$ arises from the positive weights of the Perron-Frobenius eigenvector, $f_m\,e_m$ must take the left hand path in the diamond of figure to a non zero multiple of the right end path. Again this multiple is easily computed to be $(S_{m-2}S_m)^{1/2}/\beta S_{m-1}$. We then define $u^*_m = \beta S_{m-1}/(S_{m-2}S_m)^{1/2} f_m\,e_m$, so that $u_m$ flips the path as in Figure 18.

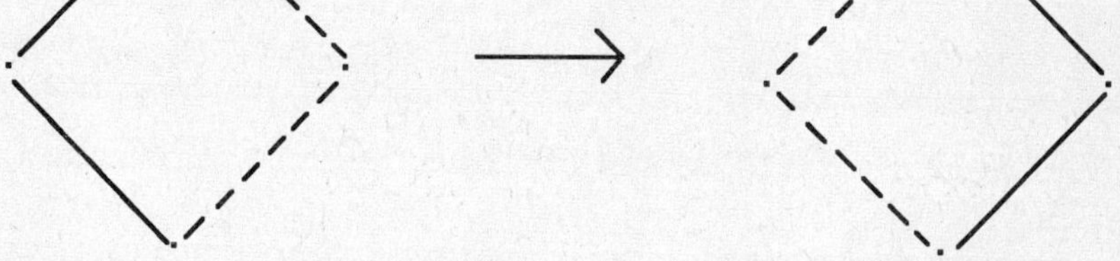

Figure 18

We combine these operations as follows. To define $e_{\mu\mu'}$ when $\mu$ is the path in Figure 19, we should take $e_1 e_3 \ldots e_{2n-1}$ shift$^{2n}$ $(f_m)$ where we need the shift by 2n units. Now the Kramers-Wannier shift by one unit $\kappa$ could be implemented by the braid elements: $\rho_i = te_i - (1-e_i)$, where $t/(1+t)^2 = \tau$, namely $\mu(x) = \lim \rho_1\rho_2 \ldots \rho_n x \rho_n^{-1} \ldots \rho_2^{-1} \rho_1^{-1}$ converges if x is in the algebra generated by $e_1, e_2, \ldots$ and $\mu(e_m) = e_{m+1} = \kappa(e_m)$.

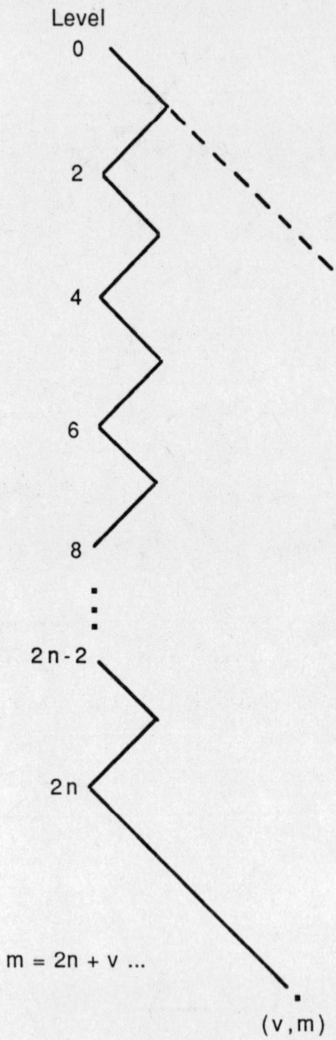


Figure 19

However, when $\tau^{-1} > 4$, the braid element $\rho_i$ is not unitary, and we would like to construct *orthogonal* idempotents. However we only need *a* shift by two units, as the shift shown in Figure 20 is what is actually neeeded in the path algebra.

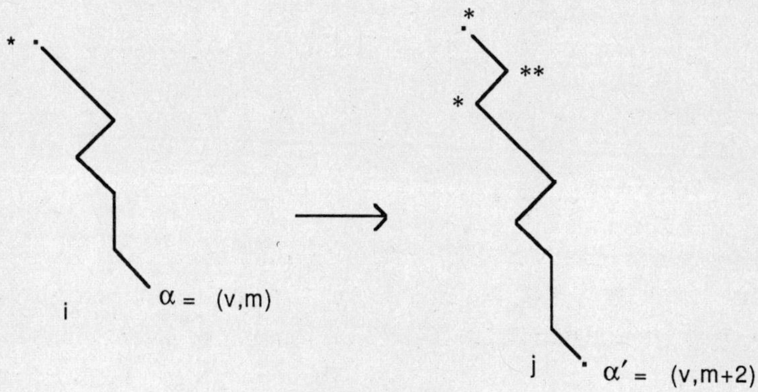


Figure 20

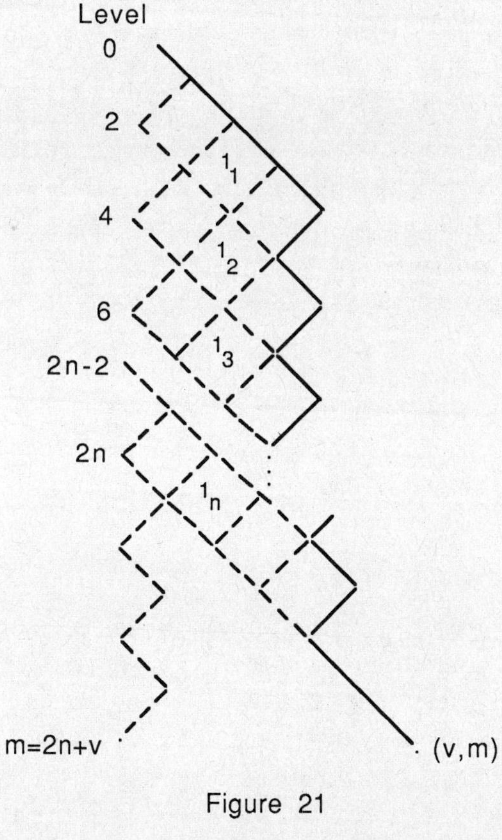


Figure 21

The path shift $\gamma$ in Figure 20 corresponds algebraically to

$$\gamma(x) = \lim_{n\to\infty} \tau^{-n} e_1 e_2 \ldots e_n x e_n \ldots e_2 e_1 \tag{97}$$

With this $\gamma$, we then define $e_{\mu\mu}$ to be

$$e_1 e_2 \cdots e_{2n+1} \gamma^n (f_m)$$

when $\mu$ is the path in Figure 19. Next consider the path $\delta$ in Figure 20. Let $\alpha = (v,m) \in \hat{A}_\infty$ and $n = (m - v)/2$, and $G_\alpha = \gamma^n(f_v)$ ($= e_{\mu\mu}$). To obtain an expression for $e_{\mu\delta}$ corresponding to the paths $\mu$ in Figure 19 and $\delta$ in Figure 21 one postmultiplies $G_\alpha$ by the operator $\gamma^{n-1}(\Delta_{i_n})\gamma^{n-2}(\Delta_{i_{n-1}}) \ldots \gamma(\Delta_{i_2})\Delta_{i_1}$. Thus $G_\alpha \gamma^{n-1}(\Delta_{i_n})$ corresponds to the projection $e_{\mu\delta'}$ where $\delta'$ is the path obtained from that in Figure 19 by flipping $i_n$ diamonds in the $n^{th}$ diagonal strip shown in Figure 21. Postmultiplying the new matrix unit by $\gamma^{n-2}(\Delta_{i_{n-1}})$, flips $i_{n-1}$ diamonds in the $(n-1)^{th}$ strip etc. If $\tau^{-1} = 4\cos^2(\pi/(m+1))$, then $S_m$ vanishes and we can only consider matrix units corresponding to paths from $\hat{A}_m$.

Embeddings $A(\Gamma_1) \to A(\Gamma_2)$ of the AF algebras associated with graphs $\Gamma_1$ and $\Gamma_2$ are discussed in [EG2] from a dimension group point of view. For the infinite T-shaped graphs of figure we have $K_o(A(\Gamma)) \simeq \mathbb{Z}[t]$, with positive cone identified with $\{0\} \cup \{P \in \mathbb{Z}(t) : P(\lambda) > 0,\ \lambda \in (0,\gamma]\}$, where $\gamma = \gamma(\Gamma) = ||\Gamma||^{-2} < \frac{1}{4}$. (That these groups $\mathbb{Z}[t]$ ordered by the interval $(0,\gamma]$ are indeed dimension groups is a consequence of the work of Bratteli, Elliott and Herman [BEH],but no general construction of the AF algebras associated to such dimension groups was known). Hence for these graphs there exists a unital homomorphism $A(\Gamma_1) \to A(\Gamma_2)$ if $||\Gamma_1|| \le ||\Gamma_2||$. For the graphs $\Gamma = T_{p,q,r}$ we first seek rational functions $\{\phi_v : v \in \Gamma^{(0)}\}$ in an indeterminate $x$ such that

$$x\phi_v = \sum_{w\in\Gamma^{(o)}} \Delta(v,w)\phi_w, \qquad \phi_* = 1 \tag{98}$$

If $d(v)$ is the distance of a minimal path from $*$ to $v$, let

$$Q_v(t) = x^{-d(v)} f_v(x) \tag{99}$$

where $t = x^{-2}$. For the graphs $T_{p,2,\infty}$, the polynomials $\{t^m Q_v(t) : m = 0,1,2,\ldots,\ v \in \Gamma^{(0)}\}$, conveniently relabelled as $\{Q_\alpha(t) : \alpha \in \hat{\Gamma}^{(0)}\}$ will generate the positive cone $K_o(A(T_{p,2,\infty}))_+$ in $K_o(A(T_{p,2,\infty})) \simeq \mathbb{Z}[t]$. When $\Gamma = A_\infty$, the functions $\{\phi_v\}$ are the Chebyshev polynomials $\phi_n = S_n \in \mathbb{Z}[x]$ of the second kind satisfying (96), and $Q_n = P_n$, where $P_r \in \mathbb{Z}[t]$, $r = 0,1,2, \ldots\ldots$ are defined by

$$P_r(t) = x^{-r} S_r(x), \qquad t = x^{-2} \tag{100}$$

and are the polynomials occuring in [Jon1]:

$$P_r = P_{r-1} - tP_{r-2}, \quad P_o = 1, \ P_{-1} = 0. \tag{101}$$

For the finite T-shaped graphs $T_{p,2,r}$ of Figure 14, $K_o(A(\Gamma)) \simeq \mathbb{Z}[t]/\langle Q\rangle$ where $\langle Q\rangle$ denotes the ideal generated by a polynomial $Q = Q(\Gamma)$ which is essentially the characteristic polynomial of the graph $\Gamma$, and positive cone identified with $\{0\} \cup \{f + \langle Q\rangle : f(\gamma) > 0\}$ where $\gamma = \gamma(\Gamma) = ||\Gamma||^{-2}$. Hence there exists a unital homomorphism $A(\Gamma_1) \to A(\Gamma_2)$ if $||\Gamma_1|| = ||\Gamma_2||$, and $Q(\Gamma_1)$ divides $Q(\Gamma_2)$. In particular we have an explicit embedding of $A(A_{11})$ in $A(E_6)$. This embedding was also described by [Pas] using ad hoc methods. Taking suitable von Neumann algebra completions, one obtains the example of Goodman, de la Harpe and Jones [GHJ] of an irreducible subfactor of the hyperfinite $II_1$ factor with index $3+\sqrt{3}$. Finding explicit matrices of multiplicities of mapping the finite dimensional subalgebras of $A(A_1)$ in those of $A(\Gamma)$ amounts to being able to write polynomials which generate $\{Q \in \mathbb{Z}[t] : Q(\lambda) > 0, \ \lambda \in (0,\frac{1}{4}]\}$, essentially Chebyshev polynomials of the second kind, as a positive linear combination of polynomials which generate $\{Q \in \mathbb{Z}[t] : Q(\lambda) > 0, \ \lambda \in (0,\gamma_p]\}$. These embeddings are best understood in terms of the polynomials $\{f_v(x) : v \in \Gamma^{(0)}\}$ rather than $\{Q_\alpha : \alpha \in \hat{\Gamma}^{(0)}\}$. For $\Gamma$ a connected locally finite bipartite graph with distinguished vertex $*$, the embedding $A(A_n) \to A(\Gamma)_n$ can be described as follows. Define $a_m$ in $V(\Gamma)$ the free module over $\mathbb{Z}$, generated by the vertices of $\Gamma$ (identifying an element $a \in V(\Gamma)$ as $a = (a_v)$, $a_v \in \mathbb{Z}$, $v \in \Gamma^{(0)}$) by $a_m = S_m(\Delta)a_o = \Delta a_{m-1} - a_{m-2}$, $a_0 = *$, $a_{-1} = 0$. Then [EG2]:

(a) If $||\Gamma|| < 2$, then $||\Gamma|| = 2\cos[\pi/(n+1)]$, $a_0 \gneqq 0, \ldots, a_{n-1} \gneqq 0$, $a_n = 0$, and there is an inclusion of graph algebras $j : A(A_n) \to A(\Gamma)$ given by the matrix $A = (a_0,\ldots,a_{n-1})$.

(b) If $||\Gamma|| \geq 2$, then $a_m \gneqq 0$ for all m, and there is an inclusion of graph algebras $j: A(A_\infty) \to A(\Gamma)$ given by the matrix $A = (a_0,a_1,a_2,\ldots.)$.

Positivity of the embedding matrix A is ensured by work of Hamachi [Ham] on the index of subfactors in an ergodic theory context. The algebraic presentation of () for these T-shaped graphs further illuminates the above embeddings.

In general the dimension group of an AF algebra, is merely an ordered group not an ordered ring as it always is for $K_o$ of a commutative C*-algebra. However, we see that for the above graphs, $K_o(A(\Gamma))$ is indeed an ordered ring. There is product structure on $K_o(A(\Gamma))$ compatible with the ordered group strucure, in particular the product of two positive elements is positive. The structure of $K_o(A(\Gamma))$ as an ordered ring is related to the fusion rules of rational conformal field theory. From a two dimensional lattice model, one gets via the transfer matrix method a field theory in two dimensions by taking a continuum or scaling limit; letting the lattice spacing $\varepsilon \to 0$, whilst simultaneously approaching the critical temperature. As the scale or correlation length becomes infinite, one obtains a scale invariant or conformal invariant theory, equipped with representations of the Virasoro algebra. In a rational conformal field theory [MS,Ver] the physical Hilbert space $\mathsf{H}$ decomposes as a finite sum

$$\mathsf{H} = \bigoplus_{r,\bar{r}=0}^{N} h_{r,\bar{r}}\, H_r \otimes H_{\bar{r}} \tag{102}$$

where $H_r$ is an irreducible representation of the chiral algebra $\mathsf{A}$ and $h_{r,\bar{r}}$ represents a multiplicity. Chiral vertex operators are intertwining operators. Given three representations i,j,k, then the chiral vertex operators of type $\begin{pmatrix} i \\ j\,k \end{pmatrix}$ will be a vector space $V^i_{jk}$ of linear transformations $H_i^\vee \otimes H_j \otimes H_k \to \mathbb{C}$. The fusion rules are $N^i_{jk} = \dim V^i_{jk}$, and with primary fields $\{\phi_i\}$ we have the formal product rule or operator product expansion:

$$\phi_j \times \phi_k = \Sigma\, N^i_{jk}\, \phi_i. \tag{103}$$

For certain graphs $\Gamma$, $K_0(A(\Gamma))$ will by the preceeding be an ordered ring, and $K_0(A(\Gamma))_+$ has generators $\{Q_\alpha : \alpha \in \hat{\Gamma}^{(0)}\}$. So from the ordered ring structure we have non-negative integers $a^\gamma_{\alpha\beta}$, $\alpha,\beta,\gamma \in \hat{\Gamma}^{(0)}$, such that:

$$Q_\alpha Q_\beta = \Sigma\, a^\gamma_{\alpha\beta}\, Q_\gamma \tag{104}$$

The fusion rules of the Verlinde algebra (103) for certain models (e.g. Wess-Zumino $SU(2)_k$) models will be related to the ordered ring structure (104) of $K_0(A(\Gamma))$ for certain graphs (e.g. $A_{k-1}$ - but the methods of [EG2] allow us to compute fusion rules for other graphs). Again, it seems that (104) is best understood in terms of the set of polynomials $\{\phi_v(x)\}$ indexed by the vertices v of the graph $\Gamma$, rather than the set $\{Q_\alpha(t)\}$ indexed by the vertices a of the Bratteli diagram $\hat{\Gamma}$.

The structure of the dimension group of $K_0(A(\Gamma))$ also sheds light on the embedding of [CoE] of the observable algebra O in $A(\Gamma)$. Motivated by multiplicative K-homology and understanding critical phenomena in some classical statistical mechanical models, actions of $GL(\infty)$ we constructed in [CoE] on Temperley-Lieb and path algebras, and were analysed in terms of U(1)-current algebras. Now the gauge invariant observable algebra O is the fixed point algebra of the Fermion algebra under the action of the gauge group $O = (\otimes_{\mathbb{Z}} M_2)^{\mathbb{T}}$ and is isomorphic to the path algebra $A(A_{\infty,\infty})$ [Bra]. Moreover $K_0(A(A_{\infty,\infty})) \simeq \mathbb{Z}[t]$ [Ren], with the positive cone identified with

$$\{0\} \cup \{P \in \mathbb{Z}[t] : P(t) > 0,\ t \in (0,1)\} \tag{105}$$

We have also noted from [EG] in the preeceding that $K_o(A(A_m)) \simeq \mathbb{Z}[t]/<P_m>$, where $P_\infty = 0$, and positive cone ordered by $1/4 \cos^2\pi/(m+1)$ when $m < \infty$, and ordered by $(0,1/4]$ when $m = \infty$. Indeed there is a commutative diagram:

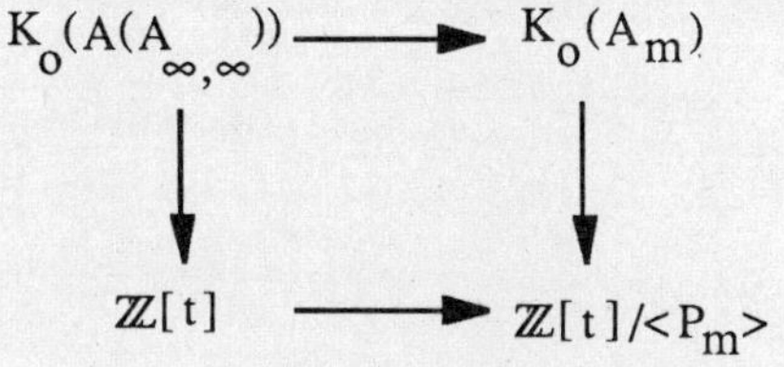


Figure 22

for $3 \le m \le \infty$, where the vertical maps are isomorphisms and $\bar{id}$ is the quotient map induced by the identity map on $\mathbb{Z}[t]$. Consider the following projections $g_i$ and partial isometries $v_i$ in the path algebra $A(A_{\infty,\infty})$. Define $g_i$ to be the projection in $A[i-1,i]$ as projection on the edges

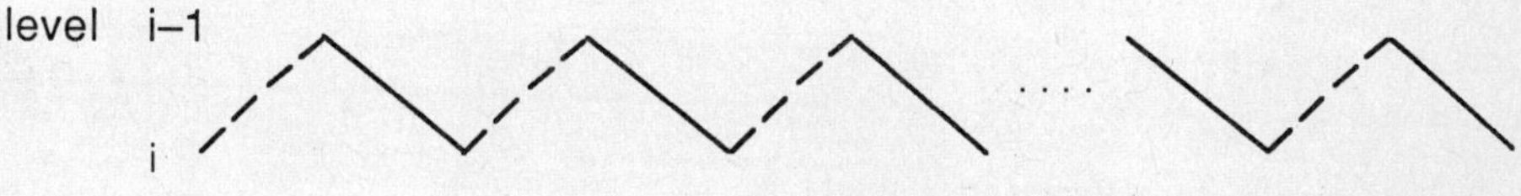


Figure 23

and $v_i$ to be the partial isometry in $A[i-1,i+1]$ which flips the diamonds by taking

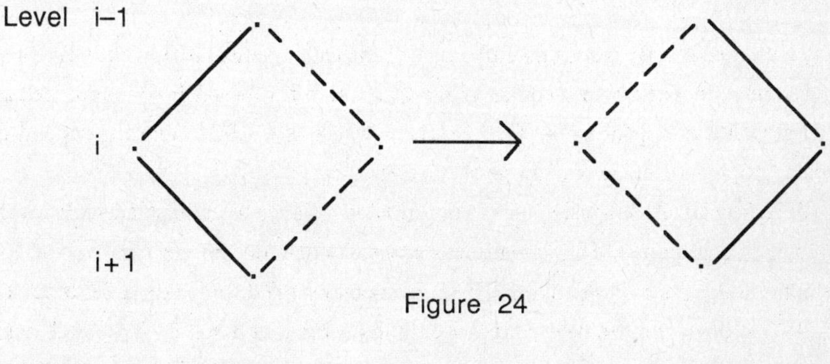


Figure 24

and is zero on other paths from level i-1 to i+1. Then it is clear that $\{g_i\}$ and $\{v_i\}$ generate $A(A_{\infty,\infty})$ and that the following relations are satisfied:

$$g_i^2 = g_i^* = g_i \tag{106}$$

$$[g_i, g_j] = 0 \tag{107}$$

$$[v_i, g_j] = 0 \qquad \text{if } j = i, i+1 \tag{108}$$

$$[v_i, v_j^{\#}] = 0 \qquad \text{if } |i-j| \geq 2. \tag{109}$$

$$v_i^* v_i = g_{i+1}(1- g_i) \ , \ v_i v_i^* = g_i(1- g_{i+1}) \tag{110}$$

The universal algebra for these generators and relations is the observable algebra $O$, [CoE] and we can indeed identify

$$a_i^* a_i = g_i \ , \quad a_i^* a_{i+1} = v_i \tag{111}$$

Note the infinitesimal braid/Temperley-Lieb relations

$$v_i^2 = 0; \qquad v_i v_{i\pm 1} v_i = 0 \tag{112}$$

which signify the Yang-Baxter equations:

$$R_i(s) R_{i+1}(s+t)R_i(t) = R_{i+1}(t) R_i(s+t)R_{i+1}(t), \text{ if } R_i(s) = 1 + sv_i. \tag{113}$$

There is a family $\{j_t: t \in \mathbb{T}\}$ of embeddings of the observable algebra O in $A(A_m)$ given by

$$a_i^* a_i \to e_{2i-1} \qquad a_i^* a_{i+1} \to t\, e_{2i-1}(1 - \tau^{-1} e_{2i})\, e_{2i+1} \tag{114}$$

for any fixed $t \in \mathbb{T}$. By working a little more in the path algebra $A(\Gamma)$, we can find a family of embeddings indexed by a compact group $G_\Gamma$, which for the ADE series (indeed any T-shaped graph) is of the form $t \in \mathbb{T}^m \oplus U(2)$ where $\varepsilon \in \{0,1\}$, and m depend on the particular graph and choice of distinguished vertex.

Polyakov [BPZ] suggested using conformal invariance to understand scale invariance of special critical points. This is particularly well understood in the Ising model [SMJ,]. Here in the scaling or continuum limit [SMJ] one obtains the disorder variable $\sigma$ as $T \downarrow T_c$: $T^n \sigma_m^x T^{-n} \to \sigma(a)$ and $(m,n) \sim (a^0, a^1) = a$ in the sense that $\varepsilon = |T - T_c| \to 0$, $\varepsilon m = a^0$, $\varepsilon n = \sqrt{-1} a^1$. Similarly one obtains the order variable $\mu$ as $T \uparrow T_c$. Moreover the free Fermions give rise to a two component Majorana Fermion field $\psi(x)$ satisfying the Dirac equation with

$$y_{\pm}(x)s(a) = \begin{cases} s(a)y_{\pm}(x) & x^1 < a^1 \\ -s(a)y_{\pm}(x) & x^1 > a^1 \end{cases} \tag{115}$$

and the signs reversed for $\mu$. Conformal invariance is understood via representations of the Virasoro algebra, and a knowledge of which representations can appear tells us something about the field theory, and the nature of the critical point. The Witt algebra of vector fields on $S^1$ is generated by $l_m = -z^{m+1}d/dz$, where $[l_m, l_n] = (m-n)l_{m+n}$. The Virasoro algebra is the central extension of the Witt algebra generated by $L_m$ and a central element $c$ with

$$[L_m, L_n] = (m-n)L_{m+n} + (c/12)m(m^2-1)\delta_{m,-n}. \tag{116}$$

Irreducible unitary highest weight representations are determined by a pair $(c,h)$ of real numbers. It is known that such representations exist if and only if either [FQS,GKO]:

(a) $c \geq 1, \quad h \geq 0$

or

(b) $c = 1 - 6/m(m+1), \quad h = h_{r,s}(c) = [((m+1)r - ms)^2 - 1]/4m(m+1)$, (117)

where $m = 2,3,4,\ldots, \quad r = 1,2,\ldots,m-1, \quad s = 1,2,\ldots,p$.

Physically, the possible values of $h$ determine the scaling dimension, and hence the critical exponents. Thus comparing with known exponents, one can match (117) [FQS,Hus] with the Ising model $(m = 3)$, tricritical Ising model $(m = 4)$, 3-state Potts model and $m \geq 5$ with the ABF models. On the other hand we have already noted how the Potts model and ABF models are related to the Jones' classification, via the transfer matrix method. It was therefore natural to ask for a direct link between the discrete sequence of index of subfactors and the discrete sequence of unitary irreducible highest weight representations of the Virasoro algebra.

The bilateral operator algebras $A(\tau)(\mathbb{Z}) = C^*(e_i, i \in \mathbb{Z})$, $A(\Gamma)(\mathbb{Z})$, the Temperley-Lieb and path algebras provide a natural setting for deforming the Fermi-Dirac quantisation of the classical fields on $S^1$. We have already noted from [CoE] how to embed a copy of an observable algebra in $A(\tau)$ or $A(\Gamma)$. The Hamiltonian $L_0 = \Sigma\, n e_{2n+1}$ generates time evolution, and has a KMS state $\omega_\beta$ for each inverse temperature $\beta$ (unique if $\tau^{-1} < 4$), and canonical ground state $\omega_\infty$ given by:

$$\omega_\infty(x) = \lim_{N\to\infty} \frac{\mathrm{Tr}\, e_{-(2N+1)} \cdots e_{-1}(1-e_1)\cdots(1-e_{2N+1})x}{\mathrm{Tr}\, e_{-(2N+1)} \cdots e_{-1}(1-e_1)\cdots(1-e_{2N+1})}.$$

Take an embedding j of the bilateral observable algebra $O = C^*(e_i, i \in \mathbb{Z})$ in $A(\tau)$ as in (114), and consider the derivations

$$\delta_k(x) = \sum_n i\,[j(a^*_{n+k}\, a_n), x] \tag{118}$$

defined on the elements of finite support in $A(\Gamma)(\mathbb{Z})$. The derivations leave the state quasi invariant, and are implemented by operators

$$T_k = \Sigma\, i :j(a^*_{n+k}\, a_n): \tag{119}$$

satisfying $[T_n, T_m] = n\delta_{m,-n}$. From the currents $T_n$, one gets a representation of the Virasoro algebra with central charge c=1 from the Sugawara formula. It remains to understand the representation of the Virasoro algebra with central charge $c = 1\text{-}6/m(m+1)$ in terms of the Temperley Lieb/path algebra $A(A_m)$.

[1]The problem of the action of the Virasoro algebra on the algebras $C^*(e_n)$ can be formulated very clearly as a problem of lifting from Out = Aut/Int (the quotient of the group of automorphisms by the inner ones) to Aut, provided one works with the algebra $A(\tau)(\mathbb{N})$ generated by the projections $e_n$, $n \geq 0$, instead of $A(\tau)(\mathbb{Z})$ generated by the $e_n$, $n \in \mathbb{Z}$. As we have seen above, one can act on $A(\tau)(\mathbb{Z})$, but then the corresponding action comes from currents and thus automatically has central charge 1. To explain the problem for finding an action on $A(\tau)(\mathbb{N})$, let us begin with the case $\tau = 1/2$. The formula for the $L_n$ acting on $A(1/2)(\mathbb{Z})$, is:

$$L_n = \sum_{k\in\mathbf{Z}} (k + n/2)\, a^*_k\, a_{n+k} \quad k\in\mathbb{Z} \tag{120}$$

Then of course $\mathrm{ad}(L_n)$ fails to leave $A(1/2)(\mathbb{N})$ globally invariant , but one can define its restriction modulo inner derivations by the formula

$$L'_n = \sum_{k\geq 0\; k+n\geq 0} (k + n/2)\, a^*_k\, a_{n+k} \tag{121}$$

since the product $a^*_k\, a_{n+k}$, $k\geq 0$, $k+n\geq 0$ belong to $A(1/2)(\mathbb{N})$. The construction of the Ramond spinor field (the critical behaviour of the Ising model with h = 1/16) provides a lifting to the above action of $\mathrm{Diff}^+(S^1)$ in $\mathrm{Out}(A(1/2)(\mathbb{N}))$ to a true action. Working at the Lie algebra level, this action is given by the following formula:

$$L''_n = \Big(\sum_{k\geq 1} + \sum_{k\geq -n}\Big)\, (k + n/2)/2\; b_{-k}\, b_{n+k} \tag{122}$$

1. The remainder of this paper is joint work with Alain Connes.

where $b_q = a_q$ for $q > 0$, $b_q = a^*_{-q}$ for $q < 0$ and $b_0 = (a_0 + a^*_0)/\sqrt{2}$, so that $b_0^2 = 1$.

Formula (122) differs from (121) only by finitely many terms, since for k large enough

$$b_{-k}\, b_{n+k} = a^*_k\, a_{n+k} \tag{123}$$

For instance the first few terms are:

$$L''_0 = \sum_{k\geq 0} k\, a^*_k\, a_k + 1/16 \tag{124}$$

$$L''_1 = \sum_{k\geq 1} (k + 1/2)\, a^*_k\, a_{k+1} + b_0 a_1/2 \tag{125}$$

$$L''_2 = \sum_{k\geq 1} (k + 1)\, a^*_k\, a_{k+2} + b_0 a_2 \tag{126}$$

$$L''_3 = \sum_{k\geq 1} (k + 3/2)\, a^*_k\, a_{k+3} + a_1 a_2/2 + 3b_0 a_3/2 \tag{127}$$

We see that the correction terms such as $b_0 a_1/2 = (a_0 + a^*_0)a_1/2\sqrt{2}$, *do not belong* to the observable algebra $O = O(\mathbb{N})$, but belong to the even Clifford algebra $A(1/2)(\mathbb{N})$. Indeed it follows from Connes and Karoubi [CoK] that there is no lifting of the above action in Aut($O$), the observable algebra. Thus the fact that correction terms belong to $A(1/2)(\mathbb{N})$ but not $O$ is a necessity. Let us now consider the general case. Using the construction of elements $e_{k,l} \in A(\tau)(\mathbb{N})$, such that $e_{k,k} = e_{2k-1}$, and that the Lie algebra relations are fulfilled, we have a replacement for $a^*_k\, a_{n+k}$ in (), namely $e_{k,k+n}$, and we can thus define as above:

$$L'_n = \sum_{k\geq 0, k+n\geq 0} (k + n/2)\, e_{k,k+n} \tag{128}$$

as derivations of $A(\tau)(\mathbb{N})$. The derivations ad($L'_n$) satisfy the Witt algebra relations *modulo inner derivations* , and the natural question is whether we can correct them by inner derivations so that the exact Virasoro algebra is fulfilled. In which case the unique full state $\omega_\tau$ of the algebra $A(\tau)$, i.e. the pure state such that $\omega_\tau(e_{2k-1}) = 1$, for all k, will be quasi invariant under $L''_n$ (as it is by construction invariant under the $L'_n$) and thus would yield a representation of the Virasoro algebra. Consider

$$L''_0 = \sum_{k\geq 0} k\,(1 - e_{2k+1}) + h \tag{129}$$

since we have taken the full state $\omega_\tau$ (rather than the vacuum state for $\tau = 1/2$), so that we also have to replace (128) by (130):

$$L'_n = \sum_{k\geq 0, k+n\geq 0} (-k+ n/2)\, e_{k,-n+k} \tag{130}$$

using the automorphism $L_n \to L_{-n}$ of the Virasoro algebra. Thus we should compute Trace($q^{L_0}$), using the Trace in $B(H_\tau)$, where $H_\tau$ is the Hilbert space of the state $\omega_\tau$, and compare with the characters of the Virasoro algebra for central charge $c = 1 - 6/m(m+1)$ when $\tau = 4\cos^2(\pi/(m+1))$. The Hilbert space $H_\tau$ can be identified with $H_R = l^2(R)$, where $R$ is the space of paths

$$\{(\sigma_k : k=1,2,...) : \sigma_{2i+1} = *, \sigma_{2i} = **, \text{ for all large } i\} \tag{131}$$

i.e., paths on the Bratteli diagram which are eventually on the solid line in Figure 25.

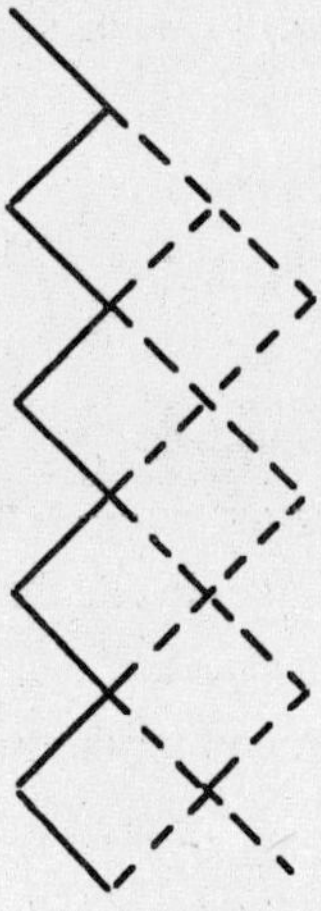

Figure 25

We can write $\pi : O \to B(H_\tau)$ as $\sum_{k\geq 0} \nu_j \pi_j$, where $\pi_j$ is the canonical representation of $O$ on $\Lambda^j(l^2)$, the space of anti-symmetric tensors on $l^2$, then

$$\text{Trace}(q^{L''_0}) = q^h \sum_{k\geq 0} \nu_k \sum_{j_1<j_2<...<j_k} q^{j_1+j_2+...+j_k}$$

$$= q^h \sum_{k\geq 0} \nu_k\, q^{k(k-1)/2} /(1-q)(1-q^2).....(1-q^k) \tag{132}$$

Here $v_j$ = dimension $(1-e_1)(1-e_3).....(1-e_{2j-1})$. To get the cut down of $A(\tau)$ by $\prod_{k\geq 1}(1-e_{2k-1})$, first take the Bratteli diagram for $A(\tau)$, and telescope two steps at a time as illustrated below for m = 4:

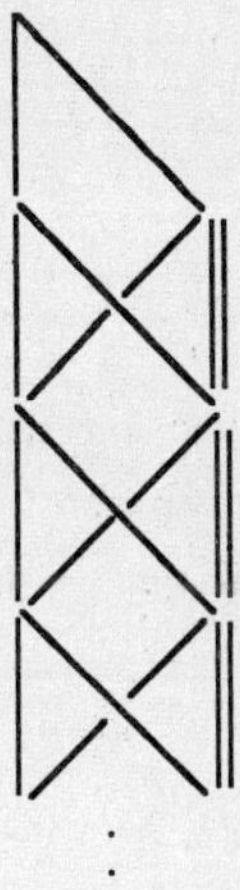

Figure 26

Then $e_{2k-1}$ is a projection of rank one down the vertical lines at the $k^{th}$ stage, and so to get the $\Pi(1-e_{2k-1})$, cutdown, we remove one vertical line from each column, again illustrated below for m = 4:

Figure 27

Hence the $v_j$ satisfy the recurrence relation $v_{j+2} = v_{j+1} + v_j$, $v_0 = 1$, $v_1 = 1$. The solution is $v_j = (\alpha^j - \beta^j)/\sqrt{5}$, if $\alpha = (1+\sqrt{5})/2$ and $\beta = (1-\sqrt{5})/2$. Thus (132) can be summed using [And]:

$$\sum_{k\geq 0} t^k q^{k(k-1)/2} /(1-q)(1-q^2).....(1-q^k) = \prod_{n\geq 0}(1+tq^n) \tag{133}$$

Note that for the Ising case $v_j = 1$, and so

$$\sum_{k\geq 0} q^{k(k-1)/2} /(1-q)(1-q^2).....(1-q^k) = \prod_{n\geq 0}(1+q^n) = \chi^3_{22}(q). \tag{134}$$

Here the character $\chi_{(c,h)} = q^{-h}\, \mathrm{tr}\, q^{h_0}$, $\chi^m_{rs} = \chi_{c,h_{rs}}$ .

Let us consider in detail the case of m = 4, c = 7/10.'' It is not hard to see, by computing the first few terms in their expansions that Trace($q^{L_0}$ ) is not an integral linear combination of the characters $\{\chi^4_{rs}\}$ of the Virasoro algebra with central chage c = 7/10. We could also consider the possibility of replacing $L_0''$ by

$$\sum_{k\geq 0} \pi_k(L_0'')\otimes 1 + 1\otimes X_k \ , \text{ where } X_k \in M_{v_k},$$

so that (132) is replaced by

$$\sum_{k\geq 0} v_k(q)\, q^{k(k-1)/2} /(1-q)(1-q^2).....(1-q^k) \tag{135}$$

where $v_k(q) = \mathrm{Trace}(q^{X_k})$, $v_k(1) = v_k$.

Let us consider in detail the characters $\chi^4_{22}$ and $\chi^4_{21}$ . Then by Rocha-Caridi [RC, page 470] we know that

$$\chi^4_{22} = \prod_{n\geq 0}(1+q^{5n+2})(1+q^{5n+3})(1+q^{5n+5})/(1-q^{5n+1})(1-q^{5n+4}) \tag{136}$$

$$\chi^4_{21} = \prod_{n\geq 0}(1+q^{5n+1})(1+q^{5n+4})(1+q^{5n+5})/(1-q^{5n+1})(1-q^{5n+4}) \tag{137}$$

Writing $(1+z) = (1-z^2)/(1+z)$, we see that

$$\chi^4_{22} = \prod_{n\geq 0} (1-q^{10n+4})(1-q^{10n+6})(1-q^{10(n+1)}) / \prod_{n\geq 1}(1-q^n)$$

$$= \prod_{n\geq 0, n\neq 0, \pm 4 \bmod 10} (1-q^n)^{-1} \tag{138}$$

$$\chi_{21}^{4} = \prod_{n\geq 0}(1-q^{10n+2})(1-q^{10n+8})(1-q^{10(n+1)}) \,/\, \prod_{n\geq 1}(1-q^{n})$$

$$= \prod_{n\geq 0,\ n\neq 0,\ \pm 2 \bmod 10} (1-q^{n})^{-1} \tag{139}$$

Moreover by Rogers (see [Sla,(44),(46)]) we know that

$$\prod_{n\geq 0\ n\neq 0,\ \pm 4 \bmod n} (1-q^{n})^{-1} =$$

$$\sum_{k\geq 0} q^{k(3k-1)/2} /(1-q)(1-q^{2})...(1-q^{k})\ (1-q)(1-q^{3})..(1-q^{2k-1}) \tag{140}$$

$$\prod_{n\geq 0\ n\neq 0,\ \pm 2 \bmod n} (1-q^{n})^{-1} =$$

$$\sum_{k\geq 0} q^{3k(k+1)/2} /(1-q)(1-q^{2})...(1-q^{k})\ (1-q)(1-q^{3})..(1-q^{2k+1}) \tag{141}$$

Comparing (135) with formula (140) does not seem too promising as one would need to take

$$\nu_{k}(q) = q^{k^{2}} /\ (1-q)(1-q^{3}).....(1-q^{2k-1}) \tag{142}$$

which is not a polynomial even, let alone $\nu_k(1) = \nu_k$. However formulae (139-140) features in Baxter's solution of the hard hexagon model [Bax1]. When computing the magnetisation in regime III of that model, Baxter considers for j = 0,1

$$F(j) = \sum_{\sigma} q^{n_j(\sigma)} \tag{143}$$

where

$$n_j(\sigma) = \sum_{k\geq 1} (k+j-1)(\sigma_{k+1} - \sigma_k\, \sigma_{k+2}) = \sum_{k\geq 1} (k+j-1)\,|\sigma_k - \sigma_{k+2}|/2 \tag{144}$$

and the summation is over all sequences

$$P = \{(\sigma_k\!: k=1,2,..)\!: \sigma_1=1,\ \sigma_k=0,1,\ \sigma_k\sigma_{k+1} = 0\} \tag{145}$$

with $\sigma_m = 0$ for all large m.

The hard hexagon model is the $A_4$ model of Andrews, Baxter and Forrester [ABF]. The constraints $\sigma_k = 0,1$, $\sigma_k\sigma_{k+1} = 0$, can be represented by the graph:

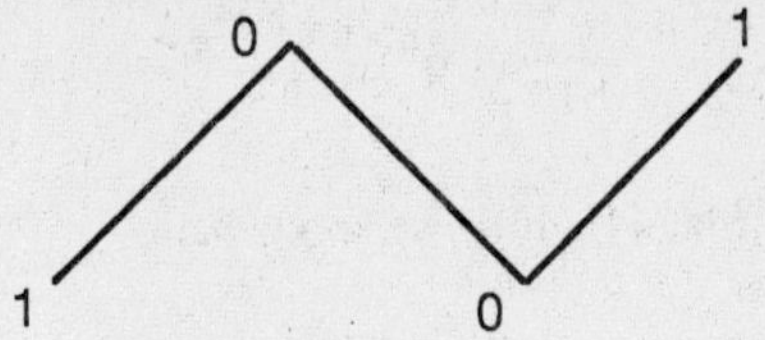


Figure 28

The only transition not allowed is 1 to 1. Thus *P* consist of paths on the Bratteli diagram $\hat{A}_4$ in Figure 29 which are eventually on the central line, starting at *.

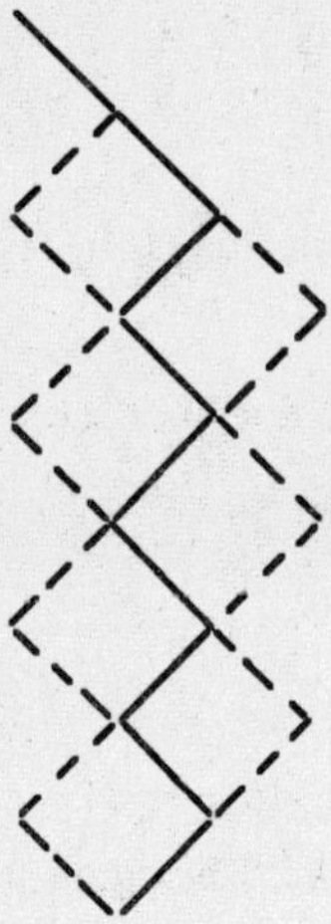

Figure 29

Then [Bax1,(79)] says that

$$F(0) = \sum_{k\geq 0} q^{k(3k-1)/2} /(1-q)(1-q^2).....(1-q^k)\ (1-q)(1-q^3).....(1-q^{2k-1}) \qquad (146)$$

$$F(1) = \sum_{k\geq 0} q^{3k(k+1)/2} /(1-q)(1-q^2).....(1-q^k)\ (1-q)(1-q^3).....(1-q^{2k+1}). \qquad (147)$$

Hence combining (138), (140), (146) we have

$$F(0) = \prod_{n\geq 0,\ n\neq 0,\ \pm 4 \bmod n} (1-q^n)^{-1}$$

$$= \sum_{k\geq 0} q^{k(3k-1)/2} /(1-q)(1-q^2)...(1-q^k)\ (1-q)(1-q^3)...(1-q^{2k-1}) = \chi^4_{22} \quad (148)$$

and similarly

$$F(1) = \prod_{n\geq 0, n\neq 0, \pm 2 \bmod n} (1-q^n)^{-1}$$

$$= \sum_{k\geq 0} q^{3k(k+1)/2} /(1-q)(1-q^2)...(1-q^k)\ (1-q)(1-q^3)...(1-q^{2k+1}) = \chi^4_{21} \quad (149)$$

We are thus guided to define projections $E_j$ living in A[j-1,j+1], by projections on the two paths of Figure 30 if j is odd and by projections on the two paths of Figure 31 if j is even.

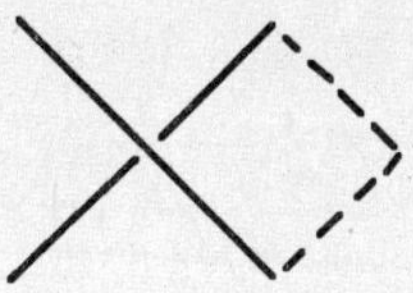

Figure 30

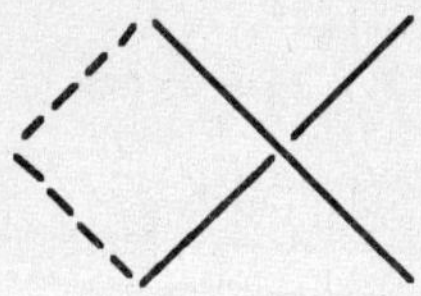

Figure 31

Then if $H_P$ denotes the Hilbert space $l^2(P)$, we can make the following identifications:

$$L_0 = \sum_{k\geq 0} jE_j/2 + 7/16 \qquad (r,s) = (2,1) \qquad (150)$$

$$L_0 = \sum_{k\geq 0} (j-1)E_j/2 + 3/80 \qquad (r,s) = (2,2). \qquad (151)$$

Although it is now clear that the Virasoro algebra acts on $H_P$ with the correct central charge c = 7/10, for (r,s) = (2,2) and (2,1), it still remains to understand how $L_n$, n ≠ 0 operate. Note that the representation (2,2) of the Virasoro algebra with central charge c = 7/10 generate the five other representations with the same central charge, through taking operator product expansions (see for example [Tod]).

**Acknowledgements**:

This work was supported by NATO, SERC, and the Center for Mathematical Analysis, Canberra.

## References


[ABF] G.E. Andrews, R.J. Baxter, J.P. Forrester. Eight vertex SOS model and generalised Rogers-Ramanujan type identities. *J. Stat. Phys.* **35** (1984) 193-266.

[AE] H. Araki, D.E. Evans. On a C*-algebra approach to phase transition in the two dimensional Ising model. *Commun. math. Phys.* **106** (1986) 241-266.

[And] G.E. Andrews. The Theory of Partitions, *Encyclopedia of Mathematics and Its Applications*. G.-C. Rota ed. Vol. 2., Addison Wesley, Reading 1976. (Reprinted Cambridge University Press, London and New York, 1984).

[Ara] H. Araki. Analyticity of correlation functions for the two dimensional Ising model. *Commun. math. Phys.* **106** (1986) 241-266.

[AS] M.F. Atiyah, I. Singer. Index theory for skew adjoint Fredholm operators. *Publ. IHES* **37** (1969) 305-326.

[AW] Y. Akutsu, M. Wadati. Knot invariants and critical statistical systems. *J. Phys. Soc. Japan* **56** (1987) 839-842.

[BakP] B.M. Baker, R.T. Powers. Product states and C*-dynamical system of product type. *J. Funct. Anal.* **50** (1983) 229-266.

[Bax1] R.J. Baxter. Rogers-Ramanujan identities in the Hard Hexagon model. *J. Stat. Phys.* **26** (1981) 427-452.

[Bax2] R.J. Baxter. Exactly solved models in statistical mechanics. Academic Press, London 1982.

[Bax3] R.J. Baxter. A simple solvable $Z_N$ Hamiltonian. *Phys. Lett. A.*

[Bax4] R.J. Baxter. The superintegrable chiral Potts model. *Phys. Lett. A.* **133** (1988) 185-189.

[Bax5] R.J. Baxter. Superintegrable Chiral Potts model: thermodynamic properties, an "inverse" model, and a simple associated Hamiltonian. Preprint ANU 1989.

[BEEK1] O. Bratteli, G.A. Elliott, D.E. Evans, A. Kishimoto. Non-commutative spheres. Preprint Trondheim 1988.

[BEEK2] O. Bratteli, G.A., Elliott, D.E. Evans, A. Kishimoto. Finite group actions on AF algebras obtained by folding the interval. Preprint Trondheim 1989.

[BEEK3] O. Bratteli, G.A. Elliott, D.E. Evans, A. Kishimoto. Non-commutative spheres II: rational rotations. Preprint Trondheim 1989.

[BEH] O. Bratteli, G.A. Elliott, R. Herman. On the possible temperatures of a C*-dynamical system. *Commun. math. Phys.* **74** (1980) 281-295.

[BeS] J. Bellisard, B. Simon. Cantor spectrum for the almost Mathieu equation. *J. Functional Analysis* **48** (1982) 408-419.

[Bla1] B. Blackadar. Symmetries of the CAR algebra. Preprint 1988.

[Bla2] B. Blackadar. K-theory for Operator Algebras, *MSRI Publication Series* v.5, Springer-Verlag, New York-Heidelberg-Berlin-Tokyo, 1986.

[BP] L. Brown, G.K. Pederson. C*-algebras of real rank zero. Preprint. Copenhagen 1989.

[BPA] R.J. Baxter, J.H.H. Perk, H. Au-Yang. New solutions of the star-triangle relations for the chiral Potts model. *Phys. Lett. A* **128** (1988) 138-142.

[BPZ] A.A.Belavin, A.M. Polyakov, A.B. Zamalodchikov. Infinite conformal symmetry in two-dimensional quantum field theory. *Nucl. Phys. B* **241** (1980) 333-380.

[BR] O. Bratteli, D.W. Robinson. Operator algebras and quantum statistical mechanics. I and II. Springer-Verlag, 1979 and 1981.

[Bra] O. Bratteli. Inductive limits of finite-dimensional C*-algebras, *Trans. Amer. Math. Soc.* **171** (1972), 195-234.

[CaE] A.L. Carey, D.E. Evans. The operator algebras of the two dimensional Ising model. in Braids, editors J. Birman, A. Libgober. *Contemporary Math.* **78** (1988) 117-165.

[Car] J.L. Cardy. Operator content of two dimensional conformally invariant theories. *Nuclear Physics. B* **270** **[FS16]** (1986) 186.

[CE] A. Connes, D.E. Evans. Embeddings of U(1)-current algebras in non-commutative algebras of classical statistical mechanics. *Commun. math. Phys.* **121** (1989) 507-525.

[CEY] M.-D. Choi, G.A. Elliott, N. Yui. Gauss polynomials and the rotation algebras.

[CIZ] A. Cappelli, C. Itzykson, J.B. Zuber. The A-D-E classification of minimal and $A^{(1)}{}_1$ conformal invariant theories. *Commun. math. Phys.* **113** (1987) 1-26.

[CK] A. Connes, M. Karoubi. Caractere multiplicatif d'un module de Fredholm. *K-theory.* **2** (1988) 431-463.

[Co] A. Connes. Classification of injective factors. *Ann. Math.* **104** (1976) 73-115.

[Dob] R.L. Dobrushin. Gibbsian random fields for lattice systems with pairwise interactions. *Func. Anal. Appl.* **2** (1968) 292-301.

[Eff1] E.G. Effros. Dimensions and C*-algebras. CBMS Lectures 46, Amer. Math. Soc. 1981.

[Eff2] E.G. Effros. On the structure of C*-algebras: some old and new problems. *Proceedings of Symposia in Pure Math.* **38** Part I (1982), 19-34.

[EG1] D.E. Evans, J. Gould. Dimension groups, embeddings and presentations of AF algebras associated to solvable lattice models. *Modern Physics Letters A* **20** (1989), 1883-1890.

[EG2] D.E. Evans, J. Gould. Embeddings and dimension groups of AF algebras associated to models in classical statistical mechanical models. Preprint Swansea 1989.

[EHS] E.G. Effros, D.E. Handelman, C.L. Shen. Dimension groups and their affine representations. *Amer. J. Math.* **102** (1980) 385-407.

[EK] D.E. Evans, A. Kishimoto. Compact group actions on UHF algebras obtained by folding the interval. *J. Funct. Anal.* (to appear).

[Ell1] G.A. Elliott. On the classification of inductive limits of sequences of semi-simple finite dimensional algebras. *J. Algebra* **38** (1976) 29-44.

[Ell2] G.A. Elliott. Gaps in the spectrum of an almost periodic Schrödinger operator. in *Geometric Methods in Operator Algebras.* eds. H. Araki, E.G. Effros. Pitman Research Notes in Mathematics 123, Longman, London 1986, 181-191.

[EL1] D.E. Evans, J.T. Lewis. The spectrum of the transfer matrix in the C*-algebra of the two-dimensional Ising model. *Commun. math. Phys.* **92** (1984) 309-327.

[EL2] D.E.Evans, J.T. Lewis. A C*-algebra approach to phase transition in the two-dimensional Ising model II. *Commun. math. Phys.***102** (1986) 521-535.

[ET] *Operator Algebras and Applications.* Editors D.E. Evans and M. Takesaki. Proceedings of UK-US joint seminar July 1987, volume 2. London Math. Soc. Lecture Notes 135, Cambridge Univ. Press 1988.

[Eva1] D.E. Evans. The C*-algebras of topological Markov chains. Lecture notes, Tokyo Metropolitan University, 1983.

[Eva2] D.E. Evans. The C*-algebras of the two-dimensional Ising model. Springer Lecture Notes in Math. 1136 (1985) 162-176.

[Eva3] D.E. Evans. Quasi-product states on C*-algebras. in *Operator Algebras and Connections with Topology and Ergodic Theory.* editors H. Araki, C.C. Moore, S. Stratila, D. Voiculescu. Springer Lecture Notes in Mathematics 1132 (1985) 129-151.

[FQS] D. Friedan, D. Qui, S. Shenker. Conformal invariance, unitarity, and two dimensional critical exponents. In *Vertex Operators in Mathematics and Physics.* Publ. MSRI No. 3. Springer-Verlag, 1984, 419-437.

[FYHLMO] P.Freyd, D. Yetter, J. Hoste, W. Lickorish, K. Millet, A. Ocneanu. A new polynomial invariant of knots and links. *Bull AMS* **12** (1985) 183-212.

[GHJ] F. Goodman, P. de la Harpe, V.F.R. Jones. Dynkin diagrams and towers of algebras. Publ. MSRI .

[GKO] P. Goddard, A. Kent, D. Olive. Unitary representations of the Virasoro and super-Virasoro algebras. *Commun. math. Phys.* **103** (1986) 105-119.

[Gol] M. Goldman. On subfactors of factors of type $II_1$. *Mich. Math. J.* **6** (1960) 167-172.
[Gou] J.D. Gould. Ph.D. thesis, University of Warwick, 1989.
[GW] D. Gepner. E. Witten. String theory on group manifolds. *Nucl. Phys. B* **278** (1986) 493-549.
[Ham] T. Hamachi. Jones index and Markov odometers. Preprint Kyushu University.
[HeS] B. Helfer, J. Sjöstrand. Analyse semi-classique pour l'equation de Harper (avec application à l'equation de Schrödinger avec champ magnetique). Preprint.
[HKS] R. Hoegh-Krohn, T. Skjelbred. Classification of C*-algebras admitting ergodic actions of the two-dimensional torus. *J. Reine Angewandtte Math.* **328** (1981) 1-8.
[Hof] A.J. Hofman. On limit points of spectral radii of non-negative symmetric integral matrices. *Lecture Notes in Math.* **303** Springer Verlag 1972, 165-172.
[HR] D.E. Handelman, W. Rossman. Actions of compact groups on C*-algebras. *Illinois J. of Math.* **29** (1985) 51-95.
[HS] U. Haagerup, J. Schou. Some new subfactors of the hyperfinite $II_1$-factor. Preprint Mittag-Leffler January 1989.
[Hus] D. Huse. Exact exponents for infinitely many new critical points. *Phys. Rev. B*.**30** (1984), 3908-3915.
[Jim1] M. Jimbo. A q-difference analogue of U(g) and the Yang-Baxter equation. *Lett. Math. Phys.* **10** (1985) 63-69.
[Jim2] M. Jimbo. A q-analog of U(gl(N+1)), Hecke algebras and the Yang-Baxter equation. *Lett. Math. Phys.* **10** (1985) 63-69.
[JMO] M. Jimbo, T. Miwa, M. Okado. Solvable lattice models related to the vector representation of classical simple Lie algebras. *Commun. math. Phys.* **116** (1988) 507-525.
[Jon1] V.F.R. Jones. Index for subfactors. *Invent. Math.* **72** (1983) 1-25.
[Jon2] V.F.R. Jones. A polynomial invariant for knots via von Neumann algebras. Bull. AMS 12 (1985) 103-111.
[Jon3] V.F.R. Jones. Braid groups, Hecke algebras and type $II_1$ factors. in *Geometric Methods in Operator Algebras*. eds. H. Araki and E.G. Effros. Longman Sci. and Tech. 1986, 242-273.
[Jon4] V.F.R. Jones. Hecke algebra representations of braid groups and link polynomials. *Ann. Math.* **126** (1987) 335-388.
[KAW] A. Kuniba, Y. Akutsu, M. Wadati. Virasoro algebra, von Neumann algebra and critical eight vertex SOS model. *J. Phys. Sci. Japan.* **55** (1986) 3285-3288.
[Kui] R. Kuik. On the q-state Potts model by means of non-commutative algebras. Thesis, Groningen 1986.
[Kum1] A. Kumjian. An involutive automorphism of the Bunce-Deddens algebra, *C.R. Math. Rep. Acad. Sci. Canada* (1988).
[Kum2] A. Kumjian. Non-commutative spherical orbifolds. in preparation.
[LaR] O.E. Lanford, D. Ruelle. Observables at infinity and states with short range correlations in statistical mechanics. *Commun. math. Phys.* **13** (1969) 194-215.
[LS] J.T. Lewis, P.N.M. Sisson. A C*-algebra of the two-dimensional Ising model. *Commun. math. Phys.* **44** (1975) 279-292.
[LW] J.T. Lewis, M. Winnink. The Ising model phase transition and the index of states on the Clifford algebra. *Coll. Math. Soc. Janes Bolyai* **27** Random Fields, Esztergom, Hungary 1979.
[Mat1] T. Matsui. A link of quantum and classical Potts models. Preprint Swansea 1989.
[Mor1] A.O. Morris. On a generalised Clifford algebra. *Quart. J. Math. Oxford (2)* **18** (1967) 7-12.
[Mor2] A.O. Morris. On a generalised Clifford algebra (II). *Quart. J. Math. Oxford (2)* **19** (1968) 289-299.

[MS1] G. Moore, N. Seiberg. Classical and quantum conformal field theory. *Commun. math. Phys.* **123** (1989) 177-254.

[MS2] G. Moore, N. Seiberg. Naturality in conformal field theory. *Nucl. Phys.* (in press).

[Nah] W.Nahm. Lie group exponents and SU(2) current algebras. *Commun. math. Phys.* **118** (1988) 171-176.

[Nes] R. Nest. Cohomology of a certain non-commutative orbifold. Preprint Copenhagen 1989.

[Ocn] A. Ocneanu. Quantized groups, string algebras and galois theory for algebras. in [ET].

[Pas1] V. Pasquier. Operator content of the ADE lattice models. *J. Phys. A:Math. Gen.* **20** (1987) 5707-5717.

[Pas2] V. Pasquier. Two-dimensional critical systems labelled by Dynkin diagrams. *Nuclear Phys. B* **285[FS19]** (1987) 162-172.

[Pas3] V. Pasquier. Etiology of IRF models. *Commun. math. Phys.* **118** (1988) 355-364.

[Pir] S. Pirogov. States associated with the two dimensional Ising model. *Theor. Math. Phys.* **11**(3), (1972) 614-617.

[Pot] R. Potts. *Proc. Camb. Phil. Soc.* **48** (1952) 106-109.

[PP] M. Pimsner, S. Popa. Entropy and index for subfactors. *Ann. Sci. Ec. Norm Sup.* **19** (1986) 57-106.

[PS] R.T. Powers, S. Sakai. Existence of ground states and KMS states for approximately inner dynamics. *Commun. math. Phys.* **39** (1975) 273-288.

[PT1] J.Palmer, C.Tracy. Two dimensional Ising correlations: convergence of the scaling limit. *Adv. in Appl. Math.* **2** (1981) 329-388.

[PT2] J. Palmer, C. Tracy. Two dimensional Ising correlations: the SMJ analysis. *Adv. in Appl. Math.* **4** (1983) 46-102.

[RC] A. Rocha-Caridi. Vacuum vector representation of the Virasoro algebra. in *Vertex Operators in Mathematical Physics.* MSRI Publ. 3. Springer, Heidelberg 1984, 451-473.

[Res] N. Yu. Reshetikhin. Quantized universal enveloping algebras, the Yang-Baxter equation and invariant of links I. LOMI Preprint, Leningrad 1987.

[Ren] J. Renault. A groupoid approach to C*-algebras. *Lecture notes in Mathematics* **793** (1980) Springer Verlag, Heidelberg.

[Rie] M. Rieffel. C*-algebras associated with irrational rotations. *Pacific J. Math.* **93** (1981), 415-429.

[Ros1] M. Rosso. Comparison des groupes SU(2) quantique de Drinfeld et de Woronowicz. *C.R. Acad. Sc. Paris* **304** (1987) 323-326.

[Ros2] M. Rosso. Finite dimensional representations of the quantum analog of the enveloping algebra of a complex simple Lie algebra. *Commun. math. Phys.*

[She] J.B. Shearer. On the distribution of the maximum eigenvalue of graphs. *Linear Algebra and its Applications.* **114/115** (1989) 17-20.

[Sla] L.J. Slater. Further identities of the Rogers-Ramanujan type. *Proc. London Math. Soc.* (2) **54** (1951) 147-167.

[Slo] P. Slodowy. In: *Algebraic Geometry.* J. Dolgachev ed. *Lecture Notes in Math.* **1008**. Springer Verlag 1983.

[SMJ] M. Sato, T. Miwa, M. Jimbo. Holonomic quantum fields. *Publ. RIMS, Kyoto Univ.* **16** (1980) 531-584.

[SO'C] R. Schor, M. O'Carroll. The scaling limit and Osterwalder-Schrader axioms for the two dimensional Ising model. *Commun. math. Phys.* **84** (1982) 153-170.

[Sun] V.S. Sunder. A model for AF algebras and a representation of the Jones projections. *J. Operator Theory* **18** (1987) 289-301.

[Tem] H.N.V. Temperley. Potts model and related problems. Preprint Swansea 1986.

[TK] A. Tsuchiya, Y. Kanie. Vertex operators on conformal field theory on $\mathbb{P}^1$ and monodromy representations of braid groups. *Adv. Stud. in P. Math.* **16** (1988) 297-372.

[TL] H.N.V. Temperley, E. Lieb. *Proc. Roy. Soc. Lond. A.***322** (1971) 251-280.

[Tod] I. Todorov. Infinite dimensional Lie algebras in conformal QFT models. in *Conformal Groups and Related Symmetries*. eds. A. Barat, H. Doebner. *Lecture Notes in Physics* **261**, Springer Berlin 1986, 387-443.

[Ver] E. Verlinde. Fusion rules and modular transformations in 2D conformal field theory.

[Was] A.J. Wassermann. Automorphic actions of compact groups on operator algebras. Ph.D. thesis, University of Pennsylvania 1981.

[Wen1] H. Wenzl. On sequences of projections. *C.R. Math. Rep. Acad. Sci. Canada*. **9** (1987) 5-9.

[Wen2] H. Wenzl. Hecke algebras of type $A_n$ and subfactors. *Invent. math.* **92** (1988) 349-383.

International Journal of Modern Physics B, Vol. 4, No. 5 (1990) 1119 – 1126
© World Scientific Publishing Company

# ENTROPY INCREASE AND K-SYSTEMS IN QUANTUM THEORY

W. Thirring

* Department of Physics and Mathematical Physics
The University of Adelaide
G.P.O. Box 498, Adelaide
South Australia 5001.

SUMMARY

It is shown that the relative entropy converges to its equilibrium value when the states converge strongly. This can happen only for subalgebras for which the clustering is uniform. It happens for K-systems for which this uniformity in clustering has the maximal extent which is possible.

* Visiting the Department of Physics and Mathematical Physics

Permanent address: Institute für Theoretische Physik,
Wien IX - Boltzmanngasse 5
Austria.

## 1. THE ENTROPY INCREASE

J. von Neumann proposed for a state over $B(H)$ given by a density matrix $\rho$ the entropy

$$S(\rho) = -\mathrm{Tr}\rho\ln\rho \ . \tag{1.1}$$

This expression can be singled out by a few plausible axioms. It does not change under a unitary time evolution $\rho \to \rho_t = U_t\rho U_t^*$ and it will become infinite upon passing to the thermodynamic limit. Then only the entropy density or the information contained in a subsystem remain meaningful. The entropy of the state $\omega$ restricted to a subalgebra $D$, $s(\omega|_D)$ can change with time but it might increase or decrease.

This is to be expected since a subsystem may exchange energy with the rest. As one learns in thermodynamics in this situation the entropy need not increase. Instead the second law requires the free energy to decrease. The difference of the free energies of a state $\rho$ and the equilibrium $\omega$ is the relative entropy

$$s(\omega|\rho) = \mathrm{Tr}\rho(\ln\rho-\ln\omega) \tag{1.2}$$

multiplied by the temperature (for $\omega = e^{-H+F_\omega}$, $s(\omega|\rho) = \mathrm{Tr}\rho H - s(\rho) - F_\omega$). Thus the second law requires

$$s(\omega|\rho_t) \le s(\omega|\rho_t{}') \ \forall t \ge t' . \tag{1.3}$$

Araki has generalized $s(\omega|\rho)$ for arbitrary states over von Neumann algebras but even for $H = C^2$ this object is discontinuous and may be $\infty$

$\left(\text{take } \rho = \begin{pmatrix} \frac{1}{2} & 0 \\ 0 & \frac{1}{2} \end{pmatrix}, \ \omega = \begin{pmatrix} 1 & 0 \\ 0 & 0 \end{pmatrix}\right)$.

To avoid these difficulties we shall in the sequel assume $\exists\lambda_{1,2}\epsilon R$ with $\lambda_1\rho \le \omega \le \lambda_2\rho$. Then we get the [1] .

Properties of $s(\omega|\rho)$ (1.4)

(i) $s(\omega\circ\gamma|\rho\circ\gamma) \le s(\omega|\rho)$ for $\gamma$ a Schwarz positive map

(i.e. $\gamma(a^*a) \ge \gamma(a^*)\gamma(a)$ $\forall a\epsilon A$

(ii) $\omega,\rho \to s(\omega|\rho)$ is weakly lower semicontinuous

(iii) $(\omega,\rho) \to s(\omega|\rho)$ is strongly continuous

(iv) $s(\omega|\rho) \ge \frac{1}{2}\|\omega-s\|^2$

Corollaries (1.5)

(i) If $\gamma$ has a Schwarz positive right inverse $\gamma_1$, $\gamma o \gamma_1 = 1$ then $s(\omega|\rho) = s(\omega o\gamma o\gamma_1|\rho o\gamma o\gamma_1) \leq s(\omega o\gamma|\rho o\gamma) \leq s(\omega|\rho)$ and thus $s(\omega|\rho)$ does not change. This happens for conditional expectations since the inclusion is the right inverse and in particular for automorphism.

(ii) If $\omega_t \to \omega_\infty, \rho_t \to \rho_\infty$ then $s(\omega_\infty|\rho_\infty) \leq \lim_{t\to\infty} s(\omega_t|\rho_t)$. For us this will be worthless since we will consider $\omega_t = \omega$, $\rho_t \to \omega$ and we know $s(\omega|\omega) = 0$ and $s(\omega_t|\rho t) \geq 0$ anyway.

(iii) and (iv) tells $s(\omega|\rho_t) \to 0$ iff $\rho_t \to \omega$.

Thus we conclude that the free energy of a state converges to the free energy of equilibrium if and only if the state converges strongly to the equilibrium state.

## 2. MIXING PROPERTIES.

Notation. We consider a von Neumann algebra $A$ with an automorphism $\alpha$, $\alpha^t(a) = a_t$ $\forall t\epsilon Z$ a faithful invariant state $\omega = \omega o\alpha$. In the GNS-representation $\omega(a) = \langle\Omega|a|\Omega\rangle$ and $a_t = U_t{}^* a U_t$ with $U_t|\Omega\rangle = |\Omega\rangle$.

Proposition (2.1)

The following properties are equivalent

(i) $\lim_{t\to\infty} \omega(a_t b) = \omega(a)\omega(b)$ $\forall a,b\epsilon A$

(ii) $\nu_t \to \omega$ for all states v over $A$

(iii) $a_t \to \omega(a)\, 1$ $\forall a\epsilon A$

(iv) $U_t \to |\Omega\rangle\langle\Omega|$

Proof: (i) $\Rightarrow$ (ii) : $\exists b\epsilon A$: $\nu(a) = \omega(ab)$ with $\omega(b) = 1$

$$\nu_t(a) = \omega(a_t b) \to \omega(a)$$

(ii) $\Rightarrow$ (iii) with the modular automorphis $\sigma_i$ of $\omega$ we have

$\omega(ab) = \omega(b\sigma_i a)$. Thus

$$\langle\Omega|ba_t c|\Omega\rangle = \omega(a_t c\sigma_i b) \to \omega(a)\omega(c\sigma_i b)$$
$$= \omega(a)\langle\Omega|b1c|\Omega\rangle.$$

(iii) $\Rightarrow$ (iv) $\langle\Omega|bU_t c|\Omega\rangle = \omega(b_t c) \to \omega(b)\omega(c)$

$$= \langle\Omega|b|\Omega\rangle\langle\Omega|c|\Omega\rangle$$

(iv) $\Rightarrow$ (i) $\omega(a_t b) = \omega(aU_t b) \to \omega(a)\omega(b)$

Definition (2.2)

A state over $A$ with the properties (2.1) is called mixing.

Remarks (2.3)

1) If $A$ has a mixing state it can have only one invariant state.

2) Since $\omega(a_t b) = \omega(\sigma_{-i}(b)a_t) = \omega(ab_{-t})$ the cluster properties also hold for $\omega(ba_t)$ and $t \to -\infty$.

So far we got only weak convergence which is of no use for the question of entropy increase and we have to look for something better.

Proposition (2.3)

Let $A_o \subset A$ be a subalgebra and $A_t = \alpha^t A_o$ then the following properties are equivalent.

(i) $\forall b \epsilon A,\ \epsilon > 0 \exists T : |\omega(a_t b) - \omega(a)\omega(b)| \leq \epsilon \|a\| \forall t > T, a \epsilon A_o$

(ii) $\forall b \epsilon A, \tau \epsilon R, \epsilon > 0 \exists T : |\omega(a_t b) - \omega(a)\omega(b)| \leq \epsilon \|a\| \forall t > T, a \epsilon A_\tau$

(iii) $v_{t \mid A_o} \xrightarrow{t \to \infty} \omega$ for a states v over $A$.

Proof

(i) $\Leftrightarrow$ (ii) is clear and if $v(a) = \omega(ab)$, $\omega(b) = 1$, both (i) and (iii) are to $\sup_{\substack{a \epsilon A_o \\ \|a\|=1}} |v_t(a) - \omega(a)| < \epsilon \forall t > T$ equivalent.

Definition (2.4)

If $(A, \alpha, A_o)$ satisfy (2.3) we say $(A, \alpha)$ is $A_o$ mixing.

Remarks (2.5)

1. The clustering is not uniform in b and thus it goes only in one time direction.

2. $A_o$=A is impossible (take $b_{-t}$=a). Therefore also $v_t \to \omega$ implies v=$\omega$.

3. $a_t \to \omega(a)1$ happens only if $a = z1$ : $a_t{}^2$ would then converge strongly to both $\omega(a^2)$ and $\omega(a)^2$ which means $\omega((a-\omega(a))^2) = 0$. Since $\omega$ is faithful we conclude $a = \omega(a)1$.

4. $U_t \to |\Omega\rangle\langle\Omega|$ is impossible because a unitarity can converge strongly only to an isometry.

## 3. K-SYSTEMS

Definition (3.1)

Denote by $\vee$ (resp $\wedge$) the algebraic union (resp. intersection). If

(i) $A_t \supset A_{t}{}' \ \forall t \geq t'$

(ii) $\bigvee_{t>0} A_t = A$

(iii) $\bigwedge_{t<0} A_t = Z1$

We call ($A$, $\alpha$, $A_o$ a K-system) [2,3].

Proposition (3.2)

Denote by $P_t$ the orthogonal projector onto $A_t|\Omega\rangle$. Then (3.1) is equivalent to

(i) $P_t \geq P_t \ \forall t \geq t'$

(ii) $\underset{t\to\infty}{\text{s-lim}}\, P_t = 1$

(iii) $\underset{t\to\infty}{\text{s-lim}}\, P_t = |\Omega\rangle\langle\Omega|$

Proof. The nontrivial point is the quality of convergence. We have strong convergence since projectors converging weakly to a projector do so strongly:

$\|(P_\mu - P_\infty)|v\rangle\|^2 = \langle v|P_\mu - P_\mu\ P_\infty - P_\infty\ P_\mu + P_\infty|v\rangle \to 0$ if $P_\mu - P_\infty$ .

Theorem (3.3)

For a K-System ($A$, $\alpha^{-1}$) is $A_o$ mixing.

Proof.

$$|\omega(a_t b) - \omega(a)\omega(b)| = |\langle\Omega|a(P_o U_t - |\Omega\rangle\langle\Omega|)b|\Omega\rangle|$$

$$= |\langle\Omega|aU_t\ (P_t - |\Omega\rangle\langle\Omega|)|b\rangle| \leq \|a\|\epsilon\ \ \forall t>T.$$

Corollary (3.4)

For a K-system $S(\omega|_{A_\tau}|\nu_t|_{A_\tau})$ converges for any $\tau$ to zero if $t \to -\infty$.

Remarks (3.5)

1. Since $\omega_t = \omega$, and $\omega_t|_{A_\tau} = \omega|_{A_{\tau+t}}$, $\nu_t|_{A_\tau} = \nu|_{A_{\tau+t}}$ we see that $s(\omega|_{A_\tau}|\nu_t|_{A_\tau})$ increases monotonically with t. Since $A_t \uparrow A$ we conclude from the weak semicontinuity that for $t \to \infty$ it increases to $s(\omega|_A|\nu|_A) = s(\omega|\nu)$.

2. The reader might be surprised that the equilibrium is reached for $t \to -\infty$ rather than $t \to +\infty$. However in the K-system appearing in physics there are also contracting algebras $D_t \subset D_{t'}$ $\forall t \geq t'$ and $s(\omega|_{D_t}|\nu|_{D_t})$ decreases to zero for $t \to +\infty$.

3. For finite subalgebras $F \subset A$ the weak and strong topologies coincide and since $\nu_t$ tends weakly to $\omega$ for $t \to \pm D$ we have $\lim_{t \to \pm\infty} s(\omega|_{F_t}|\nu|_{F_t}) = 0$. In this case the monotonic behaviour is lost.

One can consider as germ of a K-system the situation where one has an algebra $A_o$ which is isomorphic to a proper subalgebra $A_{-1}$. Continuing with this automorphism $\alpha^{-1}$ one gets a semi-infinite chain $A_o \overset{\alpha^{-1}}{\to} A_{-1} \overset{\alpha^{-1}}{\to} A_{-2} \overset{\alpha^{-1}}{\to} \ldots$ .

One can now ask three questions.

1. Can the chain be consistently continued to the left such that we get an automorphism $\alpha$ on $A = \bigvee_{i=-\infty}^{\infty} A_i$?
2. When is $(A, \alpha, A_o)$ a K-system?
3. If it is not does it contain a sub-K-system?

If $A_o$ is a $II_1$-Factor the questions can be answered as follows. [1]

1. With the Jones construction [4] one can get an $A_1$ and an isomorphism $\alpha_1 : A_1 \to A_o$ such that $\alpha_1|_{A_o} = \alpha$. Continuing one can construct $A$.

2. This leads to a K-system if $A_0$ does not have a subalgebra invariant under $\alpha^{-1}$.

3. $A$ always contains the Jones algebra as a sub-algebra on which $\alpha$ acts as the shift $e_i \to e_{i+1}$. This is a K-system.

Remarks (3.6)

K-systems cannot have finite invariant subalgebras, even quasi-periodicity is forbidden for finite sets. However there may be infinite invariant subalgebras. In particular the tensor product of two K-systems is in a natural way a K-system and each factor is an invariant subalgebra. (There are no finite K-systems).

The author is grateful to Professor C.A. Hurst for the kind hospitality extended to him during his stay in Adelaide.

REFERENCES

[1] H. Narnhofer, W. Thirring, Vienna preprint 17, 1989.

[2] G. Emch, Comm. Math. Phys. 49, 191 (1976).

[3] B. Kümmerer, W. Schröder, Comm. Math. Phys. 90, 251 (1983).

[4] V. Jones, Invent. Math. 72, 1 (1983).